Graduate Texts in Mathematics 124

B. A. Dubrovin
A. T. Fomenko
S. P. Novikov

Modern Geometry—Methods and Applications

Part III. Introduction to Homology Theory

Translated by Robert G. Burns

With 120 Illustrations

Springer

B. A. Dubrovin
c/o VAAP—Copyright Agency
of the U.S.S.R.
B. Bronnaya 6a
103104 Moscow
U.S.S.R.

A. T. Fomenko
Department of Geometry and Topology
Faculty of Mathematics and Mechanics
Moscow State University
119899 Moscow
U.S.S.R.

S. P. Novikov
L.D. Landau Institute for
Theoretical Physics
Academy of Sciences of the U.S.S.R.
Vorobevskoe Shosse, 2
117334 Moscow, U.S.S.R.

R. G. Burns (Translator)
Department of Mathematics
York University Downsview
Onatrio, M3J1P3
Canada

AMS Subject Classifications: 55-01, 57-01, 55NXX, 55PXX, 55Q22, 55Rxx, 55SXX, 55TXX, 57N65, 57015, 57020, 57R19, 57R20, 57R75, 57TXX, 58AXX, 58EXX

Library of Congress cataloging-in-Publication Data
(Revised for Vol. 2)
Dubrovin, B. A.
Modern geometry—methods and applications.
(Springer series in Soviet mathematics) (graduate texts in mathematics; 93- , 124)
"Original Russian edition: Sovremennaja geometria: Metody i Prilozenia. Moskva: Nauka, 1979"
—T.p. verso.
Includes bibliographies and indexes.
Contents: pt. 1.The geometry of surfaces, transformation groups, and fields—pt. 2. The geometry and Topology of manifolds—pt. 3. Introduction to homology theory.
1.Geomtery. I.Fomenko, A. T. II. Novikov, Sergei Petrovich. III. Title. IV. Series.
Graduate texts in mathematics; 93, etc.
QA445.D8213 1984 516 83-16851

This book is part of the Springer Series in Soviet Mathematics.
Advisors: L.D. Faddeev (Leningrad), and R. V. Gamkrelidze (Moscow).
Original Russian edition: Sovremennaja Geometria: Metody i Prilozenia. Moskva: Nauka, 1979.

Dubrovin et. al.: Modern Geometry—Methods and Applications
Part III: Introduction to Homology Theory

First Indian Reprint 2010

ISBN 978-81-8489-660-2

This edition is published by Springer (India) Private Limited, (a part of Springer Science+Business Media), Registered Office: 3rd Floor, Gandharva Mahavidyalaya, 212, Deen Dayal Upadhyaya Marg, New Delhi – 110 002, India

Printed in India by Rakmo Press (P) Ltd., New Delhi.

Preface

In expositions of the elements of topology it is customary for homology to be given a fundamental role. Since Poincaré, who laid the foundations of topology, homology theory has been regarded as the appropriate primary basis for an introduction to the methods of algebraic topology. From homotopy theory, on the other hand, only the fundamental group and covering-space theory have traditionally been included among the basic initial concepts. Essentially all elementary classical textbooks of topology (the best of which is, in the opinion of the present authors, Seifert and Threlfall's *A Textbook of Topology*) begin with the homology theory of one or another class of complexes. Only at a later stage (and then still from a homological point of view) do fibre-space theory and the general problem of classifying homotopy classes of maps (homotopy theory) come in for consideration. However, methods developed in investigating the topology of differentiable manifolds, and intensively elaborated from the 1930s onwards (by Whitney and others), now permit a wholesale reorganization of the standard exposition of the fundamentals of modern topology. In this new approach, which resembles more that of classical analysis, these fundamentals turn out to consist primarily of the elementary theory of smooth manifolds,† homotopy theory based on these, and smooth fibre spaces. Furthermore, over the decade of the 1970s it became clear that exactly this complex of topological ideas and methods were proving to be fundamentally applicable in various areas of modern physics. It was for these reasons that the present authors regarded as absolutely

† Evidently the beginning ideas of topology, which can be traced back to Gauss, Riemann and Poincaré, actually arose, historically speaking, in this order. However, at the time of Gauss and Riemann, a correspondingly organized conceptual basis for a theory of topology was unrealizable. It was Poincaré who, in creating the homology theory of simplicial complexes, was able to provide a quite different, precise foundation for algebraic topology.

essential material for a training in topology, in the first place precisely the theory of smooth manifolds, homotopy theory, and fibre spaces, and incorporated this subject matter in Part II of their textbook *Modern Geometry*. It is assumed in the present text that the reader is acquainted with that material.

On the other hand, the solution of the more complex problems arising both within topology itself (the computation of homotopy groups, the classification of smooth manifolds, etc.) and in the numerous applications of the algebro-topological machinery to algebraic geometry and complex analysis, requires a very extensive elaboration of the methods of homology theory. There is in the contemporary topological literature a complete lack of books from which one might assimilate the complex of methods of homology theory useful in applications within topology. It is part of the aim of the present book to remedy this deficiency.

In expounding homology theory we have, wherever possible, striven to avoid using the abstract terminology of homological algebra, in order that the reader continually remain cognizant of the fact that cycles and boundaries, and homologies between them, are after all concrete geometrical objects. In a few places, for instance in the section devoted to spectral sequences, this self-imposed restriction has inevitably led to certain defects of exposition. However, it is our experience that the usual expositions of the machinery of modern homological algebra lead to worse defects in the reader's understanding, essentially because the geometric significance of the material is lost from view. Certain fundamental methods of modern algebraic topology (notably those associated with spectral sequences and cohomology operations) are described without full justification, since this would have required a substantial increase in the volume of material. It must be remembered that those methods are based exclusively on the formal algebraic properties of the algebraic entities with which they are concerned, and in no way involve their explicit geometric prototypes whence they derive their *raison d'être*. In the final chapter of the book the methods of algebraic topology are applied to the investigation of deep properties of characteristic classes and smooth structures on manifilds. It is the intention of the authors that the present monograph provide a path for the reader giving access to the contemporary topological literature.

A large contribution to the final version of this book was made by the editor, Victor Matveevich Bukhshtaber. Under his guidance several sections were rewritten, and many of the proofs improved upon. We thank him for carrying out this very considerable task.

Translator's acknowledgements. Thanks are due to G. C. Burns and Abe Shenitzer for much encouragement, to several of my colleagues (especially Stan Kochman) for technical help, and to Eadie Henry for her advice, superb typing, and forbearance.

Contents

CHAPTER 1

Homology and Cohomology. Computational Recipes

§1. Cohomology Groups as Classes of Closed Differential Forms. Their Homotopy Invariance

Among the most important of the homotopy invariants of a manifold are its homology and cohomology groups, which we have already encountered (in §§19.3, 24.7, 25.5 of Part II), and which we shall now expound systematically.

There are several (equivalent) ways of defining the homology groups of a manifold; to begin with we give the definition (of the cohomology groups) in terms of differential forms on the manifold (as in §25.5 of Part II). Thus we shall initially be considering closed differential forms of rank k on our manifold M^n (where as usual the index n indicates the dimension of the manifold), given locally by

$$\omega = \sum_{i_1 < \cdots < i_k} a_{i_1 \cdots i_k}\, dx^{i_1} \wedge \cdots \wedge dx^{i_k}, \qquad d\omega \equiv 0. \tag{1}$$

(Recall that a differential k-form is *closed* if $d\omega \equiv 0$, and is *exact* if $\omega = d\omega'$ for some form ω' of rank $k - 1$, and also that $d(d\omega') \equiv 0$, so that the exact forms figure among the closed ones (see §25.2 of Part I).)

1.1 Definition.† The *kth cohomology group* $H^k(M^n; \mathbb{R})$ (actually a real vector space) of a manifold M^n is the quotient group of the group (vector space) of all closed forms of rank k on M^n by its subgroup (linear subspace) of exact

† In the sequel we shall give several different definitions of the homology and cohomology groups with coefficients from various groups. In view of the fact that these definitions all yield essentially the same concept (see §§6, 14 below), we shall refrain from introducing indices to indicate any particular version of the concept as it arises in the various contexts.

forms. Thus the elements of $H^k(M^n; \mathbb{R})$ are the equivalence classes of closed k-forms where two forms are taken as equivalent if they differ by an exact form:

$$\omega_1 \sim \omega_2 \quad \text{means} \quad \omega_1 - \omega_2 = d\omega'. \tag{2}$$

The following result gives the simplest property of the (0th) cohomology groups.

1.2. Proposition. *For any manifold M^n the* 0*th cohomology group $H^0(M^n; \mathbb{R})$ is the vector space whose dimension q is equal to the number of connected components of the manifold.*

Proof. A form of rank zero is just an ordinary scalar function $f(x)$ on the manifold. If such a form is closed, then $df(x) \equiv 0$, so that $f(x)$ is locally constant, and therefore constant on each connected component of the manifold. Hence each closed 0-form on M^n can be identified with a sequence of q constants, one for each of the q components of the manifold. In view of the fact that there are no exact 0-forms, the proposition now follows. □

Any smooth map $f: M_1 \to M_2$ between manifolds determines a map $\omega \mapsto f^*(\omega)$, the "pullback", of forms ω on M_2 to forms $f^*(\omega)$ on M_1, satisfying $df^*(\omega) = f^*(d\omega)$ (see §§22.1, 25.2 of Part I). Hence each such map f determines a map (in fact a homomorphism, or better still a linear transformation)

$$f^*: H^k(M_2; \mathbb{R}) \to H^k(M_1; \mathbb{R}) \tag{3}$$

between the cohomology groups (since under f^* closed forms are sent to closed forms, and exact to exact).

1.3. Theorem. *Let $f_1: M_1 \to M_2$, $f_2: M_1 \to M_2$ be two smooth maps of manifolds. If f_1 is homotopic to f_2 then the corresponding homomorphisms f_1^* and f_2^* of the cohomology groups, coincide:*

$$f_1^* = f_2^*: H^k(M_2; \mathbb{R}) \to H^k(M_1; \mathbb{R}).$$

Proof. Let $F: M_1 \times I \to M_2$ be a smooth homotopy between f_1 and f_2, where I is the interval $1 \le t \le 2$, $F(x, 1) = f_1(x)$, and $F(x, 2) = f_2(x)$. In terms of local co-ordinates on $M_1 \times I$ of the form $(x^1, \ldots, x^n, t) = (x, t)$, where $x^1, \ldots, x^n$ are local co-ordinates on M_1, any differential form Ω of rank k on $M_1 \times I$ can be written as

$$\Omega = \omega_1 + \omega_2 \wedge dt, \qquad \Omega|_{t=t_0} = \omega_1(t_0), \tag{4}$$

where ω_1 is a form of rank k which does not involve the differential dt (in the sense that all of its components of the form

$$b_{i_1 \ldots i_{k-1}}\, dx^{i_1} \wedge \cdots \wedge dx^{i_{k-1}} \wedge dt, \qquad i_1 < \cdots < i_{k-1},$$

are identically zero), and ω_2 is a form of rank $(k-1)$ with the same property. Let ω be any form of rank k on the manifold M_2, and write $F^*(\omega) = \Omega =$

$\omega_1 + \omega_2 \wedge dt$, with ω_1 and ω_2 as just described, i.e. given locally by

$$\omega_2 = \sum_{i_1 < \cdots < i_{k-1}} a_{i_1 \ldots i_{k-1}}(x, t)\, dx^{i_1} \wedge \cdots \wedge dx^{i_{k-1}},$$

$$\omega_1 = \sum_{j_1 < \cdots < j_k} b_{j_1 \ldots j_k}(x, t)\, dx^{j_1} \wedge \cdots \wedge dx^{j_k}.$$

We now define (locally) a form $D\Omega$ of rank $(k-1)$ on the manifold $M_1 \times I$, by means of the formula

$$D\Omega = \sum_{i_1 < \cdots < i_{k-1}} \left(\int_1^2 a_{i_1 \ldots i_{k-1}}(x, t)\, dt \right) dx^{i_1} \wedge \cdots \wedge dx^{i_{k-1}}$$
$$= (-1)^{k-1} \int_1^2 \omega_2\, dt. \tag{5}$$

At this point we require a certain property of the form $D\Omega$, to establish which we now interrupt our proof.

1.4. Lemma. *The following formula holds (cf. the defining condition for an "algebraic homotopy" in* §2(5) *below):*

$$d(D(F^*(\omega))) \pm D(d(F^*(\omega))) = f_2^*(\omega) - f_1^*(\omega). \tag{6}$$

PROOF. We shall show that in fact for any form Ω on $M_1 \times I$, the following formula is valid:

$$dD(\Omega) \pm D(d\Omega) = \Omega|_{t=2} - \Omega|_{t=1}. \tag{7}$$

To this end we calculate $dD\Omega$ and $Dd\Omega$, with $\Omega = \omega_1 + \omega_2 \wedge dt$ as before. Locally we have (by definition of the operator d and its various properties—see §25.2 of Part I)

$$dD\Omega = \sum_{i_1 < \cdots < i_{k-1}} \sum_j \left(\int_1^2 \frac{\partial a_{i_1 \ldots i_{k-1}}}{\partial x^j}\, dt \right) dx^j \wedge dx^{i_1} \wedge \cdots \wedge dx^{i_{k-1}}.$$

On the other hand

$$D d\Omega = D(d\omega_1) + D(d\omega_2 \wedge dt)$$
$$= D\left(\sum_{j_1 < \cdots < j_k} \sum_q \frac{\partial b_{j_1 \ldots j_k}}{\partial x^q}\, dx^q \wedge dx^{j_1} \wedge \cdots \wedge dx^{j_k} + \sum_{j_1 < \cdots < j_k} \frac{\partial b_{j_1 \ldots j_k}}{\partial t}\, dt \wedge dx^{j_1} \wedge \cdots \wedge dx^{j_k} \right)$$
$$+ D\left(\sum_{i_1 < \cdots < i_{k-1}} \sum_p \frac{\partial a_{i_1 \ldots i_{k-1}}}{\partial x^p}\, dx^p \wedge dx^{i_1} \wedge \cdots \wedge dx^{i_{k-1}} \wedge dt \right)$$
$$= \sum_{j_1 < \cdots < j_k} (b_{j_1 \ldots j_k}(x, 2) - b_{j_1 \ldots j_k}(x, 1))\, dx^{j_1} \wedge \cdots \wedge dx^{j_k} + (-1)^{k-1}\, dD\Omega$$
$$= \Omega|_{t=2} - \Omega|_{t=1} + (-1)^{k-1}\, dD\Omega,$$

whence the desired formula (7). Putting $\Omega = F^*(\omega)$, so that $\Omega|_{t=2} = f_2^*(\omega)$, $\Omega|_{t=1} = f_1^*(\omega)$, formula (7) then yields (6), completing the proof of the lemma. □

We now return to the proof of the theorem. Let ω be any closed form on M_2 (so that $d\omega \equiv 0$). Then, since $dF^*(\omega) = F^*(d\omega) \equiv 0$, formula (6) yields

$$dDF^*(\omega) = f_2^*(\omega) - f_1^*(\omega),$$

so that the difference of the forms $f_2^*(\omega)$ and $f_1^*(\omega)$ is exact. Since this is by definition equivalent to the statement that the homomorphisms $f_1^*, f_2^*: H^k(M_2; \mathbb{R}) \to H^k(M_1; \mathbb{R})$ coincide, the proof of the theorem is complete. □

Recall (from §17.4 of Part II) that two manifolds M_1, M_2 are said to be *homotopically equivalent* if there exist (smooth) maps $f: M_1 \to M_2$ and $g: M_2 \to M_1$, such that the composites $gf: M_1 \to M_1$ and $fg: M_2 \to M_2$ are homotopic to the respective identity maps

$$M_1 \to M_1 \quad (x \mapsto x), \qquad M_2 \to M_2 \quad (y \mapsto y).$$

(Thus, for example, Euclidean space $\mathbb{R}^n$, as also the disc

$$D^n = \left\{ \sum_{\alpha=1}^{n} (x^\alpha)^2 \le R^2 \right\},$$

is homotopically equivalent to the one-point space, or what is equivalent, is *contractible* (over itself to a point), meaning that the identity map $\mathbb{R}^n \to \mathbb{R}^n$ $(x \mapsto x)$ is homotopic to a constant map $(\mathbb{R}^n \to \{0\})$.)

1.5. Theorem. *Homotopically equivalent manifolds have isomorphic cohomology groups.*

PROOF. Let M_1, M_2 be homotopically equivalent manifolds, and let $f: M_1 \to M_2$, $g: M_2 \to M_1$ be maps satisfying the defining conditions (see above) of homotopy equivalence. Consider the corresponding homomorphisms $f^*: H^k(M_2; \mathbb{R}) \to H^k(M_1; \mathbb{R})$ and $g^*: H^k(M_1; \mathbb{R}) \to H^k(M_2; \mathbb{R})$. Since the maps fg and gf are homotopic to the appropriate identity maps, it follows from Theorem 1.3 that the homomorphisms $(fg)^* = g^*f^*$ and $(gf)^* = f^*g^*$ are actually the corresponding identity homomorphisms:

$$1 = g^*f^*: H^k(M_2) \to H^k(M_2),$$

$$1 = f^*g^*: H^k(M_1) \to H^k(M_1).$$

Hence f^* and g^* are (mutually inverse) isomorphisms, and the theorem is proved. □

Remark. This theorem suggests a way of extending the definition of the cohomology groups to any topological space X with the property that there is a manifold M in which it can be embedded ($M \supset X$) which "contracts" to

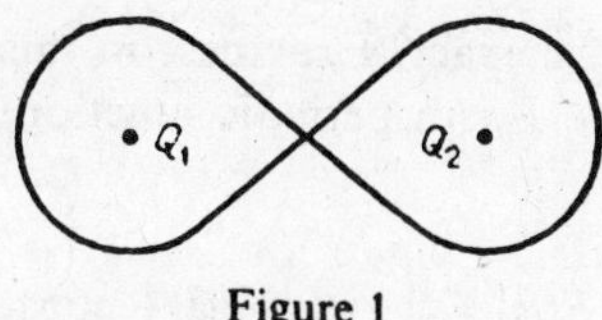

Figure 1

it, in the sense that the inclusion map $i: X \to M$ is a homotopy equivalence (so that there is a map $f: M \to X$ with the property that if and fi are homotopic to the appropriate identity maps). For such spaces X we simply define

$$H^k(X; \mathbb{R}) \equiv H^k(M; \mathbb{R}). \tag{8}$$

Thus, for instance, the "figure eight", while not a manifold, will now, according to this definition, have the same cohomology groups as $\mathbb{R}^2 \setminus \{Q_1, Q_2\}$, the plane with two points removed (see Figure 1).

1.6. Corollary. *The cohomology groups of Euclidean space $\mathbb{R}^n$ (and of the disc D^n) are isomorphic to those of a one-point space. Thus $H^k(\mathbb{R}^n)$ is trivial for $k > 0$, while $H^0(\mathbb{R}^n) \simeq \mathbb{R}$, the one-dimensional real vector space.*

This fact leads almost immediately to the so-called "Poincaré lemma": *Locally, i.e. in some neighbourhood of any point Q of a manifold M^n, every closed form ω ($d\omega \equiv 0$) of rank > 0 is exact: $\omega = d\omega'$.* To see this, we have merely to choose as the neighbourhood any disc $D^n = \{\sum_{\alpha=1}^n (x^\alpha - x_0^\alpha)^2 \leq \varepsilon\}$ with centre Q, wholly contained in some local co-ordinate neighbourhood (i.e. chart) of the manifold, and then apply the conclusion of Corollary 1.6, to the effect that $H^k(D^n) = 0$ for $k > 0$.)

The reader will no doubt recall the case $k = 1$ of the Poincaré lemma from courses in analysis: Given a 1-form $\omega = f_k\, dx^k$ with $d\omega \equiv 0$ (i.e. $\partial f_k/\partial x^i \equiv \partial f_i/\partial x^k$ in local notation), we have $\omega = dF$ where $F(P) = \int_Q^P f_k\, dx^k$, the (path-independent) line integral of the form along any smooth path in the disc from a fixed point Q to the variable point P.

What are the cohomology groups of the circle S^1?

1.7. Proposition. *The cohomology groups of the circle S^1 are as follows:*

$$\begin{aligned} H^k(S^1; \mathbb{R}) &= 0 \qquad \textit{for} \quad k > 1; \\ H^1(S^1; \mathbb{R}) \simeq \mathbb{R}; &\qquad H^0(S^1; \mathbb{R}) \simeq \mathbb{R}. \end{aligned} \tag{9}$$

Proof. The triviality of the cohomology groups of S^1 for $k > 1$ is immediate from the fact that $\dim S^1 = 1$. That $H^0(S^1) \simeq \mathbb{R}$ follows from Proposition 1.2 and the connectedness of S^1. Thus we have only to show that $H^1(S^1) \simeq \mathbb{R}$.

To this end we introduce on S^1 the usual local co-ordinate φ, where for all integers n the numbers $\varphi + 2\pi n$ represent the same point of the circle as φ. A form of rank 1 is then given by $\omega = a(\varphi)\, d\varphi$, where $a(\varphi)$ is a periodic function on $\mathbb{R}$: $a(\varphi + 2\pi) = a(\varphi)$. We always have $d\omega = 0$, again since $\dim S^1 = 1$.

When will $\omega = a(\varphi)\,d\varphi$ be exact? Exactness in this context means precisely that $a(\varphi)\,d\varphi = dF$, where F is a periodic function, or equivalently that the function defined by

$$F(\varphi) = \int_0^{\varphi} a(\psi)\,d\psi + \text{const.}$$

is periodic of period 2π or, in yet other words, that $\int_{S^1} \omega = 0$.

We see therefore that a 1-form $\omega = a(\varphi)\,d\varphi$ on S^1 is exact precisely if $\int_{S^1} \omega = 0$, i.e. $\int_0^{2\pi} a(\varphi)\,d\varphi = 0$. Hence two 1-forms $\omega_1 = a(\varphi)\,d\varphi$ and $\omega_2 = b(\varphi)\,d\varphi$ determine the same cohomology class if and only if

$$\int_{S^1} \omega_1 = \int_{S^1} \omega_2, \quad \text{i.e.} \quad \int_0^{2\pi} a(\varphi)\,d\varphi = \int_0^{2\pi} b(\varphi)\,d\varphi,$$

so that the cohomology classes are in (appropriate) one-to-one correspondence with the possible values of such integrals, i.e. with $\mathbb{R}$. This completes the proof. □

1.8. Corollary. *The cohomology groups of the Euclidean plane with one point removed* $\mathbb{R}^2\backslash Q$ *(or an annulus), being (by Theorem* 1.5) *isomorphic to those of a circle, are as follows*:

$$H^k(\mathbb{R}^2\backslash Q) = 0, \quad k > 1; \qquad H^1(\mathbb{R}^2\backslash Q) \simeq H^0(\mathbb{R}^2\backslash Q) \simeq \mathbb{R}. \tag{10}$$

Remark. We indicate another method for calculating the first cohomology group $H^1(S^1)$ of the circle. With each 1-form $\omega(\varphi) = a(\varphi)\,d\varphi$ on the circle, we associate its *average* $\hat{\omega}$ (also a form) defined by

$$\hat{\omega} = \frac{1}{2\pi}\int_0^{2\pi} \omega(\varphi + \tau)\,d\tau = \frac{1}{2\pi}\left[\int_0^{2\pi} a(\varphi + \tau)\,d\tau\right] d\varphi.$$

1.9. Proposition. *The forms* ω *and* $\hat{\omega}$ *are cohomologous.*

PROOF. For each fixed τ the form $\omega(\varphi + \tau)$ is induced from ω via the map $\varphi + \tau \mapsto \varphi$ of the circle onto itself. Since such a map is homotopic to the identity, we have (by Theorem 1.3) that $\omega(\varphi) \sim \omega(\varphi + \tau)$. For an arbitrary Riemann sum for the form $\hat{\omega}$ (as an integral) we shall therefore have

$$\frac{1}{2\pi}\sum_i \omega(\varphi + \tau_i)\Delta\tau_i \sim \omega(\varphi)\cdot\frac{1}{2\pi}\sum_i \Delta\tau_i = \omega(\varphi). \tag{11}$$

Since any Riemann sum for $\hat{\omega}$ is thus cohomologous to ω, it follows that $\hat{\omega}$ will also be cohomologous to ω, as required. □

Continuing with our remark, we note next that $\hat{\omega}$ is given by

$$\hat{\omega}(\varphi) = \alpha\,d\varphi, \qquad \text{where} \quad \alpha = \text{const.} = \frac{1}{2\pi}\int_0^{2\pi} a(\psi)\,d\psi,$$

since

$$\hat{\omega}(\varphi) = \frac{1}{2\pi}\left[\int_0^{2\pi} a(\varphi + \tau)\,d\tau\right] d\varphi = \frac{1}{2\pi}\left[\int_{\varphi}^{2\pi+\varphi} a(\psi)\,d\psi\right] d\varphi$$
$$= \frac{1}{2\pi}\left[\int_0^{2\pi} a(\psi)\,d\psi\right] d\varphi.$$

(Thus the form $\hat{\omega}(\varphi)$ is, as they say, "rotation-invariant": $\hat{\omega}(\varphi + \varphi_0) = \hat{\omega}(\varphi)$.) From this and the above proposition, we see that the correspondence $\omega \mapsto \hat{\omega}$ essentially associates (in what is clearly an appropriate one-to-one manner) a real number, namely α, with each 1-form ω on the circle, whence $H^1(S^1) \simeq \mathbb{R}$. In the sequel we shall use a generalization of this method to calculate the cohomology groups of compact homogeneous spaces.

1.10. Proposition. *An orientable, closed, Riemannian manifold M^n of dimension n has non-trivial nth cohomology group $H^n(M^n)$.*

Proof. As usual we denote by Ω the volume element on M; thus locally

$$\Omega = \sqrt{|g|}\,dx^1 \wedge \cdots \wedge dx^n,$$

where $g = \det(g_{ij})$, (g_{ij}) being the Riemannian metric with which we are assuming our manifold endowed. If the local co-ordinates on the charts of M^n are all arranged to agree in orientation (i.e. so that the Jacobians of the transition functions on the regions of overlap are all positive), then (see Part I, §18.2) Ω can be regarded as a differential form of rank n on M^n, which can therefore be integrated over M, yielding its volume $\int_{M^n} \Omega > 0$.

Since M^n has dimension n and $d\Omega$ has rank $n + 1$, we must of course have $d\Omega = 0$, i.e. Ω is closed. If Ω were in fact exact, say $\Omega = d\omega$, then by the general Stokes formula (Part I, §26.3) we should have

$$\int_{M^n} \Omega = \int_{M^n} d\omega = \int_{\partial M^n} \omega = 0, \tag{12}$$

since by hypothesis M^n is without boundary. Hence we have found a closed n-form which is not exact (namely Ω), whence the proposition. □

Remark. It will be shown below (in §3) that on the other hand for every non-orientable closed manifold M^n (for example, $M^2 = \mathbb{R}P^2$, the projective plane) the group $H^n(M^n; \mathbb{R})$ is trivial. (Of course, the above proof fails for such manifolds since the volume element does not behave like a differential form under co-ordinate changes with negative Jacobian.)

For any manifold M^n we write

$$H^*(M) = \sum_{k=0}^{n} H^k(M^n), \tag{13}$$

the direct sum of (all) the cohomology groups of M. The following proposition shows that the wedge (or exterior) product of forms can be used to define a "multiplicative" operation on $H^*(M)$, thereby turning it into a ring.

1.11. Proposition. *For any closed forms ω_1, ω_2 on M^n, the forms $\omega_1 \wedge \omega_2$ and $(\omega_1 + d\omega') \wedge \omega_2$ are also closed, and moreover cohomologous.*

PROOF. By Leibniz' formula (see Part I, Theorem 25.2.4) we have

$$d(\omega' \wedge \omega_2) = d\omega' \wedge \omega_2 \pm \omega' \wedge d\omega_2 = d\omega' \wedge \omega_2. \tag{14}$$

Hence

$$(\omega_1 + d\omega') \wedge \omega_2 = \omega_1 \wedge \omega_2 + d(\omega' \wedge \omega_2),$$

so that $\omega_1 \wedge \omega_2$ and $(\omega_1 + d\omega') \wedge \omega_2$ are cohomologous, as required. (The closure of $\omega_1 \wedge \omega_2$ is immediate from Leibniz' formula.) □

In view of this proposition the exterior-product operation on $H^*(M)$ is well defined. It is easy to see that with this as its multiplicative operation $H^*(M)$ becomes a ring (in fact, an algebra), called the *cohomology ring* of the manifold M^n. Note that if $\omega_1 \in H^p(M^n)$, $\omega_2 \in H^q(M^n)$, then $\omega_1\omega_2 \in H^{p+q}(M^n)$, and that the multiplication in $H^*(M)$ is skew-commutative in the sense that (see Part I, Lemma 18.3.1)

$$\omega_2\omega_1 = (-1)^{pq}\omega_1\omega_2. \tag{15}$$

We shall now describe the geometric significance of the cohomology groups. (More precise considerations will be left to later sections.)

Given any manifold M^n we define "periods", or "integrals over cycles", of any closed form ω (of rank k) on M^n, as follows. As a preliminary, we define a *cycle* in M^n to be a pair (M^k, f), where M^k is any k-dimensional manifold (of dimension equal to the rank of ω) and $f: M^k \to M^n$ is any smooth map.

1.12. Definition. The *period* of a k-form ω on M^n with respect to a cycle (M^k, f) is the integral $\int_{M^k} f^*(\omega)$.

Let N^{k+1} be any oriented manifold-with-boundary. Its boundary $\partial N^{k+1} = M^k$ say, is then a closed, oriented manifold (which may have several connected components). We define a *film* (see Appendix 2 for an explanation of this name) to be a map $F: N^{k+1} \to M^n$ from the manifold-with-boundary N^{k+1} to the manifold M^n under consideration.

1.13. Theorem

(i) *The period of an exact form ω on M^n with respect to any cycle (M^k, f) is zero.*
(ii) *The period of a closed form ω on M^n is zero with respect to any cycle (M^k, f) in M^n which is the boundary of a film (N^{k+1}, F) (i.e. is such that $M^k = \partial N^{k+1}$ and $F|_{M^k} = f$).*

PROOF. (i) Writing $\omega = d\omega'$, we have by the general Stokes formula

$$\int_{M^k} f^*\omega = \int_{M^k} f^*(d\omega') = \int_{M^k} d(f^*\omega') = \int_{\partial M^k} f^*\omega' = 0, \tag{16}$$

where the last equality is a consequence of the fact that the manifold M^k is without boundary.

(ii) Since M^k is the boundary of N^{k+1} (with orientation induced from that of N^{k+1}), and $F|_{M^k} = f$, the general Stokes formula yields

$$\int_{M^k} f^*\omega = \int_{N^{k+1}} dF^*(\omega) = \int_{N^{k+1}} F^*(d\omega) = 0, \tag{17}$$

where in the last equality we have used the hypothesis $d\omega = 0$. □

We note without proof the following important fact (a partial converse to part (i) of the above theorem): *If the period of a closed form is zero with respect to every cycle, then the form is exact.* (See §14 below.)

Example. *For the n-dimensional sphere S^n we have $H^k(S^n) = 0$ for $k \neq 0, n$.*

PROOF. For $k > n$ it is trivial that $H^k(S^n) = 0$, so we may assume $0 < k < n$. If (M^k, f) is any cycle in S^n (where $0 < k < n$), then by Sard's theorem (Theorem 10.2.1 of Part II), there are certainly points of S^n outside $f(M^k)$. If $Q \in S^n$ is such a point, then the cycle (M^k, f) may be regarded as a cycle in $S^n \backslash Q \cong \mathbb{R}^n$. Now, essentially by Poincaré's lemma (see above), every closed form on $\mathbb{R}^n$ is exact, so that by Theorem 1.13(i) the period of every closed k-form with respect to the cycle (M^k, f) is zero. Since the cycle (M^k, f) was arbitrary, it follows from the above-mentioned partial converse of Theorem 1.13(i) that every closed k-form on S^n is exact, whence (for $0 < k < n$) $H^k(S^n) = 0$. □

This fact can also be established by means of an argument analogous to that used above for calculating $H^1(S^1)$ (in the remark following Corollary 1.8): one first shows that each cohomology class of closed k-forms on S^n contains a form ω invariant under the group $SO(n+1)$ of (proper) isometries of S^n. Such a form is of course determined by its components at a single point of the sphere, and these components will be invariant under the stationary group $SO(n) \subset SO(n+1)$ fixing that point (i.e. under the stabilizer of that point). We leave it to the reader to deduce that if $0 < k < n$ then these components must all be zero. (Consider to begin with the case of a 1-form on $S^2 \cong \mathbb{R}^2 \cup \{\infty\}$ whose components at the origin of $\mathbb{R}^2$ are rotation-invariant.)

We shall now show how an analogue of this method can be used for calculating the cohomology groups of Lie groups and symmetric spaces.

Recall (from §6 of Part II) that a homogeneous space (see §5.1 of Part II) M of a Lie group G, with isotropy group H, is said to be *symmetric* if there is an involutory Lie automorphism of G, i.e. a Lie automorphism $I: G \to G$ such that $I^2 = 1$ and $I|_H = 1$ (so that the automorphism I fixes H pointwise); it is also required that all points fixed by I that are sufficiently close to the identity element of G, should lie in the subgroup H. Corresponding to each point x of such a manifold M there is then a naturally determined "symmetry" s_x of M, whose effect on an arbitrary point y of M is defined as follows: since

G acts transitively on M we must have $y = g(x)$ for some $g \in G$; we set

$$s_x(y) = s_x(g(x)) = I(g)(x). \tag{18}$$

It is easily verified that s_x is a well-defined self-diffeomorphism of M, and that $s_x^2 = 1$ and $s_x(x) = x$. It is also not difficult to see that $(s_x)_*$, the induced map of the tangent space to M at x, is the reflection in the origin (cf. Definition 6.1.1 of Part II, in which this figured as a defining property of a symmetric space).

Note that in particular every compact Lie group G is a symmetric space of the group $G \times G$, where the action of $G \times G$ on G is defined by

$$T_{(g,h)}(x) = gxh^{-1}. \tag{19}$$

Here the isotropy group H is the diagonal $\{(g, g)\}$, the involution I is given by $I(g, h) = (h, g)$, and the symmetry s_x corresponding to the identity element e of G is easily verified to be given by

$$s_e(g) = g^{-1}.$$

Of particular importance among the differential forms on an arbitrary homogeneous space of a Lie group G are those that are *invariant* under the action of G, i.e. satisfy $g^*\omega = \omega$ for all $g \in G$. It is easy to see that the differential of such an invariant form is again invariant:

$$g^*d\omega = d(g^*\omega) = d\omega; \tag{20}$$

and that the wedge product preserves invariance:

$$g^*(\omega_1 \wedge \omega_2) = g^*\omega_1 \wedge g^*\omega_2 = \omega_1 \wedge \omega_2.$$

It follows that the set of all invariant forms on a homogeneous space M forms a ring. It turns out that in the computation of the cohomology ring of any homogeneous space of a compact, connected Lie group, the invariant forms play an essential part. Indeed, if the homogeneous space is symmetric then its cohomology ring coincides (in essence) with its ring of invariant forms; that is the import of the following

1.14. Theorem. *If M is a compact, symmetric space of a compact, connected Lie group G, then:*

(i) *every invariant form on M is closed;*
(ii) *every closed form on M is cohomologous to some invariant form; and*
(iii) *no non-zero invariant form on M is cohomologous to zero.*

PROOF. (i) Let ω be any invariant form on M, of rank k say, and consider the form $s_x^*\omega = \hat{\omega}$. We shall show first that $\hat{\omega}$ also is invariant. From (18) it follows, writing T_g for the transformation of M determined by $g \in G$, that for all $x \in M$, $g \in G$,

$$s_x T_g = T_{Ig} s_x. \tag{21}$$

(To see this let $y = T_h(x)$ be any element of M; then on the one hand

$$s_x T_g(y) = s_x T_g T_h(x) = s_x T_{gh}(x) = T_{I(gh)}(x),$$

where the last equality is a consequence of (18), and on the other hand

$$T_{Ig} s_x(y) = T_{Ig} s_x T_h(x) = T_{Ig} T_{Ih}(x) = T_{I(gh)}(x).)$$

Hence

$$T_g^* \hat{\omega} = T_g^* s_x^* \omega = (s_x T_g)^* \omega = s_x^* T_{Ig}^* \omega = s_x^* \omega = \hat{\omega},$$

where the third and fourth equalities come respectively from (21) and the assumed invariance of ω. Thus $\hat{\omega}$ is invariant, as claimed.

Since the transformation s_x induces the reflection in its origin of the tangent space at the point x, it follows that $\hat{\omega}|_x = (-1)^k \omega|_x$. The forms ω and $\hat{\omega}$ being invariant under the action of G (which is transitive by definition of a homogeneous space), we infer that this equality must hold at every point of M, so that in fact

$$\hat{\omega} = (-1)^k \omega. \tag{22}$$

It is then immediate that also $d\hat{\omega} = (-1)^k \, d\omega$. However, the forms $d\omega$ and $d\hat{\omega}$ are also invariant and related by $s_x^* \, d\omega = d\hat{\omega}$, so that the above argument applied to them (in place of ω and $\hat{\omega}$) yields, since their rank is $k + 1$,

$$d\hat{\omega} = (-1)^{k+1} \, d\omega.$$

From this and (22) we conclude that $d\omega = 0$, establishing statement (i) of the theorem.

(ii) Let ω be any closed form on M: $d\omega = 0$. Since G is compact, an invariant metric can be defined on it (see Part I, §24.4 and Part II, §8.3). Such a metric determines an invariant volume-element on G, which we shall denote by $d\mu(g)$; thus, for all $g, h \in G$,

$$d\mu(hg) = d\mu(g). \tag{23}$$

We may suppose this volume element normalized so that the volume of the whole group G is 1:

$$\int_G d\mu(g) = 1. \tag{24}$$

This assumed, we define a form $\tilde{\omega}$, in terms of our arbitrary closed form ω on M, by

$$\tilde{\omega} = \int_G T_g^* \omega \, d\mu(g). \tag{25}$$

We claim that the form $\tilde{\omega}$ is invariant and cohomologous to ω. Its invariance follows directly by computing the form $T_h^*\tilde{\omega}$ for arbitrary $h \in G$:

$$T_h^*\tilde{\omega} = \int_G T_{hg}^*\omega \, d\mu(g) = \int_G T_{hg}^*\omega \, d\mu(hg)$$
$$= \int_G T_{g'}^*\omega \, d\mu(g') = \tilde{\omega},$$

where in the second equality we have used (23), and in the third we have put $g' = hg$ (such a change of variables being of course smooth and invertible).

It remains to show that the form $\tilde{\omega}$ is cohomologous to ω. Observe first that for each $g \in G$, the map T_g of the manifold M to itself is homotopic to the identity map on M; for if $g(t)$ is any curve in the group G joining the point g to the identity element, then $T_{g(t)}$ will serve as a homotopy of the desired kind. (Remember that by hypothesis G is connected.) Hence by Theorem 1.3 the forms $T_g^*\omega$ and ω are cohomologous: $T_g^*\omega \sim \omega$. Consequently

$$\tilde{\omega} = \int_G T_g^*\omega \, d\mu(g) \sim \int_G \omega \, d\mu(g) = \omega \int_G d\mu(g) = \omega,$$

where in the last equality we have invoked (24). This completes the proof of statement (ii).

(iii) We wish to show (finally) that a non-zero invariant form ω on a compact symmetric space is never cohomologous to zero. Recall first that on the manifold M there can be defined a Riemannian metric (h_{ij}) invariant under the action of G (see Part II, §8.3), and also that in terms of such a metric one can then define the operator $*$ on forms ω on M (see Part I, §19.3). Since the rank of $*\omega$ is equal to (dim M − rank ω), this enables us to define in turn a "scalar square" of ω by setting

$$\langle \omega, \omega \rangle = \int_M \omega \wedge *\omega. \tag{26}$$

This scalar is always positive (provided $\omega \neq 0$), since if in local co-ordinates

$$\omega = \sum_{i_1 < \cdots < i_k} a_{i_1 \ldots i_k} \, dx^{i_1} \wedge \cdots \wedge dx^{i_k},$$

then

$$\langle \omega, \omega \rangle = \int \omega \wedge *\omega = \int h^{i_1 j_1} \ldots h^{i_k j_k} a_{i_1 \ldots i_k} a_{j_1 \ldots j_k} \sqrt{h} \, dx^1 \wedge \cdots \wedge dx^n > 0. \tag{27}$$

(Here as usual (h^{ij}) denotes the matrix inverse to (h_{ij}), $h = \det(h_{ij})$, and $n =$ dim M.)

In view of the invariance of the metric (h_{ij}), the operator $*$ commutes with every operator T_g^*, $g \in G$. From this, together with the invariance of our form ω, we infer that the form $*\omega$ is also invariant, and therefore, by virtue of the already-established statement (i), is closed: $d(*\omega) = 0$.

Suppose now that, contrary to statement (iii), ω is cohomologous to zero, i.e. is exact: $\omega = d\omega'$ for some form ω'. Then

$$d(\omega' \wedge *\omega) = d\omega' \wedge *\omega \pm \omega' \wedge d(*\omega) = \omega \wedge *\omega.$$

Hence

$$\langle \omega, \omega \rangle = \int_M \omega \wedge *\omega = \int_M d(\omega' \wedge *\omega) = 0$$

by Stokes' theorem (since M is without boundary). However this contradicts (27). This completes the proof of the theorem. □

We now consider some examples.

Examples

(a) The torus $T^n = \mathbb{R}^n/\Gamma$, where Γ is the integral lattice in $\mathbb{R}^n$ spanned by n independent vectors, is a compact commutative Lie group, and can be considered as a symmetric space of the Lie group $\mathbb{R}^n$, or alternatively of itself.

Let $x^1, \ldots, x^n$ be Euclidean co-ordinates in $\mathbb{R}^n$. Each basic form $dx^{i_1} \wedge \cdots \wedge dx^{i_k}$ on $\mathbb{R}^n$ is invariant under the action of $\mathbb{R}^n$ on itself (i.e. under translations), and therefore defines an invariant form on the torus T^n. If, on the other hand, a form

$$\omega = a_{i_1 \ldots i_k}(x)\, dx^{i_1} \wedge \cdots \wedge dx^{i_k}$$

on T^n (in terms of local co-ordinates on T^n induced from $x^1, \ldots, x^n$ on $\mathbb{R}^n$) is invariant, then clearly we must have for all y that

$$a_{i_1 \ldots i_k}(x + y) = a_{i_1 \ldots i_k}(x),$$

i.e. that the coefficients of the form ω are constant:

$$a_{i_1 \ldots i_k} = \text{const}.$$

We deduce that the invariant forms on T^n are just the linear combinations with constant coefficients of exterior products of the forms $dx^1, \ldots, dx^n$. From Theorem 1.14 we therefore conclude that: *The cohomology ring $H^*(T^n)$ of the torus T^n is (isomorphic to) the free exterior algebra $\bigwedge[e_1, \ldots, e_n]$ with free generators $e_1, \ldots, e_n$ (of degree* 1). (Here e_i denotes the cohomology class of the form dx^i.)

(b) By way of a second example we examine the situation where the manifold is actually a compact Lie group G (considered as a symmetric space of the direct product $G \times G$; see above). Here the invariant forms on G will be those that are two-sided invariant in the sense of being unaffected by left and right multiplicative by the elements of G.

We consider, to begin with, left-invariant 1-forms ω on G which are vector-valued (rather than scalar-valued), taking their values in the Lie algebra $\mathfrak{g}$ of the Lie group G. For an arbitrary Lie group G a vector-valued, left-invariant 1-form can be constructed as follows (and in fact every such 1-form arises in this way): the 1-form is of course determined by its value at each tangent vector ξ to G at each point g; the value we assign is that vector in the tangent space to G at the identity (i.e. in the Lie algebra of G), obtained by translating ξ back to the identity, i.e. by applying to it the left-translation operator $(L_{g^{-1}})_*$ (cf. Part II, §25.1). In view of this construction we denote such a form ω by $\omega = \omega(g) = g^{-1}\,dg$. (If G is a matrix Lie group then we shall have $g = (g_{ik})$, $dg = (dg_{ik})$, and ω will literally be the matrix product $g^{-1}\,dg = (\omega_{ik})$, a matrix whose entries are ordinary (scalar-valued) 1-forms.) It is immediate that a form defined in this way is indeed left-invariant, i.e., in the above notation,

$$\omega(hg) = g^{-1}h^{-1}d(hg) = g^{-1}\,dg = \omega(g). \tag{28}$$

We now turn our attention to the ordinary (i.e. scalar-valued) forms on a Lie group G. Let $\theta^1, \ldots, \theta^N$ form a basis for the vector space of scalar-valued left-invariant 1-forms on G. (For a matrix Lie group G one may (essentially) take as the θ^i the elements of a maximal linearly independent subset of the set of entries of the form $\omega = (\omega_{ik}) = g^{-1}\,dg$. For instance, if $G = SO(n)$, then the Lie algebra $\mathfrak{g} = so(n)$ consists of all skew-symmetric matrices, so that $\omega = (\omega_{ik})$ is also skew-symmetric, and as a basis for the left-invariant 1-forms on $SO(n)$ we may take the ω_{ik} with $i < k$.)

1.15. Lemma. *The dimension N of the vector space of (scalar-valued) left-invariant 1-forms on a Lie group is equal to the dimension of the group.*

Proof. It follows directly from the left-invariance that each left-invariant 1-form is completely determined by the values it takes on the tangent space at the identity of the group. Since such a form may clearly be constructed so as to coincide at the identity with any element of the dual space of the tangent space at the identity, the lemma follows. □

Since as far as its vector-space structure is concerned, the Lie algebra of a Lie group is the same as the tangent space at the identity of the group, we deduce the

1.16. Corollary. *The vector space of all left-invariant 1-forms on a Lie group G is (naturally) isomorphic to the space $\mathfrak{g}^*$ of linear functionals on the Lie algebra $\mathfrak{g}$ of G.*

1.17. Lemma. *Every left-invariant k-form ω on a Lie group G has the form*

$$\omega = \sum_{i_1 < \cdots < i_k} a_{i_1 \ldots i_k} \theta^{i_1} \wedge \cdots \wedge \theta^{i_k}, \tag{29}$$

where the $a_{i_1 \ldots i_k}$ are constants (and the θ^i form a basis for the space of left-invariant 1-forms on G).

PROOF. By Lemma 1.16, at the identity e of G each k-form ω (whether or not it is invariant) can be represented as

$$\omega(e) = \sum_{i_1 < \cdots < i_k} a_{i_1 \ldots i_k} \theta^{i_1}(e) \wedge \cdots \wedge \theta^{i_k}(e). \tag{30}$$

The invariance of the forms $\theta^1, \ldots, \theta^N$ and ω then implies that (30) holds at every point $g \in G$. □

1.18. Corollary. *The algebra of left-invariant forms on a Lie group G is isomorphic to the exterior algebra $\bigwedge[\mathfrak{g}^*]$ generated by the space $\mathfrak{g}^*$ of linear functionals on the Lie algebra $\mathfrak{g}$, i.e. to the algebra of skew-symmetric* (*or "alternating"*) *multilinear maps on the Lie algebra $\mathfrak{g}$.*

Since we are interested in the two-sided invariant forms on our Lie group G, it is now appropriate to ask for those among the left-invariant forms on G which are also right-invariant. Clearly, a left-invariant 1-form ω on G will also be invariant under a right translation $y \mapsto yh^{-1}$ ($y \in G$, $g \in G$) if and only if it is invariant under the inner automorphism $y \mapsto hyh^{-1}$, i.e. if ω is invariant under the operator $\mathrm{Ad}(h)$ (which as usual we shall sometimes also indicate by the notation $\mathrm{Ad}(h)\colon X \mapsto hXh^{-1}$, $X \in \mathfrak{g}$). Hence we have

1.19. Lemma. *A skew-symmetric multilinear map $\varphi(X_1, \ldots, X_k)$ belonging to $\bigwedge[\mathfrak{g}^*]$ corresponds (as in the preceding corollary) to a two-sided invariant form if and only if it is Ad-invariant, i.e.*

$$\varphi(hX_1h^{-1}, \ldots, hX_kh^{-1}) = \varphi(X_1, \ldots, X_k) \qquad \textit{for all} \quad h \in G. \tag{31}$$

From Theorem 1.14 we immediately conclude that:

The cohomology ring of a connected, compact Lie group G is (*naturally*) *isomorphic to the ring $\bigwedge_{\mathrm{inv}}[\mathfrak{g}^*]$ of Ad-invariant multilinear alternating maps on the Lie algebra $\mathfrak{g}$.*

We shall now indicate (by way of an application) how one can deduce from this the non-triviality of $H^3(G)$ for a large class of Lie groups G. Thus let $\langle\ ,\ \rangle$ denote the Killing form on the Lie algebra $\mathfrak{g}$ of a Lie group G (see Definition 3.1.3 of Part II). We define a 3-linear map $\Omega(X, Y, Z)$ on the Lie algebra $\mathfrak{g}$ by setting

$$\Omega(X, Y, Z) = \langle [X, Y], Z \rangle. \tag{32}$$

The skew-symmetry of this map in the variables X, Y, Z represents simply a basic property of the Killing form (see Part I, §24.4 or Part II, §6.4). Since, moreover, $\mathrm{Ad}(h)$ is for each $h \in G$ a Lie algebra automorphism of $\mathfrak{g}$, so that in particular

$$[hXh^{-1}, hYh^{-1}] = h[X, Y]h^{-1},$$

it follows that the 3-linear map Ω is also Ad-invariant. Hence we can infer immediately from the preceding statement the following

1.20. Proposition. *The group $H^3(G)$ is non-trivial for every compact Lie group G for which the form Ω defined by* (32) *is non-zero (and so in particular if the Killing form of G is non-degenerate, i.e. for semisimple G; see the concluding remark of* §3.1 *of Part II).*

(Note in connexion with this proposition the fact that a compact Lie group is abelian if and only if its Lie algebra is commutative; see Part II, Corollary 3.1.2.)

(c) In this our third example, we take M once more to be a general symmetric space of a Lie group G, with corresponding isotropy group H. Fixing arbitrarily on a point x of M, we obtain a corresponding map $p: G \to M$, defined by $p(g) = T_g(x)$. Using this map p, any form ω given on the manifold M can be pulled back to the form $p^*\omega$ on G. It follows from the definition of the pullback operator (see Part I, §22.1), that the form $p^*\omega$ takes the value zero on the tangent space to the subgroup H (a closed, and therefore Lie, subgroup of G). Since the map p sends each left coset gH of the isotropy group to a single point of M, it follows also that the form $p^*\omega$ is invariant under right translations by elements of H. If, furthermore, ω is invariant under the action of G on M, then the form $p^*\omega$ on G will be left-invariant, i.e. invariant under left translations by elements of G. Since right H-invariance and left G-invariance of $p^*\omega$ are together equivalent to left G-invariance plus invariance under inner automorphism of G determined by elements of H, and since the correspondence $\omega \leftrightarrow p^*\omega$ is an exterior-algebra isomorphism, we have as the upshot of the foregoing (together with Corollary 1.18) the following

1.21. Theorem. *The ring of invariant differential forms on a homogeneous space M of a group G, with isotropy group H, is isomorphic to the exterior algebra $\Lambda_{\mathrm{inv}}[(\mathfrak{g}/\mathfrak{h})^*]$, the algebra of multilinear alternating maps on $\mathfrak{g}$ which vanish on the Lie algebra $\mathfrak{h}$ of H, and are invariant under the operators* $\mathrm{Ad}(h)$ *for all h in H.*

Invoking Theorem 1.14 yet again we can now infer an isomorphism between the cohomology ring of a (suitable) symmetric space and the exterior algebra figuring in this result.

(d) In this our final example we compute (partially) the cohomology ring of the n-dimensional complex projective space

$$\mathbb{C}P^n \cong U(n+1)/(U(1) \times U(n)). \tag{33}$$

Now since $\mathbb{C}P^n$ is compact (see Part II, §2.2), and the group $U(n+1)$ is compact (being identifiable with a closed surface in $\mathbb{R}^{2(n+1)^2}$) and connected (since $U(n+1)/U(n) \cong S^{2n+1}$), we have, once again by Theorem 1.14, that the cohomology ring of $\mathbb{C}P^n$ is isomorphic to the ring of differential forms on $\mathbb{C}P^n$ invariant under $U(n+1)$.

Let $(z^0, \ldots, z^n)$ be, as usual, homogeneous co-ordinates on $\mathbb{C}P^n$ (i.e. $(n+1)$-

tuples from $\mathbb{C}^{n+1}\setminus\{0\}$, where two $(n+1)$-tuples are regarded as equivalent if one is a (complex) multiple of the other), and consider in $\mathbb{C}^{n+1}$ the real differential 2-form

$$\Omega = \frac{i}{2}\sum_k dz^k \wedge d\bar{z}^k. \tag{34}$$

It is straightforward to verify (directly from the fact that $U(n+1)$ preserves the Hermitian scalar product; see Part I, §11.2) that the form Ω on $\mathbb{C}^{n+1}$ is invariant under the action of $U(n+1)$. We shall now show that as a $U(n+1)$-invariant 2-form on S^{2n+1}, Ω is the pullback of some (invariant) form ω on $\mathbb{C}P^n$, i.e. that $\Omega = p^*\omega$ where $p\colon S^{2n+1} \to \mathbb{C}P^n$ is the natural projection. For this it clearly suffices to show that the form Ω is preserved by transformations of S^{2n+1} of the form

$$\begin{aligned} z^k &\to e^{i\varphi}z^k \qquad (dz^k \to e^{i\varphi}(dz^k + iz^k\, d\varphi)),\\ \bar{z}^k &\to e^{-i\varphi}\bar{z}^k \qquad (d\bar{z}^k \to e^{-i\varphi}(d\bar{z}^k - i\bar{z}^k\, d\varphi)). \end{aligned} \tag{35}$$

Now since $\sum_{k=0}^n z^k\bar{z}^k = 1$ on the sphere S^{2n+1}, we have

$$\sum (z^k\, d\bar{z}^k + \bar{z}^k\, dz^k) = 0.$$

Hence under the transformation (35) our form Ω transforms as follows:

$$\begin{aligned} \frac{i}{2}\sum dz^k \wedge d\bar{z}^k \to \frac{i}{2}\sum dz^k \wedge d\bar{z}^k + i\, d\varphi \wedge \sum (z^k\, d\bar{z}^k + \bar{z}^k\, dz^k)\\ = \frac{i}{2}\sum dz^k \wedge d\bar{z}^k. \end{aligned}$$

Thus we have found an invariant 2-form ω on $\mathbb{C}P^n$. Its powers ω^k are all non-zero for $k \le n$, since the corresponding powers of Ω are non-zero (verify this!). We conclude that:

The cohomology algebra $H^*(\mathbb{C}P^n)$ of the complex projective space $\mathbb{C}P^n$ contains as subalgebra the algebra of polynomials over $\mathbb{C}$ in the rank-2 form ω (which of course satisfies $\omega^{n+1} = 0$).

(It follows from results to be established in §4 (see Example (g) there, and also the first example in §7) that in fact there are no other elements of $H^*(\mathbb{C}P^n)$.)

§2. The Homology Theory of Algebraic Complexes

2.1. Definition. An additively written abelian group C is called an (algebraic) *complex* (*chain or cochain complex*) if:

(i) The group C is given in the form of a direct sum $\sum_{k\ge 0} C_k$ of subgroups C_k, $k = 0, 1, 2, \ldots$, whose indices are called the *dimensions* or *degrees* of the corresponding summands. (A group given in this way is said to be *graded*.)

(ii) There is prescribed a $\mathbb{Z}$-linear operator (i.e. abelian group endomorphism) ∂ on C, which satisfies $\partial\partial \equiv 0$, and which lowers the dimension of every C_k by 1 (i.e. $\partial C_k \subset C_{k-1}$ for every $k > 0$, and $\partial C_0 = 0$), or raises the dimension of every C_k by 1 (i.e. $\partial C_k \subset C_{k+1}$). In the former case the elements of C are termed *chains*, and in the latter *cochains*. (We shall sometimes refer to ∂ as a "differential operator".)

2.2. Definition

(i) The *k-dimensional homology group* $H_k(C)$ of a chain complex (C, ∂), is the quotient group of the group $Z_k = \operatorname{Ker}(\partial|_{C_k})$ of *k-dimensional cycles* (i.e. those elements of C_k sent to 0 by ∂) by its subgroup $B_k = \operatorname{Im}(\partial|_{C_{k+1}}) = \partial C_{k+1}$ of *boundaries* ($B_k \subset Z_k$):

$$H_k(C) = Z_k/B_k. \tag{1}$$

(ii) The *k-dimensional cohomology group* $H^k(C)$ of a cochain complex is the quotient group of the group $Z^k = \operatorname{Ker}(\partial|_{C_k})$ of *k-dimensional cocycles* by its subgroup $B^k = \partial C_{k-1}$ of *coboundaries* ($B^k \subset Z^k$):

$$H^k(C) = Z^k/B^k. \tag{2}$$

(iii) *The full homology group* $H_*(C)$ (resp. *full cohomology group* $H^*(C)$) is then the direct sum $H_*(C) = \sum_{k\geq 0} H_k(C)$ (resp. $H^*(C) = \sum_{k\geq 0} H^k(C)$).

Example

(a) With each manifold M^n there is associated the cochain complex $C = \sum_{k=0}^{n} C_k$ consisting of the (smooth) differential forms on M^n; here C_k is the subgroup of all (smooth) k-forms on the manifold, and the operator $\partial: C_k \to C_{k+1}$ is just the usual differential operator d on forms. The cohomology groups of this complex were earlier called the cohomology groups of the manifold M^n (see Definition 1.1).

(b) On a Lie group, or more generally a symmetric space, there is defined the complex of invariant differential forms (see the latter part of §1). Since, essentially by Theorem 1.14(i), all such forms are closed, the operator $\partial = d$ is in this situation trivial, i.e. the zero operator. From the full Theorem 1.14 it follows that, at least under certain conditions, the cohomology groups of this complex are in essence the same as the cohomology groups of the complex of all differential forms on the space.

We shall in the sequel meet with various further examples of complexes.

We resume our exposition with the definition of a homomorphism between complexes. (In what follows we shall frequently state definitions and theorems for chain complexes only, leaving it to the reader to formulate their cohomological duals.)

2.3. Definition. Given two complexes $(C^{(1)}, \partial^{(1)})$ and $(C^{(2)}, \partial^{(2)})$ (both chain complexes or both cochain complexes) we call a homomorphism $f: C^{(1)} \to C^{(2)}$ a *homomorphism of complexes* if it preserves the grading and commutes with the differential operators:

$$f(C_k^{(1)}) \subset C_k^{(2)} \quad \text{for} \quad k = 0, 1, 2, \ldots; \qquad f\partial^{(1)} = \partial^{(2)} f. \tag{3}$$

2.4. Proposition. *A homomorphism f of algebraic complexes induces homomorphisms (which we also denote by f) between the corresponding homology groups*

$$f: H_k(C^{(1)}, \partial^{(1)}) \to H_k(C^{(2)}, \partial^{(2)}), \qquad k = 0, 1, 2, \ldots. \tag{4}$$

PROOF. It is easy to verify directly from (3) that for every k a homomorphism of complexes maps k-cycles (i.e. elements of $Z_k^{(1)}$) to k-cycles (i.e. into $Z_k^{(2)}$), and k-boundaries (i.e. elements of $B_k^{(1)}$) to k-boundaries (i.e. into $B_k^{(2)}$). It follows that a homomorphism of complexes induces, for every k, a homomorphism between the respective kth homology groups, as claimed. □

An important example is the following one. As we saw early on in §1, a smooth map $f: M \to N$ between manifolds determines a map f^* between the corresponding complexes of differential forms on the manifolds, but in the opposite sense:

$$f^*: C(N) \to C(M).$$

Since this map is linear and commutes with the differential operator d (i.e. $f^* d\omega = df^*\omega$ for every form ω) we conclude that f^* is a homomorphism of complexes in the sense of Definition 2.3.

2.5. Definition. Let $f: C^{(1)} \to C^{(2)}$ and $g: C^{(1)} \to C^{(2)}$ be two homomorphisms of algebraic complexes. We say that the homomorphisms f and g are *(algebraically) homotopic* if there exists an ordinary homomorphism $D: C^{(1)} \to C^{(2)}$, such that (cf. Lemma 1.4)

$$D\partial^{(1)} \pm \partial^{(2)} D = f - g. \tag{5}$$

We also impose on D the condition (natural in view of (5)) that if $\partial^{(1)}$ and $\partial^{(2)}$ "raise the grading" then D lowers it, and vice versa; thus either

$$D(C_k^{(1)}) \subset C_{k-1}^{(2)} \qquad \text{or} \qquad D(C_k^{(1)}) \subset C_{k+1}^{(2)}. \tag{6}$$

2.6. Proposition. *If f and g are homotopic maps (i.e. homomorphisms) from a complex $C^{(1)}$ to a complex $C^{(2)}$, then the corresponding induced homomorphisms of the homology groups coincide*:

$$f = g: H_k(C^{(1)}, \partial^{(1)}) \to H_k(C^{(2)}, \partial^{(2)}). \tag{7}$$

PROOF. Let $c_k \in C_k^{(1)}$ be an arbitrary k-cycle: $\partial^{(1)} c_k = 0$. Then

$$f(c_k) - g(c_k) = D\partial^{(1)} c_k \pm \partial^{(2)} D c_k = \pm \partial^{(2)} D c_k,$$

whence we see, in view of the second inclusion in (6), that $f(c_k) - g(c_k)$ is a

k-boundary, so that $f(c_k) \sim g(c_k)$ in the homology group $H_k(C^{(2)}, \partial^{(2)})$. This completes the proof. □

The reader may recall that a specific algebraic homotopy was used in the proof of Theorem 1.3 concerning the homotopy invariance of induced maps between the cohomology groups of manifolds. Further examples will appear subsequently.

We shall now assume that the summands C_k of a complex are finitely generated.

2.7. Definition. Let b_k denote the (torsion-free) rank of the group $H_k(C, \partial)$ (i.e. the number of infinite cyclic direct summands in a decomposition of the abelian group H_k as a direct product of cyclic groups). We call the alternating sum

$$\chi(C, \partial) = \sum_{k \geq 0} (-1)^k \operatorname{rank} H_k = \sum_{k \geq 0} (-1)^k b_k \tag{8}$$

(if it exists) the *Euler characteristic of the complex* (C, ∂).

2.8. Proposition. *The Euler characteristic of a complex* (C, ∂) *is also given by*

$$\chi(C, \partial) = \sum_{k \geq 0} (-1)^k \operatorname{rank} \cdot C_k. \tag{9}$$

Proof. Denote by z_k the rank of the group Z_k of k-cycles, and by β_k the rank of the group B_k of k-boundaries. From the theory of finitely generated, free abelian groups we have

$$b_k = z_k - \beta_k, \qquad \beta_k = \operatorname{rank} C_{k+1} - z_{k+1}, \tag{10}$$

where in the second equation we are assuming that we are dealing with a chain complex, so that ∂ "lowers the grading". Hence

$$b_k = z_k + z_{k+1} - \operatorname{rank} C_{k+1},$$

and therefore

$$\sum_{k \geq 0} (-1)^k b_k = z_0 + \sum_{k \geq 0} (-1)^{k+1} \operatorname{rank} C_{k+1}.$$

Since $z_0 = \operatorname{rank} C_0$, the proposition follows. (We leave it to the reader to carry out the very similar proof when (C, ∂) is a cochain complex.) □

Given any (additively written) abelian group G and any complex C, we can form the new complex $C \otimes G = \sum_{k \geq 0} C_k \otimes G$, called a *complex with coefficients from* G. (We remind the reader briefly of the definition of the tensor product $A \otimes B$ of two abelian groups A and B: The abelian group $A \otimes B$ consists of all possible finite sums $\sum a_i \otimes b_i$, $a_i \in A$, $b_i \in B$ (i.e. has the symbols $a \otimes b$, $a \in A$, $b \in B$, as generators) with the following defining relations imposed:

$$\begin{aligned} (a_1 + a_2) \otimes b &= a_1 \otimes b + a_2 \otimes b, \\ a \otimes (b_1 + b_2) &= a \otimes b_1 + a \otimes b_2. \end{aligned} \tag{11}$$

One immediately deduces from these the following useful relation: $ma \otimes b = a \otimes mb$ for every integer m. (Thus $A \otimes B$ is in general algebraic parlance the "tensor product of the $\mathbb{Z}$-modules A and B".)

Exercise

Show that for every abelian group G we have $G \otimes \mathbb{Z} = G$. Identify the tensor product $\mathbb{Z}_m \otimes \mathbb{Z}_n$ of two finite cyclic groups. Prove that the tensor product of any finite abelian group with the group of reals (or even rationals) is zero.

We have yet to define appropriately the boundary operator on the complex $C \otimes G$. If ∂ denotes the boundary operator on C, we define the action of the boundary operator on $C \otimes G$, also denoted by ∂, as follows: For each of the generators $c_k \otimes g$, $c_k \in C_k$, $g \in G$, we put

$$\partial(c_k \otimes g) = \partial c_k \otimes g,$$

and then extend this action via linearity to the whole group $C \otimes G$. Clearly, we shall then have $\partial\partial = 0$, as required. We call the homology groups of the complex $C \otimes G$ the *homology groups of C with coefficients from* G, and denote them by

$$H_k(C; G) \equiv H_k(C \otimes G).$$

(Thus the "ordinary" homology groups of C can now be identified with the homology groups of C with coefficients from $\mathbb{Z}$.)

With G as before an additive abelian group, and (C, ∂) a (chain) complex, we now define its *dual cochain complex* C^* to consist of the $\mathbb{Z}$-linear maps (i.e. homomorphisms) from C to G with the obvious addition; thus $C^* = \mathrm{Hom}(C, G)$ in the notation in general use in algebra. The group C^* comes with the natural grading

$$C^* = \sum_{k \geq 0} C_k^*, \tag{12}$$

where $C_k^* = \mathrm{Hom}(C_k, G)$, and with boundary operator ∂^* defined in terms of ∂ by

$$\begin{aligned} &\partial^*\colon C_k^* \to C_{k+1}^*, && \partial\colon C_k \to C_{k-1}, \\ &(\partial^* x, c) = (x, \partial c); && c \in C, \quad x \in C^*, \end{aligned} \tag{13}$$

where (x, c) denotes the value in G taken by $x \in C^*$ at $c \in C$. It is immediate that $\partial^*\partial^* = 0$. The cohomology groups $H^k(C^*, \partial^*)$ are usually denoted simply by $H^k(C; G)$, and are called the *cohomology groups of the chain complex C with values in G.*

Suppose now that G is the additive group of a field $\mathbf{k}$ (which might be, for instance, any of the fields $\mathbb{R}$, $\mathbb{C}$ or $\mathbb{Q}$ of real numbers, complex numbers or rational numbers, respectively, or the finite field $\mathbb{Z}_p$ of p elements, p prime) and let C be a (chain) complex whose summands are finite-dimensional vector spaces over $\mathbf{k}$.

2.9. Theorem. *The vector spaces $H^k(C; \mathbf{k})$ and $H_k(C)$ are mutually dual; in pariticular, they have the same dimension.*

PROOF. We shall assume that C is a chain complex, so that the operator ∂ lowers dimension, and, as usual, leave the formulation and proof of the cohomological counterpart of the theorem to the reader.

We first show that an element c^k of C_k^* is a k-cocycle in the complex C^* precisely if $(c^k, B_k) = 0$, where $B_k (\subset C_k)$ is the subspace of k-boundaries. Since $B_k = \partial C_{k+1}$, each element of B_k has the form ∂c_{k+1} for some $c_{k+1} \in C_{k+1}$. Hence a typical element of (c^k, B_k) has the form

$$(c^k, \partial c_{k+1}) = (\partial^* c^k, c_{k+1}) = 0,$$

if c^k is a cocycle. Conversely, if $c^k \in C_k^*$ is such that $(c^k, \partial c_{k+1}) = 0$ for all $c_{k+1} \in C_{k+1}$, then $(\partial^* c^k, C_{k+1}) = 0$, whence $\partial^* c^k = 0$, i.e. c^k is a cocycle.

We have thus established that each subspace Z^k of cocycles of the complex C^* coincides with the subspace of C_k^* consisting of those linear functionals which vanish on B_k. We may therefore identify Z^k with the space of linear functionals on C_k/B_k. Since each space C_k is finite dimensional, so that $(C_k^*)^*$ is naturally isomorphic to C_k, the last statement but one can be dualized to yield a natural identification of each space Z_k of k-cycles of C with the space of those linear functionals on C_k^* which vanish on B^k, the space of k-dimensional coboundaries of C^*. This translates easily (via the natural isomorphism $C_k \simeq (C_k^*)^*$) into the equivalent assertion that B^k consists of just those linear functionals on C_k which vanish on Z_k. Combining this with the aforementioned identification of Z^k with the space of linear functionals on C_k/B_k, we obtain finally a natural identification of $Z^k/B^k = H^k(C; \mathbf{k})$ with $Z_k/B_k = H_k(C)$, which clearly fulfils the claim of the theorem. □

We next define the *tensor product* $C = C^{(1)} \otimes C^{(2)}$ *of a pair of (chain) complexes*. We first recall for the reader the concept of the tensor product $A \otimes B$ of two vector spaces A and B over a field $\mathbf{k}$, as the "tensor product over the field $\mathbf{k}$ of A and B regraded as k-modules"; thus the vector space $A \otimes B$ is once again generated, as an abelian group, by the symbols $a \otimes b$, $a \in A$, $b \in B$, subject to the relations (11) supplemented by $\lambda a \otimes b = a \otimes \lambda b$, where λ is any scalar. (Note that if the scalars come instead from $\mathbb{Z}$, then we are back in the case of the tensor product of abelian groups; see above.) Scalar multiplication in $A \otimes B$ is then defined by $\lambda(a \otimes b) = \lambda a \otimes b$. It follows readily that if $\{a_1, \dots, a_r\}$ and $\{b_1, \dots, b_s\}$ are bases for A and B respectively, then the rs elements $a_i \otimes b_j$ form a basis for $A \otimes B$.

Returning to the tensor product $C = C^{(1)} \otimes C^{(2)}$ (of abelian groups or vector spaces, depending on the provenance of the scalars), we define the grading of C by $C = \sum_{k \geq 0} C_k$, where

$$C_k = (C^{(1)} \otimes C^{(2)})_k = \sum_{p+q=k} C_p^{(1)} \otimes C_q^{(2)}, \tag{14}$$

and the boundary operator by

$$\partial(c_p^{(1)} \otimes c_q^{(2)}) = (\partial^{(1)} c_p^{(1)}) \otimes c_q^{(2)} + (-1)^p c_p^{(1)} \otimes \partial^{(2)} c_q^{(2)}. \tag{15}$$

It is easy to verify that $\partial\partial = 0$.

2.10. Theorem. *For any pair $C^{(1)}$, $C^{(2)}$ of (chain) complexes of vector spaces over any field* $\mathbf{k}$ *(the cases $\mathbf{k} = \mathbb{R}, \mathbb{C}, \mathbb{Q}, \mathbb{Z}_p$ being for us the important ones), the homology groups of the tensor product satisfy*

$$H_k(C^{(1)} \otimes C^{(2)}) = \sum_{p+q=k} H_p(C^{(1)}) \otimes H_q(C^{(2)}), \tag{16}$$

i.e. there is a natural vector-space isomorphism between these spaces.

For the proof we need the following ancillary result:

2.11. Lemma. *If $C = \sum_{n \geq 0} C_n$ is any chain complex of vector spaces over the field* $\mathbf{k}$*, then there is for each C_n a "canonical" basis $(x_{n,i}, y_{n,j}, h_{n,l})$ on which the operator ∂ acts as follows:*

$$\partial x_{n,i} = y_{n-1,i}, \qquad \partial y_{n,j} = 0, \qquad \partial h_{n,l} = 0. \tag{17}$$

PROOF. From (17) and the requirement that $(x_{n,i}, y_{n,j}, h_{n,l})$ be a basis for C_n, it is immediate that the $y_{n,j}$ will have to form a basis for the space B_n of n-boundaries, the $h_{n,l}$ a basis for a complement H_n of B_n in the space Z_n of n-cycles: $Z_n = B_n \oplus H_n$ (so that the $h_{n,l}$ represent a basis for the homology group $H_n(C)$), and finally that the $x_{n,i}$ must form a basis for a complement X_n of Z_n in C_n, i.e. $C_n = X_n \oplus Z_n = X_n \oplus B_n \oplus H_n$. Starting at C_0 (for which $X_0 = 0$), it is not difficult to construct bases of the desired kind using induction on n. (Thus if bases satisfying (17) have already been chosen for $C_0, C_1, \ldots, C_{n-1}$, then as the basis elements $x_{n,i}$ for $X_n \subset C_n$ choose any preimages under $\partial: C_n \to C_{n-1}$ of the $y_{n-1,i}$, and choose the basis elements $y_{n,j}$ for B_n and $h_{n,l}$ for H_n arbitrarily.) □

PROOF OF THE THEOREM. Choose canonical bases $(x_p^{(1)}, y_p^{(1)}, h_p^{(1)})$ and $(x_q^{(2)}, y_q^{(2)}, h_q^{(2)})$ for the subspaces $C_p^{(1)}$ and $C_q^{(2)}$ respectively. (The second subscript index, enumerating the members of each basis, will be omitted.) From these bases we construct as follows canonical bases for the subspaces $C_k = \sum_{p+q=k} C_p^{(1)} \otimes C_q^{(2)}$ of $C^{(1)} \otimes C^{(2)}$. The basis vectors for C_k grouped first, i.e. the non-cycles, spanning a complement in C_k of the space of k-cycles, are given by

$$\begin{aligned} x_{pq} &= x_p^{(1)} \otimes x_q^{(2)}; \qquad & a_{pq} &= \tfrac{1}{2}[x_p^{(1)} \otimes x_q^{(2)} + (-1)^{p-1} y_{p-1}^{(1)} \otimes x_{q+1}^{(2)}], \\ \alpha_{pq} &= x_p^{(1)} \otimes h_q^{(2)}; \qquad & \beta_{pq} &= (-1)^p h_p^{(1)} \otimes x_q^{(2)}, \end{aligned} \tag{18}$$

where throughout (p, q) ranges over all ordered pairs of non-negative integers satisfying $p + q = k$. The basis vectors for B_k are defined by

$$\begin{aligned} &b_{pq} = y_{p-1}^{(1)} \otimes x_{q+1}^{(2)} - (-1)^{p+1} x_p^{(1)} \otimes y_q^{(2)}, \\ &y_{pq} = y_p^{(1)} \otimes y_q^{(2)}; \qquad \gamma_{pq} = y_p^{(1)} \otimes h_q^{(2)}; \qquad \delta_{pq} = h_p^{(1)} \otimes y_q^{(2)}, \end{aligned} \tag{19}$$

where (p, q) varies as before. We leave to the reader the verification of the linear independence of the vectors in (18) and (19). It remains to specify basis vectors for a complement of B_k in Z_k; the vectors of the form $h_p^{(1)} \otimes h_q^{(2)}$, $p + q = k$, clearly linearly independent of the vectors in (18) and (19), are the ones we use for this purpose. Finally, we check that the operator ∂ on C (defined in (15)) acts in accordance with (17). From (15) and (17) (as it applies to the canonical bases chosen for the $C_p^{(1)}$ and $C_q^{(2)}$) we calculate that

$$\partial x_{pq} = b_{p,q-1}, \qquad \partial a_{pq} = y_{p-1,q}, \qquad \partial \alpha_{pq} = \gamma_{p-1,q}, \qquad \partial \beta_{pq} = \delta_{p,q-1},$$

$$\partial b_{pq} = \partial y_{pq} = \partial \gamma_{pq} = \partial \delta_{pq} = \partial(h_p^{(1)} \otimes h_q^{(2)}) = 0,$$

so that the bases we have constructed (one for each C_k) are indeed canonical. Hence, by the parenthetical observation in the proof of Lemma 2.11, the vectors $h_p^{(1)} \otimes h_q^{(2)}$ with $p + q = k$ represent a basis for $H_k(C^{(1)} \otimes C^{(2)})$, and this is what we wished to prove. □

§3. Simplicial Complexes. Their Homology and Cohomology Groups. The Classification of the Two-Dimensional Closed Surfaces

We now describe a different approach to the definition (and investigation) of homology and cohomology groups, which considerably extends their field of application.

A (closed) *n-dimensional simplex* (or *n-simplex*) in $\mathbb{R}^k$, $k \geq n$, is defined inductively as follows: A 0-simplex is just a point $[\alpha_0]$; a 1-simplex is a (closed) straight-line segment $[\alpha_0\alpha_1]$; a 2-dimensional simplex is a triangle $[\alpha_0\alpha_1\alpha_2]$ (including its interior); and a 3-dimensional simplex is a solid tetrahedron $[\alpha_0\alpha_1\alpha_2\alpha_3]$ (see Figure 2). Proceeding inductively, if we have an n-dimensional simplex $\sigma^n = [\alpha_0\alpha_1 \ldots \alpha_n]$ already defined in $\mathbb{R}^n$, then to construct from it an $(n+1)$-simplex we take any point α_{n+1} outside the hyperplane $\mathbb{R}^n \subset \mathbb{R}^{n+1}$ and join that point by means of straight-line segments to every point of σ^n; the geometrical figure made up of the totality of the points on these line segments is then an $(n+1)$-simplex $\sigma^{n+1} = [\alpha_0\alpha_1 \ldots \alpha_{n+1}]$.

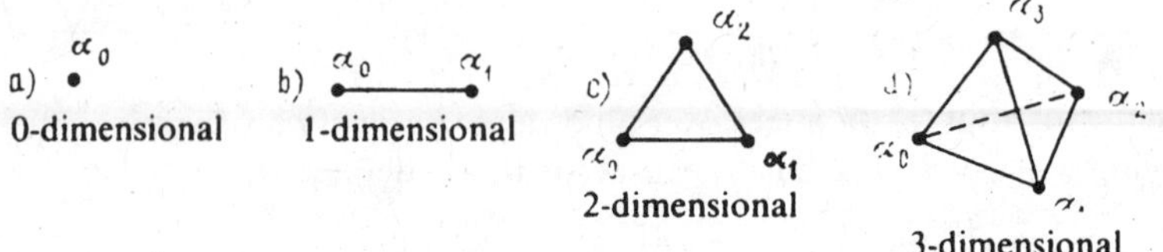

Figure 2. Simplexes.

Alternatively, an n-dimensional simplex may be defined as the convex hull of any $n+1$ points in $\mathbb{R}^k$, $k \geq n$ (called the *vertices* of the simplex), which are not contained in any $(n-1)$-dimensional hyperplane in $\mathbb{R}^k$.

The boundary of an n-simplex $[\alpha_0 \dots \alpha_n]$ determined by its vertices $\alpha_0, \dots, \alpha_n$, is made up of the $(n-1)$-simplexes (called $(n-1)$-*dimensional faces*) $[\alpha_0\alpha_1 \dots \alpha_{n-1}]$, $[\alpha_0\alpha_1 \dots \alpha_{n-2}\alpha_n]$, ..., $[\alpha_1 \dots \alpha_n]$; in other words, the ith face $\sigma_{(i)}^{n-1}$ is the $(n-1)$-simplex determined by all of the vertices $\alpha_0, \dots, \alpha_n$ but the ith; it is the face "opposite" the vertex α_i:

$$\sigma_{(i)}^{n-1} = [\alpha_0 \dots \hat{\alpha}_i \dots \alpha_n] \tag{1}$$

(where as usual the hat indicates that α_i is understood as deleted). More generally, the faces of dimensions $(n-1)$, $(n-2)$, ..., 0, of our simplex $[\alpha_0 \dots \alpha_n]$ are (formally) obtained by deleting any $1, 2, \dots, n$ vertices respectively.

3.1. Definition. The *oriented boundary* $\partial\sigma^n$ of a simplex $\sigma^n = [\alpha_0 \dots \alpha_n]$ (whose points are given in the order indicated) is the following formal linear combination of its $(n-1)$-dimensional faces:

$$\partial\sigma^n = \partial[\alpha_0 \dots \alpha_n] = \sum_{i=0}^{n} (-1)^i [\alpha_0 \dots \hat{\alpha}_i \dots \alpha_n] = \sum_{i=0}^{n} (-1)^i \sigma_{(i)}^{n-1}. \tag{2}$$

For example, for 0-, 1-, and 2-simplexes we have:

$$\begin{aligned} \partial[\alpha_0] &= 0, \\ \partial[\alpha_0\alpha_1] &= [\alpha_1] - [\alpha_0], \\ \partial[\alpha_0\alpha_1\alpha_2] &= [\alpha_1\alpha_2] - [\alpha_0\alpha_2] + [\alpha_0\alpha_1]. \end{aligned} \tag{3}$$

In the last case at least it is clear from Figure 2 that the faces (here "edges") enter with appropriate signs.

3.2. Lemma. *For any n-dimensional simplex we have*

$$\partial\partial[\alpha_0 \dots \alpha_n] = 0. \tag{4}$$

This follows essentially by direct calculation. For instance, in the case $n = 2$, we have

$$\partial[\alpha_0\alpha_1\alpha_2] = [\alpha_1\alpha_2] - [\alpha_0\alpha_2] + [\alpha_0\alpha_1],$$

$$\partial\partial[\alpha_0\alpha_1\alpha_2] = \{[\alpha_2] - [\alpha_1]\} - \{[\alpha_2] - [\alpha_0]\} + \{[\alpha_1] - [\alpha_0]\} = 0.$$

For general n the calculation is analogous: we have

$$\partial\partial\sigma^n = \partial\left(\sum_{i=0}^{n} (-1)^i \sigma_{(i)}^{n-1}\right) = \sum_{i=0}^{n} (-1)^i \partial\sigma_{(i)}^{n-1},$$

where in this sum the face $\sigma_{(ij)}^{n-2}$ (obtained by omitting the vertices α_i and α_j) occurs twice, namely in the expressions for the boundaries $\partial\sigma_{(i)}^{n-1}$ and $\partial\sigma_{(j)}^{n-1}$, with opposite signs. □

3.3. Definition. A *simplicial complex* is a collection of simplexes of arbitrary dimensions with the following two properties:

(i) together with every simplex in the collection, all of its faces (of all dimensions) are also in the collection;
(ii) any two simplexes in the collection are either disjoint or intersect in a single whole face (of some dimension). The *dimension* of a simplicial complex is the largest (if there is a largest) of the dimensions of the simplexes of which it is composed.

If the collection of simplexes is finite, we speak of a *finite simplicial complex*.

Given a simplicial complex M, we fix arbitrarily on an enumeration of its vertices: $\alpha_0\alpha_1\ldots$. The r-dimensional simplexes of the complex are then taken to be of the form $[\alpha_{i_0}\alpha_{i_1}\ldots\alpha_{i_r}]$ for certain subsets $\{\alpha_{i_0},\ldots,\alpha_{i_r}\}$ of size r of the set of vertices, with the given ordering imposed. (Note that we shall usually require of a simplicial complex that it have only a finite number of simplexes of each dimension.)

Suppose now that we are also given an (arbitrary) abelian group G with the group operation denoted by the addition symbol $+$. A *chain of dimension k with coefficients from G* of the simplicial complex M is defined to be a formal finite linear combination of the distinct k-simplexes of the complex, of the form

$$c_k = \sum_i g_i\sigma_i, \tag{5}$$

where the k-simplexes σ_i have their vertices written in the order determined by the prescribed ordering of the vertices of the complex, and the g_i are arbitrary elements of the group G. *Addition* of a pair of such k-chains is then defined as follows: if $c_k = \sum g_i\sigma_i$ and $c_k' = \sum g_i'\sigma_i$, then $c_k + c_k' = \sum (g_i + g_i')\sigma_i$. With this additive operation, the set of all k-chains clearly forms an abelian group.

The (*oriented*) *boundary* ∂c_k of a k-chain c_k given by (5), is defined by the formula

$$\partial c_k = \sum_i g_i\partial\sigma_i, \tag{6}$$

where $\partial\sigma_i$ is given by (2). It is immediate from Lemma 3.2 that $\partial\partial c_k = 0$. The *k-cycles of the simplicial complex M* are now defined to be those k-chains c_k satisfying $\partial c_k = 0$; they form a subgroup which we denote by Z_k. The *boundary k-cycles* are those that are "homologous to zero", i.e. are of the form ∂c_{k+1} for some $(k+1)$-cycle c_{k+1}; the group they comprise is denoted by B_k (cf. the more abstract analogues of these concepts introduced in the preceding section).

3.4. Definition (cf. Definition 2.2). The *k-dimensional homology group $H_k(M; G)$ of a simplicial complex M, with coefficients from G*, is the quotient of the group Z_k of k-cycles by the subgroup B_k of boundary k-cycles. (Thus we say that two k-cycles c_k', c_k'' are *homologous* if in combination they form a bounding k-cycle, i.e., more precisely, if $c_k' - c_k'' = \partial c_{k+1}$ for some $(k+1)$-chain c_{k+1} of the complex M.)

The interesting cases are: $G = \mathbb{Q}$, the additive rational numbers; $G = \mathbb{C}$; $G = \mathbb{Z}$; $G = \mathbb{Z}_2$ (the integers modulo 2), and more generally $G = \mathbb{Z}_m$ (the integers modulo m) especially when m is prime, i.e. $\mathbb{Z}_m$ is a field. When $G = \mathbb{R}$ the homology group $H_i(M; \mathbb{R})$ is for every i a real vector space; its dimension is in this case called the *ith Betti number of the simplicial complex M*.

Given a finite simplicial complex M, we define its *Euler characteristic* $\chi(M)$ as follows: denoting by γ_i the number of i-simplexes in the complex M, we set

$$\chi(M) = \sum_{i \geq 0} (-1)^i \gamma_i. \tag{7}$$

The following result makes it easier to perceive the analogy between this definition and the earlier one (Definition 2.7) for an algebraic complex with coefficients from $\mathbb{Z}$.

3.5. Theorem. *Denoting by b_i the dimension of the vector space $H_i(M; \mathbb{R})$ (the ith Betti number), we have*

$$\chi(M) = \sum_{i \geq 0} (-1)^i \gamma_i = \sum_{i \geq 0} (-1)^i b_i. \tag{8}$$

PROOF. It follows directly from the definition of an i-chain with coefficients from $\mathbb{R}$, that the group of i-chains of the simplicial complex M is a vector space of dimension γ_i. This noted, the desired conclusion is obtained by imitating the proof of Proposition 2.8 with the word "rank" replaced throughout by "dimension" (and "free abelian group" by "vector space over $\mathbb{R}$").

Remark. In Part II, §15, the Euler characteristic $\chi(M)$ (where M is now a "triangulated" smooth manifold; see below) was defined (equivalently) as the sum of the indices of the critical points of a smooth function on M (see also Part II, Theorem 15.2.7). We have now the option of computing $\chi(M)$ instead by homological means.

We now turn to the duals of the above concepts. We define a *k-dimensional cochain c^k of a simplicial complex M* to be a linear map from the group of k-dimensional integral chains of M to a group G of coefficients; in other words a k-cochain associates with each k-simplex σ an element $c^k(\sigma)$ of the group G, and extends to all integral k-chains via linearity:

$$c^k(a\sigma_1 + b\sigma_2) = ac^k(\sigma_1) + bc^k(\sigma_2), \qquad a, b \in \mathbb{Z}.$$

Clearly the totality of k-cochains of M forms a group under addition of maps.

We next define the *coboundary* δc^k of an arbitrary cochain c^k to be the $(k + 1)$-dimensional cochain given (on $(k + 1)$-simplexes) by

$$\delta c^k(\sigma) = c^k(\partial\sigma), \tag{9}$$

where σ is any $(k + 1)$-simplex. (Note that in §2 we used the alternative notation ∂^* for the operator δ.) That $\delta\delta = 0$ follows from $\partial\partial = 0$:

$$\delta\delta c^k(\sigma) = \delta c^k(\partial\sigma) = c^k(\partial\partial\sigma) = 0.$$

A *k-dimensional cocycle of the complex M* is, as might be expected, a k-cochain c^k for which $\delta c^k = 0$, and a k-cocycle c^k is *cohomologous to zero* if $c^k = \delta c^{k-1}$ for some $(k-1)$-cocycle c^{k-1}, i.e. if it is a coboundary.

3.6. Definition. The *kth cohomology group $H^k(M; G)$ of a simplicial complex M, with coefficients from G*, is the quotient group of the group of k-cocycles by the group of k-dimensional coboundaries. Thus $c_1^k \sim c_2^k$ in the cohomology group precisely if $c_1^k - c_2^k$ has the form δc^{k-1}, i.e. is a coboundary.

The (algebraic) complex of cochains of the simplicial complex M is the dual (in the sense of §2) of the (algebraic) complex of chains of M. Hence from Theorem 2.9 we infer immediately the following result (pertaining to the case where G is a field $\mathbf{k}$).

3.7. Corollary. *If M is a finite simplicial complex and* $\mathbf{k}$ *any field, then for each i the vector spaces* $H_i(M; \mathbf{k})$ and $H^i(M; \mathbf{k})$ *have the same dimension.*

We now consider in more detail the homology and cohomology of a simplicial complex M when the group G of coefficients is $\mathbb{Z}_m$ (in particular, the important case $G = \mathbb{Z}_p$, the field of p elements, p prime). Let $x \in H_q(M; \mathbb{Z}_m)$, i.e. let x be (the coset of) a q-cycle, and let $\bar{x}$ be any integral q-chain such that $x = \bar{x} (\text{mod } m)$. Then since x is a q-cycle we have $\partial x = 0$, or, in terms of the integral chain $\bar{x}$,

$$\partial \bar{x} = mu \qquad \left(\text{or} \quad u = \frac{\partial \bar{x}}{m}\right)$$

for some integral $(q-1)$-chain u. Any other integral q-cycle representing x will have the form $\bar{x} + \partial y + mz$ for some integral chains y and z. Applying the boundary operator to such a representative, we get

$$\partial(\bar{x} + \partial y + mz) = \partial \bar{x} + \partial\partial y + m\partial z = mu + m\partial z,$$

so that the replacement $\bar{x} \to \bar{x} + \partial y + mz$ entails $u = \partial \bar{x}/m \to u + \partial z$. Since $\partial u = 0$ and ∂z is a boundary we deduce that the "Bockstein homomorphism"

$$\partial_1 \colon H_q(M; \mathbb{Z}_m) \to H_{q-1}(M; \mathbb{Z}) \tag{10}$$

is well defined by

$$x \to \frac{\partial \bar{x}}{m}, \qquad \text{where} \quad \bar{x} \ (\text{mod } m) \sim x \in H_q(M; \mathbb{Z}_m).$$

One defines analogously the Bockstein homomorphisms between the mod m and integral cohomology groups:

$$\delta_1 \colon H^q(M; \mathbb{Z}_m) \to H^{q+1}(M; \mathbb{Z}). \tag{11}$$

3.8. Proposition. *For any element* $x \in H_q(M; \mathbb{Z}_m)$, *we have* $\partial_1 x = 0$ *in* $H_{q-1}(M; \mathbb{Z})$ *if and only if x is obtained from an element y of* $H_q(M; \mathbb{Z})$ *by* *reduction modulo m:*

$$x = y \,(\text{mod } m) \quad \Leftrightarrow \quad \partial_1 x = 0.$$

(The analogous statement is valid for the corresponding cohomology groups of M: For any $x \in H^q(M; \mathbb{Z}_m)$,

$$\delta_1 x = 0 \quad \Leftrightarrow \quad x = y \,(\text{mod } m) \qquad \text{for some} \quad y \in H^q(M; \mathbb{Z}).)$$

PROOF. If $x = y(\text{mod } m)$ where $y \in H_q(M; \mathbb{Z})$, then we can represent x modulo m by an integral q-cycle $\bar{x}$, and then since $\partial\bar{x} = 0$, we have $\partial_1 x = \partial\bar{x}/m = 0$ in $H_{q-1}(M; \mathbb{Z})$.

Conversely, if $\partial_1 x = 0$ in $H_{q-1}(M; \mathbb{Z})$, then for every integral q-chain $\bar{x}$ representing x we must have that $\partial\bar{x}/m$ is a boundary, so that $\partial\bar{x}/m = \partial z$ for some integral chain z. If we then take $y = \bar{x} - mz$, we get $\partial y = 0$ and $x = y\,(\text{mod } m)$, as required. This completes the proof. □

From this result we see that knowledge of ∂_1 (and δ_1) allows us to identify in "mod m homology" (and cohomology) those cycles which come from integral cycles via reduction modulo m. As a further application of the proposition we have the following fact: *The image $\partial_1 H_q(M; \mathbb{Z}_m)$ in $H_{q-1}(M; \mathbb{Z})$ consists exactly of the m-torsion elements of $H_{q-1}(M; \mathbb{Z})$, i.e. of those elements u such that $mu = 0$.* (This can be seen as follows: on the one hand, for any element x of $H_q(M; \mathbb{Z}_m)$, we have $m(\partial_1 x) = \partial_1(mx) = \partial_1(0) = 0$. For the converse, let v be an integral $(q-1)$-cycle such that mv is homologous to zero in $H_{q-1}(M; \mathbb{Z})$; then $mv = \partial\bar{x}$ for some integral q-chain $\bar{x}$. Reducing modulo m the coefficients in $\bar{x}$, we obtain a q-chain x with coefficients in $\mathbb{Z}_m$ which, in view of $mv = \partial\bar{x}$, satisfies $\partial x = 0$, i.e. is a cycle. Hence $v = \partial\bar{x}/m$ is the image under ∂_1 of a q-cycle, namely x.)

Example. For $M = \mathbb{R}P^2$ (triangulated as described below) and $G = \mathbb{Z}_2$, it turns out that (see Proposition 3.11 below, and §4, Example (f))

$$H_2(\mathbb{R}P^2; \mathbb{Z}_2) \simeq \mathbb{Z}_2 \simeq H_1(\mathbb{R}P^2; \mathbb{Z}).$$

Hence, in view of the preceding application of Proposition 3.8, the map $\partial_1 \colon H_2(\mathbb{R}P^2; \mathbb{Z}_2) \to H_1(\mathbb{R}P^2; \mathbb{Z})$ must in fact be an isomorphism.

EXERCISE

Show that for every non-orientable manifold M^n (triangulated—see below) the group $H_n(M^n; \mathbb{Z}_2)$ contains a cycle $[M^n] = x$ (see Corollary 3.12 below) such that $\partial_1 x$ has order 2 in $H_{n-1}(M^n; \mathbb{Z})$ (and so, in particular, $\partial_1 x \neq 0$). (For the coholomogy groups of M^n the analogous assertion is valid: There is an element u in $H^1(M^n; \mathbb{Z}_2)$ such that $\delta_1 u$ has order 2 in $H^2(M^n; \mathbb{Z})$.)

We turn now to the application of the above theory to an arbitrary manifold M. Since the homology theory of simplicial complexes is essentially combinatorial (or algebraic) we can apply it to any "triangulation" of M, that is, any subdivision (if such exists) of M into "smooth simplexes" in such a way that it becomes, in essence, a simplicial complex.

Thus we define a *smooth k-dimensional simplex* σ^k to be a smooth embedding of an ordinary k-simplex, together with some open neighbourhood of it in $\mathbb{R}^k$, into M. We then say that the manifold M^n is *triangulated* if it is subdivided into smooth simplexes in such a way as to form a simplicial complex (of dimension n) (i.e. if the smooth simplexes of the subdivision satisfy the conditions (i) and (ii) of Definition 3.3).

Here are two important facts:

(i) The homology and cohomology groups of a triangulable manifold are independent of the triangulation, and homotopically invariant (i.e. invariant under homotopy equivalences). (See §6 for the proof.)
(ii) For $G = \mathbb{R}$ the simplicial cohomology groups of a manifold M determined by any triangulation, coincide with the cohomology groups of M defined (in §1) in terms of differential forms on M. (See the conclusion of §14.)

We now give some indication as to why the latter statement holds. To this end, let σ^k be any smooth k-simplex of the (triangulated) manifold M, and let ω_k be any differential form of rank k on M. We denote the integral of the form ω_k over the simplex σ^k by $\langle \omega_k, \sigma^k \rangle$:

$$\langle \omega_k, \sigma^k \rangle = \int_{\sigma^k} \omega_k. \tag{12}$$

Extending via linearity we then define $\langle \omega_k, c_k \rangle$, for any k-chain $c_k = \sum_i r_i \sigma_i^k$ with real coefficients, by setting

$$\langle \omega_k, c_k \rangle = \sum_i r_i \int_{\sigma_i^k} \omega_k = \int_{c_k} \omega_k, \tag{13}$$

where the last equality constitutes the definition of the integral of the form ω_k over the k-chain c_k. From the general Stokes formula (see Part I, §26.3) it follows that for any $(k+1)$-chain c we have

$$\int_c d\omega_k = \int_{\partial c} \omega_k, \qquad \text{i.e.} \qquad \langle d\omega_k, c \rangle = \langle \omega_k, \partial c \rangle. \tag{14}$$

Thus by means of (13) we can associate with each k-form ω on M, a linear functional on the space of k-chains of the triangulation of M. If c and $\hat{c}$ are homologous k-cycles, say $c = \hat{c} + \partial c'$, and ω is closed, then in view of (14),

$$\langle \omega, c \rangle = \langle \omega, \hat{c} \rangle + \langle d\omega, c' \rangle = \langle \omega, \hat{c} \rangle.$$

Furthermore, if ω is exact, then it easily follows from (14) that $\langle \omega, c \rangle = 0$ for every k-cycle c, so that exact k-forms are associated with the zero linear functional on the space of k-cycles. We conclude that: *For each k, equation* (13) *determines a linear map from the space $H^k(M; \mathbb{R})$, defined in terms of the closed k-forms on the manifold M, to the space of linear functionals on the simplicial homology group $H_k(M; \mathbb{R})$.*

To obtain statement (ii) in full, one needs of course to show that this linear map in bijective, i.e. that every linear functional on $H_k(M; \mathbb{R})$ can be expressed as in (13), and that a closed but not exact k-form on M gives rise via (13) to a non-zero linear functional on $H_k(M; \mathbb{R})$. We shall not however pursue this further here.

Let M^n be a closed, connected manifold. It is not difficult to see that in any triangulation of M^n, i.e. subdivision as an n-dimensional complex, each $(n-1)$-simplex is a face of precisely two n-simplexes; we exploit this fact in proving the following three assertions.

3.9. Theorem. *If M^n is a closed and connected manifold of dimension n, admitting a finite triangulation, then*

$$H_n(M^n; \mathbb{Z}_2) \simeq \mathbb{Z}_2$$

(where $\mathbb{Z}_2$ is, as usual, the 2-element group of residues modulo 2).

PROOF. Clearly, a k-chain over $\mathbb{Z}_2 = \{0, 1\}$ is simply a finite sum of k-simplexes, i.e. has the form $\sum_i \sigma_i^k$, and furthermore the orientation of simplexes plays no role in this case. Hence over $\mathbb{Z}_2$ we have, for any k-simplex σ^k,

$$\partial\sigma^k = \sum_{i=0}^{k} \sigma_i^{k-1},$$

where the σ_i^{k-1}, $i = 0, \dots, k$, are the faces of σ^k. If now $z = \sum_j \sigma_j^n$ is the sum of all the (finitely many) n-simplexes, then in the sum $\partial z = \sum_j \partial\sigma_j^n$, each $(n-1)$-simplex of the complex will occur exactly twice (by the remark immediately preceding the statement of the theorem). Hence $\partial z = 0$. Since z is (for similar reasons and since M is connected) the only non-zero n-cycle, the theorem is proved. □

For manifolds which are in addition orientable, we have the following stronger result:

3.10. Proposition. *For every closed, connected and orientable n-dimensional manifold M^n (admitting a finite triangulation), the nth homology group $H_n(M^n; G)$ is isomorphic to G for every abelian group G.*

PROOF. Choose an orientation of the manifold, i.e. at each point of M^n choose one of the two possible orientation classes of tangent frames in such a way that the class varies continuously with the points (see Part II, §2.1). Orient the n-simplexes in accordance with the chosen orientation of M^n. (This will determine, up to cyclic permutations, an ordering of the vertices of each n-simplex.) Then if two n-simplexes σ_1^n and σ_2^n intersect in an $(n-1)$-simplex σ^{n-1} (as depicted for the case $n = 2$ in Figure 3), that $(n-1)$-simplex will occur as a term in the boundaries $\partial\sigma_1^n$ and $\partial\sigma_2^n$ with opposite signs prefixed. Hence the n-chain $[M^n] = \sum_i \sigma_i^n$, in which all n-simplexes occur, each with coefficient 1, will again be a cycle. It is easy to see (again by the connectedness assumption

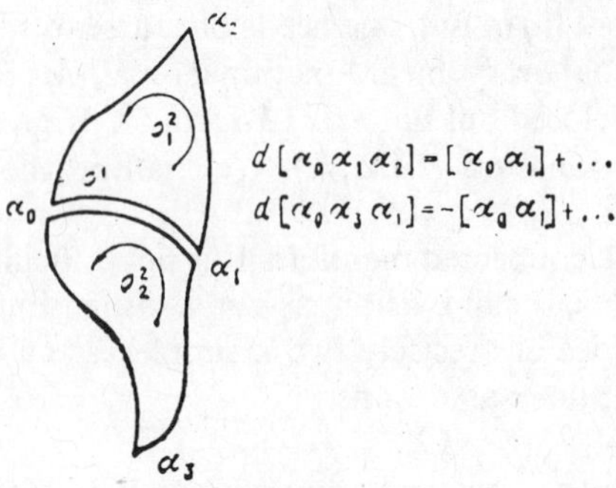

Figure 3

together with the remark preceding Theorem 3.9) that in fact the n-cycles are just the chains of the form $g[M^n]$, $g \in G$. Since there are no n-dimensional boundaries, the proposition follows. □

3.11. Proposition. *For every non-orientable, connected, closed, n-dimensional manifold M^n (admitting a finite triangulation) we have*

$$H_n(M^n; \mathbb{Z}) = 0, \qquad H_n(M^n; \mathbb{Z}_2) \simeq \mathbb{Z}_2.$$

PROOF. We have already seen, in Theorem 3.9, that $H_n(M^n; \mathbb{Z}_2) \simeq \mathbb{Z}_2$. We need to show here, therefore, only that the nth integral homology group of M^n is zero. The connectedness assumption together with the remark preceding Theorem 3.9 imply as before that every non-zero n-cycle has the form $z = \lambda\sum_i \sigma_i^n$, where λ is a non-zero integer, and the sum is over all n-simplexes (suitably oriented). If two n-simplexes σ_1^n and σ_2^n intersect in an $(n-1)$-simplex, then this simplex enters into the boundaries $\partial\sigma_1^n$ and $\partial\sigma_2^n$ with opposite signs if and only if the simplexes σ_1^n and σ_2^n happen to agree in their orientations. Hence $\partial z = 0$ if and only if the orientations chosen on all n-simplexes agree, i.e. if and only if the manifold M^n is oriented. This contradiction completes the proof. □

3.12. Corollary (cf. the exercise above). *Let $[M^n] = \sum_i \sigma_i^n$ denote the sum of all n-simplexes of (some finite triangulation of) a non-orientable manifold M^n (the "fundamental class" of M^n). Then $[M^n]$ generates the group $H_n(M^n; \mathbb{Z}_2)$, and in the group $H_{n-1}(M^n; \mathbb{Z})$ we have*

$$\partial_1[M^n] \neq 0, \qquad 2\partial_1[M^n] = 0.$$

We shall, for the remainder of this section, concentrate our attention on 2-dimensional smooth manifolds. We shall, to be more specific, be concerned with the question as to whether such manifolds can be triangulated, and with the problem of classifying them. We shall in fact show how to classify completely all connected, closed, 2-dimensional smooth manifolds. (Recall that "closed" means "compact and without boundary".)

3.13. Lemma. *Every connected, closed, 2-dimensional smooth manifold M^2 admits a finite triangulation, i.e. can be subdivided by means of smooth curves into finitely many smooth triangles (smooth 2-simplexes) in such a way that any two of these triangles intersect either not at all, or in a single common vertex (0-face), or in a single common edge (1-dimensional face).*

PROOF. By §9 of Part I, we may assume our manifold M^2 embedded in some finite-dimensional Euclidean space, and then M^2 will have a Riemannian metric induced on it by the Euclidean metric on that Euclidean space. There then exists a (small) number $\varepsilon > 0$ such that whenever two points $x, y \in M^2$ are such that $\rho(x, y) < \varepsilon$ (where ρ is the distance function on M^2 determined by the induced Riemannian metric—see Part II, §1.2), there is a unique shortest geodesic arc $\gamma_{x,y}$ in M^2 joining them (cf. Part I, §§29.2, 36.2). About each point x of M^2 we can find an open neighbourhood N_x of radius $< \varepsilon/2$ diffeomorphic to an open disc, and then inside N_x we can find a closed region D_x containing x, diffeomorphic to a closed disc, and with its boundary made up of geodesic arcs of length $< \varepsilon$ (verify this!). Since M^2 is compact we can cover it with finitely many of these discs, say $D_1, \ldots, D_N$. Clearly, each of these discs can be finitely triangulated by means of geodesic arcs. On the regions of overlap of two or more of the D_i the geodesic arcs representing the edges of their respective triangulations can intersect in at most finitely many points (unless they coincide for all or part of their length); it follows that by admitting finitely many additional points to the triangulations of the discs D_i, we shall obtain a finite triangulation of M^2, as required. □

Remark. In fact every differentiable manifold, and every 2- or 3-dimensional (continuous, but not necessarily differentiable) manifold can be triangulated. These results are more difficult, however, and we shall omit their proofs.

We now turn to the classification theorem for closed, connected 2-manifolds. We shall show that, up to a diffeomorphism, every such manifold is a member of one of the following two series. The first series is made up of the manifolds M_g^2, $g = 0, 1, 2, \ldots$, the *spheres-with-g-handles* (or "orientable closed surfaces of genus g"). These manifolds can be smoothly embedded in $\mathbb{R}^3$ as surfaces; see Figure 4. As we saw in Part II, §4.2, they arise in particular as the Riemann surfaces of complex algebraic functions of the form $w = \pm\sqrt{P_n(z)}$, where $P_n(z)$ is a polynomial without repeated roots; thus the manifold in $\mathbb{C}P^2$ whose points are the solutions (z, w) (in homogenized form) of the equation $w^2 = P_n(z)$, is (essentially) diffeomorphic to M_g^2 where $n = 2g + 1$ or $2g + 2$.

The members of the second series of manifolds N_μ^2, $\mu = 1, 2, \ldots$, are obtained as follows. From the sphere S^2, μ pairwise non-intersecting open

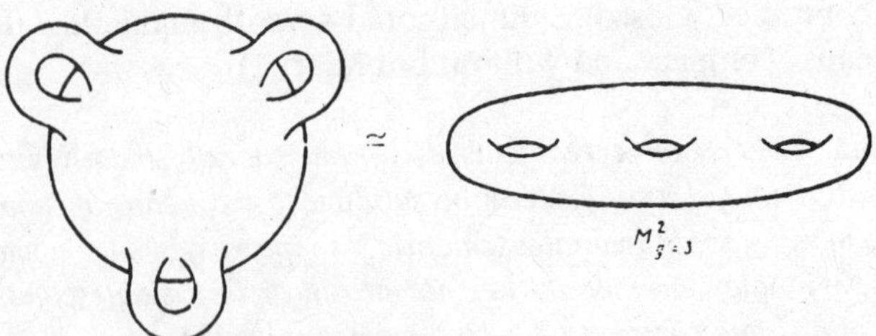

Figure 4. The sphere-with-g-handles $S^2 + (g) = M_g^2$. (In the figure, $g = 3$.)

discs D^2 are removed, and then each of the resulting μ holes is closed up by identifying each pair of diametrically opposite points of its boundary (see Figure 5(a)). (This process is called "attaching μ Möbius bands to S^2", to obtain a "sphere-with-μ-crosscaps".)

In particular, the surface $M_{\mu=1}^2$ is the real projective plane (Figure 5(b)), and $M_{\mu=2}^2$ is the Klein bottle. (Recall that in Part II, §§16.2, 18.2, the Klein bottle was defined as the orbit space of the action of a certain discrete group of motions of the plane; that in this realization the Klein bottle is diffeomorphic to $M_{\mu=2}^2$ can be seen from Figure 5(c).)

A priori one might expect there to be an independent third "mixed" type of surface, obtained by attaching g handles and μ crosscaps to S^2. However, each such "mixed" type of surface is in fact already accounted for among the M_μ^2. This can be seen as follows: Consider, to begin with, the surface obtained by attaching a single handle and a single crosscap; by moving one end of the handle until the circle S^1, in which it meets the sphere, is on the attached

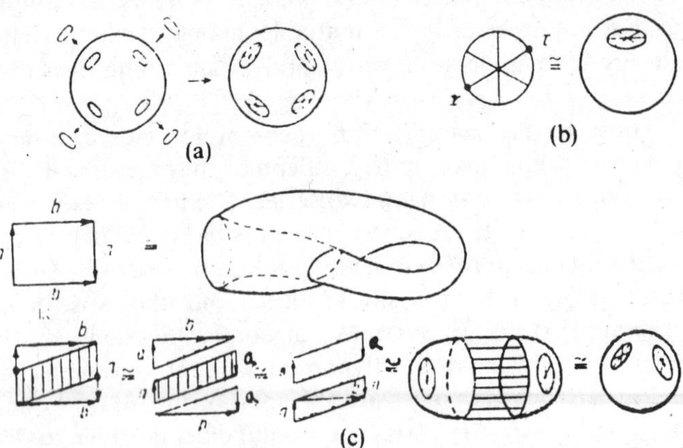

Figure 5. (a) The manifold $M_\mu^2 = S^2 + (\mu)$ (where in the figure $\mu = 4$) obtained by attaching μ crosscaps (Möbius bands) to the sphere S^2. (b) The real projective plane $M_{\mu=1}^2 \cong \mathbb{R}P^2$. (c) The Klein bottle $M_{\mu=2}^2$.

Möbius band, then moving it once around the Möbius band and then off it, the orientation of S^1 is reversed, so that the end of the handle is now attached to the inside of the sphere (as in the Klein bottle; see Figure 5(c)). (Figure 6 is intended to indicate this process. Note that the handle does not actually meet the sphere; that it seems to do so in the last of the four diagrams comprising Figure 6, is a consequence of the fact that the Klein bottle is not embeddable in $\mathbb{R}^3$.) Thus do we see that a sphere-with-one-handle-and-one-crosscap is diffeomorphic to a Klein bottle-with-crosscap, and therefore, in view of Figure 7 (or, equivalently, Figure 5(c)) to the surface $M^2_{\mu=3}$, the sphere-with-three-crosscaps (see Figure 8). The upshot is that in the presence of at least one crosscap, each handle can be replaced, via a diffeomorphism, by two crosscaps.

Having quelled this potential initial doubt, we shall now prove rigorously that every closed, compact 2-manifold M^2 is indeed accounted for in one or the other of the infinite lists $\{M^2_g\}$, $\{M^2_\mu\}$.

Thus consider any such manifold M^2; by Lemma 3.13 we may assume it to be finitely triangulated. We make cuts in M^2 along every edge of the triangulation, inscribing beforehand, however, the same identifying letter on both sides of each edge (i.e. on the two faces abutting at each edge), using

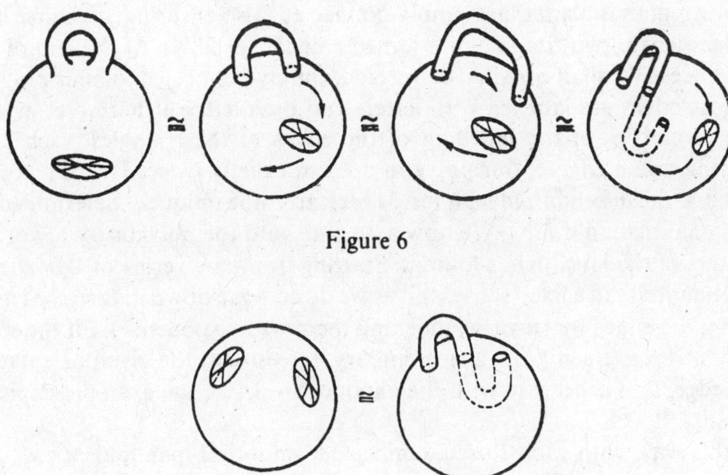

Figure 6

Figure 7. The sphere S^2 with handle going from outside to inside.

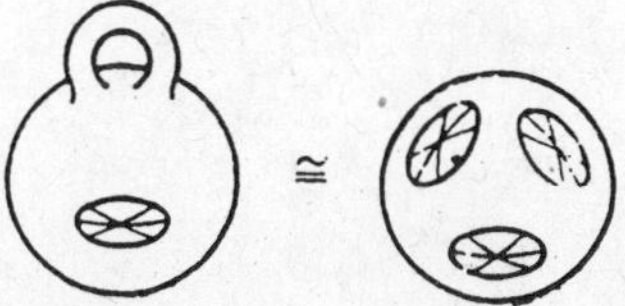

Figure 8

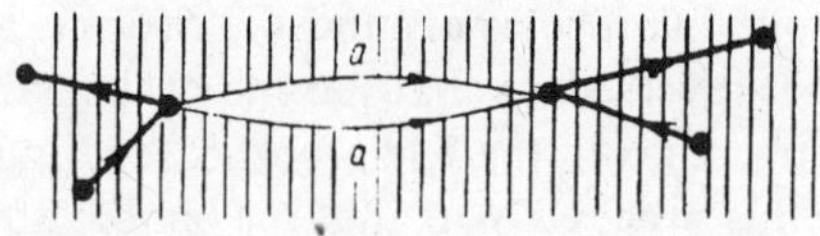

Figure 9

different letters for different edges, and also indicating on those sides, by means of arrows, an orientation of the edge (see Figure 9). Our manifold is in this way transformed into a finite collection of triangles (i.e. 2-simplexes) with their edges oriented, and labelled with letters in such a way that each letter which occurs as a label, does so exactly twice, and then on edges of distinct triangles. Clearly, the labelling and orientation of the edges of the triangles (without regard to their shape or size) is enough to determine the original manifold M^2.

We now flatten out the triangles (and allow their size and shape to vary) so that the following reverse process, of partial reconstruction of M^2, can be carried out. We begin with one triangle, attach to it in the prescribed way a second, and then to the resulting polygon a third trigangle, and so on, always adjusting the sizes and shapes of the triangles we are adjoining, so that the resulting region is planar and simply-connected (also ensuring, of course, that the edges where two triangles are joined coincide exactly). At the end of this finite process we shall obtain a connected, simply-connected, planar polygon W, whose edges are labelled with letters and oriented. Furthermore, in view of the properties of the labelling of the edges of the triangles, each label appearing on an edge of our polygon does so exactly twice. The polygon W is called a *fundamental polygon* for M^2 (clearly not uniquely determined by the original manifold M^2). We now associate with the polygon W a word in the letters of the labelling as follows: Starting from any vertex of W we trace out its boundary in a fixed sense, and as we do so write down in turn the letters labelling the edges we traverse, indexing them with exponent $+1$ if the sense in which we are tracing out the boundary agrees with the given orientation of the edge, and otherwise with the exponent -1 (see the example depicted in Figure 10).

In this way, with each closed, connected, smooth 2-manifold M^2 we can associate (non-uniquely) a (finite) word

$$w = a_{i_1}^{\varepsilon_1} a_{i_2}^{\varepsilon_2} \dots a_{i_k}^{\varepsilon_k},$$

Figure 10

where $\varepsilon_\alpha = \pm 1$, k is the (necessarily even) number of sides of the fundamental polygon W whence this word arose, and each letter a_α occurs exactly twice. We now take a wider view and consider for each such manifold M^2 all finite, planar, connected, simply-connected polygons with boundary edges labelled and oriented so that the corresponding identifications yield a manifold diffeomorphic to M^2. (From the foregoing we know that for each M^2 there are infinitely many such polygons and associated words.) Each of the corresponding words in the labels on the edges of such a polygon therefore "encodes" M^2, in the sense that the manifold can be, up to a diffeomorphism, reconstructed from it. We shall now describe "elementary operations" on these words, by means of which they can be brought into one of three canonical forms, determining a classification of the class of 2-manifolds we are considering (see Theorem 3.19 below).

3.14. Lemma. *Every closed, connected, smooth 2-manifold M^2 is determined by a finite planar, connected, simply-connected polygon (via identification of its edges according to their labels and orientation) in which all vertices of the polygon correspond to (i.e. are identified with) a single point of M^2.*

(We note that such a polygon, or word arising from it, is said to be *reduced*; clearly, it cannot arise as a fundamental polygon from a triangulation of M^2.)

PROOF. Suppose that we have a polygon W determining M^2, with at least two equivalence classes of vertices (under identification). We may then clearly find two distinct such classes $\{P\}$ and $\{Q\}$ say, such that there is an edge (labelled a) of W joining a point P of $\{P\}$ and a point Q of $\{Q\}$. We then carry out the "elementary operation" on our polygon indicated in Figure 11, introducing thereby two new edges labelled d, and eliminating the edges labelled with c. (In that figure and subsequent ones the heavy line segments are intended to represent parts of the boundary of the polygon not relevant to the issue at hand.) It is intuitively clear that such an operation leaves unaltered (to within a diffeomorphism) the manifold determined by the polygon.

From Figure 11 it can be seen that the net effect of that elementary operation is to reduce by one the number of vertices in the class $\{P\}$, while increasing by one the number in the vertex-class $\{Q\}$, without affecting any

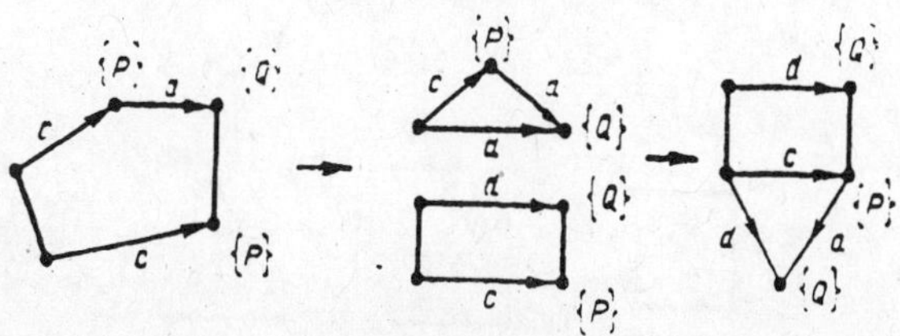

Figure 11

other vertex-classes (or introducing additional ones). The procedure is now clear: we concentrate on a particular vertex class $\{P\}$, reducing the number of vertices in it by one at each step (without thereby altering, in essence, the manifold determined by the polygons) till there is only one remaining, at the same time enlarging various of the other vertex-classes; and, finally, when there is but one vertex left in $\{P\}$, we eliminate it as indicated in Figure 12. We are then left with fewer vertex-classes than before, so that the desired result follows by induction. □

3.15. Lemma. *Any segment of the form cc^{-1} (or $c^{-1}c$) of a word arising from a polygon representing M^2, may be deleted (yielding an "equivalent" word, determining a diffeomorphic manifold).*

PROOF. This is clear from Figure 12. □

3.16. Lemma. *Any word of the form — a — a — is equivalent to one of the form — aa — of the same length.*

PROOF. This follows as in Figure 13. (Note that subsequent to the final step the letter c is to be replaced by a, which no longer appears as a boundary label.) □

3.17. Lemma. *Any word of the form — a — b — a^{-1} — b^{-1} — is equivalent to one of the form — $aba^{-1}b^{-1}$ — of the same length.*

PROOF. See Figure 14. □

3.18. Lemma. *Let w be a reduced word determining as above a manifold M^2 (see Lemma 3.14). If w has the form — a —a^{-1} —, where the indicated symbols a and a^{-1} are not adjacent, i.e. the set α of symbols between them is non-empty, then there is a boundary label b such that one of the symbols b, b^{-1} is in α, but not the other. It follows that w has the following form (after possibly inter-*

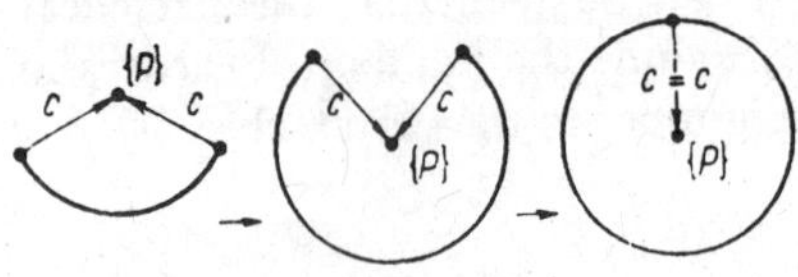

Figure 12

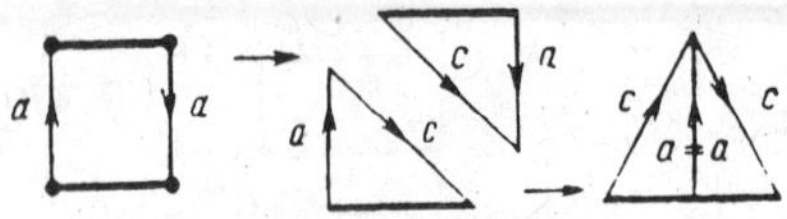

Figure 13

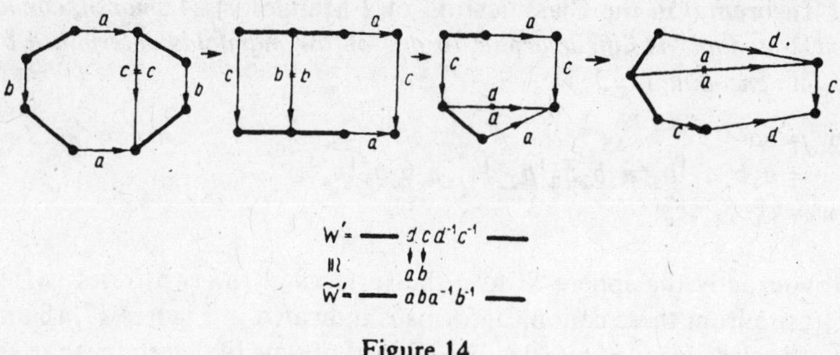

Figure 14

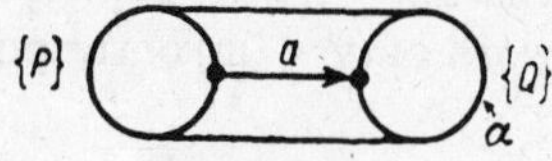

Figure 15

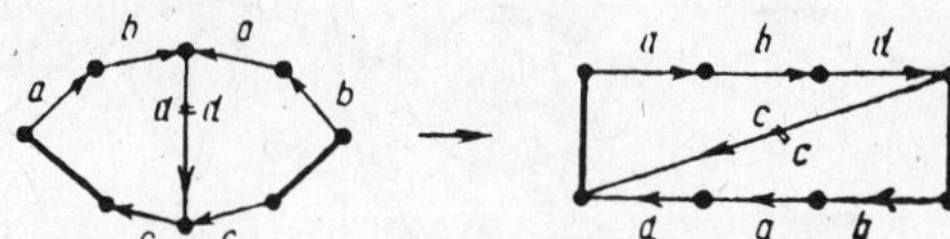

Figure 16

changing the symbols b, b^{-1} *and performing a cyclic permutation*):

$$w = -a \underbrace{- b -}_{\alpha} a^{-1} - b^{-1} -$$

PROOF. Suppose on the contrary that together with every c (or c^{-1}) in α, we also have c^{-1} (or c) in α. In this case all of the vertices of the edges in α on the one hand, and the initial vertices of the edges labelled with a on the other, would be identified with distinct points of M^2 (see Figure 15). This however contradicts our assumption that w is reduced. □

3.19. Lemma. *A word* w *of the form* $- aba^{-1}b^{-1} - cc -$ *is equivalent to one of the form* $- a^2 - b^2 - c^2$.

PROOF. This follows by rearranging the polygon corresponding to w as indicated in Figure 16, and then applying Lemma 3.16 (three times). □

From these lemmas (and their proofs) we infer immediately the promised clasification of words (or polygons) determining manifolds M^2 according to three canonical forms.

3.20. Theorem (On the Classification of 2-Manifolds). *A smooth, connected, closed 2-manifold is diffeomorphic to one of the manifolds determined by the following encoding words w:*

(i) $w = aa^{-1}$;
(ii) $w = a_1 b_1 a_1^{-1} b_1^{-1} a_2 b_2 a_2^{-1} b_2^{-1} \ldots a_g b_g a_g^{-1} b_g^{-1}$;
(iii) $w = c_1^2 c_2^2 \ldots c_\mu^2$.

How exactly the sphere S^2 and the surfaces in the two series $\{M_g^2\}$ and $\{M_\mu^2\}$ arise from these canonical forms is indicated in Figures 17, 18 and 19. Thus it is clear from Figure 17 that a word of type (i) determines the sphere S^2. How a word of type (ii) yields the sphere-with-g-handles may be inferred from Figure 18, where the cases $g = 1, 2$ are depicted. Finally, Figure 19 shows how (in the case $\mu = 2$) a word of type (iii) corresponds to the sphere-with-μ-crosscaps.

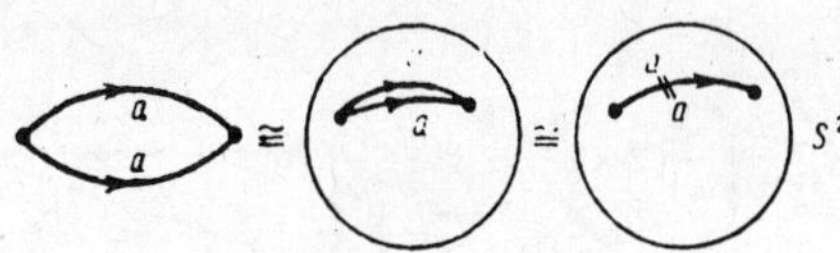

Figure 17

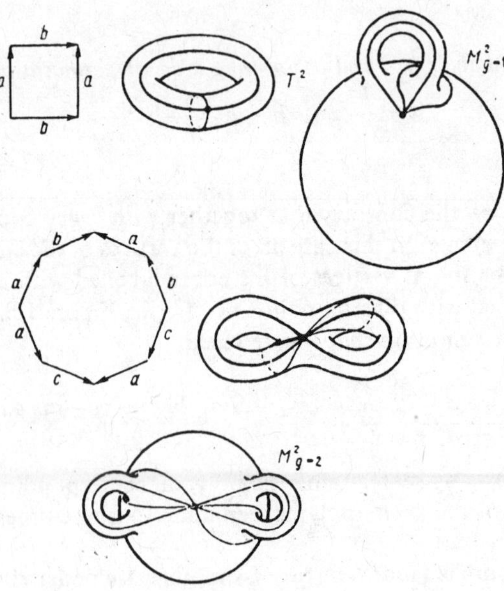

Figure 18

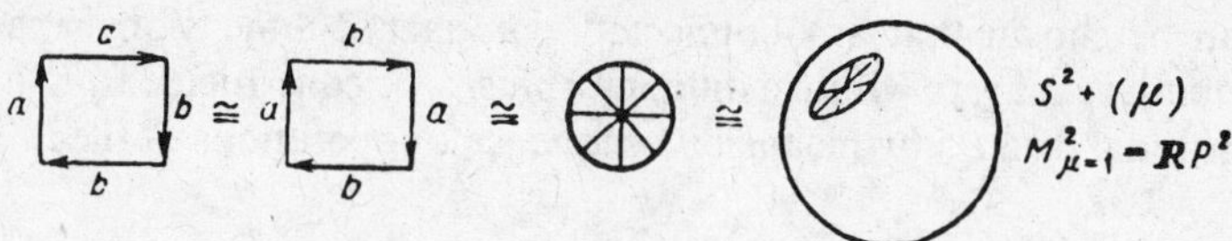

Figure 19

Remarks

1. It is not difficult to compute the homology groups, with integer coefficients for instance, of the manifolds associated with the canonical forms (i), (ii), (iii) (see §4 below). From the fact that, as it turns out, those groups are distinct, it follows (in view of the above-mentioned homotopy invariance of the homology groups) that distinct canonical words determine non-homeomorphic manifolds M^2.

2. There exist alternative canonical forms for the words encoding closed, connected 2-manifolds. For instance, each such M^2 can be encoded as a word w of the following form (omitting the easy case of the sphere S^2):

$$w = a_1 a_2 \ldots a_N a_1^{-1} a_2^{-1} \ldots a_{N-1}^{-1} a_N^{\varepsilon}, \qquad \varepsilon = \pm 1,$$

where N is even if $\varepsilon = -1$, and then $M^2 \cong M_g^2$ with $g = N/2$ (the orientable case), while if $\varepsilon = +1$ (and N arbitrary) we get $M^2 \cong M_\mu^2$ (the non-orientable case). This can be seen as follows. Consider the orientable case: $w = a_1 \ldots a_N a_1^{-1} \ldots a_N^{-1}$, $N = 2g$. By means of elementary transformations (cf. the above lemmas) we separate off one by one standard handles corresponding to word-segments of the form $aba^{-1}b^{-1}$; the way to achieve this is clear from Figure 20 (in which Q stands for $a_3 \ldots a_N$ and P for $a_3^{-1} \ldots a_N^{-1}$), where the procedure for separating out the first handle (represented by the segment $d^{-1}cdc^{-1}$) is indicated. By repeating this procedure (taking into account that $N = 2g$) we shall after g steps arrive at the canonical form (ii) of Theorem 3.20. This establishes the claim in the orientable case. We leave the (analogous) proof of the non-orientable case to the reader.

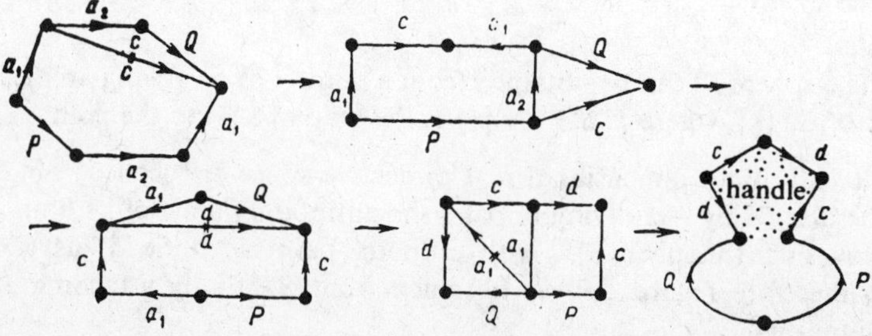

Figure 20

It can be shown that any connected, compact, 2-dimensional manifold-with-boundary may, to within a diffeomorphism, be constructed from a closed 2-dimensional disc D^2 by means of a sequence of operations of the following types:

(i) removal of a finite number of open discs of small radius;
(ii) attachment of finitely many handles;
(iii) attachment of finitely many crosscaps.

(Note that these operations are to be carried out away from the boundary of D^2.)

§4. Attaching Cells to a Topological Space. Cell Spaces. Theorems on the Reduction of Cell Spaces. Homology Groups and the Fundamental Groups of Surfaces and Certain Other Manifolds

As usual we denote by D^n the (closed) n-dimensional disc (or ball) and by S^{n-1} its boundary ∂D^n, the $(n-1)$-sphere. We shall in what follows assume a fixed orientation given on D^n, which then of course induces a certain orientation on its boundary S^{n-1}. Let X be any topological space, and suppose we are given a map f from the boundary S^{n-1} of D^n to X:

$$f: S^{n-1} \to X.$$

From these data we form a new space, denoted by $D^n \cup_f X$, by identifying each point x of the sphere S^{n-1} with the point $f(x)$ of X, and we say that this new space has been obtained by *attaching to X the cell* (D^n, f). (Thus $D^n \cup_f X$ is a quotient space of $D^n \cup X$ via the obvious identification map $D^n \cup X \to D^n \cup_f X$; it follows that its closed sets, which of course determine its topology, are just those subsets $K \subset D^n \cup_f X$ with the property that the complete inverse image of K under the identification map intersects each of D^n and X in a closed set.)

Examples

(α) The sphere S^n can be obtained from a point $*$ by attaching an n-cell to it: $S^n \cong D^n \cup_f \{*\}$, where $f: S^{n-1} \to \{*\}$ is the map of S^{n-1} to the point.

(β) Recall that n-dimensional real projective space $\mathbb{R}P^n$ can be obtained from the disc D^n by identifying in pairs the antipodal points of its boundary S^{n-1}. Now since the sphere S^{n-1} with antipodal points identified is just $\mathbb{R}P^{n-1}$, we can therefore regard $\mathbb{R}P^n$ as obtained from $\mathbb{R}P^{n-1}$ by attaching an n-dimensional cell:

$$\mathbb{R}P^n \cong D^n \cup_{f_n} \mathbb{R}P^{n-1}, \tag{1}$$

where $f_n: S^{n-1} \to \mathbb{R}P^{n-1}$ is the standard covering.

4.1. Lemma. *Homotopic maps* $f, g: S^{n-1} \to X$, *determine homotopically equivalent spaces* $D^n \cup_f X$ *and* $D^n \cup_g X$.

PROOF. Let $F: S^{n-1} \times I \to X$ be a homotopy between f and g, where as usual I denotes the unit interval. We attach the product space $D^n \times I$ to X by means of the map F of part of the boundary of $D^n \times I$, obtaining thereby the space

$$\hat{X} = (D^n \times I) \cup_F X.$$

Clearly the space $\hat{X}$ contains the spaces $D^n \cup_f X$ and $D^n \cup_g X$, if we make the obvious identifications:

$$D^n \cup_f X = ((D^n \times 0) \cup_F X) \subset \hat{X}, \qquad D^n \cup_g X = ((D^n \times 1) \cup_F X) \subset \hat{X}.$$

Consider the homotopy φ_t (of the identity map on $D^n \times I$) which in effect collapses $D^n \times I$ onto $(D^n \times 0) \cup (S^{n-1} \times I)$ by sliding the points of $D^n \times I$ along rays emanating from a suitable point, as indicated in Figure 21. Since $(D^n \times 0) \cup (S^{n-1} \times I)$ is (clearly) left pointwise fixed throughout, this homotopy determines a homotopy equivalence between $(D^n \times 0) \cup (S^{n-1} \times I)$ and $D^n \times I$, with the property that the mutual "homotopically inverse" maps between these spaces both fix $(D^n \times 0) \cup (S^{n-1} \times I)$ pointwise. It follows that this homotopic equivalence "extends" to the pair of spaces $D^n \cup_f X$ and $\hat{X}$. A completely analogous argument shows that $\hat{X}$ is also homotopically equivalent to $D^n \cup_g X$, whence the lemma. □

4.2. Definition. A *cell space* is a topological space obtained from a finite set of points (i.e. from a finite discrete space) by iterating the procedure of attaching cells of arbitrary dimensions, with the condition that only finitely many cells of each dimension are attached. (The initial discrete points may be regarded as 0-dimensional cells.)

4.3. Definition. A cell space X is called a *cell complex* (or *CW-complex*) if each cell is attached to cells of lower dimension. The subcomplex made up of all cells of dimension $k \leq n$ is then the *n-skeleton* X_n of X; thus a cell complex X is the union of the ascending chain of its n-skeletons:

$$X_0 \subset X_1 \subset \cdots \subset X_n \cdots \subset X.$$

Remark. A simplicial complex (see Definition 3.3) is a particular case of a cell complex; its n-skeleton is the union of its simplexes of all dimensions up to and including the nth.

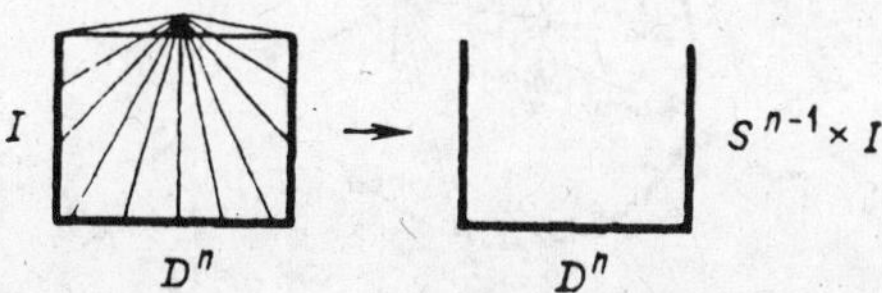

Figure 21

4.4. Theorem. *Every cell space is homotopically equivalent to some cell complex.*

PROOF. It suffices to show that every map of the sphere S^k to a cell complex Y is homotopic to a map of S^k to its k-skeleton Y_k, since the theorem will then follow by repeated appeal to Lemma 4.1.

Thus let $f: S^k \to Y$ by any map. Being compact, the image $f(S^k)$ of S^k under f will intersect the interiors of only finitely many cells. Suppose that $f(S^k)$ intersects the interior D^n of some cell of dimension $n > k$; we shall show, in essence, that this intersection can be deformed out onto the boundary (in Y) of that n-cell. By Part II, Theorem 12.1.3, the restriction of the map f to the complete inverse image $f^{-1}(D^n)$ of that open n-cell is homotopic to a smooth map $\hat{f}$ say, and then by Sard's theorem (Part II, Theorem 10.2.1) there is certainly a point P in that open n-cell which is not an image point under $\hat{f}$. Hence by projecting $D^n \backslash P$ along rays emanating from P (as in Figure 22), we can push the image under $\hat{f}$ out onto the boundary, which is of course part of the $(n-1)$-skeleton of Y. By repeating this proceduce for each open cell of dimension exceeding k which intersects with $f(S^k)$, we shall, after finitely many steps, have deformed f to a map of S^k whose image is contained in the k-skeleton Y_k *of* Y. This completes the proof. □

4.5. Definition. A map $f: X \to Y$ of cell complexes is *cellular* if for every k it sends the k-skeleton X_k of the complex X to the k-skeleton Y_k of Y.

4.6. Theorem (Cellular Approximation Theorem). *Every continuous map between cell complexes is homotopic to a cellular map.*

(The proof of this theorem is completely analogous to that of the previous one; by way of an (easy) exercise, we leave it to the reader to verify this.)

Let X be a cell complex with its cells oriented arbitrarily, and let X_{k-1}, X_{k-2} be its $(k-1)$-skeleton and $(k-2)$-skeleton. It is clear that the quotient space X_{k-1}/X_{k-2}, obtained from X_{k-1} by identifying its subspace X_{k-2} with a single point, is just a bouquet of finitely many oriented $(k-1)$-spheres, one arising from each $(k-1)$-cell D^{k-1} of X. The adjunction of an oriented k-cell (D^k, f) to X_{k-1} then gives rise to a corresponding adjunction of a k-cell to this

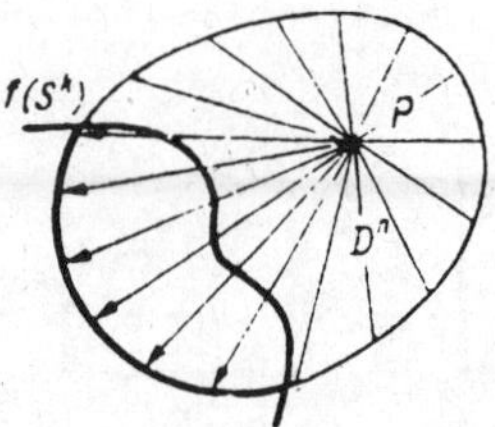

Figure 22

bouquet of $(k-1)$-spheres, via the composite map

$$S^{k-1} \xrightarrow{f} X_{k-1} \to X_{k-1}/X_{k-2}. \tag{2}$$

Now let $\sigma_i^{k-1} = (D^{k-1}, f_i)$, $i = 1, \ldots, N$, be the $(k-1)$-cells of the cell complex X, and let $\sigma^k = (D^k, f)$ be any k-cell of X. The *incidence number* $[\sigma^k : \sigma_i^{k-1}]$ of the pair of (oriented) cells σ^k and σ_i^{k-1} is defined as the degree (see Part II, §13.1) of the composite of the map (2) with the projection of the bouquet X_{k-1}/X_{k-2} onto its ith component sphere S_i^{k-1}, arising from the cell σ_i^{k-1}. (In the case $k = 1$, $S^{k-1} = S^0$ is a pair of points with different signs $\pm$ attached; we set $[\sigma^1, \sigma_i^0] = \pm 1$ if just one of the points of S^0 is mapped to σ_i^0, otherwise $[\sigma^1, \sigma_i^0] = 0$.)

We are now in a positive to define the cell-chain complex $C(X; G)$ of a cell-complex X over an arbitrary additively written abelian group G. We first define a *cell-chain* c_k *of dimension* k of X to be a formal linear combination of the k-cells with coefficients from G:

$$c_k = \sum_{i=1}^{N} g_i \sigma_i^k,$$

where $\sigma_1^k, \ldots, \sigma_N^k$ are the k-cells of X and $g_i \in G$. We then write $C_k(X; G)$ for the additive group of all k-chains, and define the full *cell-chain complex of the cell complex* X as the direct sum of all $C_k(X; G)$ (cf. the definitions of algebraic and simplicial chain complexes in §§2, 3):

$$C(X; G) = \sum_{k \geq 0} C_k(X; G).$$

The action of the *boundary operator* ∂ on k-chains, which (as usual) we first specify on the k-cells, and then extend via G-linearity, is given by

$$\begin{gathered} \partial \sigma^k = \sum_i [\sigma^k : \sigma_i^{k-1}] \sigma_i^{k-1} \qquad (\partial \sigma^0 = 0), \\ \partial\colon C_k(X; G) \to C_{k-1}(X; G). \end{gathered} \tag{3}$$

(Note that in extending, it is understood that $g(n\partial^k) = (ng)\sigma^k$, $g \in G$, $n \in \mathbb{Z}$.)

Remarks

1. If X is a simplicial complex, then the boundary operator defined by (3) coincides with that defined in §3. (Verify this!)

2. In the case $G = \mathbb{Z}$, there is a natural epimorphism

$$\alpha\colon \pi_k(X_k, X_{k-1}) \to C_k(X; \mathbb{Z}),$$

from the relative homotopy group (see Part II, §21.2) onto the group of integral k-chains. This can be seen as follows: each map $(D^k, S^{k-1}) \to (X_k, X_{k-1})$ (representing an element of $\pi_k(X_k, X_{k-1})$) determines a map $(D^k, S^{k-1}) \to (X_k/X_{k-1}, *)$ or $S^k \to X_k/X_{k-1}$; the desired epimorphism now follows by using the fact that X_k/X_{k-1} is a bouquet of k-spheres, one for each k-cell, together with Theorem 13.3.1 of Part II, which tells us, essentially, that for $k > 1$ a map $S^k \to X_k/X_{k-1}$ is, up to a homotopy, determined by its degree on each k-sphere of the bouquet.

4.7. Lemma. *For the operator defined by* (3), *we have* $\partial\partial = 0$.

PROOF. Clearly each k-cell $\sigma^k: D^k \to X_k$, represents an element $[\sigma^k]$ of the relative homotopy group $\pi_k(X_k, X_{k-1})$. It is not difficult to see, using Remark 2 above and the definition of the incidence numbers, that the effect of the boundary operator (3) on a k-cell σ^k (regarded as an integral k-chain) is equivalently given by

$$\partial(\sigma^k) = \alpha(j\hat{\partial}[\sigma^k]) \in C_{k-1}(X; \mathbb{Z}), \tag{4}$$

where α is as in Remark 2, and the homomorphisms

$$\hat{\partial}: \pi_k(X_k, X_{k-1}) \to \pi_{k-1}(X_{k-1})$$

(the "boundary homomorphism") and $j: \pi_{k-1}(X_{k-1}) \to \pi_{k-1}(X_{k-1}, X_{k-2})$ are as defined in connection with the "exact sequence of a pair" in Part II, §21.2 (with a slight modification). Since by exactness $\operatorname{Im} j = \operatorname{Ker} \hat{\partial}$, whence $\hat{\partial} j = 0$, it follows that $\partial\partial(\sigma^k) = 0$. Since every k-chain over an arbitrary group G is a linear combination with coefficients from G of the k-cells σ^k, and the boundary operator is G-linear, it follows that $\partial\partial = 0$, as claimed. □

One now defines in the usual way (see §2) the *cellular homology* and *cohomology groups* of a cell-chain complex $C(X; G)$. It is clear from the above remarks that if the complex X is simplicial, then these groups will coincide with the simplicial homology and cohomology groups introduced in §3.

We now consider examples of cell-complexes.

Examples

(a) The sphere S^n. We noted in Example (α) above that for $n \geq 1$ the n-sphere S^n can be obtained as the result of attaching a single n-cell σ^n to a 0-cell σ^0. Now $\partial\sigma^0 = 0$ by definition, and for $n > 1$, $\partial\sigma^n = 0$ is clear. In the case $n = 1$, the boundary of the cell $\sigma^n = \sigma^1$ is the 0-dimensional sphere S^0 (a pair of points), and the map (2) is the map $S^0 \to \sigma^0$; hence by the definition of the incidence number $[\sigma^k, \sigma^{k-1}]$ in the case $k = 1$ (see above), we have here $[\sigma^1, \sigma^0] = 0$, so that $\partial\sigma^1 = 0$ also. We conclude immediately that:

$$\begin{aligned} &H_0(S^n; G) \simeq G, \qquad n \geq 1; \\ &H_n(S^n; G) \simeq G; \\ &H_k(S^n; G) = 0, \qquad k \neq 0, n. \end{aligned} \tag{5}$$

Consider more generally the bouquet K_q^n of q spheres S_j^n, $n \geq 1, j = 1, \ldots, q$, joined as usual at a single point. Clearly, this space can be realized as the cell complex obtained by attaching q n-cells $\sigma_1^n, \ldots, \sigma_q^n$ to a single 0-cell σ^0. As in the case of a single sphere, we have $\partial\sigma^0 = 0$, $\partial\sigma_i^n = 0$, whence it follows that:

$$\begin{aligned} &H_0(K_q^n; G) \simeq G; \\ &H_n(K_q^n; G) \simeq G \oplus G \oplus \cdots \oplus G \quad (n \text{ direct summands}); \\ &H_k(K_q^n; G) = 0, \qquad k \neq 0, n. \end{aligned} \tag{6}$$

We note that the space obtained from the Euclidean space $\mathbb{R}^{n+1}$ by removing q points has the same homology groups as in (6), since that space contracts to the bouquet of q n-spheres (i.e. is homotopically equivalent to it). (The homotopy invariance of the cellular homology groups will be estabished in §6 below.)

(b) The torus T^2 has the cellular decomposition with cells $\sigma^0, \sigma_1^1, \sigma_2^1, \sigma^2$, as indicated in Figure 23. We have $\partial\sigma^0 = \partial\sigma_1^1 = \partial\sigma_2^1 = \partial\sigma^2 = 0$, whence

$$H_0(T^2) \simeq G, \qquad H_1(T^2) \simeq G \oplus G, \qquad H_2(T^2) \simeq G. \tag{7}$$

(c) The Klein bottle K^2 has the cellular decomposition into cells $\sigma^0, \sigma_1^1, \sigma_2^1, \sigma^2$, as in Figure 24. Here we have $\partial\sigma^0 = \partial\sigma_1^1 = \partial\sigma_2^1 = 0$, $\partial\sigma^2 = 2\sigma_1^1$, so that

$$H_0(K^2; \mathbb{Z}) \simeq \mathbb{Z}, \qquad H_2(K^2; \mathbb{Z}) = 0, \qquad H_1(K^2; \mathbb{Z}) \simeq \mathbb{Z} \oplus \mathbb{Z}_2,$$
$$H_2(K^2; \mathbb{Z}_2) \simeq \mathbb{Z}_2. \tag{8}$$

(d) The projective plane $\mathbb{R}P^2$ has the cell decomposition into three cells σ^0, σ^1, σ^2, as in Figure 25. Here $\partial\sigma^0 = \partial\sigma^1 = 0$, $\partial\sigma^2 = 2\sigma^1$, whence

$$H_0(\mathbb{R}P^2; \mathbb{Z}) \simeq \mathbb{Z}, \qquad H_1(\mathbb{R}P^2; \mathbb{Z}) \simeq \mathbb{Z}_2, \qquad H_2(\mathbb{R}P^2; \mathbb{Z}) = 0. \tag{9}$$

(e) Consider, more generally than (b), the orientable surface of genus g. We saw in Theorem 3.20(ii) that this surface can be obtained by identifying the

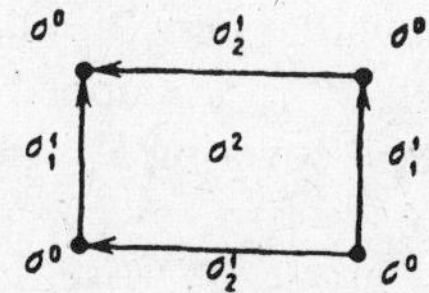

Figure 23. The torus T^2.

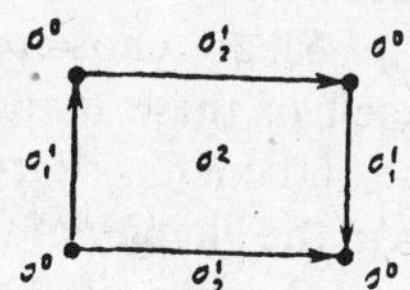

Figure 24. The Klein bottle K^2.

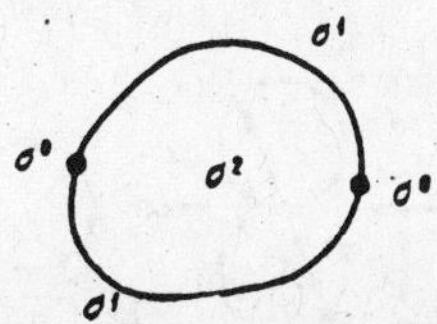

Figure 25. The projective plane $\mathbb{R}P^2$.

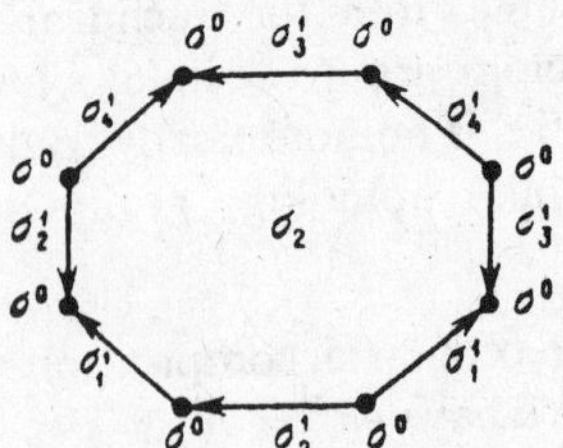

Figure 26. The pretzel.

edges of a $4g$-gon in pairs, as indicated (in the case $g = 2$) in Figure 26. Thus we may regard the surface as a cell complex with cells $\sigma^0, \sigma_1^1, \ldots, \sigma_{2g}^1, \sigma^2$. Since each of these cells has zero boundary, the homology groups are (taking $G = \mathbb{Z}$):

$$H_0(M_g^2; \mathbb{Z}) \simeq \mathbb{Z}; H_1(M_g^2; \mathbb{Z}) \simeq \mathbb{Z} \oplus \cdots \oplus \mathbb{Z} \text{ (2g summands)}; H_2(M_g^2; \mathbb{Z}) \simeq \mathbb{Z}. \tag{10}$$

(f) Real projective space $\mathbb{R}P^n$ (cf. Example (d)). In Example (β) at the beginning of the section, we realized $\mathbb{R}P^n$ as $D^n \cup_{f_n} \mathbb{R}P^{n-1}$, i.e. as resulting from the adjunction of an n-cell to $\mathbb{R}P^{n-1}$ via the standard covering map $f_n: S^{n-1} \to \mathbb{R}P^{n-1}$. Iterating this decomposition for $\mathbb{R}P^{n-1}$, $\mathbb{R}P^{n-2}$, ... in turn, we obtain a cell complex with exactly one cell $\sigma^m = (D_m, f_m)$ of each dimension $m = 0, 1, \ldots, n$. We shall now show that $\partial\sigma^{2k+1} = 0$, $\partial\sigma^{2k} = 2\sigma^{2k-1}$. Since the boundary S^m of each $(m+1)$-cell σ^{m+1} is mapped to the m-skeleton ($\mathbb{R}P^m$) of the cell complex via the standard covering $S^m \to \mathbb{R}P^m$, we need to compute the degree of the composite map

$$S^m \to \mathbb{R}P^m \to \mathbb{R}P^m/\mathbb{R}P^{m-1} \cong S^m.$$

It is easy to see that this map is the sum (in the group $\pi_m(S^m)$) of two maps from S^m to S^m, as indicated in Figure 27. A little thought reveals that, provided local co-ordinates on the copies of S^m are chosen appropriately, for m odd (consider, for instance, $m = 1$) each of these maps has degree $+1$ so that $\partial\sigma^{m+1} = 2\sigma^m$, while for m even one of them has degree $+1$, the other -1. (For instance, if $m = 2$, a pair of charts in the vicinity of the north and south poles, determining the same orientation of the domain 2-sphere, will map to charts determining different orientations of the codomain.) Hence the homology

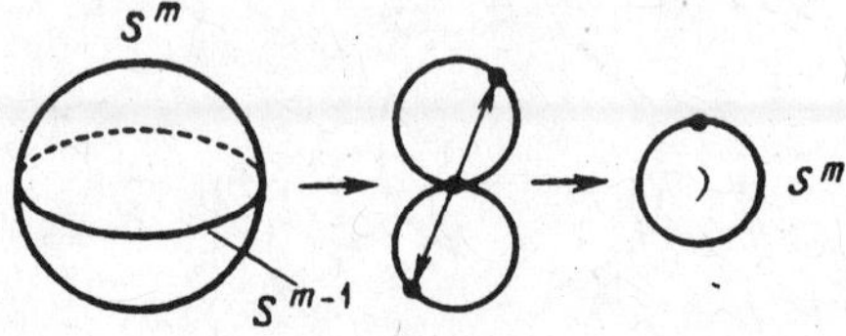

Figure 27

groups (over $\mathbb{Z}$ and $\mathbb{Z}_2$) of n-dimensional real projective space $\mathbb{R}P^n$ are as follows:

$$\text{(i)}\quad H_0(\mathbb{R}P^n;\mathbb{Z})\simeq\mathbb{Z};\qquad H_n(\mathbb{R}P^n;\mathbb{Z})\simeq\begin{cases}0 & \text{for } n \text{ even},\\ \mathbb{Z} & \text{for } n \text{ odd};\end{cases}$$

$$\text{and for } 0<k<n,\ H_k(\mathbb{R}P^n;\mathbb{Z})\simeq\begin{cases}0 & \text{for } k \text{ even},\\ \mathbb{Z}_2 & \text{for } k \text{ odd}.\end{cases}\tag{11}$$

$$\text{(ii)}\quad H_k(\mathbb{R}P^n;\mathbb{Z}_2)\simeq\mathbb{Z}_2,\qquad k=0,1,\ldots,n.\tag{12}$$

(g) Complex projective space $\mathbb{C}P^n$. Let $z^0, \ldots, z^n$ be homogeneous co-ordinates for $\mathbb{C}P^n$. The equation $z^0 = 0$ then determines a submanifold of $\mathbb{C}P^n$ identifiable with $\mathbb{C}P^{n-1}$, whose complement $\mathbb{C}P^n \backslash \mathbb{C}P^{n-1}$ is a submanifold identifiable with ordinary n-dimensional complex space $\mathbb{C}^n$ (co-ordinatized by $z^1/z^0, \ldots, z^n/z^0$) which can therefore be regarded as a $2n$-dimensional cell σ^{2n}. Iterating this decomposition, we ultimately arrive at a cell decomposition of $\mathbb{C}P^n$ into $n + 1$ even-dimensional cells $\sigma^0, \sigma^2, \ldots, \sigma^{2n}$. Since the boundaries of these cells are all trivially zero (there being no odd-dimensional cells), it follows immediately that

$$H_{2k}(\mathbb{C}P^n;G)\simeq G,\qquad H_{2k+1}(\mathbb{C}P^n;G)=0,\qquad 0\le k\le n.$$

Cell complexes are also used for computing homotopy groups. In this connexion the following theorem is useful. (A topological space is called *n-connected* if it is path-connected and all its homotopy groups $\pi_i(X)$ up to the nth (inclusive) are trivial.)

4.8. Theorem. *Every n-connected cell complex K is homotopically equivalent to a cell complex $\tilde{K}$ with exactly one vertex σ^0 and without cells of dimensions* $1, 2, \ldots, n$.

Before giving the proof we consider two examples.

Examples

(a) Consider the case $n = 0$ of the theorem. The reduction of a path-connected cell complex K to a one-vertex complex $\tilde{K}$ of the same homotopy type, is effected as follows: If there is in the complex K an edge σ^1 (i.e. a 1-cell) whose boundary consists of two distinct vertices $\sigma_1^0 \neq \sigma_2^0$, we then form the quotient space K' of K obtained by identifying the whole edge σ^1 with a single point $\tilde{\sigma}_1^0 = \tilde{\sigma}_2^0$ (which will then figure as a vertex of the new complex K') and apart from this leaving all other cells unchanged. The resulting complex K' is then homotopically equivalent to the original one K (as will be shown below, in the proof of Theorem 4.8). Iterating this procedure finitely may times, we end up with a complex $\tilde{K}$ with exactly one 0-cell $\tilde{\sigma}^0$. Its 1-skeleton $\tilde{K}_1$ will then of course be a bouquet of N circles, where the circles are just the 1-cells $\tilde{\sigma}_i^1$, $i = 1, \ldots, N$. Hence the fundamental group $\pi_1(\tilde{K}_1)$ of the 1-skeleton is free, with free generators the homotopy classes $[\tilde{\sigma}_i^1] = a_i$ say (see Part II, §19.2).

Each 2-cell $\tilde{\sigma}_j^2$ of $\tilde{K}$ is attached to the 1-skeleton $\tilde{K}_1$ by means of a map of its boundary $S_j^1 = \partial\tilde{\sigma}_j^2 \to \tilde{K}_1$, which corresponds in the obvious way to a word w_j in the generators $a_1, \ldots, a_N$ (and their inverses) of the free group $\pi_1(\tilde{K}_1)$. It is then at least very intuitive that the group $\pi_1(\tilde{K})$, the fundamental group of the whole complex $\tilde{K}$, and therefore also $\pi_1(K)$, has a presentation with generators $a_1, \ldots, a_N$ and relations $w_j = 1$, one for each 2-cell $\tilde{\sigma}_j^2$ of the complex $\tilde{K}$. On the other hand, the homology group $H_1(\tilde{K}; \mathbb{Z})$ (and therefore the isomorphic group $H_1(K; \mathbb{Z})$—see §6 below) is clearly generated as an abelian group by the basic cycles $\tilde{\sigma}_i^1$ with relations corresponding to the w_j, i.e. is obtained from the relations $w_j = 1$ by replacing each a_i in them by $\tilde{\sigma}_i^1$, and reinterpreting the results as relations in an additively written abelian group. This line of argument provides the justification for the definition of the first integral homology group, given in Part II, §19.3, as the abelianization of the fundamental group:

$$H_1(K; \mathbb{Z}) \simeq \pi_1(K)/[\pi_1(K), \pi_1(K)]. \tag{13}$$

(b) If $n > 0$, then the groups $H_{n+1}(\tilde{K}; \mathbb{Z})$ and $\pi_{n+1}(\tilde{K})$ (with $\tilde{K}$ as in the theorem) are both commutative. Analogously to the case $n = 0$ considered above, they also have essentially the same generators, namely the $(n + 1)$-dimensional cells $\tilde{\sigma}_i^{n+1}$ of K $(i = 1, \ldots, N)$, subject to the same relations (as we shall show below), namely those corresponding to the boundaries of the $(n + 2)$-cells $\tilde{\sigma}_j^{n+2}$. Hence we have the following

4.9. Corollary (Hurewicz Isomorphism Theorem). *For an n-connected cell complex K, where* $n > 0$, *the groups* $\pi_{n+1}(K)$ *and* $H_{n+1}(K; \mathbb{Z})$ *are (naturally) isomorphic.*

PROOF. By Theorem 4.6 each map of an $(n + 1)$-sphere S^{n+1} to the cell complex K is homotopic to a map sending S^{n+1} to the $(n + 1)$-skeleton K_{n+1} of K. Since K is n-connected, we may by Theorem 4.8 (yet to be proved) assume for the purposes of the corollary, that its $(n + 1)$-skeleton K_{n+1} is a bouquet of $(n + 1)$-spheres (gathered at the unique vertex σ^0), so that the homotopy classes of maps $S^{n+1} \to K_{n+1}$ are in one-to-one correspondence with the formal integral linear combinations of the $(n + 1)$-cells σ_i^{n+1} of K (see Part II, §13.3). Each relation $\sum_i \lambda_i \sigma_i^{n+1} \sim 0$ on the σ_i^{n+1}, regarded as generators of the group $\pi_{n+1}(K)$, then corresponds to a map of a disc D^{n+2} to K, whose restriction to the boundary $S^{n+1} = \partial D^{n+2}$ corresponds in the other direction (as a map $S^{n+1} \to K_{n+1}$) to the linear combination $\sum_i \lambda_i \sigma_i^{n+1}$. Now such a map $D^{n+2} \to K$ is, again by Theorem 4.6, homotopic to one sending D^{n+2} to the $(n + 2)$-skeleton, moreover via a homotopy which is constant on the boundary $\partial D^{n+2} = S^{n+1}$ (this can be secured by arguing much as in the proof of Theorem 4.4). Hence each relation "$\sum \lambda_i \sigma_i^{n+1}$ is homotopic to zero" in $\pi_{n+1}(K)$, has its counterpart "$\sum \lambda_i \sigma_i^{n+1}$ is homologous to zero" in $H_{n+1}(K; \mathbb{Z})$ (and vice versa). This completes the proof of the corollary. □

EXERCISE

1. Prove the following converse statement: If K is a connected and simply-connected cell complex such that $H_k(K; \mathbb{Z}) = 0$ for $0 < k < n$, then $\pi_k(K) = 0$ for all these k, and $\pi_n(K) \simeq H_n(K; \mathbb{Z})$.

PROOF OF THEOREM 4.8. Fix on one vertex σ^0 of our n-connected cell complex K, and join it to the other vertices σ_i^0 by means of paths γ_i, one for each σ_i^0. We may suppose (by Theorem 4.6 if you like) that these paths are all contained in the 1-skeleton of K. For each i we attach a half-disc to K along γ_i (see Figure 28), thereby obtaining a larger cell complex $\hat{K}$ containing, in addition to the cells of K, the cells σ_i^1 and σ_i^2 (for each i), as in Figure 28. Since each pair of these new 1-cells intersect only in σ^0, their union $\bigcup_i \sigma_i^1$ is contractible in $\hat{K}$ (to the vertex σ^0). It follows easily that the resulting quotient space

$$\tilde{K} = \hat{K} \Big/ \bigcup_i \sigma_i^1$$

is homotopically equivalent to $\hat{K}$. On the other hand, since the complex $\hat{K}$ clearly contracts to K (by shrinking each attached half-disc onto its diameter), we have similarly that $\hat{K}$ is homotopically equivalent to K. Hence K is homotopically equivalent to $\tilde{K}$, a cell complex with exactly one 0-cell.

Now suppose inductively that we have a cell complex K with just one vertex (0-cell), and without cells of dimensions $1, 2. \ldots, k - 1$, where $k < n$. If there are any cells of dimension k, then the k-skeleton of K will be a bouquet of k-spheres S_i^k. Since each of these spheres S_i^k is, in view of the assumed n-connectedness of K, homotopic to zero in K, there is for each i a map of a $(k + 1)$-dimensional disc D_i^{k+1} to K which sends the boundary onto S_i^k (and which, in essence by the proof of Theorem 4.4, we may assume to send D_i^{k+1} to the $(k + 1)$-skeleton of K). For each i we now attach a $(k + 2)$-dimensional disc D_i^{k+2} to K by identifying half of its boundary with the image of D_i^{k+1} in K. This gives us a cell complex $\hat{K}$ homotopically equivalent to K, containing for each k-cell S_i^k an additional pair of cells σ_i^{k+1} and σ_i^{k+2} (analogously to the first part of the proof). We see much as before (where we had $k = 0$) that the union of the new $(k + 1)$-cells σ_i^{k+1} is contractible in $\hat{K}$, whence it follows that

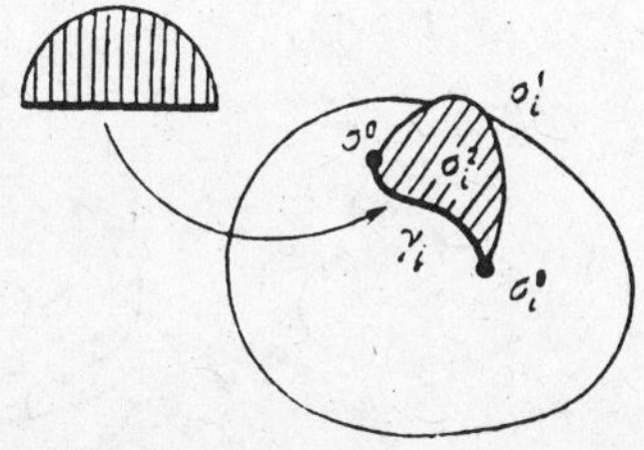

Figure 28

the quotient space

$$\tilde{K} = \hat{K} \Big/ \bigcup_i \sigma_i^{k+1}$$

is homotopically equivalent to $\hat{K}$. Thus we have constructed a cell complex $\tilde{K}$ which is homotopically equivalent to K, and has no cells of dimensions $1, 2, \ldots, k-1, k$. This completes the inductive step, and thence the proof of the theorem. □

The classification theorem for closed surfaces (Theorem 3.20) affords us a decomposition of each such surface M^2 as a cell complex of the form

$$M^2 = \sigma^0 \cup \left(\bigcup_\alpha \sigma_\alpha^1 \right) \cup \sigma^2,$$

with just one 0-cell σ^0 and one 2-cell σ^2; here the 1-skeleton is a bouquet of circles $\bigvee_\alpha S_\alpha^1$, to which a disc D^2 (corresponding to the 2-cell σ^2) is attached (see Figure 29) in a manner determined by the appropriate canonical word w of Theorem 3.20. (The case of $M_{g=1}^2$ is illustrated in Figure 30.) More precisely, if the circles S_α^1 of the bouquet are denoted by $a_1, \ldots, a_n$ (where $n = 2g$ in the case of M_g^2, and $n = \mu$ for M_μ^2), then as the disc we take a plane polygon with sides labelled and oriented in accordance with the appropriate word w in the letters $a_1, \ldots, a_n$ (and their inverses) and then make the obvious identifications of the circles a_i with the correspondingly labelled edges of the polygon, matching the orientations of the edges with arbitrarily prescribed orientations of the circles. Since the fundamental group of the bouquet of circles is free of rank n on $a_1, \ldots, a_n$, the adjunction of the cell σ^2 in this way introduces the single relation $w = 1$ (and its consequences of course). Hence the fundamental groups $\pi_1(M^2)$ have the following presentations in terms of generators and

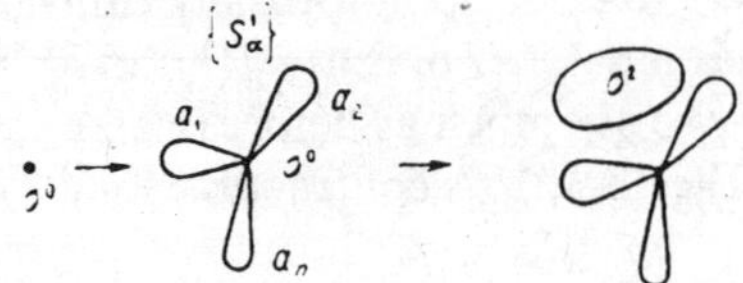

Figure 29

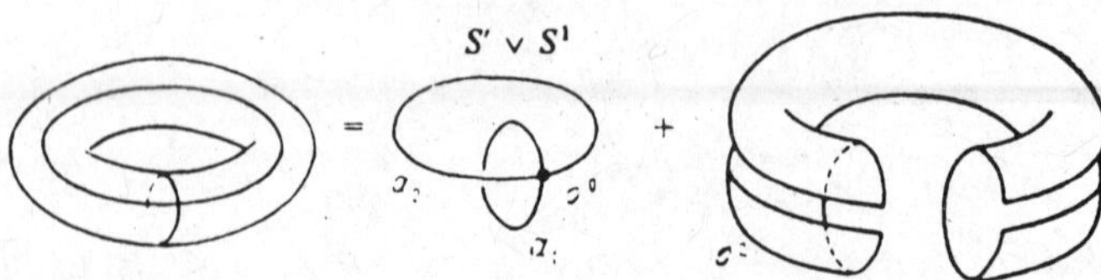

Figure 30

defining relations:

$$\pi_1(M^2) \simeq \begin{cases} \{1\} & \text{for } S^2; \\ a_1, b_1, \ldots, a_g, b_g; & \\ w = a_1 b_1 a_1^{-1} b_1^{-1} \ldots a_g b_g a_g^{-1} b_g^{-1} = 1 & \text{for } M_g^2; \\ a_1, \ldots, a_\mu;\ w = a_1^2 a_2^2 \ldots a_\mu^2 = 1 & \text{for } M_\mu^2. \end{cases} \tag{14}$$

EXERCISES (continued)

2. Show directly that the following group presentations (of $\pi_1(M_g^2)$ and $\pi_1(M_\mu^2)$) define isomorphic groups (cf. §3, Remark 2):

$$\text{(i)} \quad \begin{cases} \langle a_1, b_1, \ldots, a_g, b_g;\ w = a_1 b_1 a_1^{-1} b_1^{-1} \ldots a_g b_g a_g^{-1} b_g^{-1} = 1\rangle; \\ \langle \bar{a}_1, \bar{b}_1, \ldots, \bar{a}_g, \bar{b}_g;\ \bar{w} = \bar{a}_1 \ldots \bar{a}_g \bar{b}_1 \ldots \bar{b}_g \bar{a}_1^{-1} \bar{b}_1^{-1} \ldots \bar{a}_g^{-1} \bar{b}_g^{-1} = 1\rangle. \end{cases}$$

$$\text{(ii)} \quad \begin{cases} \langle a_1, \ldots, a_\mu;\ w = a_1^2 \ldots a_\mu^2 = 1\rangle; \\ \langle \bar{a}_1, \bar{b}_1, \ldots, \bar{a}_k, \bar{b}_k;\ \bar{w} = \bar{a}_1 b_1 \bar{a}_1^{-1} \bar{b}_1 \ldots \bar{a}_k \bar{b}_k \bar{a}_k^{-1} \bar{b}_k = 1\rangle, & \mu \text{ even, } k = \mu/2; \\ \langle \bar{\bar{a}}_1, \ldots, \bar{\bar{a}}_\mu;\ \bar{\bar{w}} = \bar{\bar{a}}_1 \ldots \bar{\bar{a}}_\mu \bar{\bar{a}}_1^{-1} \ldots \bar{\bar{a}}_{\mu-1}^{-1} \bar{\bar{a}}_\mu = 1\rangle, & \mu \text{ arbitrary.} \end{cases}$$

3. Show that distinct surfaces M^2 have non-isomorphic fundamental groups $\pi_1(M^2)$ (in fact non-isomorphic first homology groups $H_1(M^2) \simeq \pi_1/[\pi_1, \pi_1]$).

4. Give a classification of all connected, smooth 2-manifolds (not necessarily compact).

5. Show that a necessary and sufficient condition for an orientable, connected, smooth 2-manifold (with or without boundary) to be realizable as a planar region is that the intersection index (see Part II, §15.1) of any two 1-dimensional cycles vanish. (Note that a planar 2-manifold must of course be orientable.)

6. Show that every non-compact, connected 2-manifold is homotopically equivalent to a finite or countably infinite bouquet of circles:

$$\bigvee_{i=1}^{k} S_i^1 \quad (k < \infty) \qquad \text{or} \qquad \bigvee_{i=1}^{\infty} S_i^1,$$

and therefore has free fundamental group.

Remark. Every closed, connected, smooth 2-manifold can be endowed with a Riemannian metric of constant curvature. On the sphere S^2 and the projective plane $\mathbb{R}P^2$ metrics of constant positive curvature can easily be defined (for S^2 see Part I, §9; the required metric on $\mathbb{R}P^2$ is induced from that on S^2 via the standard projection $S^2 \to \mathbb{R}P^2$), while the torus and Klein bottle can be endowed with metrics of zero curvature. (A metric of zero curvature on the torus T^2 is induced from the Euclidean metric on $\mathbb{R}^2$ via the realization of T^2 as an orbit space $\mathbb{R}^2/\Gamma$, where $\Gamma = \mathbb{Z}(a) \oplus \mathbb{Z}(b)$ is generated by translations (and so isometries) a and b of $\mathbb{R}^2$. For the Klein bottle the construction is analogous since it can be realized as an orbit space $\mathbb{R}^2/\Gamma$, where now Γ is the subgroup of the isometry group of $\mathbb{R}^2$ generated by the transformations

$$T_1(x, y) = (x, y + 1), \qquad T_2(x, y) = (x + \tfrac{1}{2}, -y)$$

(which, as we saw in Part II, §18.2, satisfy the relation $T_2^{-1}T_1T_2T_1 = 1$).) On all other connected, closed, smooth 2-manifolds, a Riemannian metric of constant negative curvature can be defined, since they are all realizable as orbit spaces of the form $M^2 = L^2/\Gamma$, where L^2 denotes as always the Lobachevskian plane with its usual metric (of constant negative curvature; see Part I, §10), and Γ is a subgroup of the isometry group of L^2 isomorphic to $\pi_1(M^2)$ (see Part II, §20).

We shall now consider a useful topological construction which generalizes that of attaching handles and Möbius bands to a sphere. The details of this construction, yielding the "connected sum" of two manifolds of the same dimension, are as follows. Let M_1^n and M_2^n be two manifolds of the same dimension n. By Part II, §11.1, for sufficiently large N we can embed these manifolds in the Euclidean space $\mathbb{R}^N$. By taking N even larger, if necessary, we can clearly so position the manifolds in $\mathbb{R}^N$, that they lie on opposite sides of a hyperplane $\mathbb{R}^{N-1} \subset \mathbb{R}^N$ (so that they certainly do not intersect), and that for any given $\varepsilon > 0$ and any pair of arbitrarily prescribed points $x \in M_1$, $y \in M_2$, the distance between the points x, y is ε, the tangent planes T_x (to M_1 at x) and T_y (to M_2 at y) are parallel, and finally that the straight-line segment $[x, y]$ joining x and y is perpendicular to T_x and T_y (see Figure 31). (Thus T_x, T_y and the straight-line segment $[x, y]$ will be contained in an $(n + 1)$-dimensional Euclidean subspace $\mathbb{R}^{n+1}$ of $\mathbb{R}^N$.) Consider now the cylinder of sufficiently small radius $\delta > 0$, with cross-section the $(n - 1)$-sphere, and with the line segment $[x, y]$ as axis. It we remove from M_1 and M_2 the n-dimensional open neighbourhoods of radius δ and centres x and y respectively, and identify the boundaries of these neighbourhoods with the ends of the cylinder (see Figure 32), then we obtain a new n-manifold, the *connected sum* of M_1^n and M_2^n, denoted by $M_1 \# M_2$.

It is important to note that, as is easy to see intuitively, the manifold $M_1 \# M_2$ is uniquely determined by M_1 and M_2 (at least provided the latter are connected), in the sense that replacement of the points x, y by other points $x' \in M_1$, $y' \in M_2$, in the above-described construction, yields a diffeomorphic manifold. It is clear also that the operation $\#$ is associative (i.e.

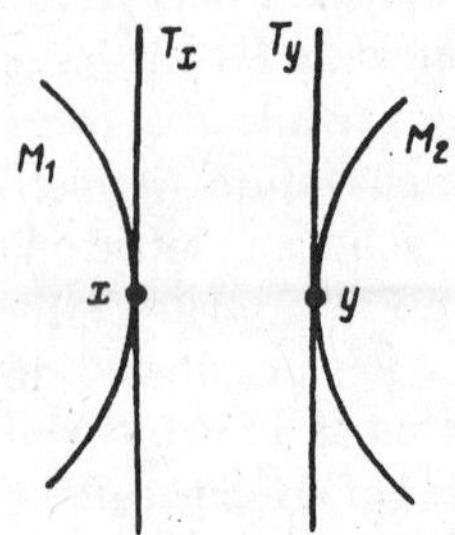

Figure 31

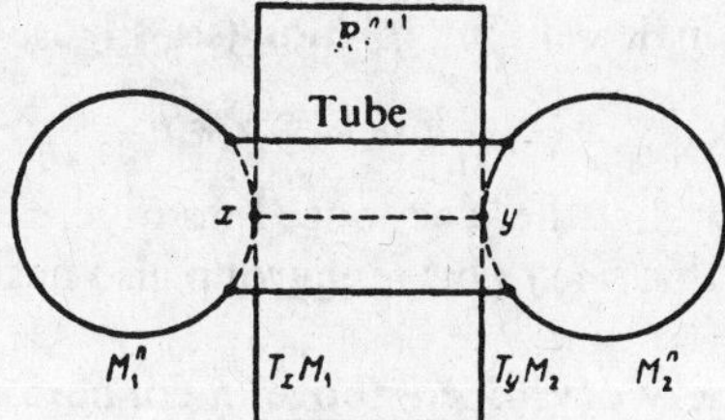

Figure 32

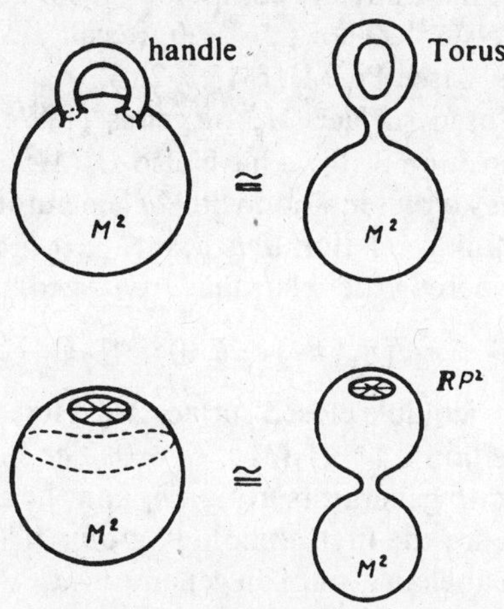

Figure 33

$(M \# N) \# Q \cong M \# (N \# Q)$ where $\cong$ indicates diffeomorphism) and commutative.

Equipped with the operation of taking the connected sum, we now reconsider the operations of attaching a handle or Möbius band to a 2-manifold, introduced in §3 in connexion with the classification of closed surfaces. It is clear from Figure 33 that the operation of attaching a standard handle (corresponding to a word segment of the form $aba^{-1}b^{-1}$) is equivalent to the formation of the connected sum of M^2 and the torus T^2, and that attaching a Möbius band to M^2 amounts to taking the connected sum of M^2 and the projective plane $\mathbb{R}P^2$. Similarly we have: $M^2 \# S^2 \cong M^2$; $M_{g_1}^2 \# M_{g_2}^2 \cong M_{g_1+g_2}^2$; $M_{\mu_1}^2 \# M_{\mu_2}^2 \cong M_{\mu_1+\mu_2}^2$; and $M^2 \# M_{g=1}^2 \# M_{\mu=1}^2 \cong M^2 \# M_{\mu=3}^2$ (see Figure 8). Note also that the Klein bottle is the connected sum of two projective planes (see Figure 5). Thus the introduction of the connected-sum construction yields as a by-product the result that the diffeomorphism classes of connected, closed 2-manifolds M^2, form under that operation a commutative semigroup P with two generators a (the torus T^2) and b (the projective plane

$\mathbb{R}P^2$), and with the single defining relation (see Figure 8)

$$a \# b = b \# b \# b.$$

(As an exercise, show that there ae indeed no other relations, i.e. other than the consequences of this one.) This semigroup also has a zero, represented by the sphere S^2.

Using the machinery developed above in connexion with cell complexes, we can now readily compute the (integral) homology groups of the connected, closed 2-manifolds.

1. The sphere S^2. We have already computed the homology groups of S^2 (see (5)): $H_0(S^2; \mathbb{Z}) \simeq H_2(S^2; \mathbb{Z}) \simeq \mathbb{Z}$; $H_1 = 0$. Recall also that $\pi_1(S^2) = 0$ and $\pi_2(S^2) \simeq \pi_3(S^2) \simeq \mathbb{Z}$ (see Part II, §§13.3, 23.3).
2. The orientable closed surface M_g^2 of genus g. We have $H_0(M_g^2; \mathbb{Z}) \simeq \mathbb{Z}$ trivially. By Proposition 3.10 we have also $H_2(M_g^2; \mathbb{Z}) \simeq \mathbb{Z}$. Since the fundamental group has a presentation with $2g$ generators $a_1, \ldots, a_g, b_1, \ldots, b_g$, and the single defining relation $a_1 b_1 a_1^{-1} b_1^{-1} \ldots a_g b_g a_g^{-1} b_g^{-1} = 1$, we obtain by abelianizing (whereby the relation is trivialized):
$$H_1(M_g^2; \mathbb{Z}) \simeq \pi_1/[\pi_1, \pi_1] \simeq \mathbb{Z} \oplus \cdots \oplus \mathbb{Z} \quad (2g \text{ summands}).$$
3. The general non-orientable closed surface M_μ^2. Here we have $H_0(M_\mu^2; \mathbb{Z}) \simeq \mathbb{Z}$ and, by Proposition 3.11, $H_2(M_\mu^2; \mathbb{Z}) = 0$. The fundamental group has the presentation with generators $a_1, \ldots, a_\mu$ and the single defining relation $a_1^2 a_2^2 \ldots a_\mu^2 = 1$; hence the first homology group $H_1(M_\mu^2; \mathbb{Z}) \simeq \pi_1/[\pi_1, \pi_1]$ is presented (as an abelian group) on generators $\hat{a}_1, \ldots, \hat{a}_\mu$, with the relation $2(\hat{a}_1 + \cdots + \hat{a}_\mu) = 0$, whence
$$H_1(M_\mu^2; \mathbb{Z}) \simeq \underbrace{\mathbb{Z} \oplus \cdots \oplus \mathbb{Z}}_{\mu - 1} \oplus \mathbb{Z}_2.$$
(Here as generators of the summands $\mathbb{Z}$ we may take $\hat{a}_1, \ldots, \hat{a}_{\mu-1}$; the generator of $\mathbb{Z}_2$ is $\hat{a}_1 + \cdots + \hat{a}_\mu$.)

We next consider the 3-dimensional *lens space* L_m (already encountered in Part II, §24.4), which is defined as the orbit space of the 3-sphere S^3: $|z^1|^2 + |z^2|^2 = 1$, regarded as embedded in $\mathbb{C}^2(z^1, z^2)$, under the action of the group $\mathbb{Z}_m$, given by

$$(z^1, z^2) \mapsto (z^1 e^{2\pi i/m}, z^2 e^{2\pi i/m}). \tag{15}$$

(It is easy to see that when $m = 2$ this space is just the 3-dimensional real projective space $\mathbb{R}P^3$.)

To obtain a cell decomposition of the lens space L_m, we first realize S^3 as a cell complex by specifying for each $q = 0, 1, \ldots, m - 1$, one cell of each of the dimensions 0, 1, 2, 3, as follows: we take σ_q^3 to consist of those points $(z^1, z^2) \in S^3$ such that $z^2 = \rho e^{i\varphi}$ where $\rho > 0$, and

$$\frac{2\pi q}{m} < \varphi < \frac{2\pi(q + 1)}{m}; \tag{16}$$

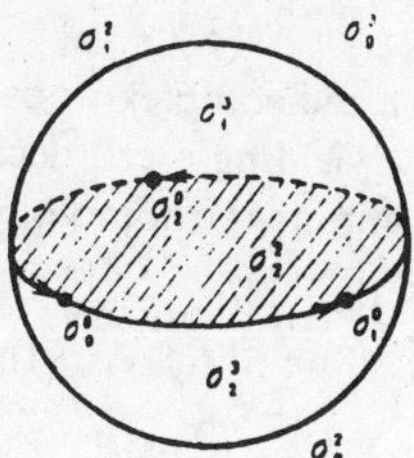

Figure 34

we take σ_q^2 to consist of the points $(z^1, z^2) \in S^3$ such that $z^2 = \rho e^{i\varphi}$ where $\rho > 0$ and $\varphi = 2\pi q/m$; we take σ_q^1 to consist of the points $(z^1, 0)$ where $z^1 = e^{i\varphi}$, with φ again satisfying (16); and finally σ_q^0 to be the point $(e^{2\pi i q/m}, 0)$. The disposition of these cells is indicated schematically in Figure 34 in the case $m = 3$. (In that figure the sphere S^3 is taken in its guise as the 1-point compactification of ordinary Euclidean 3-space). It is not difficult to see that, after orienting these cells appropriately, we shall have

$$\partial\sigma_q^3 = \sigma_{q+1}^2 - \sigma_q^2, \qquad \partial\sigma_q{}^2 = \sigma_0^1 + \cdots \sigma_m^1, \qquad \partial\sigma_q^1 = \sigma_{q+1}^0 - \sigma_q^0, \tag{17}$$

where $(q + 1)$ is understood to be reduced modulo m. Upon identifying points in each orbit of the action of $\mathbb{Z}_m$, the cells σ_q^3, σ_q^2, σ_q^1, σ_q^0 of each dimension clearly become identified, so that the above cell decomposition of S^3 yields a decomposition of L_m as a cell complex with just four cells $\sigma^3, \sigma^2, \sigma^1, \sigma^0$, which moreover by (17) satisfy $\partial\sigma^3 = 0$, $\partial\sigma^2 = m\sigma^1$, $\partial\sigma^1 = 0$. It follows that the integral homology groups of the lens space L_m are given by

$$H_3(L_m; \mathbb{Z}) \simeq \mathbb{Z} \simeq H_0(L_m; \mathbb{Z}), \quad H_2(L_m; \mathbb{Z}) = 0, \quad H_1(L_m; \mathbb{Z}) \simeq \mathbb{Z}_m, \tag{18}$$

and over $\mathbb{Z}_m$ by

$$H_i(L_m; \mathbb{Z}_m) \simeq \mathbb{Z}_m, \qquad i = 0, 1, 2, 3. \tag{19}$$

EXERCISE

7. Identify the cohomology groups $H^i(L_m; \mathbb{Z})$.

The *general lens space* $L_m^{2n-1}(q_1, \ldots, q_{n-1})$ of dimension $2n - 1$, is defined as the orbit space of the sphere $S^{2n-1} \subset \mathbb{C}^n$ under the action of the group $\mathbb{Z}_m$, given by

$$(z^1, \ldots, z^n) \mapsto (e^{2\pi i/m} z^1, e^{2\pi i q_1/m} z^2, \ldots, e^{2\pi i q_{n-1}/m} z^n), \tag{20}$$

where it is required that each q_i be relatively prime to m in order to ensure that the orbit space is a manifold (verify this!); thus with this action of $\mathbb{Z}_m$, we set

$$L_m^{2n-1}(q_1, \ldots, q_{n-1}) = S^{2n-1}/\mathbb{Z}_m.$$

From covering space theory (see Part II, §19.2) it follows that $\pi_1(L_m^{2n-1}) \simeq \mathbb{Z}_m$. Obviously, two general lens spaces $L_m^{2n-1}(q_1, \ldots, q_{n-1})$, $L_m^{2n-1}(q_1', \ldots, q_{n-1}')$ are the same if $q_i \equiv q_i' \pmod m$ for $i = 1, \ldots, n - 1$.

EXERCISES (continued)

8. Realize the sphere S^{2n-1} as a cell complex whose cells are freely permuted by the above action of $\mathbb{Z}_m$, thereby yielding a cell decomposition of L_m^{2n-1} (cf. the case $n = 2$ considered above). Compute the homology groups of the general lens space.

9. Show that the general lens space with $q_1 = q_2 = \cdots = q_{n-1} = 1$, is a smooth fibre bundle with base $\mathbb{C}P^{n-1}$ and fibre S^1 (cf. Part II, §24.4):
$$L_m^{2n-1}(1, \ldots, 1) \xrightarrow{p} \mathbb{C}P^{n-1}, \qquad F = S^1.$$

10. Compute the cohomology ring of the general lens space (with coefficients from $\mathbb{Z}_m$).

Cell decomposition of various smooth fibre bundles are also of interest. We shall briefly consider here one of the simplest situations only, namely that where the fibre F is the sphere S^{n-1}, decomposed as before into two cells: $S^{n-1} = F = \sigma_F^0 \cup \sigma_F^{n-1}$. An important instance of this kind is the unit tangent bundle (see Part II, §7.1):
$$p\colon M^{2n-1} \to M^n, \qquad F = S^{n-1},$$
and in fact we shall confine ourselves to this example. A cell decomposition of the base M^n into cells σ_j^q, gives rise to a cell decomposition of the bundle space M^{2n-1} since above the interior $\hat{\sigma}_j^q$ of each σ_j^q we have
$$p^{-1}(\hat{\sigma}_j^q) \cong \hat{\sigma}_j^q \times F = \hat{\sigma}_j^q \times (\sigma_F^0 \cup \sigma_F^{n-1}),$$
a consequence of the result that any bundle over a disc is trivial, i.e. is merely the (direct) product (see Part II, §24.4). Thus we obtain a cell decomposition of M^{2n-1} into cells
$$\sigma_j^q \times \sigma_F^0, \qquad \sigma_j^q \times \sigma_F^{n-1},$$
one such pair of cells arising from each cell σ_j^q of the base M^n. It turns out, however, that the computation of the homology groups of M^{2n-1} by means of this cell decomposition is, in general, difficult (see §8). We therefore specialize our example further by taking the base to be the surface M_g^2 of genus $g > 0$, in its standard realization as a cell complex (see above):
$$M_g^2 = \sigma^0 \cup \left(\bigcup_{j=1}^{2g} \sigma_j^1 \right) \cup \sigma^2.$$

This cell decomposition then gives rise, in the manner just described, to a cell decomposition of the unit tangent bundle M^3 over M_g^2 into cells
$$\sigma^0 \times \sigma_F^0, \quad \sigma_j^1 \times \sigma_F^0, \quad \sigma^2 \times \sigma_F^0, \quad \sigma^0 \times \sigma_F^1, \quad \sigma_j^1 \times \sigma_F^1, \quad \sigma^2 \times \sigma_F^1. \tag{21}$$

Since M^3 is orientable, the single 3-cell $\sigma^2 \times \sigma_F^1$ must, by Proposition 3.10, be a cycle. We leave it to the reader to verify that each cell $\sigma_j^1 \times \sigma_F^1$ is also a cycle (although in the fundamental group $\pi_1(M^3)$ its boundary represents the commutator of the paths σ_j^1 and σ_F^1 (cf. Figure 23)). The cell $\sigma^2 \times \sigma_F^0$, on the other hand, is not a cycle; it can be shown that in fact
$$\partial(\sigma^2 \times \sigma_F^0) = [(\partial\sigma^2) \times \sigma_F^0] \cup [\sigma^0 \times (\sigma_F^1)^{2-2g}], \tag{22}$$

where the symbol $(\sigma_F^1)^{2-2g}$ signifies that in the boundary of the cell $\sigma^2 \times \sigma_F^0$ the 1-cycle σ_F^1 is traced out $2 - 2g$ times (and we assume given beforehand appropriate orientations of the cells).

EXERCISE

11. Prove the formula (22) by considering a vector field on M_g^2 with exactly one singular point, of index $2 - 2g$ (see Part II, §14.4).

In terms of one of the standard cell decompositions of the base M_g^2 considered earlier in the section, we shall have

$$\partial\sigma^2 = \prod_{i=1}^{g} (a_i b_i a_i^{-1} b_i^{-1}),$$

where the paths a_i correspond to the cells σ_i^1, and b_i to σ_{i+g}^1, in M_g^2 $(i = 1, \ldots, g)$. Using this (and a little more besides—see Part II, §24.3, Example (d)), we obtain a presentation of $\pi_1(M^3)$ on generators $a_1, \ldots, a_g, b_1, \ldots, b_g, \gamma$ (where γ is the path class represented by any fibre $F = S^1$), with defining relations

$$\begin{gathered} [a_i, \gamma] = a_i\gamma a_i^{-1}\gamma^{-1} = 1, \qquad [b_i, \gamma] = b_i\gamma b_i^{-1}\gamma^{-1} = 1, \\ \gamma^{2-2g} = \prod_{i=1}^{g} (a_i b_i a_i^{-1} b_i^{-1}) = \prod_{i=1}^{g} [a_i, b_i]. \end{gathered} \tag{23}$$

Abelianization of this presentation then yields $H_1(M^3)$, while the structure of the other homology groups can be inferred from the knowledge, elicited above, of which cells in the list (21) are cycles; thus the homology groups of M^3 are as follows:

$$H_0 \simeq \mathbb{Z}; \quad H_1 \simeq \underbrace{\mathbb{Z} \oplus \cdots \oplus \mathbb{Z}}_{2g} \oplus \mathbb{Z}_{2g-2}; \quad H_2 \simeq \underbrace{\mathbb{Z} \oplus \cdots \oplus \mathbb{Z}}_{2g}; \quad H_3 \simeq \mathbb{Z}. \tag{24}$$

§5. The Singular Homology and Cohomology Groups. Their Homotopy Invariance. The Exact Sequence of a Pair. Relative Homology Groups

We shall now consider the most general approach to the definition, in homotopically invariant fashion, of the homology and cohomology groups, requiring neither the concept of a manifold nor of a simplicial or cell complex.

5.1. Definition. A *singular k-dimensional simplex* of an arbitrary topological space X, is a pair (σ^k, f), where $f: \sigma^k \to X$ is a continuous map from a standard (oriented) k-simplex $\sigma^k = [\alpha_0 \ldots \alpha_k]$ to X. A *singular k-dimensional chain* is then a formal, finite linear combination of the form

$$c_k = \sum_i g_i(\sigma_i^k, f_i),$$

where the coefficients g_i come from some additively written abelian group G, and the $(\sigma_i^k, f_i)(f_i: \sigma_i^k \to X)$ are singular k-simplexes of X. The *boundary of a singular simplex* (σ^k, f) is the formal linear combinaton

$$\partial(\sigma^k, f) = \sum_q (-1)^q(\sigma_q^{k-1}, f|_{\sigma_q^{k-1}}), \tag{1}$$

where $\sigma_q^{k-1} = [\alpha_0 \dots \hat{\alpha}_q \dots \alpha_k]$ is the qth face of the standard k-simplex σ^k (determining the *singular* $(k-1)$-*dimensional face* $(\sigma_q^{k-1}, f|_{\sigma_q^{k-1}})$ of the singular k-simplex (σ^k, f)). (Note that the σ_q^{k-1} are to be identified with a single standard $(k-1)$-simplex.) The boundary of a singular k-chain c_k is then defined as usual by extending linearly:

$$\partial c_k = \sum_i g_i \partial(\sigma_i^k, f_i).$$

(It then follows from Lemma 3.2 that $\partial\partial c_k = 0$.) A *singular k-cycle* is, naturally, a k-chain c_k for which $\partial c_k = 0$, and a *singular k-boundary* a k-chain c_k such that $c_k = \partial c_{k+1}$ for some singular $(k+1)$-chain c_{k+1}. (Clearly the singular k-boundaries figure among the singular k-cycles.) The kth *singular* (*simplicial*) *homology group* $H_k(X; G)$ is then the quotient group of the group of singular k-cycles by its subgroup of singular k-boundaries. Finally, the *singular cohomology groups* $H^k(X; G)$ are also defined as in §2; thus, a *singular k-cochain* is a linear functional on the k-chains, and the *singular coboundary operator* δ is the dual operator to ∂.

The usefulness of singular homology and cohomology derives largely from the fact that every continuous map $\varphi: X \to Y$ between topological spaces induces in a very transparent manner homomorphisms φ_* and φ^* between the corresponding homology and cohomology groups:

$$\varphi_*: H_k(X; G) \to H_k(Y; G);$$

$$\varphi^*: H^k(Y; G) \to H^k(X; G).$$

(Thus a singular k-chain $c_k = \sum g_i(\sigma_i^k, f_i)$ is mapped by φ_* to the singular k-chain $\varphi_*(c_k) = \sum g_i(\sigma_i^k, \varphi \circ f_i)$. For the cohomology groups the map φ^* operates in the reverse direction: a singular k-cochain c^k has image $\varphi^*(c^k)$ defined by $(\varphi^*(c^k), \tilde{c}_k) = (c^k, \varphi_*(\tilde{c}_k))$ where, for instance, $(\varphi^*(c^k), \tilde{c}_k)$ denotes as before the value of the functional $\varphi^*(c^k)$ on the k-chain $\tilde{c}_k$. Since the maps φ_* and φ^*, as is easy to see, commute with the boundary operators ∂ and δ, it follows that they are well defined on the homology and cohomology groups respectively. That they are indeed homomorphisms is also easily verified.

It is clear from this (or more immediately) that homeomorphic topological spaces will have naturally isomorphic homology and cohomology groups. We shall now establish a stronger result, namely the homotopy invariance of the singular homology (and cohomology) groups. (It is a routine matter to verify that in what follows the analogous assertions (and proofs) are valid also for the singular cohomology groups.)

5.2. Theorem. *If $\varphi_0: X \to Y$ and $\varphi_1: X \to Y$ are homotopic maps of topological spaces, then the induced homomorphisms of the singular simplicial homology groups $\varphi_{0*}, \varphi_{1*}: H_k(X; G) \to H_k(Y; G)$ coincide: $\varphi_{0*} \equiv \varphi_{1*}$ and, analogously, for the cohomology groups $\varphi_0^* \equiv \varphi_1^*$).*

PROOF. Let Φ be a homotopy between the maps φ_0 and φ_1:

$$\Phi(x, t): X \times I \to Y, \qquad \Phi|_{t=0} = \varphi_0, \quad \Phi|_{t=1} = \varphi_1, \quad 0 \le t \le 1, \quad x \in X.$$

With each singular simplex (σ, f), $f: \sigma \to X$, there is then associated the map $\Phi(f \times 1) = \tilde{f}$ says, from the cylinder $\sigma \times \mathrm{I}$ to Y:

$$\Phi(f \times 1)(\sigma, t) = \Phi(f(\sigma), t): \sigma \times I \to X \times I \to Y.$$

We decompose the cylinder $\sigma \times I$ into simplexes as follows: If $\sigma = [\alpha_0 \ldots \alpha_k]$, then as vertices of the simplicial decomposition we take the points $\alpha_i^0 = \alpha_i \times 0$ in the base of the cylinder, and $\alpha_i^1 = \alpha_i \times 1$ in the lid, and as oriented $(k + 1)$-simplexes those of the form

$$\sigma_q = [\alpha_0^0 \ldots \alpha_q^0 \alpha_q^1 \alpha_{q+1}^1 \ldots \alpha_k^1], \qquad q = 0, \ldots, k. \tag{2}$$

(It is not difficult to verify that this does represent a decomposition of $\sigma \times \mathrm{I}$ as a simplicial complex; see Figue 35 for the cases $k = 1, 2$.) In terms of this simplicial decomposition, for each k (the dimension of σ), with each singular k-simplex (σ, f), $f: \sigma \to X$ we can now associate via the map $\Phi(f \times 1) = \tilde{f}$, the singular simplicial $(k + 1)$-chain $D(\sigma, f)$ given by

$$D(\sigma, f) = (-1)^{k-1} \sum_{q=0}^{k} (-1)^q (\sigma_q, \tilde{f}). \tag{3}$$

Extended linearly to all of $C_k(X)$, this yields a homomorphism

$$D: C_k(X) \to C_k(Y), \tag{4}$$

where, for instance, $C_k(X)$ denotes the additive group of all singular k-chains of X. To complete the proof of the theorem we shall need the following fact concerning the homomorphism D.

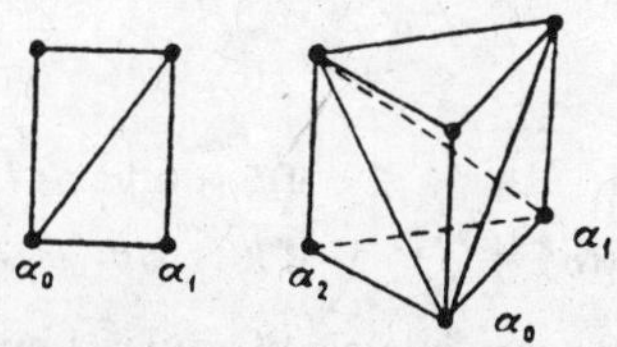

Figure 35. Decompositions of cylinders over simplexes

5.3. Lemma (cf. Lemma 1.4 and Definition 2.5). *The homomorphism D defined by* (3) *and* (4) *satisfies the equation*

$$D \circ \partial + (-1)^{k-1} \partial \circ D = \varphi_{1*} - \varphi_{0*}. \tag{5}$$

PROOF. Denoting by $d[\alpha_0 \dots \alpha_k]$ the following sum of simplexes of the form (2):

$$d[\alpha_0 \dots \alpha_k] = (-1)^{k-1} \sum_{q=0}^{k} (-1)^q [\alpha_0^0 \dots \alpha_q^0 \alpha_q^1 \dots \alpha_k^1], \tag{6}$$

and recalling (from Definition 3.1) that

$$\partial[\alpha_0 \dots \alpha_k] = \sum_{i=0}^{k} (-1)^i [\alpha_0 \dots \hat{\alpha}_i \dots \alpha_k],$$

(where as usual the hat indicates omission of a symbol), we obtain by direct calculation that

$$d\partial[\alpha_0 \dots \alpha_k] + (-1)^{k-1} \partial d[\alpha_0 \dots \alpha_k] = [\alpha_0^1 \dots \alpha_k^1] - [\alpha_0^0 \dots \alpha_k^0]. \tag{7}$$

(Since $d[\alpha_0 \dots \alpha_k]$ corresponds to a simplicial decomposition of the cylinder $[\alpha_0 \dots \alpha_k] \times I$, equation (7) is easy to understand geometrically as a consequence of the simple fact that the boundary of the cylinder consists of its lid and base (appropriately signed) together with the cylinder $(\partial[\alpha_0 \dots \alpha_k]) \times I$.) The lemma is now immediate from this equation. □

The theorem now follows easily (as in the proof of Proposition 2.6) since on applying equation (5) to any cycle z_k we obtain $\varphi_{1*} z_k - \varphi_{0*} z_k = \pm \partial D z_k$, a boundary. □

5.4. Corollary. *Homotopically equivalent spaces have naturally isomorphic singular (simplicial) homology and cohomology groups.*

Examples

(a) Any contractible space (i.e. contractible over itself to a point) is homotopically equivalent to the one-point space. The singular simplicial homology groups of the latter space are easily computed: Since there is just one map $f: \sigma^k \to \{*\}$ from each standard simplex σ^k to a singleton set $\{*\} = X$, we shall have but one singular simplex of each dimension k. Since the boundary of the simplex σ^k has the form

$$\partial(\sigma^k) = \sum_{q=0}^{k} (-1)^q (\sigma_q^{k-1}),$$

it follows that

$$\partial(\sigma^k, f) = \begin{cases} 0 & \text{if } k \text{ is odd or } k = 0, \\ (\sigma^{k-1}, f) & \text{if } k \text{ is even, } k > 0, \end{cases}$$

whence we see that for $k > 0$ every k-cycle (if any) is a boundary. We conclude that

$$H_0(*; G) \simeq G, \qquad H_k(*; G) = 0 \quad \text{for} \quad k > 0. \tag{8}$$

(b) For a path-connected topological space X, we have $H_0(X; G) \simeq G$. To see this note first that of course all 0-chains are (trivially) cycles. Now let (σ^0, f) and (σ^0, g), where $f(\sigma^0) = x_1$, $g(\sigma^0) = x_2$, be any two 0-simplexes; if $\varphi: [0, 1] \to X$ is a path in X joining x_1 to x_2, then clearly

$$(\sigma^0, g) - (\sigma^0, f) = \partial(\sigma^1, \varphi).$$

Hence any 0-chain $\sum g_i(\sigma^0, f_i)$ is homologous to a 0-chain of the form $(\sum g_i)(\sigma^0, f)$ for some $f: \sigma^0 \to X$, so that such a chain is a boundary precisely if $\sum g_i = 0$. It follows readily that $H_0(X; G) \simeq G$, as claimed. (This argument readily extends to show that for a topological space X with n path-connected components, the group $H_0(X; G)$ is isomorphic to the direct sum of a n copies of G.)

Singular *cubic* homology theory is for certain purposes more convenient than the simplicial variety. We begin with the relevant definitions. The *standard n-dimensional unit cube* I^n is the set of points $(x^1, \ldots, x^n)$ in $\mathbb{R}^n$ satisfying $0 \le x^i \le 1$. (The 0-dimensional unit cube I^0 is taken to be a single point.) We denote by $\lambda_i^\varepsilon I^n$ that $(n-1)$*-dimensional face* of I^n consisting of all points of I^n with $x^i = \varepsilon (\varepsilon = 0$ or $1)$; thus each face $\lambda_i^\varepsilon I^n$ is an $(n-1)$-dimensional unit cube, and the n-cube I^n has altogether $2n$ such faces.

A *singular n-cube* in a topological space X is a pair (I^n, f) where $f: I^n \to X$ is a continuous map. The *faces* of a singular cube (I^n, f) are then defined by restricting f

$$\lambda_i^\varepsilon(I^n, f) = (\lambda_i^\varepsilon I^n, f|_{\lambda_i^\varepsilon I^n}), \qquad i = 1, \ldots, n, \quad \varepsilon = 0, 1; \tag{9}$$

for $\varepsilon = 0$ this is called the ith *lower face*, and for $\varepsilon = 1$ the ith *upper face* of the singular cube. The following equation for the operators λ_i^ε is easily verified:

$$\lambda_i^\varepsilon \lambda_j^\eta = \lambda_{j-1}^\eta \lambda_i^\varepsilon, \qquad i < j, \quad \varepsilon, \eta = 0, 1. \tag{10}$$

A *singular cubic chain of dimension n* (with coefficients from a group G) is then a formal finite linear combination of the form

$$c_n = \sum g_i(I^n, f_i), \qquad g_i \in G;$$

we denote the additive group of all such n-chains by $\hat{C}_n(X; G)$. The *boundary of a singular cube* is defined by

$$\partial(I^n, f) = \sum_{i=1}^n (-1)^i[\lambda_i^1(I^n, f) - \lambda_i^0(I^n, f)]; \tag{11}$$

the boundary operator ∂ is then, as usual, extended linearly to the whole of $\hat{C}_n(X; G)$. It follows from equation (10) that $\partial\partial(I^n, f) = 0$ (verify this!), so that $\partial\partial = 0$.

A singular n-cube (I^n, f) is said to be *degenerate* if the map $f: I^n \to X$ decomposes as the canonical projection $I^n \to I^{n-1}$ of I^n onto one of its faces, followed by a map $g: I^{n-1} \to X$. Denoting by $D_n(X; G)$ the subgroup of $\hat{C}_n(X; G)$ consisting of all formal finite linear combinations over G of degenerate singular n-cubes, we now form the quotient group

$$C_n(X; G) = \hat{C}_n(X; G)/D_n(X; G)$$

of *normalized* singular cubic n-chains. Since, as is readily verified, the boundary operator ∂ preserves $D_n(X; G)$, it follows that it induces an operator (which we also denote by ∂)

$$\partial: C_n(X; G) \to C_{n-1}(X; G),$$

which will of course satisfy $\partial\partial = 0$. Finally, we define the nth *singular cubic homology group* to be the quotient of the group of normalized singular cubic n-cycles by its subgroup of normalized boundaries (and as usual define the singular cubic cohomology groups dually).

The singular cubic homology (and cohomology) groups thus defined are, like their predecessors, homotopically invariant.

5.5. Theorem. *Homotopic maps $\varphi_0, \varphi_1: X \to Y$ of topological spaces, induce the same homomorphism $\varphi_{0*} \equiv \varphi_{1*}: H_n(X; G) \to H_n(Y; G)$ between the respective singular cubic homology groups (and the same homomorphism $\varphi_0^* \equiv \varphi_0^*$ between the singular cubic cohomology groups).*

(The proof of this theorem is analogous to that of the corresponding assertion for the singular simplicial homology groups (Theorem 5.2). In the present "cubic" context the operator D, with the defining property of an algebraic homotopy (Definition 2.5), associates with each singular n-cube in the space X a singular $(n + 1)$-cube in Y: if $\Phi: I \times X \to Y$ is a homotopy between the maps φ_0 and φ_1, then D is defined by

$$D(I^n, f) = (I^{n+1}, \Phi(1 \times f)),$$

where we are taking $I^{n+1} = I \times I^n$, $1 \times f: I^{n+1} \to I \times X$. It is easy to verify that the image under D of a degenerate singular cube is again degenerate, so that D can be regarded as operating on the group of normalized chains. One then shows, somewhat as in the proof of Lemma 5.3, that

$$D\partial \pm \partial D = \varphi_{1*} - \varphi_{0*},$$

whence the theorem.)

Example. The computation of the singular cubic homology groups of the one-point space $X = \{*\}$ (and therefore, by the above theorem, of any contractible space) is easy. Clearly, there is exactly one singular cube $(I^n, f_n)(f_n(I^n) = \{*\})$ of each dimension n, and for $n > 0$ these are obviously degenerate. Hence the groups of normalized cubic n-chains, and also the cubic homology groups, are given by

$$C_0(*; G) \simeq G \simeq H_0(*; G); \qquad C_n(*; G) = 0 \qquad \text{for} \quad n > 0, \tag{12}$$

from which we observe that the singular cubic homology groups of a contractible space are the same as the singular simplicial versions.

Remark. In contrast with (12), the homology groups $\hat{H}(X; G)$, obtained by using instead the full group $\hat{C}_n(X; G)$ of singular cubic chains (rather than its quotient of normalized chains), turn out to be non-trivial for the one-point space $X = \{*\}$. We leave the precise details to the reader as an

EXERCISE

(i) Find the group $\hat{H}(*; \mathbb{Z})$.

(ii) Prove that for any space X we have

$$\hat{H}(X; \mathbb{Z}) \simeq \sum_{k \geq 0} H_{n-k}(X; \hat{H}(*; \mathbb{Z})). \tag{13}$$

We now define the *relative* singular homology groups of a topological space. (The definition is essentially the same in the cubic as in the simplicial case.) If Y is a subspace of the space X, then the group $C_k(Y)$ of singular chains can be regarded in the obvious way as a subgroup of the group $C_k(X)$; we can therefore form the quotient $C_k(X, Y) = C_k(X)/C_k(Y)$, the *group of relative singular k-chains*. (The group G of coefficients, which is left implicit here, is arbitrary.) Since the boundary operator ∂ maps $C_k(Y)$ to $C_{k-1}(Y)$, it induces a boundary operator (also denoted by ∂) on the quotient group

$$\partial: C_k(X, Y) \to C_{k-1}(X, Y), \tag{14}$$

and this is the final ingredient of the *singular (simplicial or cubic) relative chain complex*; the dual *relative cochain complex* is defined in the usual way.

The *relative k-cycles* are, as expected, those relative k-chains c_k for which $\partial c_k = 0$; we denote the group they form by $Z_k(X, Y)$. The *relative k-boundaries*, forming a subgroup $B_k(X, Y) \subset Z_k(X, Y)$, are those relative k-chains c_k such that $c_k = \partial c_{k+1}$ for some relative $(k+1)$-chain c_{k+1}. Finally, the kth *relative homology group* $H_k(X, Y)$ is defined as the quotient $Z_k(X, Y)/B_k(X, Y)$.

Essentially by regarding each k-cycle as a relative k-cycle, we obtain natural isomorphisms

$$H_k(X) \xrightarrow{j} H_k(X\ Y), \qquad H^k(X, Y) \xrightarrow{j} H^k(X). \tag{15}$$

(Thus j is induced by the natural map $C_k(X) \to C_k(X)/C_k(Y) = C_k(X, Y)$.) We have also, corresponding to the inclusion map $i: Y \to X$, the *inclusion homomorphisms* (see p. 60)

$$i_*: H_k(Y) \to H_k(X), \qquad i^*: H^k(X) \to H^k(Y). \tag{16}$$

Thirdly and lastly, we define (as follows) the *boundary homomorphism* ∂_* from $H_k(X, Y)$ to $H_{k-1}(Y)$ (with cohomological dual $\partial_*: H^{k-1}(Y) \to H^k(X, Y)$). A relative k-cycle $c_k \in C_k(X, Y)$ may be regarded as an ordinary (i.e. "absolute") k-chain, i.e. as an element (not necessarily a cycle) of $C_k(X)$, provided it is borne in mind that it is defined only up to the addition of an element of $C_k(Y)$. Its boundary $c_{k-1} = \partial c_k$ will then of course be a $(k-1)$-cycle from $Z_{k-1}(Y)$, determined up to addition of an element of $\partial C_k(Y)$, i.e. of a boundary, and therefore corresponding to a well-defined element of $H_{k-1}(Y)$. We define

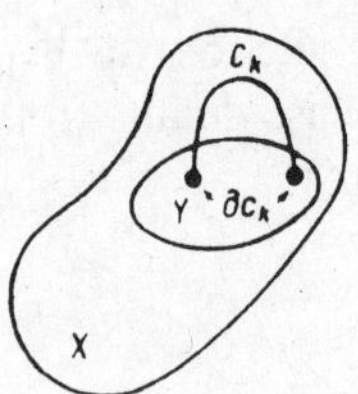

Figure 36

$\partial_* c_k$ to be this element (see Figure 36). It is clear that, as defined, the map

$$\partial_*: H_k(X, Y) \to H_{k-1}(Y) \tag{17}$$

is a homomorphism.

In combination the maps j, i_* and ∂_* yield a sequence of homomorphisms

$$\begin{aligned} \cdots \xrightarrow{\partial_*} H_k(Y) \xrightarrow{i_*} H_k(X) \xrightarrow{j} H_k(X, Y) \xrightarrow{\partial_*} H_{k-1}(Y) \xrightarrow{i_*} \cdots \to H_0(Y) \\ \to H_0(X) \to H_0(X, Y) \to 0 \end{aligned} \tag{18}$$

called the *exact* (*homology*) *sequence of the pair* (X, Y) (cf. Part II, §21.2). In calling it this we are anticipating the following

5.6. Theorem. *The sequence of homomorphisms* (18) *is exact; i.e.*

(i) Ker $i_* = \operatorname{Im} \partial_*$;
(ii) Ker $j = \operatorname{Im} i_*$;
(iii) Ker $\partial_* = \operatorname{Im} j$.

PROOF. (i) A $(k-1)$-cycle $c_{k-1} \in Z_{k-1}(Y)$ represents an element in the kernel of $i_*: H_{k-1}(Y) \to H_{k-1}(X)$ precisely if $i_*(c_{k-1})$ is a boundary in $C_{k-1}(X)$, i.e. if there is a chain c_k in $C_k(X)$ such that $\partial c_k = i_* c_{k-1}$, i.e. $\partial c_k = c_{k-1}$ if we regard c_{k-1} as a chain in $C_{k-1}(X)$. Now it is easy to see that the k-chains c_k of $C_k(X)$ with the property that ∂c_k is a cycle in $C_{k-1}(Y)$ are precisely those representing k-cycles of $C_k(X, Y)$. The desired conclusion now follows (the case $k = 0$ being easily verified separately).

(ii) Let c_k be a k-cycle in the space X such that $j(c_k) = 0$ (with the usual abuse of language). Thus $\partial c_k = 0$, and there are chains c_{k+1} in X and $\tilde{c}_k$ in Y such that

$$c_k = \tilde{c}_k + \partial c_{k+1}.$$

Hence a cycle c_k in X belongs to Ker j precisely if there is a cycle $\tilde{c}_k$ in Y homologous to it, i.e. precisely if c_k represents an elements of Im i_*.

(iii) Let c_k be a chain in X representing an element of $H_k(X, Y)$ in the kernel of the homomorphism $\partial_*: H_k(X, Y) \to H_{k-1}(Y)$. From the definition of ∂_* it is immediate that this is equivalent to the existence of a cycle $\hat{c}_k$ in X and a chain $\tilde{c}_k$ in Y such that

$$c_k = \hat{c}_k + \tilde{c}_k,$$

i.e. to the existence of an "absolute" cycle $(c_k - \tilde{c}_k)$ in X representing the same element of $H_k(X, Y)$ as does c_k. Since this is just the condition that c_k represent an element of Im j, the equality (iii) follows, and the proof of the theorem is complete. □

Remark. If Y is a simplicial (or cell) subcomplex of a simplicial (or cell) complex X, then the analogues of the above three homomorphisms are defined in the obvious way, and the exactness of the corresponding sequence of homomorphisms is established by an argument similar to the above. (We leave the details to the reader as an exercise.)

5.7. Corollary. *From the exactness of the sequence* (18) *it follows that for any path-connected space X we have*

$$\begin{aligned} &H_0(X, *) = 0, \\ &H_k(X, *) \simeq H_k(X), \qquad k > 0. \end{aligned} \tag{19}$$

Proof. From (18) with $Y = \{*\}$ we obtain in particular the exact sequence

$$\begin{array}{ccccc} H_0(*) & \xrightarrow{i_*} & H_0(X) & \rightarrow & H_0(X; *) \rightarrow 0. \\ \wr\!\| & & \wr\!\| & & \\ G & & G & & \end{array}$$

That $H_0(X; *) = 0$ follows from the exactness of this sequence by taking into account the easily verified fact that i_* is here an isomorphism (cf. Example (b) above).

For $k > 0$, putting $Y = \{*\}$ in (18) we obtain the exact sequence

$$H_k(*) \rightarrow H_k(X) \rightarrow H_k(X, *) \rightarrow H_{k-1}(*) \rightarrow H_{k-1}(X) \rightarrow \cdots.$$

Since by Example (a) above we have $H_k(*) = 0$ for $k > 0$, we obtain the exact sequence

$$0 \rightarrow H_k(X) \xrightarrow{j} H_k(X; *) \rightarrow 0 \tag{20}$$

for all $k > 1$, and in fact we have such an exact sequence also in the case $k = 1$, by virtue of the above-noted fact that the map $H_0(*) \rightarrow H_0(X)$ is an isomorphism. Hence we have an exact sequence (20) for all $k > 0$, so that for these k the homomorphism j is in fact an isomorphism. □

An important property of the relative homology and cohomology groups is their so-called "naturality", by which is meant the following. Any map of pairs of topological spaces

$$f: (X, X') \rightarrow (Y, Y')$$

(where $X' \subset X$, $Y' \subset Y$ and f is a map from X to Y such that $f(X') \subset Y'$) induces in the obvious way the homomorphisms

$$f_*: H_k(X) \rightarrow H_k(Y), \qquad f^*: H^k(Y) \rightarrow H^k(X), \tag{21}$$

$$f_*: H_k(X, X') \to H_k(Y, Y'), \qquad f^*: H^k(Y, Y') \to H^k(X, X'), \tag{22}$$
$$f_*: H_k(X') \to H_k(Y') \qquad f^*: H^k(Y') \to H^k(X'). \tag{23}$$

The various homomorphisms figuring in the exact sequence (18) are then "natural" in the sense that they commute with the maps f_*, so that we obtain a "morphism of exact homology sequences" given by the following commutative diagram:

$$\begin{array}{ccccccccc} \xrightarrow{\partial_*} & H_k(X') & \xrightarrow{i_*} & H_k(X) & \xrightarrow{j} & H_k(X, X') & \xrightarrow{\partial_*} & H_{k-1}(X') & \to \\ & \downarrow f_* & & \downarrow f_* & & \downarrow f_* & & \downarrow f_* & \\ \xrightarrow{\partial_*} & H_k(Y') & \xrightarrow{i_*} & H_k(Y) & \xrightarrow{j} & H_k(Y, Y') & \xrightarrow{\partial_*} & H_{k-1}(Y') & \to \end{array} \tag{24}$$

and for the exact cohomology sequences we have analogously:

$$\begin{array}{ccccccccc} \xrightarrow{\delta^*} & H^k(X, X') & \xrightarrow{j} & H^k(X) & \xrightarrow{i^*} & H^k(X') & \xrightarrow{\delta^*} & H^{k+1}(X, X') & \to \\ & \uparrow f^* & & \uparrow f^* & & \uparrow f^* & & \uparrow f^* & \\ \xrightarrow{\delta^*} & H^k(Y, Y') & \xrightarrow{j} & H^k(Y) & \xrightarrow{i^*} & H^k(Y') & \xrightarrow{\delta^*} & H^{k+1}(Y, Y') & \to \end{array} \tag{25}$$

The following proposition exemplifies the great usefulness of the "naturality" property.

5.8. Proposition. *Suppose we have a map of pairs*

$$f: (X, X') \to (Y, Y')$$

for which for all k the induced homomorphisms

$$H_k(X) \xrightarrow{f_*} H_k(Y), \qquad H_k(X') \xrightarrow{f_*} H_k(Y')$$

are isomorphisms. Then the corresponding homomorphisms (also denoted by f_*, as in (24)) between the relative homology groups $H_k(X, X')$ and $H_k(Y, Y')$ are also isomorphisms (and dually for the cohomology groups).

PROOF. We first prove injectivity, by showing that the kernel of the map $f_*: H_k(X, X') \to H_k(Y, Y')$, is trivial. Let $\alpha \in H_k(X, X')$ be any element of this kernel; thus $f_*\alpha = 0$, and then from the commutative diagram (24) we obtain $f_*\partial_*\alpha = \partial_*f_*\alpha = 0$, i.e. $f_*(\partial_*\alpha) = 0$, where $\partial_*\alpha \in H_{k-1}(X')$. Since by assumption the map $f_*: H_{k-1}(X') \to H_{k-1}(Y')$ is an isomorphism, it then follows that $\partial_*\alpha = 0$, i.e. $\alpha \in \operatorname{Ker} \partial_*$. Since by exactness $\operatorname{Ker} \partial_* = \operatorname{Im} j$ (see Theorem 5.6), we infer that $\alpha = j(\beta)$ for some $\beta \in H_k(X)$, whence $0 = f_*(\alpha) = f_*j(\beta) = jf_*(\beta)$, again by the commutativity of diagram (24). Hence $f_*(\beta) \in \operatorname{Ker} j$ (where j maps $H_k(Y)$ to $H_k(Y, Y')$). Since we have $\operatorname{Ker} j = \operatorname{Im} i_*$ by exactness, it follows that $f_*(\beta) = i_*(\gamma)$ for some $\gamma \in H_k(Y')$. Since by hypothesis $f_*: H_k(X') \to H_k(Y')$ is an isomorphism, we can write (unambiguously) $\delta = f_*^{-1}(\gamma) \in H_k(X')$. The commutativity of diagram (24) then gives us that $\beta = f_*^{-1}i_*(\gamma) = i_*f_*^{-1}(\gamma) = i_*(\delta)$, whence $\alpha = j(\beta) = ji_*(\delta) = 0$ by exactness. Hence $f_*(\alpha) = 0$ implies that $\alpha = 0$, as we wished to show.

For the surjectivity we must show that every element γ of $H_k(Y, Y')$ has the form $\gamma = f_*(\delta)$ for some $\delta \in H_k(X, X')$. Consider first the case that $\partial_*(\gamma) = 0$. By exactness we shall then have $\gamma = j(\beta)$ for some $\beta \in H_k(Y)$. The commutativity of diagram (24) and the assumption that $f_*: H_k(X) \to H_k(Y)$ is an isomorphism, together give us then that $f_* j f_*^{-1}(\beta) = f_* f_*^{-1} j(\beta) = j(\beta) = \gamma$, so that γ is indeed an image under f_*, namely of $\delta = j f_*^{-1}(\beta)$. Suppose, on the other hand, that $\partial_* \gamma \neq 0$. Then $f_*^{-1}\partial_*(\gamma) = \partial_* \beta$ for some $\beta \in H_k(X, X')$ (in view of the commutativity of diagram (24) and the hypothesis that $f_*: H_{k-1}(X') \to H_{k-1}(Y')$ is an isomorphism). Hence $\partial_*(f_*(\beta) - \gamma) = 0$, and then from the case already treated we know that $f_*(\beta) - \gamma$ has a preimage under f_*, whence so does γ. This completes the proof of the proposition. □

Remark. A similar line of argument can be used to show that, more generally, if f_* is an isomorphism as applied to any two of the groups $H_*(X)$, $H_*(X')$, $H_*(X, X')$, then it will automatically be an isomorphism of the third (and dually for the cohomology groups).

In the next section (§6) the fact that for simplicial and, more generally, cell complexes X the singular homology groups coincide with the simplicial and cellular homology groups, will be established using the above formal properties of the homology groups together with the following important further property.

5.9. Theorem ("Excision Theorem"). *If K is a cell complex and L a subcomplex, then there is a (natural) isomorphism (of singular homology groups)*

$$H_k(K, L) \simeq H_k(K/L), \qquad k > 0,$$

where K/L denotes the quotient space obtained from K by identifying L with a point. (Note that K/L is homotopically equivalent to the cell complex $K \cup CL$, where CL is the cone over L obtained from $L \times I$ by identifying the lid of the cylinder with a point; see Figure 37.)

We shall give the proof for simplicial complexes only. As a preliminary, we define an operator β on singular simplicial chains determined by *barycentric subdivision* of simplexes. The barycentric subdivision of a simplex $[\alpha_0 \ldots \alpha_k] = \sigma^k$, $k > 0$, is achieved as follows: for a 1-simplex it is the subdivision into two 1-simplexes by means of a new vertex introduced at its mid-point; for a 2-simplex $[\alpha_0\alpha_1\alpha_2]$ (i.e. triangle) one first subdivides the edges and then

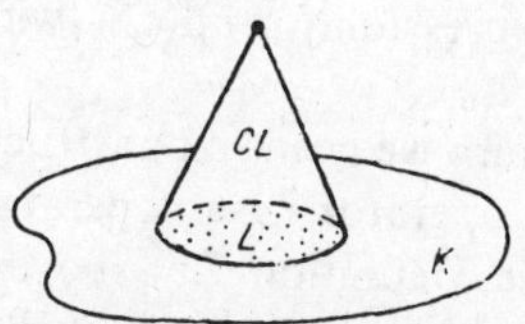

Figure 37. $K \cup CL$.

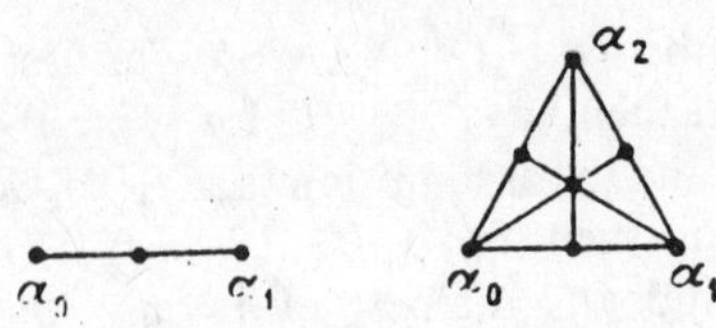

Figure 38

introduces a new vertex at the centroid of the triangle to be joined by new edges to the six vertices on its boundary (three old ones and three new ones; see Figure 38), yielding a subdivision of the triangle into six new 2-simplexes. For the general k-simplex σ^k the procedure is analogous: one chooses a new vertex σ^0 at the centroid of σ^k and introduces new edges by joining it to all vertices on the faces of σ^k, assumed (inductively) to be already barycentrically subdivided; in this way σ^k is subdivided into new simplexes σ_i^k each determined by the new vertex σ^0 and a $(k-1)$-simplex forming part of the (already subdivided) boundary of σ^k. The simplicial complex comprised of these simplexes together with their faces of all dimensions is then the barycentric subdivision of σ^k.

Now let (σ^k, f) be a singular simplex in the topological space X, and let $\sigma_1^k, \ldots, \sigma_N^k$ by the k-simplexes of the barycentric subdivision of the simplex σ^k. We define the *barycentric operator* β on singular simplexes in X by means of the formula

$$\beta(\sigma^k, f) = \sum_{i=1}^{N} (\sigma_i^k, f|_{\sigma_i^k}) \tag{26}$$

(where the sum is taken over all k-simplexes of the subdivision of σ^k), and as usual extend it linearly to the whole group $C_k(X)$ of singular simplicial k-chains:

$$\beta\colon C_k(X) \to C_k(X), \qquad k = 0, 1, \ldots. \tag{27}$$

5.10. Lemma. *The barycentric operator β commutes with the boundary homomorphism ∂, and is algebraically homotopic to the identity operator.*

PROOF. The boundary of a simplicial k-chain of the simplicial complex represented by the barycentric subdivision of a simplex σ^k is defined (as in §3) in terms of some ordering of the vertices of the subdivision. The equality $\partial\beta = \beta\partial$ is then not difficult to see, essentially since each "interior" face of the subdivision enters into the expression for $\partial\beta(\sigma^k)$ twice and with opposite signs, therefore cancelling out.

To verify the second claim we need to construct an algebraic homotopy D between β and the identity operator, i.e. an operator D such that $D(C_k) \subset C_{k+1}$ and $\partial D \pm D\partial = \beta - 1$ (see Definition 2.5). To this end, we triangulate the product $\sigma^k \times I$ of the simplex σ^k with the unit interval I, in such a way that

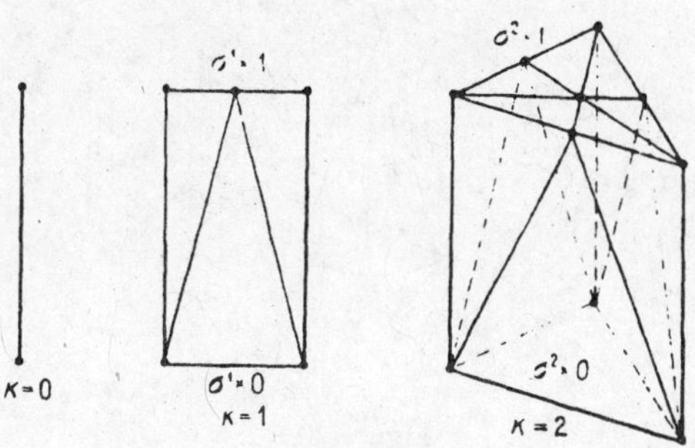

Figure 39

$\sigma^k \times 0$ is a single simplex of the triangulation, while the other end $\sigma^k \times 1$ of the cylinder is barycentrically subdivided. The triangulations in the cases $k = 0, 1, 2$ are shown in Figure 39; in the general case the triangulation in question is produced as follows: Suppose inductively that the desired triangulation has already been defined for $\sigma^{k-1} \times I$, and triangulate the "vertical" faces of the cylinder $\sigma^k \times I$ accordingly. We next leave the base $\sigma^k \times 0$ as it is, but barycentrically subdivide the lid $\sigma^k \times 1$. At this stage the whole boundary of $\sigma^k \times 1$ has been triangulated; by now joining the centroid of the lid $\sigma^k \times 1$ to every vertex in the triangulation of $\partial(\sigma^k \times I)$ we obtain finally the required triangulation of $\sigma^k \times I$.

Given any singular simplex (σ^k, f) in X, we define the corresponding "trivial" map $\hat{f}$ by

$$\hat{f}: \sigma^k \times I \to X, \qquad \hat{f}(x, t) = f(x), \tag{28}$$

and then define the operator D we are seeking by taking $D(\sigma^k, f)$ to be the $(k+1)$-chain $(\sigma^k \times I, \hat{f})$ where $\sigma^k \times I$ is triangulated as above. It is then easy to verify that the operator D has the requisite properties. □

Proof of the Excision Theorem. In view of the homotopy invariance of the singular homology groups (Corollary 5.4) we have

$$H_k(K \cup CL, CL) \simeq H_k(K \cup CL, *),$$

since the cone CL is contractible (over itself to a point). On the other hand, we know from Corollary 5.7 that for $k > 0$

$$H_k(K \cup CL, *) \simeq H_k(K \cup CL) \simeq H_k(K/L).$$

Hence it suffices to show that

$$H_k(K \cup CL, CL) \simeq H_k(K, L). \tag{29}$$

We realize this isomorphism by constructing, for each relative k-cycle c_k in $C_k(K \cup CL, CL)$, a k-cycle in $C_k(K, L)$ homologous to it. To this end, we

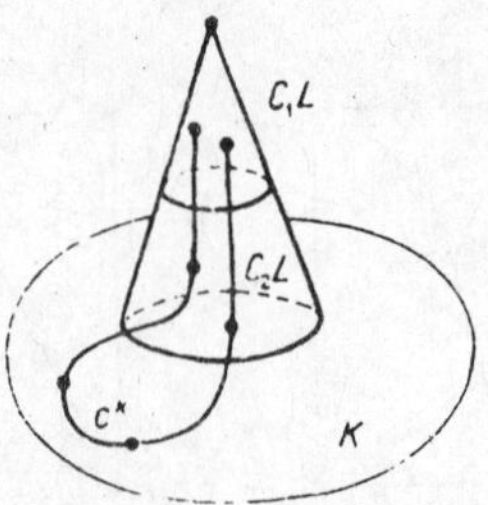

Figure 40

subdivide the cone CL into two parts C_1L and C_2L, as indicated in Figure 40. By Lemma 5.10 (and Proposition 2.6) we may replace c_k by the homologous cycle $\beta^N c_k$, and then by taking N sufficiently large (i.e. by means of sufficiently many iterated barycentric subdivisions) we can ensure that any simplex of $\beta^N c_k$ which meets the upper portion C_1L of the cone CL, is contained entirely in that cone. It follows that we can delete from our cycle $\beta^N c_k$ (N sufficiently large) all simplexes meeting C_1L, without thereby altering the relative homology class (modulo CL) of the cycle $\beta^N c_k \sim c_k$. The resulting cycle $\hat{c}_k$ say, may then be regarded as belonging to the group $H_k(K \cup C_2L, C_2L)$ which, since C_2L contracts to L, is naturally isomorphic to $H_k(K, L)$. The image of $\hat{c}_k$ under this isomorphism is then the relative cycle in $H_k(K, L)$ that we associate with the arbitrary original (relative) cycle c_k.

If the original cycle c_k is homologous to zero in $H_k(K \cup CL, CL)$, then by means of a similar argument we can "remove it" from the vicinity of the apex of the cone, obtaining thereby a bounding cycle $\hat{c}_k$ in $C_k(K \cup C_2L, C_2L)$. We leave the precise details to the reader. □

§6. The Singular Homology of Cell Complexes. Its Equivalence with Cell Homology. Poincaré Duality in Simplicial Homology

We begin by calculating the singular homology groups of the spheres S^n, $n = 1, 2, \ldots$. (We shall throughout this section take the integers as coefficients.)

6.1. Theorem. *For $n > 0$ the singular homology groups of the n-sphere S^n are given by*

$$H_i(S^n; \mathbb{Z}) \simeq \begin{cases} \mathbb{Z}, & i = 0, n, \\ 0, & i \neq 0, n. \end{cases} \tag{1}$$

PROOF. Consider first the case $n = 1$. We compute the singular homology groups of the circle S^1 by exploiting the exact sequence of the pair $(D^1, \partial D^1)$, where $\partial D^1 = S^0$ is the 2-point discrete space; thus the exact sequence (18) of §5 yields, in this context, the exact sequence

$$H_1(D^1) \to H_1(D^1, S^0) \to H_0(S^0) \to H_0(D^1) \to H_0(D^1, S^0) \to 0.$$

Since $H_1(D^1) = 0$, $H_0(D^1) \simeq \mathbb{Z} \simeq H_0(D^1, S^0)$, $H_0(S^0) \simeq \mathbb{Z} \oplus \mathbb{Z}$ (S^0 having two connected components), and $H_1(D^1, S^0) \simeq H_1(S^1)$ by Theorem 5.9, this simplifies to

$$0 \to H_1(S^1) \to \mathbb{Z} \oplus \mathbb{Z} \to \mathbb{Z} \to 0,$$

whence $H_1(S^1) \simeq \mathbb{Z}$. For $k > 1$ we have (again from Theorem 5.6) the exact sequence

$$H_k(D^1) \to H_k(D^1, S^0) \to H_{k-1}(S^0).$$

Since $H_k(D^1) = 0$ (by §5, Example (a)), and $H_{k-1}(S^0) = 0$ (verify!), it follows that $H_k(D^1, S^0) = 0$, whence $H_k(S^1) = 0$ by Theorem 5.9. This concludes the computation of the singular homology groups of the circle.

Now suppose inductively that the singular homology groups of the sphere S^{n-1} are as claimed in the theorem, where $n > 1$. The exact sequence of the pair (D^n, S^{n-1}) is as follows:

$$\cdots \to H_k(D^n) \to H_k(D^n, S^{n-1}) \to H_{k-1}(S^{n-1}) \to H_{k-1}(D^n) \to \cdots. \tag{2}$$

For $k > 1$ this becomes (in view of Theorem 5.9 and the triviality of $H_k(D^n)$ and $H_{k-1}(D^n)$):

$$0 \to H_k(S^n) \to H_{k-1}(S^{n-1}) \to 0,$$

whence $H_k(S^n) \simeq H_{k-1}(S^{n-1})$ for $k > 1$. For $k = 1$, on the other hand, the exact sequence (2) takes the form

$$H_1(D^n) \to H_1(D^n, S^{n-1}) \to H_0(S^{n-1}) \to H_0(D^n) \to 0$$

or

$$0 \to H_1(D^n, S^{n-1}) \to \mathbb{Z} \to \mathbb{Z} \to 0.$$

The exactness of this sequence implies that the homomorphism $\mathbb{Z} \to \mathbb{Z}$ is onto, and therefore an isomorphism, so that $H_1(D^n, S^{n-1}) = 0$, whence $H_1(S^n) = 0$. The assertion of the theorem now follows. □

Remark. The identity map $\sigma^n \to \sigma^n$ of an n-simplex, with the image n-simplex regarded as an n-dimensional disc, defines a relative singular cycle in $C_n(D^n, S^{n-1})$, and thence an element of $H_n(D^n, S^{n-1}) \simeq H_n(S^n)$. This cycle in fact corresponds to a generator of $H_n(S^n)$.

EXERCISE

Each permutation P of the vertices of an n-simplex $\sigma^n = [\alpha_0 \ldots \alpha_n]$ determines in the obvious way a map $\sigma^n \to \sigma^n$. Compute the element of $H_n(S^n)$ defined by this map.

6.2. Corollary. *The singular homology groups of the bouquet of N n-dimensional spheres $S_1^n, \dots, S_N^n$ $(n > 0)$ are given by*

$$H_0\Big(\bigvee_i S_i^n\Big) \simeq \mathbb{Z}, \qquad H_n\Big(\bigvee_i S_i^n\Big) \simeq \underbrace{\mathbb{Z} \oplus \cdots \oplus \mathbb{Z}}_{N},$$
$$H_k\Big(\bigvee_i S_i^n\Big) = 0 \quad \text{for} \quad k \neq 0, n. \tag{3}$$

PROOF. Consider the pair (K, L) where $K = \bigcup_i D_i^n$, $L = \bigcup_i \partial D_i^n$. It is clear that

$$H_k(K, L) \simeq \sum_i H_k(D_i^n, \partial D_i^n) \simeq \sum_i H_k(S_i^n),$$

where the latter isomorphism comes from Theorem 5.9. From this the corollary now essentially follows, using Theorem 6.1. □

6.3. Corollary. *The homomorphism $f_*: H_n(S^n) \to H_n(S^n)$ induced by a map $f: S^n \to S^n$ $(n > 0)$ is just multiplication of the elements of $H_n(S^n) \simeq \mathbb{Z}$, by the degree of f:*

$$f_*(m) = m \times \deg f, \qquad m \in H_n(S^n).$$

PROOF. A map $f: S^n \to S^n$ of prescribed (positive) degree k can be constructed as shown in Figure 41; thus S^n is first gathered into a bouquet of k n-spheres and then each of these n-spheres is mapped "identically" to the image n-sphere. By Part II, Theorem 13.3.1, every map $S^n \to S^n$ of degree k is homotopic to the map thus defined, so that for the purposes of the corollary we may restrict attention to this particular one. Now under the homomorphism of homology groups induced by the map of the sphere to the bouquet, a generator of $H_n(S^n) \simeq \mathbb{Z}$ is sent to the sum of generators of the k spheres of the bouquet; the subsequent map of the bouquet then causes each term of this sum to be sent to the same generator of the final image sphere. Hence the homomorphism $H_n(S^n) \to H_n(S^n)$ induced by the composite simply multiplies a generator, and hence every element, of $H_n(S^n)$ by $k = \deg f$, as claimed. □

6.4. Corollary. *Let K be a cell complex and let K_j denote its j-skeleton, $j = 0, 1, 2, \dots$. Then the relative homology groups of each pair (K_n, K_{n-1}) are*

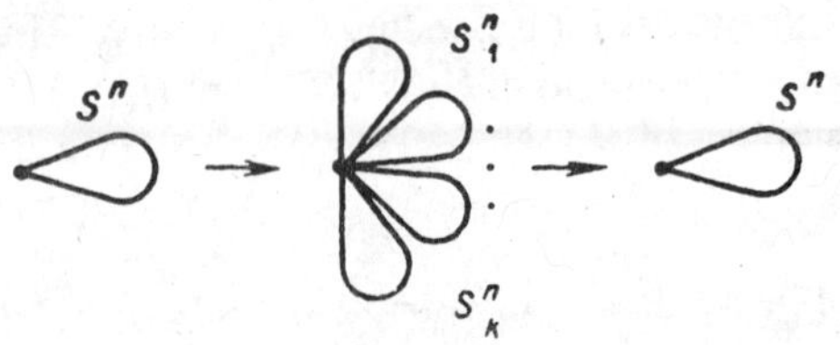

Figure 41

given by

$$H_k(K_n, K_{n-1}) = \begin{cases} 0, & k \neq n, \\ \mathbb{Z} \oplus \cdots \oplus \mathbb{Z}, & k = n, \end{cases} \tag{4}$$

where the number of summands $\mathbb{Z}$ *is equal to the number of n-cells in* K_n.

(This is an immediate consequence of Theorem 5.9, Corollary 6.2 and the fact that the quotient space K_n/K_{n-1} is homeomorphic to a bouquet of n-spheres, one for each n-cell (see §4).)

We now turn to the equivalence for cell complexes of singular homology and cell homology.

6.5. Theorem. *The singular (simplicial) homology (and cohomology) groups of a cell complex are (naturally) isomorphic to the corresponding cellular homology (and cohomology) groups.*

6.6. Corollary. *The cellular homology (and cohomology) groups are homotopy invariants. Since simplicial homology is a special case of cell homology, the simplicial homology groups of a simplicial complex are isomorphic to the corresponding singular homology groups of the complex, and are consequently homotopically invariant (and likewise for the cohomology groups).*

Before giving the full proof of the theorem we sketch a proof for simplicial complexes, since this important special case follows relatively easily from some of the results established above. Thus to begin with, observe that each simplex σ^k of our given simplicial complex K can naturally be regarded as the singular simplex (σ^k, f) where f is the inclusion map of σ^k in K; in this way we obtain an embedding of the group of simplicial chains into the group of singular chains

$$C^{\text{simp}}(K) \to C^{\text{sing}}(K). \tag{5}$$

Since this map clearly commutes with th boundary operator ∂, it induces a map of the homology groups

$$H_k^{\text{simp}}(K) \to H_k^{\text{sing}}(K), \tag{6}$$

and, given a simplicial subcomplex L of K, a map of the relative homology groups

$$H_k^{\text{simp}}(K, L) \to H_k^{\text{sing}}(K, L), \tag{7}$$

(where $H_k^{\text{simp}}(K, L)$ is defined analogously to $H_k^{\text{sing}}(K, L)$; see §5). We are thus led to a "map" (or "morphism") between the corresponding exact homology sequences (simplicial and singular) of the pair (K, L) (cf. §5(24)).

For complexes K of dimension zero, the statement of the theorem is readily verified (and so we leave the details to the reader). Let $n > 1$ and suppose inductively that the conclusion of the theorem is valid for complexes of dimension $\leq n - 1$. For a simplicial complex K_n of dimension n we have the following morphism, determined by (6) and (7), between the respective exact homology sequences of the pair (K_n, K_{n-1}) (where K_{n-1} is the $(n-1)$-skeleton of K_n):

$$\begin{array}{ccccccccccc} \to H_{j+1}^{\text{simp}}(K_n, K_{n-1}) & \xrightarrow{\partial_*} & H_j^{\text{simp}}(K_{n-1}) & \xrightarrow{i_*} & H_j^{\text{simp}}(K_n) & \to & H_j^{\text{simp}}(K_n, K_{n-1}) & \xrightarrow{\partial_*} & H_{j-1}^{\text{simp}}(K_{n-1}) \to \\ \downarrow & & \downarrow & & \downarrow & & \downarrow & & \downarrow \\ \to H_{j+1}^{\text{sing}}(K_n, K_{n-1}) & \to & H_j^{\text{sing}}(K_{n-1}) & \to & H_j^{\text{sing}}(K_n) & \to & H_j^{\text{sing}}(K_n, K_{n-1}) & \to & H_{j-1}^{\text{sing}}(K_{n-1}) \to \end{array} \tag{8}$$

Now it is not difficult to verify (essentially by using Theorem 5.9 and the fact that K_n/K_{k-1} is a bouquet of spheres) that the homomorphism

$$H_j^{\text{simp}}(K_n, K_{n-1}) \to H_j^{\text{sing}}(K_n, K_{n-1}),$$

given by (7) with $K = K_n$, $L = K_{n-1}$, is in fact an isomorphism. (Thus these groups are, by Corollary 6.4, trivial for $j \neq n$, and isomorphic to $\mathbb{Z} \oplus \cdots \oplus \mathbb{Z}$ for $j = n$, with one summand for each sphere of the bouquet K_n/K_{n-1}.) Since by inductive hypothesis the homomorphism

$$H_j^{\text{simp}}(K_{n-1}) \to H_j^{\text{sing}}(K_{n-1}),$$

given by (6) with $K = K_{n-1}$, is an isomorphism, we deduce, essentially via the remark following Proposition 5.8, that for all j the third homomorphism

$$H_j^{\text{simp}}(K_n) \to H_j^{\text{sing}}(K_n)$$

is also an isomorphism, thus concluding the inductive step. This completes our sketch of an argument in the simplicial case of the theorem. □

PROOF OF THEOREM 6.5. Let K be a general cell complex with n-skeleton K_n; then as we have noted several times, the quotient space K_n/K_{n-1} is a bouquet of n-spheres, one n-sphere arising from each n-cell of K. Hence by Theorem 5.9 and Corollary 6.2 we have that

$$H_n^{\text{sing}}(K_n, K_{n-1}) \simeq C_n(K), \qquad H_i(K_n, K_{n-1}) = 0, \qquad i \neq n,$$

where $C_n(K)$ is the group of cellular n-chains of the cell complex K. We can therefore define a boundary homomorphism $\partial\colon C_n(K) \to C_{n-1}(K)$ as the composite of the following homomorphisms:

$$C_n(K) \simeq H_n(K_n, K_{n-1}) \xrightarrow{\partial_*} H_{n-1}(K_{n-1}) \xrightarrow{j} H_{n-1}(K_{n-1}, K_{n-2}) \simeq C_{n-1}(K). \tag{9}$$

6.7. Lemma. *The homomorphism $\partial\colon C_n(K) \to C_{n-1}(K)$ defined by* (9) *coincides with the usual boundary operator on the complex of cellular chains of the cell complex K.*

PROOF. Denote by σ^n any n-cell of the cell complex K. We can regard σ^n as a generator of a direct summand of $H_n(K_n, K_{n-1})$, by identifying this group with $C_n(K)$ via the above-noted isomorphism. With this understanding, we denote by $\hat{\sigma}^n$ an absolute singular simplex representing the appropriate element of $H_n(K_n, K_{n-1})$. We shall then have that this element of $H_n(K_n, K_{n-1})$ is mapped by the boundary homomorphism ∂_* to $\partial\hat{\sigma}^n$ (see §5(17)), which by §5(1) is given by $\sum_i(-1)^i\hat{\sigma}_i^{n-1}$ where the $\hat{\sigma}_i^{n-1}$ are the $(n-1)$-dimensional faces of the singular simplex $\hat{\sigma}^n$. Under the identification map $K_{n-1} \to K_{n-1}/K_{n-2}$, the singular simplexes $\hat{\sigma}_i^{n-1}$ have their boundaries identified with a single point; it follows that the map

$$H_{n-1}(K_{n-1}) \xrightarrow{j} H_{n-1}(K_{n-1}, K_{n-2}) \simeq H_{n-1}(K_{n-1}/K_{n-2}),$$

restricted to each simplex $\hat{\sigma}_i^{n-1}$, determines a map from S^{n-1} to the bouquet K_{n-1}/K_{n-2}. On each sphere S_j^{n-1} of this bouquet, we shall have, in view of Corollary 6.3, that j is essentially just multiplicaton by the degree of the map to S_j^{n-1} it determines. Since each sphere S_j^{n-1} arises from a unique $(n-1)$-cell σ_j^{n-1} of K, we obtain, by combining the maps to the S_j^{n-1} determined by the maps j (with the orientations of the $\hat{\sigma}_i^{n-1}$ taken into account), and applying the final isomorphism in (9), that the image of the original n-cell σ^n of K under the homomorphism ∂ given by (9) is

$$\partial\sigma^n = \sum_j [\sigma^n:\sigma_j^{n-1}]\sigma_j^{n-1},$$

where $[\sigma^n:\sigma_j^{n-1}]$ is the incidence number of the indicated cells, and the summation is taken over all $(n-1)$-cells of K. Since this formula is the same as that defining the boundary operator on cellular chains (see §4(3)), the lemma follows. □

Resuming the proof of the theorem, we first note the following properties of the cellular homology groups:

(i) They are all zero in dimensions greater than that of the complex K (assuming the latter finite-dimensional):

$$H_j^{\text{cell}}(K_n) = 0 \qquad \text{for} \quad j > n.$$

(ii) The group $H_n^{\text{cell}}(K_n)$ coincides with the group $Z_n^{\text{cell}} \subset C_n(K_n)$ of n-cycles since there are no (non-zero) n-boundaries.

(iii) The group $H_j^{\text{cell}}(K)$ is fully determined by the $(j+1)$-skeleton K_{j+1} of K; thus the jth cellular homology groups of $K_{j+1}, K_{j+2}, \ldots,$ coincide.

Suppose inductively that for complexes of dimension $\leq n-1$ the equivalence of the singular and cellular homology theories is already established. (Their equivalence in the case $n = 0$ in easily seen.) Consider the homology exact sequence of the pair (K_n, K_{n-1}):

$$\to H_{j+1}^{\text{sing}}(K_n, K_{n-1}) \xrightarrow{\partial_*} H_j^{\text{sing}}(K_{n-1}) \xrightarrow{i_*} H_j^{\text{sing}}(K_n) \xrightarrow{j} H_j^{\text{sing}}(K_n, K_{n-1}) \to. \tag{10}$$

Since by Corollary 6.2 and Theorem 5.9 we have

$$H_{j+1}^{\text{sing}}(K_n, K_{n-1}) = 0 = H_j^{\text{sing}}(K_n, K_{n-1}) \qquad \text{for} \quad j \neq n-1, n,$$

it follows from the exactness of the sequence (10) that

$$H_j^{\text{sing}}(K_n) \simeq H_j^{\text{sing}}(K_{n-1}) \qquad \text{for} \quad j \neq n-1, n, \tag{11}$$

whence by the inductive hypothesis and (i) above we have

$$H_j^{\text{sing}}(K_n) \simeq \begin{cases} 0 & \text{for } j \geq n+1, \\ H_j^{\text{sing}}(K_{n-1}) \simeq H_j^{\text{cell}}(K_{n-1}) & \text{for } j \leq n-2, \end{cases}$$

whence the desired result, via the property (iii), for $j \neq n-1, n$.

For the cases $j = n-1, n$, we exploit the following exact sequences:

$$\begin{array}{ccccccccc} 0 \to H_n^{\text{sing}}(K_n) \to & H_n^{\text{sing}}(K_n, K_{n-1}) & \xrightarrow{\partial_*} & H_{n-1}^{\text{sing}}(K_{n-1}) & \xrightarrow{i_*} & H_{n-1}^{\text{sing}}(K_n) & \to & H_{n-1}^{\text{sing}}(K_n, K_{n-1}) \\ \| & \wr\wr & & \wr\wr & & \| & & \| \\ 0 \to H_n^{\text{sing}}(K_n) \to & C_n^{\text{cell}}(K_n) & \xrightarrow{\partial} & Z_{n-1}^{\text{cell}}(K_{n-1}) & \to & H_{n-1}^{\text{sing}}(K_n) & \to & 0, \end{array}$$

where the second sequence is obtained from the first by means of the indicated isomorphisms, and where in denoting the homomorphism $C_n^{\text{cell}}(K_n) \to Z_{n-1}^{\text{cell}}(K_{n-1})$ by ∂, we are invoking, in essence, Lemma 6.7, together with the inductive hypothesis (in the form $H_{n-1}^{\text{sing}}(K_{n-1}) \simeq H_{n-1}^{\text{cell}}(K_{n-1})$), and property (ii) above to the effect that $H_{n-1}^{\text{cell}}(K_{n-1}) = Z_{n-1}^{\text{cell}}$. From the resulting exactness of the second of these sequences it is immediate that $H_n^{\text{sing}}(K_n)$ is mapped one-to-one onto the kernel of ∂, so that $H_n^{\text{sing}}(K_n) \simeq Z_n^{\text{cell}}(K_n)$, whence by (ii),

$$H_n^{\text{sing}}(K_n) \simeq H_n^{\text{cell}}(K_n).$$

Similarly, looking at the right-hand half of the same exact sequence, we see that

$$H_{n-1}^{\text{sing}}(K_n) \simeq \frac{Z_{n-1}^{\text{cell}}}{\operatorname{Im} \partial} = H_{n-1}^{\text{cell}}(K_n),$$

completing the inductive step, and thence the proof of the theorem. (The dual cohomological result is proved analogously.) □

Important Remark. In the above proof of the equivalence of cell homology with singular simplicial homology, only certain formal properties of these two theories are significant, the details of their respective constructions playing no essential role. By isolating these properties in their purest form, one is led to the following set of axioms (due to Steenrod and Eilenberg) for an abstract homology theory:

(i) By an (*abstract*) *homology theory* we shall mean a "functor" (i.e. a generalized function in the categorical sense) which associates with each cell complex K (or more generally each pair (K, L) where $L \subset K$ is a subcomplex) a sequence of abelian groups $H_i(K)$ (or $H_i(K, L)$), $i = 0, 1, 2, \ldots$, and with each continuous map of complexes (which may in fact be assumed cellular—see Axiom 1 below) $f: K \to K'$ (or $f: (K, L) \to (K', L')$, where $f(L) \subset L'$) a sequence of homomorphisms

$$\begin{gathered} f_*: H_i(K) \to H_i(K'), \\ (f_*: H_i(K, L) \to H_i(K', L')), \end{gathered} \tag{12}$$

with the property that the composite fg of two maps corresponds to the composite of the corresponding homomorphisms

$$(fg)_* = f_* g_*,$$

and the identity map corresponds to the identity homomorphism: $1_* = 1$.

(ii) To qualify as a "homology theory" such a functor is required to satisfy the following conditions (the "axioms of homology theory"):

1. *Homotopy invariance.* If two maps f, g are homotopic then the corresponding homomorphisms should coincide:

$$f \sim g \quad \Rightarrow \quad f_* = g_*.$$

2. *Boundary homomorphisms* should be specified:

$$\partial\colon H_m(K, L) \to H_{m-1}(L), \qquad m = 1, 2, \ldots,$$

(where L is a subcomplex of the complex K) which commute with continuous maps of pairs of complexes:

$$\partial f_* = f_* \partial; \qquad f\colon (K, L) \to (K', L'), \qquad f(L) \subset L'.$$

3. *Exactness.* Denote by i, j the obvious inclusions

$$L \overset{i}{\subset} K = (K, \varnothing) \overset{j}{\subset} (K, L).$$

It is required that the following sequence of groups and homomorphisms be exact:

$$\cdots \to H_{m+1}(K, L) \xrightarrow{\partial_*} H_m(L) \xrightarrow{i_*} H_m(K) \xrightarrow{j} H_m(K, L) \xrightarrow{\partial_*} H_{m-1}(L) \to \cdots.$$

4. *Excision axiom.* It is required that

$$H_m(K, L) = H_m(K/L, *),$$

where L is any subcomplex of K, and K/L is the quotient complex in which L has been identified with a point.

5. *Normalization axiom.* For a one-point space $\{*\}$

$$H_m(*) = 0 \qquad \text{for} \quad m > 0.$$

Exercise

Show that once the group $H_0(*) = G$ is prescribed, these axioms determine a unique homology theory (up to isomorphism).

As we saw in §§4, 5, all of these axioms are fulfilled in the cellular and singular homology theories: this is the essential reason for their equivalence. In §5 we also noted in passing (in a remark and exercise) the un-normalized version of singular cubic homology theory, in which the normalization axiom is not satisfied, the integral homology groups of the one-point space being in that theory nontrivial in positive dimensions.

If we omit the normalization requirement from the above definition of an abstract homology theory, we obtain the more general concept of a *generalized* (or *extraordinary*) *homology theory*, of which the unreduced singular cubic homology theory furnishes a rather "trivial" instance (see the exercise in §5). Another, much more complex (and more important), example of a generalized homology theory is afforded by "bordism theory" which will be dicussed in Chapter 3 (see also the end of the present section).

An analogous list of axioms for an abstract cohomology theory can also be given. (We leave to the reader, as an exercise, the precise formulation of these axioms, and the proof of the appropriate uniqueness theorem for the resulting theory.) By using this axiomatic approach one can establish the equivalence of the cohomology theory of manifolds defined in §1 in terms of differential forms, with the various other cohomology theories encountered; for this purpose one, so to speak, simply transforms an arbitrary (finite-dimensional) cell complex into a manifold by considering instead a small neighbourhood (i.e. "thickening") of its image under an embedding into some Euclidean space. We shall not pursue this line of argument in detail here, however, since in §14 a different, more constructive, approach to establishing this equivalence will be outlined.

We now give an application of the operation of barycentric subdivision of a simplicial complex representing a manifold, namely to the elicitation of the phenomenon known as "Poincaré duality" (see also §18). Thus suppose we are given a manifold M^n triangulated into smooth simplexes, i.e. represented as a "simplicial complex" made up of smooth simplexes. We may suppose that the subdivision is as fine as we please (by carrying out a succession of barycentric subdivisions if necessary). Corresponding to each k-simplex σ_α^k of the manifold M^n we define its dual "polyhedron" $D(\sigma_\alpha^k) = \bar{\sigma}_\alpha^{n-k}$, a cell of dimension $n - k$, as follows:

(i) the dual polyhedron $D\sigma_\beta^n$ of an n-simplex σ_β^n is taken to be the vertex of the barycentric subdivision of the given simplicial decomposition of M^n, at the "centre" of σ_α^n (i.e. in its interior);

(ii) the dual polyhedron $D\sigma_\gamma^0$ of a 0-simplex σ_γ^0 is defined to be the n-cell obtained as the sum of all simplexes of the barycentric subdivision of M^n, having σ_γ^0 as a vertex (see Figure 42 for the case $n = 2$);

(iii) the dual polyhedron of an edge σ_δ^1 of the given triangulation of M^n, is the $(n - 1)$-cell $D\sigma_\delta^1$ obtained as the sum of all simplexes of dimension $(n - 1)$ of the barycentric subdivision with the point in the middle of the edge σ_δ^1 as a vertex, and transverse to that edge (see Figure 42);

(iv) the polyhedron dual to an $(n - 1)$-face σ_μ^{n-1} of M^n is taken to be that comprised of all (in fact, in this case, there are only two) 1-simplexes of the barycentric subdivision having the point at the "centre" of σ_μ^{n-1} as a vertex, and transverse to σ_μ^{n-1} (see Figure 42).

(v) The general definition is now clear: the polyhedron dual to a k-simplex σ_α^k of M^n is the cell $D\sigma_\alpha^k$ of dimension $(n - k)$, obtained by forming the

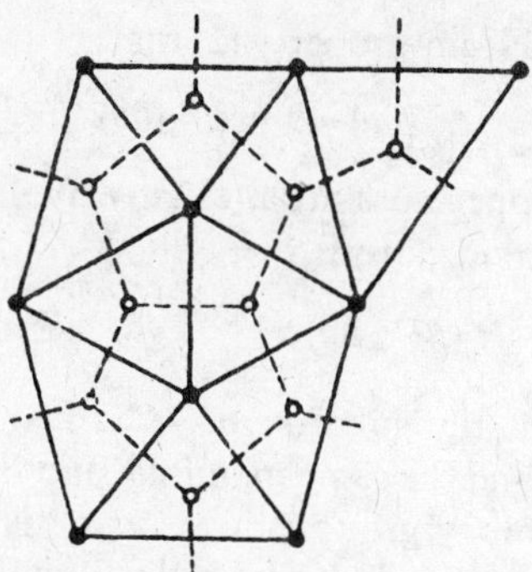

Figure 42. The initial triangulation of M^2 is indicated by the solid lines, and its dual cell decomposition by the dotted lines.

sum of all simplexes of the barycentric subdivision which have as a vertex the point of that subdivision at the centre of σ_α^k and are transverse to σ_α^k.

Clearly the "polyhedra" $D\sigma_\alpha^k$, $k = 0, 1, \ldots, n$, are then the cells of a decomposition of M^n as a cell complex, the *dual complex* of the given triangulation of M^n.

Note the following two properties of the operator D:

(1) The intersection $\sigma_s^k \cap D\sigma_s^k$ consists of a single point (namely the point at the "centre" of σ_s^k);
(2) Neglecting signs (i.e. orientations), we have

$$(\partial\sigma_s^k) \cap D\sigma_t^{k-1} = \sigma_s^k \cap (\partial D\sigma_t^{k-1}) \quad (\text{mod } 2). \tag{13}$$

(Property (1) is clear from the definition of D, while (2) is easy to see in dimensions $n = 1, 2, 3$, and not difficult to verify generally.)

Property (1) allows us to define a bilinear scalar product $a \circ b$ ($\in \mathbb{Z}_2$) of chains $a \in C_j(M^n)$ with cochains $b \in C_D^{n-j}(M^n)$ of the dual complex comprised of the cells $D\sigma_i^j$, in the following way. We write

$$a = \sum_i \lambda_i \sigma_i^j, \qquad b = \sum_k \mu_k \widehat{D\sigma_k^j}, \qquad \lambda_i, \mu_k \in \mathbb{Z}_2,$$

where in the latter sum $\widehat{D\sigma_k^j}$ denotes the functional on the $\mathbb{Z}_2$-vector space spanned by the $D\sigma_s^j$, which takes the value 1 on $D\sigma_k^j$ and vanishes on the rest. We then set

$$\sigma_i^j \circ \widehat{D\sigma_k^j} = \delta_{ik} \quad (\text{mod } 2), \tag{14}$$

whence, extending bilinearly, we obtain

$$a \circ b = \sum_{i,k} \lambda_i \mu_k \delta_{ik}. \tag{15}$$

It is then easily inferred from property (2) that

$$(\partial a) \circ b = a \circ (\partial b), \tag{16}$$

i.e. that the boundary operators are mutually dual. It follows that the map $\sigma_i^j \mapsto D\sigma_i^j$ induces an isomorphism

$$H_j(M^n; \mathbb{Z}_2) \overset{D}{\simeq} H^{n-j}(M^n; \mathbb{Z}_2), \tag{17}$$

since both the original triangulation of M^n and its dual complex are cell decompositions of one and the same manifold, and therefore yield isomorphic homology and cohomology groups; this is essentially a consequence of the homotopy invariance of the cellular homology and cohomology groups (see Theorem 4.6). The relationship between the homology and cohomology groups over $\mathbb{Z}_2$ expressed by (17) is termed "Poincaré duality". (For orientable manifolds (16) and (17) can be shown to hold over $\mathbb{Z}$.) In §18 below we shall establish the Poincaré-duality isomorphism using other means.

We end this section by considering briefly "bordism theory". (A fuller treatment will be give in §27.) In the present work we have more than once, and before even defining the homology groups, used in connexion with a given manifold M^n the terms "k-dimensional cycle" (or "singular bordism") and "$(k + 1)$-dimensional film", with the following meanings: A *cycle* in M^n is a pair (M^k, f), where M^k is a closed, oriented manifold, and $f: M^k \to M^n$ a smooth map of manifolds. A *film* is a pair (W^{k+1}, f) made up of a compact, oriented manifold-with-boundary W^{k+1}, and a smooth map $f: W^{k+1} \to M^n$; a film in M^n has therefore a *boundary*

$$\partial(W^{k+1}, f) = (\partial W^{k+1}, f|_{\partial W^{k+1}}). \tag{18}$$

We can now form the group of k-cycles in M^n, i.e. the additive group of finite formal sums of k-cycles in M^n:

$$\sum_i (M_i^k, f_i). \tag{19}$$

(Note that such a linear combination (over $\mathbb{Z}$) may be regarded as a cycle (or "singular bordism") in the above sense by thinking of it as the disjoint union $(\bigcup_i M_i^k, \bigcup_i f_i)$, and similarly for $\mathbb{Z}$-linear combinations of k-boundaries.) By factoring out the k-boundaries (18), i.e. by forming the quotient of the group of k-cycles by the subgroup generated by the k-boundaries (18), we obtain a group, analogous to the kth homology group, called the kth *bordism group* of M^n, denoted by $\Omega_k(M^n)$. Clearly, this definition extends to any cell complex, and, by analogy with relative singular homology, we can define the relative bordism groups $\Omega_k(X, Y)$ for pairs of cell complexes X, Y where Y is a subcomplex of X. It can be shown that for the bordism groups the theorem on homotopy invariance is valid, there is a corresponding exact sequence of each pair (X, Y), and even that the property $\Omega_k(X, Y) \simeq \Omega_k(X/Y)$ holds. On the other hand, as mentioned above, for contractible spaces (or equivalently the one-point space $\{*\}$), the bordism groups are, as it turns out (see §27.1), not all trivial in positive dimensions. The reason for this is simple: it is far from being the case that every manifold M^k can be realized as the boundary of a $(k+1)$-dimensional manifold-with-boundary. For instance, if a 4-manifold M^4 can be realized as the boundary of a film W^5, then it can be shown that its "Pontrjagin characteristic class" $p_1(M^4) = 0$, so that in particular $\mathbb{C}P^2$ cannot be such a boundary (see §§9.3, 27 for the relevant definitions and results).

One defines similarly the "bordism groups modulo 2" or "nonorientable bordism groups", denoted by $N_k(X)$, by admitting as cycles all pairs (M^k, f) with M^k any closed manifold (i.e. by dropping the requirement that M^k be oriented) and likewise for films.

EXERCISES

1. Prove that $\mathbb{R}P^2$ cannot be realized as the boundary of any 3-dimensional manifold-with-boundary. Draw the analogous conclusion for every product $\mathbb{R}P^2 \times \cdots \times \mathbb{R}P^2$ of $\mathbb{R}P^2$ by itself k times.
2. Prove that if a manifold M^k is a boundary, i.e. $M^k = \partial W^{k+1}$, then its Euler characteristic $\chi(M^k)$ is even (cf. §27).

There are natural homomorphisms

$$\Omega_k(X) \to H_k(X; \mathbb{Z}) \to H_k(X; \mathbb{R}), \qquad N_k(X) \to H_k(X; \mathbb{Z}_2), \tag{20}$$

which the reader will easily be able to define. The homology classes which happen to be images under these homomorphisms, are termed "cycles realizable as continuous images of manifolds"; they are the prototypes of cycles, or cycles in a simpler intuitive sense. However, bordism theory and the questions associated with it are more complex than this might suggest.

§7. The Homology Groups of a Product of Spaces. Multiplication in Cohomology Rings. The Cohomology Theory of H-Spaces and Lie Groups. The Cohomology of the Unitary Groups

The product $K_1 \times K_2$ of cell complexes K_1 and K_2 is again a cell complex, with cells the products of the cells of the complexes K_1 and K_2. This is the reason behind the following isomorphism of the group $C_n(K_1 \times K_2; \mathbb{Z})$ of integral cellular n-chains:

$$C_n(K_1 \times K_2; \mathbb{Z}) \simeq \sum_{k+l=n} C_k(K_1; \mathbb{Z}) \otimes C_l(K_2; \mathbb{Z}).$$

The boundary of the product $\sigma^i \times \sigma^j$ of two cells is given by the formula

$$\partial(\sigma^i \times \sigma^j) = (\partial\sigma^i) \times \sigma^j \cup (-1)^i \sigma^i \times (\partial\sigma^j) \tag{1}$$

(where the sign $(-1)^i$ ensures that the boundary is oriented appropriately; verify this for $\sigma^1 \times \sigma^1$, for instance). Since this accords (via the above isomorphism) with the definition of the boundary operator on the tensor product of two algebraic complexes (see §2(15)), we deduce immediately the following

7.1. Proposition. *The complex of integral chains of a product $K_1 \times K_2$ of cell complexes can be identified with the tensor product of the complexes $C(K_1; \mathbb{Z})$ and $C(K_2; \mathbb{Z})$:*

$$C(K_1 \times K_2; \mathbb{Z}) \simeq C(K_1; \mathbb{Z}) \otimes C(K_2; \mathbb{Z}).$$

This proposition clearly remains valid if, instead of the integers as coefficients, we take any commutative ring with multiplicative identity, in particular a field. Hence in view of Theorem 2.10 we have the following

7.2. Corollary. *For the homology groups of $K_1 \times K_2$ with coefficients from a field* **k** *we have the following (natural) isomorphism:*

$$H_n(K_1 \times K_2; \mathbf{k}) \simeq \sum_{l+m=n} H_l(K_1; \mathbf{k}) \otimes H_m(K_2; \mathbf{k}). \tag{2}$$

While in the general situation of an arbitrary ring G with 1 this result may not hold, the cell decomposition of $K_1 \times K_2$ via products of cells of K_1 and K_2 still yields a homomorphism

$$\sum_{k+l=m} H_k(K_1; G) \otimes H_l(K_2; G) \to H_m(K_1 \times K_2; G). \tag{3}$$

Here if $c_1 = \sum_i a_i \sigma_i^k$ and $c_2 = \sum_j b_j \sigma_j^l$ are cycles in K_1 and K_2 respectively, then as before we associate with the element $c_1 \otimes c_2$ of $C_k(K_1; G) \otimes C_l(K_2; G)$ the cycle

$$\sum_{i,j} a_i b_j (\sigma_i^k \times \sigma_j^l) \in C_{k+l}(K_1 \times K_2; G).$$

To see that this does indeed induce a well-defined map from

$$H_k(K_1; G) \otimes H_l(K_2; G)$$

to $H_{k+l}(K_1 \times K_2; G)$, it suffices to verify that if we replace c_1 by a homologous cycle $c_1 + \partial c$, then the element $c_1 \otimes c_2$ (actually a cycle in view of §2(15)) is replaced by

$$(c_1 + \partial c) \otimes c_2 = c_1 \otimes c_2 + \partial(c \otimes c_2)$$

(using once again §2(15)), which is mapped to a cycle in $C_{k+l}(K_1 \times K_2; G)$ homologous to the image of $c_1 \otimes c_2$. Thus, for an arbitrary commutative ring G, we obtain the homomorphism (3), which by the corollary is an isomorphism if $G = \mathbf{k}$, a field.

For the cohomology groups we obtain the analogous homomorphism

$$\sum_{k+l=m} H^k(K_1; G) \otimes H^l(K_2; G) \to H^m(K_1 \times K_2; G), \tag{4}$$

which is again an isomorphism if the coefficients are from a field. (The isomorphisms (2) and (4) are often referred to as the "Künneth formulae.")

Given a cell complex K, the diagonal map $\Delta: K \to K \times K$, sending x to (x, x), induces in the usual way a homomorphism of the respective (singular) cohomology groups

$$H^k(K \times K; G) \xrightarrow{\Delta^*} H^k(K; G).$$

7.3. Theorem. *Let G be an (associative), commutative ring with identity; then the composite homomorphism*

$$\Delta^*(a \otimes b) = ab: H^k(K; G) \otimes H^l(K; G) \to H^{k+l}(K \times K; G) \xrightarrow{\Delta^*} H^{k+l}(K; G), \quad (5)$$

defines on the direct sum $H^*(K; G) = \sum_{i \geq 0} H^i(K; G)$ a multiplication satisfying $ab = (-1)^{kl}ba$, thereby turning it into an associative and skew-commutative ring with a multiplicative identity $1 \in H^0(K; G)$.

Proof. (i) Associativity. If $c_1 \in H^k(K_1; G)$, $c_2 \in H^l(K_2; G)$ and $c_3 \in H^m(K_3; G)$, then the elements $(c_1 \otimes c_2) \otimes c_3$ and $c_1 \otimes (c_2 \otimes c_3)$ are easily seen to be mapped to the same element of $H^{k+l+m}(K_1 \times K_2 \times K_3; G)$.

(ii) Skew-commutativity. If $c \in H^k(K; G)$, $c' \in H^l(K; G)$, then under the map (4) (with $K_1 = K = K_2$) the element $c \otimes c'$ is sent to a G-linear combination of product functionals $\hat{\sigma}^k \times \hat{\sigma}^l$, while $c' \otimes c$ is sent under the map (4) (with k and l interchanged) to the same G-linear combination of the reversed products $\hat{\sigma}^l \times \hat{\sigma}^k$. The skew-commutativity is then a consequence of the fact that under the further homomorphism Δ^* to $H^{k+l}(K; G)$, the images of $\hat{\sigma}^k \times \hat{\sigma}^l$ and $\hat{\sigma}^l \times \hat{\sigma}^k$ differ only by the factor $(-1)^{kl}$, which arises essentially from the difference in orientations between the two product cells $\sigma^k \times \sigma^l$ and $\sigma^l \times \sigma^k$.

(iii) The identity element. Assuming for simplicity that K is connected (the general case being similar), we may, by Theorem 4.8, suppose that it has just one vertex, so that there is, up to a homotopy, just one singular 0-simplex $*$ of K. Let c be any basic singular l-cycle in K. Then since the composite map

$$K \xrightarrow{\Delta} K \times K \xrightarrow{p} K, \qquad p(x, y) = x,$$

is the identity map, we have

$$\hat{c} = (p \circ \Delta)^* \hat{c} = \Delta^* p^* \hat{c} = \Delta^*(\hat{c} \otimes \hat{*}),$$

where the hatted symbols denote the appropriate basic functionals. This completes the proof of the theorem. □

Remarks

1. If we take $K = M$, a manifold, and take the cohomology groups (of M and $M \times M$) to be defined in terms of differential forms, as in §1, then the above multiplication in the ring $H^*(M; \mathbb{R})$ takes on the following aspect. Suppose first that we have two manifolds M_1 and M_2 on which there are defined forms $\bar{\omega}$ and $\bar{\bar{\omega}}$ respectively:

$$\bar{\omega} = \sum f_{i_1 \ldots i_k} dx^{i_1} \wedge \cdots \wedge dx^{i_k}, \qquad \bar{\bar{\omega}} = \sum g_{j_1 \ldots j_l} dy^{j_1} \wedge \cdots \wedge dy^{j_l};$$

then the tensor product $\bar{\omega} \otimes \bar{\bar{\omega}}$ (at each point of the product $M_1 \times M_2$) defines a form ω on $M_1 \times M_2$ given locally by

$$\omega = \bar{\omega} \otimes \bar{\bar{\omega}} = p_1^*(\bar{\omega}) \wedge p_2^*(\bar{\bar{\omega}}) = \left(\sum f_{i_1 \dots i_k} dx^{i_1} \wedge \cdots \wedge dx^{i_k}\right) \wedge \left(\sum g_{j_1 \dots j_l} dy^{j_1} \wedge \cdots \wedge dy^{j_l}\right),$$

where $p_1: M_1 \times M_2 \to M_1$ and $p_2: M_1 \times M_2 \to M_2$ are the projections. (It can be shown that any smooth form on $M_1 \times M_2$ can be expressed as a convergent series of products of forms on the factors M_1 and M_2. We note also that the tensor product of two closed forms is a closed form on $M_1 \times M_2$, and that the product of a closed by an exact form is exact.) The above-defined product on $H^*(M; \mathbb{R})$ in this context turns out to be just the exterior product of forms: if $M_1 = M_2 = M$ say, and as before $\Delta: M \to M \times M$ is defined by $x \mapsto (x, x)$, then $\Delta^*(\bar{\omega} \otimes \bar{\bar{\omega}}) = \bar{\omega} \wedge \bar{\bar{\omega}}$ (on M), and the relation $ab = (-1)^{kl}ba$ appears as the familiar skew-commutativity of the exterior product of forms: $\bar{\omega} \wedge \bar{\bar{\omega}} = (-1)^{kl}\bar{\bar{\omega}} \wedge \bar{\omega}$.

2. If K is a finite simplicial complex then the multiplication of simplicial cochains may be defined equivalently as follows: Recall (from Definition 3.3 *et seqq.*) that if the vertices of K are $\alpha_1, \alpha_2, \dots, \alpha_N$ in the prescribed order then the k-dimensional simplexes σ^k of the complex K are given by certain correspondingly ordered subsets of $k + 1$ of these vertices:

$$\sigma^k = (\alpha_{j_0} \dots \alpha_{j_k}), \qquad j_0 < j_1 < \cdots < j_k.$$

Now let α and β be cochains of dimensions k and l respectively, i.e. functionals on the G-spaces of k-chains and l-chains of K, with values in the commutative ring G with 1. A cochain of dimension $k + l$, called the *cup product* of α and β is then defined by

$$(\alpha \smile \beta, \sigma^{k+l}) = (\alpha, \sigma_1^k)(\beta, \sigma_2^l), \tag{6}$$

where $\sigma^{k+l} = (\alpha_{j_0}\alpha_{j_1} \dots \alpha_{j_{k+l}})$ is an arbitrary $(k + l)$-simplex of K, in terms of which the simplexes σ_1^k and σ_2^l are given by

$$\sigma_1^k = (\alpha_{j_0}\alpha_{j_1} \dots \alpha_{j_k}), \qquad \sigma_2^l = (\alpha_{j_k}\alpha_{j_{k+1}} \dots \alpha_{j_{k+l}}).$$

The identity element for this multiplication of cochains is clearly the cochain, easily seen to be a cocycle, taking the value $1 \in G$ on each vertex of K (cf. part (iii) of the proof of Theorem 7.3). However the cup product is not generally skew-commutative.

Exercises

1. Verify the equality (Leibniz' formula)

$$\delta(\alpha \smile \beta) = (\delta\alpha) \smile \beta + (-1)^{\dim \alpha} \alpha \smile (\delta\beta).$$

2. Show that the following difference of cup products is cohomologous to zero if α and β are cocycles:

$$\alpha \smile \beta - (-1)^{kl}\beta \smile \alpha = \delta\gamma, \qquad k = \dim \alpha, \quad l = \dim \beta, \quad \delta\alpha = \delta\beta = 0.$$

(It follows that the cup product furnishes $H^*(K; G) = \sum_{q \geq 0} H^q(K; G)$ with a multiplicative operation, turning it into a skew-commutative ring with identity $1 \in H^0(K; G)$.)

3. Prove that this multiplication is equivalent to that given by Theorem 7.3.

In terms of the cup product of integral cochains, one can define a related product operation, the *cap product* of a k-cochain and a $(k+l)$-chain: if α^k is a cochain in $C^k(K; \mathbb{Z})$ and z_{k+l} is a chain in $C_{k+l}(K; \mathbb{Z})$, then the cap product $\alpha^k \frown z_{k+l}$ of α^k and z_{k+l} is the functional on l-cochains $\beta^l \in C^l(K; \mathbb{Z})$ given by

$$(\alpha^k \frown z_{k+l}, \beta^l) = (\alpha^k \smile \beta^l, z_{k+l}), \tag{7}$$

and regarded as an l-chain (via the natural identification of $C_l(K; \mathbb{Z})$ with its double dual).

EXERCISES (continued)

4. Prove that the cap product determines a well-defined product of elements of $H^k(K; \mathbb{Z})$ by elements of $H_{k+l}(K; \mathbb{Z})$:

$$H^k(K; \mathbb{Z}) \frown H_{k+l}(K; \mathbb{Z}) \subset H_l(K; \mathbb{Z}).$$

5. Show that if $f: K \to L$ is a continuous map of complexes, then the cap product (operating between the appropriate cohomology and homology groups) satisfies:

$$f_*(f^*(\alpha) \frown z) = \alpha \frown f_*(z).$$

6. Show that the barycentric operator D, defined in §6 in terms of a given triangulation of a manifold M^n, is given by the map $\alpha \mapsto \alpha \frown [M^n]$ where $[M^n] = z$ is the sum of all the n-simplexes of the triangulation.

Note that if the coefficient ring G is a field then by Theorem 2.9 the vector spaces H^k and H_k are mutually dual, so that the cap product can in fact the regarded as reducing to a product operation between the various homology groups; even in the integral case, however, the cap product is useful.

Example. We compute the cohomology ring of n-dimensional complex projective space $\mathbb{C}P^n$, taking the real numbers as coefficients. From our calculation of the homology groups of $\mathbb{C}P^n$ in §4, Example (g), it follows that $H_{2k+1} = 0 = H^{2k+1}$, and

$$H^{2k}(\mathbb{C}P^n; \mathbb{R}) \simeq H_{2k}(\mathbb{C}P^n; \mathbb{R}) \simeq \mathbb{R}, \qquad k \leq n. \tag{8}$$

In §1, Example (d), a 2-form c_1 was exhibited, generating (via the wedge product) a subalgebra of $H^*(\mathbb{C}P^n; \mathbb{R})$ isomorphic to the algebra of polynomials $\mathbb{R}[x]$ with the relation $x^{n+1} = 0$ imposed. In view of (8) (together with the fact that $H^{2k+1} = 0$) we now see that this subalgebra is actually the full cohomology algebra $H^*(\mathbb{C}P^n; \mathbb{R})$; thus $H^*(\mathbb{C}P^n; \mathbb{R})$ is isomorphic to the algebra of "truncated" polynomials in a single variable (representing the 2-form c_1):

$$H^*(\mathbb{C}P^n; \mathbb{R}) \simeq \mathbb{R}[c_1]/(c_1^{n+1} = 0), \qquad \text{rank } c_1 = 2. \tag{9}$$

We now turn to the problem of determining the structure of the cohomology algebras of Lie groups and H-spaces. For this purpose the following preliminary observation will be useful. Let $f: K \to L$ be an arbitrary continuous map of cell complexes (By Theorem 4.6 we may assume, if need be, that f is in fact cellular.) The map $F = f \times f$, defined in terms of f by

$$F = f \times f: K \times K \to L \times L, \qquad (x, y) \mapsto (f(x), f(y)),$$

is clearly diagonal-preserving ($F(\Delta) \subset \Delta$), and has the further property that (assuming the identification (1)) the homomorphism which it induces between the nth homology groups of $K \times K$ and $L \times L$, maps each tensor product of homology classes to the tensor product of the images of these classes (and analogously for the respective cohomology groups). From this we draw the important conclusion that, since by Theorem 7.3 multiplication in the cohomology rings of each of the complexes K, L is given by the formula $ab = \Delta^*(a \otimes b)$, the continuous map f commutes with that multiplication, i.e. more precisely,

$$f^*(ab) = f^*(a)f^*(b),$$

so that the induced homomorphism $f^*: H^*(L) \to H^*(K)$ is in fact a ring homomorphism.

We shall now exploit this fact in order to determine the structure of the cohomology ring of a Lie group or, more generally, of an H-space. (Recall (from Part II, §22.4) that a *(generalized) H-space* is by definition a topological space X equipped with a continuous multiplication $x \circ y = \psi(x, y) \in X$ (i.e. $\psi: X \times X \to X$) with respect to which there is a "homotopic identity element", i.e. an element $x_0 \in X$ with the property that the maps

$$\psi(x_0, x): X \to X, \qquad \psi(x, x_0): X \to X,$$

defining multiplication by x_0, are both homotopic to the identity map.) For our present purpose the following algebraic concept will be crucial.

7.4. Definition. A graded, skew-symmetric (associative) algebra $H = \sum_{k \geq 0} H^k$, with identity element

$$H^k H^l \subset H^{k+l}, \qquad yx = (-1)^{kl}xy \qquad \text{for all} \quad x \in H^k, \quad y \in H^l,$$

is called a *(connected) Hopf algebra* if each H^k has finite vector-space dimension and there is specified a dimension-preserving (or, in other terminology, degree-preserving) algebra homomorphism $\lambda: H \to H \otimes H$, which for $x \in H^k$, $k > 0$, has the form

$$\lambda(x) = x \otimes 1 + 1 \otimes x + x_1 \otimes y_1 + \cdots + x_k \otimes y_k, \tag{10}$$

where $0 < \deg x_i, \deg y_i < \deg x_i + \deg y_i = \deg x$. (The homomorphism λ is often termed "diagonal". Note also that the multiplication in $H \otimes H$ is defined by $(a \otimes b)(a' \otimes b') = aa' \otimes bb'$.)

Examples

(a) Consider the algebra $H = \mathbb{R}[x]$ of polynomials over the reals in the indeterminate x. If we postulate x as having positive even dimension, then we obtain a graded algebra, which is "trivially", as it were, skew-commutative. We can turn H into a Hopf algebra by specifying the value taken by the homomorphism λ on the indeterminate x to be $\lambda(x) = x \otimes 1 + 1 \otimes x$, whence it follows that

$$\lambda(x^k) = x^k \otimes 1 + 1 \otimes x^k + \sum_{i=1}^{k-1} \binom{i}{k} (x^i \otimes x^{k-i}).$$

(b) The "free" exterior algebra $H = \bigwedge[y]$ on a single generator y of odd positive degree, clearly forms a graded, skew-symmetric algebra, which can be endowed with the structure of a Hopf algebra by setting $\lambda(y) = y \otimes 1 + 1 \otimes y$.

(c) The preceding two examples represent the simplest (non-trivial) instances of a *free, skew-symmetric, graded algebra over* $\mathbb{R}$; the general definition of such an entity $H = \sum_{k \geq 0} H_k$ is as follows: The H_k are required to be finite dimensional as real vector-spaces, with, in particular, $H_0 = \mathbb{R}$; apart from this H is to have the form

$$\mathbb{R}[x_1, \ldots, x_k, \ldots] \otimes \bigwedge[y_1, \ldots, y_s, \ldots],$$

where the degrees (or "dimensions") of the x_i are even, and of the y_j odd. Thus H is generated by the pairs $(x_i, y_j) = x_i \otimes y_j$, amongst which there are no (non-trivial) relations other than those implied by skew-commutativity, namely

$$y_j^2 = -y_j^2 = 0,$$

$$y_j y_s = -y_s y_j, \qquad y_j x_i = x_i y_j, \qquad x_i x_k = x_k x_i,$$

and the number of x_i and y_j of each degree is finite. There are in general uncountably many ways of endowing such an algebra with the structure of a Hopf algebra. To see this, it suffices to observe that in view of the freeness of the algebra H, any map of the generators x_i, y_j of the form

$$\lambda(x_i) = x_i \otimes 1 + 1 \otimes x_i + \sum_s \bar{u}_i^{(s)} \otimes \bar{v}_i^{(s)},$$

$$\lambda(y_j) = y_j \otimes 1 + 1 \otimes y_j + \sum_r \bar{\bar{u}}_j^{(r)} \otimes \bar{\bar{v}}_j^{(r)},$$

where

$$\deg \bar{u}_i^{(s)}, \deg \bar{v}_i^{(s)}, \deg \bar{\bar{u}}_j^{(r)}, \deg \bar{\bar{v}}_j^{(r)} > 0,$$

and

$$\deg \bar{u}_i^{(s)} + \deg \bar{v}_i^{(s)} = \deg x_i, \qquad \deg \bar{\bar{u}}_j^{(r)} + \deg \bar{\bar{v}}_j^{(r)} = \deg y_j,$$

and the $\bar{u}_i^{(s)}, \bar{v}_i^{(s)}, \bar{\bar{u}}_j^{(r)}, \bar{\bar{v}}_j^{(r)}$ are otherwise arbitrary, extends to a unique algebra homomorphism $H \to H \otimes H$.

7.5. Theorem (Hopf). *The cohomology algebra of an H-space K is a Hopf algebra; that is, there is a homomorphism*

$$\lambda\colon H^*(K;\mathbb{R}) \to H^*(K;\mathbb{R}) \otimes H^*(K;\mathbb{R}),$$

such that for every element x of $H^q(K;\mathbb{R})$ (q arbitrary positive),

$$\lambda(x) = x \otimes 1 + 1 \otimes x + \sum_i x^{(i)} \otimes y^{(i)},$$

where $0 < \dim x^{(i)}, \dim y^{(i)} < \dim x^{(i)} + \dim y^{(i)} = \dim x$. *(We assume that the H-space is given as a (connected) cell complex.)*

Proof. The H-space multiplication $\psi\colon K \times K \to K$ induces an (algebra) homomorphism

$$\psi^*\colon H^*(K;\mathbb{R}) \to H^*(K \times K;\mathbb{R}),$$

which, in view of the identification afforded by Corollary 7.2, we may regard as a homomorphism

$$\psi^*\colon H^*(K;\mathbb{R}) \to H^*(K;\mathbb{R}) \otimes H^*(K;\mathbb{R}).$$

This latter map in fact will serve as the homomorphism we are seeking. To see this note first that for any element x of $H^*(K;\mathbb{R})$, we may write

$$\psi^* x = x^{(0)} \otimes 1 + 1 \otimes y^{(0)} + \sum x^{(i)} \otimes y^{(i)}, \tag{11}$$

where $\dim x^{(i)}, \dim y^{(i)} > 0$ (since in view of the connectedness of K every element of $H^* \otimes H^*$ has this form) and $\dim x^{(i)} + \dim y^{(i)} = \dim x^{(0)} = \dim y^{(0)} = \dim x$, since of course the induced homomorphism ψ^* is dimension-preserving. Denoting by $1 \times i$ the inclusion map $K \times a_0 \subset K \times K$, where a_0 is the homotopic identity element of the H-space K, we have, since the map $\psi(a, a_0)\colon K \to K$ is homotopic to the identity map,

$$(1 \times i)^* \psi^* x = x \otimes 1,$$

and then comparison with (11) gives $x^{(0)} = x$. A similar argument shows that $y^{(0)} = x$, completing the proof. □

The applications of this theorem depend on the following algebraic structure theorem for Hopf algebras over (in particular) the reals.

7.6. Theorem. *A Hopf algebra over a field of characteristic zero (e.g. the rational, real or complex fields) is a free skew-commutative graded algebra (as defined in Example* (c) *above).*

7.7. Corollary. *The cohomology algebra of a (finite-dimensional) connected Lie group is a finitely generated free exterior algebra* $\Lambda[y_1, \ldots, y_n]$.

Proof of the Corollary. Since a Lie group is certainly an H-space, its cohomology algebra is a Hopf algebra by Theorem 7.5, and therefore a free, skew-commutative, graded algebra by Theorem 7.6. If there were any free generators x_i of (positive) even dimension, then there would be elements of arbitrarily high dimension, an impossibility for a finite-dimensional manifold. □

Before proving Theorem 7.6 we consider two examples.

Examples

(α) Since the circle S^1 is a Lie group, we have, in view of the above corollary and Proposition 1.7, that

$$H^*(S^1; \mathbb{R}) \simeq \bigwedge [y_1], \qquad \deg y_1 = 1. \tag{12}$$

(β) *The structure of the cohomology algebra of the unitary group $U(n)$ is given by*

$$H^*(U(n); \mathbb{R}) \simeq \bigwedge [y_1, y_3, \ldots, y_{2n-1}], \qquad \deg y_i = i.$$

PROOF. Since the unitary group $U(n)$ is topologically equivalent to the direct product $S^1 \times SU(n)$: $U(n) \cong S^1 \times SU(n)$ (this was noted also in Part II, §24.4), it suffices in view of Corollary 7.2 to show that

$$H^*(SU(n); \mathbb{R}) \simeq \bigwedge [y_3, \ldots, y_{2n-1}]. \tag{13}$$

Since $SU(2) \cong S^3$, the case $n = 2$ of this isomorphism is immediate from Hopf's theorem and the facts that $H^n(S^n; \mathbb{R}) \simeq \mathbb{R}$, $H^k(S^n; \mathbb{R}) = 0$ for $k \neq 0, n$ (which can easily be established—for general n—using the methods of §4); cf. also Corollary 1.18. In the general case we exploit the standard fibre bundle

$$SU(n) \to S^{2n-1}, \tag{14}$$

with fibre $SU(n-1)$, where S^{2n-1} ($\subset \mathbb{C}^n$) is a homogeneous space for the group $SU(n)$ acting in the obvious way, with the fibre $SU(n-1)$ as isotropy group. We shall use this fibre bundle to realize $SU(n)$ as a cell complex built up (inductively) from cell decompositions of the base S^{2n-1} and the fibre $SU(n-1)$.

Consider to begin with the case $n = 3$, where the above fibre bundle is given by the projection p: $SU(3) \to S^5$, with fibre $SU(2)$. Choose any point σ^0 of the base S^5, and take the standard cell decomposition $\sigma^0 \cup \sigma^3$ of the fibre $SU(2) \cong S^3$ above that point. By Lemma 24.4.2 of Part II, above the complement $S^5 \backslash \{\sigma^0\}$ (the 5-dimensional disc) the bundle is trivial, so that on this bundle co-ordinates of the direct product may be introduced:

$$p^{-1}(S^5 \backslash \{\sigma^0\}) \cong (S^5 \backslash \{\sigma^0\}) \times SU(2) \cong D^5 \times S^3.$$

This product space can then be realized as a cell complex (in the standard way for products) as follows:

$$D^5 \times S^3 = \sigma^5 \cup \sigma^8 \qquad \text{where} \quad \sigma^8 \cong D^5 \times D^3.$$

The upshot is a cell decomposition of $SU(3)$ into four cells:

$$SU(3) = \sigma^0 \cup \sigma^3 \cup \sigma^5 \cup \sigma^8,$$

whence it follows immediately that

$$H^0(SU(3); \mathbb{R}) \simeq H^3 \simeq H^5 \simeq H^8 \simeq \mathbb{R},$$

while all other cohomology groups are zero. In view of Corollary 7.7 therefore, there are (basic) elements

$$y_3 \in H^3(SU(3); \mathbb{R}), \qquad y_5 \in H^5(SU(3); \mathbb{R}),$$

such that $y_3^2 = y_5^2 = 0$, and $y_3 y_5 = -y_5 y_3$ is basic (i.e. non-zero) for $H^8(SU(3); \mathbb{R})$.

Proceeding now to the inductive step, suppose inductively that the isomorphism (13) holds for the cohomology algebra $H^*(SU(n-1); \mathbb{R})$, i.e.

$$H^*(SU(n-1); \mathbb{R}) \simeq \bigwedge [y_3, \ldots, y_{2n-3}]. \tag{15}$$

The cell decomposition of $SU(n)$ is constructed, as in the case $n = 3$, using the fibre bundle (14). Thus the base S^{2n-1} is given the cell decomposition $\sigma^0 \cup \sigma^{2n-1}$, and we assume inductively a decomposition of the fibre $F \cong SU(n-1)$ into cells σ_F^α in natural one-to-one correspondence with a linear basis for $H^*(SU(n-1); \mathbb{R})$, including one 0-dimensional cell and representatives of the generators $y_3, \ldots, y_{2n-3}$ in (15). Since $p^{-1}(\sigma^0) = F$ and $p^{-1}(\sigma^{2n-1}) = \sigma^{2n-1} \times F$ (for the same reason as in the case $n = 3$), we obtain a cell decomposition of $SU(n)$ into cells

$$\sigma_F^\alpha \times \sigma^0, \qquad \sigma_F^\alpha \times \sigma^{2n-1}. \tag{16}$$

We shall now show that these cells all correspond to cocycles. For those of the form $\sigma_F^i \times \sigma^0$, $i = 3, \ldots, 2n-3$, corresponding to the exterior-algebra generators y_i, this is obvious since the cell of next higher dimension appearing in (16) is $\sigma_F^0 \times \sigma^{2n-1}$, of dimension $2n-1$. (The remainder of the cells $\sigma_F^\alpha \times \sigma^0$ in the fibre $SU(n-1)$ will, by inductive hypothesis, correspond to the various products of the y_i, $i = 3, \ldots, 2n-3$.) Denote by y_{2n-1} the cocycle taking the value 1 on the cell $\sigma_F^0 \times \sigma^{2n-1}$, and zero on the remaining cells in (16). We need to show that indeed y_{2n-1} is a cocycle; this follows by identifying y_{2n-1} with the tensor product of the cocycles corresponding to σ_F^0 and σ^{2n-1} (via the isomorphism (4)), and noting that this is again a cocycle by definition of the coboundary operator on tensor products (cf. §2(15)). It follows from this, together with Hopf's theorem (and Theorem 7.6), that the algebra $H^*(SU(n); \mathbb{R})$ embeds the free exterior algebra $\bigwedge [y_3, \ldots, y_{2n-1}]$. Since, however, the vector-space dimension of this algebra coincides with the number of cells in (16), we infer that in fact $H^*(SU(n); \mathbb{R})$ is isomorphic to $\bigwedge [y_3, \ldots, y_{2n-1}]$, as claimed. □

PROOF OF THEOREM 7.6. Let H denote an arbitrary Hopf algebra, and let $x_1, x_2, \ldots$ be homogeneous elements of H with $x_i \in H^{\deg x_i}$, and $0 < \deg x_i \le \deg x_j$ for $i \le j$; we may suppose further that the x_i form a minimal generating set for the Hopf algebra in the sense that every element of H should be expressible (not necessarily uniquely) as a polynomial $P(x_1, x_2, \dots)$ in the x_i, but that no generator x_k be so expressible in terms of the x_j with $j < k$:

$$x_k \neq P(x_1, \dots, x_{k-1}).$$

For each i denote by s_i the smallest positive integer such that $x^{s_i} = 0$, unless every positive power of x_i is non-zero, in which case write $s_i = \infty$. (Thus, in particular, if x_i has odd degree then in view of the skew-commutativity of H we shall have $s_i = 2$.) We first show that there are no relations between the x_i save those which are a consequence of the relations $x_i^{s_i} = 0$ (and skew-commutativity).

7.8. Lemma. *The monomials of the form*

$$x_k^{r_k} x_{k-1}^{r_{k-1}} \dots x_1^{r_1}, \qquad \textit{where} \quad 0 \leq r_i < s_i, \tag{17}$$

form a basis for the Hopf algebra H regarded as a vector space.

PROOF. We shall call a monomial of the form (17) *normal*, and a linear combination (over the ground field) of such monomials a *normal polynomial*. We define in the usual way the degree n of the normal monomial (17) to be given by

$$n = r_k \deg x_k + \cdots + r_1 \deg x_1.$$

Clearly any monomial in the x_i can be brought into the form (17) by means of the skew-commutativity relation and the relations $x_i^{s_i} = 0$. Thus we have only to show linear independence of the normal monomials, i.e. that a non-trivial normal polynomial cannot represent zero in the Hopf algebra H.

Clearly this is the case for normal polynomials of degree 1 in view of the minimality of the generating set $\{x_i\}$ of H. Suppose inductively that for polynomials of degree $< n$ the desired conclusion holds. Given any normal polynomial $P(x_k, \dots, x_1)$ of degree n, we can group the terms in which x_k occurs to highest degree (r say), and then take out the common factor x_k^r to obtain

$$P(x_k, \dots, x_1) = x_k^r Q(x_{k-1}, \dots, x_1) + R(x_k, \dots, x_1), \tag{18}$$

where in the terms of the normal polynomial R the generator x_k occurs to powers less than the rth. Suppose that there are non-trivial relations of the form $P(x_k, \dots, x_1) = 0$, where $\deg P = n$ (and $k \geq 1$ is arbitrary), and consider such a relation where the r in (18) is least positive among all such relations (with k allowed to vary). If P_n is the homogeneous part of P of degree n, then it follows from the definition of the (degree-preserving) diagonal homomorphism λ (see (10)) that the projection of $\lambda(P(x_k, \dots, x_1))$ on $H^n \otimes H^0$ is just $P_n \otimes 1$, so that since $\lambda(P) = \lambda(0) = 0$, we must have $P_n = 0$ in H. Thus we may suppose that P is homogeneous.

This assumed, we shall now show that $Q =$ const. and $r = 1$. Denoting by I_{k-1} the ideal of the algebra H generated by $x_1, \dots, x_{k-1}$, we readily deduce from the definition of λ that if $\deg Q > 0$, then

$$\lambda(x_k^r Q(x_{k-1}, \dots, x_1))$$
$$\equiv \left(\sum_{i=0}^{r} \binom{i}{r} x_k^i \otimes x_k^{r-i}\right)(1 \otimes Q + Q \otimes 1) \pmod{I_{k-1} \otimes I_{k-1}}. \quad (19)$$

Since $P = x_k^r Q + R = 0$, the terms $x_k^r Q \otimes 1$ and $1 \otimes x_k^r Q$ in (19) are respectively equal to $-(R \otimes 1)$ and $-(1 \otimes R)$. With this replacement made in (19), we see that (if $\deg Q > 1$) the terms explicitly appearing there are all linearly independent (and independent of the terms absorbed in $I_{k-1} \otimes I_{k-1}$) in view of the inductive hypothesis and the minimality of r. (We are here using the fact that, given a linearly independent set S in H, the elements $a \otimes b$, $a, b \in S$ are linearly independent in $H \otimes H$.) Since $\lambda(R(x_k, \dots, x_1))$ has no terms of the form $x_k^i a \otimes x_k^j b$ with $i + j = r$, it follows that $\lambda(P(x_k, \dots, x_1)) \neq 0$, contradicting the assumption that $P(x_k, \dots, x_1) = 0$ in H. Hence we must have $\deg Q = 0$, so that we may take $Q = 1$. We then have

$$\lambda(x_k^r Q) = \lambda(x_k^r) = \sum_{i=0}^{r} \binom{i}{r} x_k^i \otimes x_k^{r-i} \pmod{I_{k-1} \otimes I_{k-1}}, \quad (20)$$

whence, if $r > 1$, we again obtain a contradiction (in view of the inductive hypothesis and the minimality of r, and after replacing the terms $x_k^r \otimes 1$ and $1 \otimes x_k^r$ by $-R \otimes 1$ and $-1 \otimes R$ respectively). We must therefore have $r = 1$. However, then $x_k = -R(x_k, \dots, x_1)$ and x_k does not actually appear in R, contradicting the minimality of the generating set $\{x_i\}$. This completes the proof of the lemma. □

Continuing the proof of the theorem, we show that every generator x_k of even degree has the property that x_k^s is non-zero for all s. Suppose, on the contrary, that $x_k^s = 0$ and that s is the smallest positive integer for which this happens. Then from (20) (with $r = s$, and setting $x_k^s = 0$) we see that

$$\lambda(x_k^s) = \sum_{i=0}^{r-1} \binom{i}{r} x_k^i \otimes x_k^{r-i} \pmod{I_{k-1} \otimes I_{k-1}},$$

i.e. $\lambda(x_k^s)$ is a non-trivial linear combination of linearly independent elements modulo $I_{k-1} \otimes I_{k-1}$, contradicting the assumption that $x_k^s = 0$.

Thus we have shown that between the generators of a minimal generating system of our Hopf algebra there are no non-trivial relations (i.e. other than those resulting from skew-commutativity and the defining properties of an associative algebra). Hence the generators of even degree generate a subalgebra isomorphic to a polynomial algebra $\mathbb{R}[x_1', x_2', \dots]$ and the odd-degree generators generate a free exterior algebra $\bigwedge[x_1'', x_2'', \dots]$, the whole algebra H being, as is easy to see, isomorphic to the tensor product of these two algebras. This completes the proof of the theorem. □

We now consider further examples of (generalized) H-spaces.

Examples (continued)

(γ) For any complex K one can define the *loop space* $\Omega(K, x_0) = X$, with points the loops (i.e. paths) beginning and ending at $x_0 \in K$ (see Part II, §22.4). (We assume on $\Omega(K, x_0)$ the "compact-open" topology, where as a subbasis for the open sets one takes the sets of paths γ of the form $\{\gamma | \gamma(C) \subset U\}$ where C is any compact subset of $[0, 1]$ and U is any open set of K.) Recall that here the multiplicative operation is just the path-product, with the constant path as the homotopic identity, and that furthermore "homotopic inverses" exist (i.e. for each $x \in X$ there is an element $\bar{x}$ (the path inverse to x) such that the map $X \to X$ defined by $x \mapsto x \circ \bar{x}$ is homotopic to the constant map from X to the homotopic identity), and the operation is "homotopically associative" (i.e. the maps $X \times X \times X \to X$ defined by $(x, y, z) \mapsto (x \circ y) \circ z$ and $(x, y, z) \mapsto x \circ (y \circ z)$ are homotopic).

(δ) A multiplication with an identity element can be defined on the 7-sphere S^7 via that of the *Cayley numbers* (or "octonians"). These constitute a (non-associative) 8-dimensional real division algebra $\mathbb{R}^8$ with its bilinear product defined as follows: regarding each element as a pair (q_1, q_2) of quaternions (see Part I, §14.3), one multiplies such pairs according to the rule

$$(q_1, q_2)\cdot(q_1', q_2') = (q_1 q_1' - \bar{q}_2' q_2, q_2' q_1 + q_2 \bar{q}_1').$$

It follows that the inverse of a Cayley number (q_1, q_2) is

$$(q_1, q_2)^{-1} = \frac{(\bar{q}_1, -q_2)}{|q_1|^2 + |q_2|^2}.$$

We denote this division algebra by $\mathbb{K}$. The multiplication on S^7 is then obtained, analogously to that on S^0, S^1, S^3, by identifying S^7 in the usual way with the unit Cayley numbers, i.e. those with $|q_1|^2 + |q_2|^2 = 1$.

Apart from Lie groups G and products of the form $G \times S^7 \times \cdots \times S^7$ (with S^7 made into an H-space as above) there are no other examples of simply-connected, finite-dimensional H-spaces known. We indicate briefly why no continuous multiplication with homotopic identity element exists on the spheres S^{n-1} for $n \neq 1, 2, 4, 8$.

Suppose S^{n-1} can be endowed with such a multiplication

$$\psi: S^{n-1} \times S^{n-1} \to S^{n-1}, \qquad (x, y) \mapsto x \circ y.$$

Since $D^n \times D^n \cong D^{2n}$, we have

$$S^{2n-1} = \partial D^{2n} \cong (D^n \times S^{n-1}) \cup (S^{n-1} \times D^n),$$

with identification along the common boundary $S^{n-1} \times S^{n-1}$. Using this representation of S^{n-1} it is not difficult to show (via the assumed existence of a homotopic identity element) that the multiplication map ψ can be extended to a map

$$f(\psi): S^{2n-1} \cong (D^n \times S^{n-1}) \cup (S^{n-1} \times D^n) \to S^n,$$

sending $S^{n-1} \times S^{n-1}$ to the equator of S^n according to the multiplication map ψ (verify this!). We now use this map $f(\psi)$ to attach a $2n$-dimensional disc D^{2n} to the n-sphere S^n, obtaining the complex

$$K_n = D^{2n} \cup_{f(\psi)} S^n,$$

with three cells $\sigma^0, \sigma^n, \sigma^{2n}$. It follows for $n > 1$ (and also for $n = 1$ more directly, since $K_1 \cong \mathbb{R}P^2$) that

$$H^j(K_n) = \begin{cases} 0, & j \neq 0, n, 2n, \\ \mathbb{Z}, & j = 0, n, 2n. \end{cases}$$

EXERCISES (continued)

7. Let $u_n \in H^n(K_n; \mathbb{Z}_2)$, $u_{2n} \in H^{2n}(K_n; \mathbb{Z}_2)$ denote the basic mod 2 cohoomology classes. Deduce from the assumed existence of a homotopic identity element with respect to the multiplication $\psi: S^{n-1} \times S^{n-1} \to S^{n-1}$ (which is equivalent to this map's having degree 1 on each factor), that $u_n^2 = u_{2n}$ in the cohomology algebra $H^*(K_n; \mathbb{Z}_2)$.

In the cases $n = 1, 2, 4, 8$, where we know that a multiplication exists on S^{n-1} (obtained resectively from the multiplications in $\mathbb{R}$, $\mathbb{C}$, $\mathbb{H}$, and the Cayley numbers $\mathbb{K}$), we have $K_1 \cong \mathbb{R}P^2$, $K_2 \cong \mathbb{C}P^2$, $K_4 \cong \mathbb{H}P^2$, $K_8 \cong \mathbb{K}P^2$. According to a difficult result of Adams, such complexes K_n do not exist for $n \neq 1, 2, 4, 8$.

To conclude this section we sketch a proof, which uses cohomological multiplication, of the fact that for even n the groups $\pi_{2n-1}(S^n)$ are infinite.

It is not difficult to verify using the standard cell decomposition of S^n that (cf. §4, Example (a))

$$H^0(S^n; \mathbb{Z}) \simeq H^n(S^n; \mathbb{Z}) \simeq \mathbb{Z}, \qquad H^k(S^n; \mathbb{Z}) = 0 \quad \text{for} \quad k \neq 0, n.$$

If u is a generator of $H^n(S^n; \mathbb{Z})$, then in view of the cell decomposition in (21) below, and the homomorphism (2), the elements $1 = 1 \otimes 1$, $1 \otimes u$, $u \otimes 1$ and $u \otimes u$ may be regarded as forming an additive basis for the ring $H^*(S^n \times S^n; \mathbb{Z})$. Consider the bouquet

$$S^n \vee S^n = (S^n \times x_0) \cup (x_0 \times S^n) \subset S^n \times S^n,$$

and a map

$$\varphi: S^n \vee S^n \to S^n.$$

If φ has degree λ on the first n-sphere of the bouquet and degree μ on the second, then (cf. Corollary 6.3)

$$\varphi^*(u) = \lambda(u \otimes 1) + \mu(1 \otimes u).$$

Since $u^2 = 0$ in $H^*(S^n; \mathbb{Z})$, if $\lambda, \mu \neq 0$ the map φ cannot be extended to a map $\hat{\varphi}: S^n \times S^n \to S^n$; for if there were such an extension $\hat{\varphi}$, we should have $\hat{\varphi}^*(u^2) = 0$, whereas in fact $\hat{\varphi}^*(u^2) = 2\lambda\mu(u \otimes u) \neq 0$. In the cell decomposition of $S^n \times S^n$ arising from the standard decomposition of S^n, namely

$$S^n \times S^n = \sigma^0 \cup \sigma_1^n \cup \sigma_2^n \cup \sigma^{2n} = (S^n \vee S^n) \cup D^{2n}, \tag{21}$$

the cell $\sigma^{2n} = D^{2n}$ is attached via a map $\partial D^{2n} = S^{2n-1} \to S^n \vee S^n$. The composite of this map with φ:

$$S^{2n-1} \to S^n \vee S^n \xrightarrow{\varphi} S^n, \tag{22}$$

cannot be null-homotopic if $\lambda, \mu \neq 0$, since if it were then it could be extended to the disc D^{2n}, and so to $S^n \times S^n$.

EXERCISES (continued)

8. Prove that for n even the integer $\lambda\mu$ is an additive invariant of the homotopy class of the map $S^{2n-1} \to S^n$ defined in (22).

9. For each odd positive integer n construct a map $\varphi: S^n \times S^n \to S^n$ with $\lambda = 2$, $\mu = -1$.

10. Let x_0, x_1 be regular values of a map $f: S^{2n-1} \to S^n$ (see Part II, §10.2), and denote by γ the linking coefficient $\{M_1^{n-1}, M_2^{n-1}\}$ of the closed submanifolds $M_1^{n-1} = f^{-1}(x_0)$, $M_2^{n-1} = f^{-1}(x_1)$. (cf. Part II, Definition 15.4.1).† Prove that for the map $f = f(\varphi)$ given by (22) we have $\gamma = 2\lambda\mu$. Prove that for the complex $K = D^{2n} \cup_{f(\varphi)} S^n$, where the disc is attached by means of the map $f(\varphi)$, we have $u_n^2 = \gamma u_{2n}$ in the ring $H^*(K; \mathbb{Z})$.

11. Show that if f is the projection map p of a smooth fibre bundle $S^{2n-1} \to S^n$, then $\gamma = \pm 1$.

§8. The Homology Theory of Fibre Bundles (Skew Products)

For a fibre bundle (or "skew product") the relationship between the homology groups of the bundle space on the one hand, and those of the base and fibre on the other, is generally speaking incomparably more complex than in the case of the direct product (the trivial bundle). For this reason we shall assume throughout (except at one or two points), without further mention, that the coefficients are from a field. Consider a fibre bundle with projection $p: E \to B$ and fibre F, where E, B and F are cell complexes (or at least homotopic equivalents thereof). A cell decomposition of the bundle space E can be obtained from given decompositions of the base and fibre as in the case of the direct product (see §7). Thus if we denote the cells of the fibre by σ_F^j and of the base by σ_B^q, then since the interiors $\hat{\sigma}_B^q$ of the σ_B^q are (open) discs, their complete preimages $p^{-1}(\hat{\sigma}_B^q)$ are the (direct) products $\hat{\sigma}_B^q \times F$, whence as the cells of a decomposition of E we obtain

$$\sigma_E^{j+q} = \sigma_B^q \times \sigma_F^j.$$

† The linking coefficient was defined in Part II, §15.4 only for closed curves in $\mathbb{R}^3$. For submanifolds M_1^k, M_2^k of $\mathbb{R}^{2k+1}$ (or S^{2k+1}) the linking coefficient is defined analogously as the intersection index of either submanifold with a film spanning the other, i.e. with a $(k+1)$-submanifold-with-boundary having the k-submanifold in question as boundary.

Thus the cell decomposition of the bundle space E afforded by the given decompositions of the base B and fibre F is, formally at least, the same as that we utilized for the (direct) product $B \times F$. However, the boundary operator is in general much more complicated. (This complexity was evident in the example of a fibre bundle considered at the end of §4, the unit tangent bundle over a surface of genus g.) The following two properties are consequences of the definition of the boundary operator on a cell-chain complex (see §4(3)):

(i) For cells of the total space E of the form $\sigma_E^j = \sigma_B^0 \times \sigma_F^j$, where σ_B^0 is a vertex in the base of the fibre bundle, we have

$$\partial\sigma_E^j = \sigma_B^0 \times (\partial\sigma_F^j).$$

(ii) More generally, for a cell of the total space E of the form $\sigma_E^{q+j} = \sigma_B^q \times \sigma_F^j$, we have

$$\partial\sigma_E^{q+j} = \sigma_B^q \times (\partial\sigma_F^j) + \Delta, \tag{1}$$

where Δ is a linear combination of cells in the complete inverse image $p^{-1}(\overline{\partial\sigma_B^q})$, $\overline{\partial\sigma_B^q}$ denoting the smallest subcomplex of the $(q-1)$-skeleton B^{q-1} of the base B, containing the image of the boundary $\partial\sigma_B^q \cong S^{q-1}$ under the map by means of which σ_B^q is attached to B^{q-1}. Thus it is certainly the case that the cells involved in Δ come from $p^{-1}(B^{q-1})$.

EXERCISE

1. If the base B is simply-connected, and has only one 0-cell σ_B^0 and no 1-cells, then the following formula holds:

$$\partial\sigma_E^{q+j} = \sigma_B^q \times (\partial\sigma_F^j) + (-1)^j(\partial\sigma_B^q) \times \sigma_F^j + \Delta_1, \tag{2}$$

where Δ_1 only involves cells from $p^{-1}(B^{q-2})$, the complete inverse image of the $(q-2)$-skeleton B^{q-2} of the given cell decomposition of B.

We shall assume for the remainder of this section that the formula (2) holds for the fibre bundle under consideration. That formula is obviously valid also if there are no $(q-1)$-dimensional cells in the given cell decomposition of the base B; such decompositions exist for $B = S^n$ $(n > 1)$, $\mathbb{C}P^n$, $\mathbb{H}P^n$, and also if B is a complex Grassmannian manifold (Part II, §5.2), a bouquet of spheres, a product of spheres, and in several other cases.

Remark. In fact the ensuing discussion an the conclusions it leads us to, will be valid (with appropriate qualifications) in the more general situation where the action of the group $\pi_1(B)$ on the fibre F (cf. Part II, §§19.1, 22.3, 24.3) induces the trivial action of $\pi_1(B)$ on the homology groups $H_k(F)$. (For the unit tangent bundle over a manifold B, for instance, this is equivalent to orientability of B; here the fibre is of course spherical.) If the fibre is a sphere S^n, then over $\mathbb{Z}_2$ this condition will automatically be fulfilled since every homology group $H_k(S^n; \mathbb{Z}_2)$ has trivial automorphism group. (The changes in our conclusions, necessary in cases where the actions of $\pi_1(B)$ on the $H_k(F)$ are *not* all trivial, will be indicated in §11 below.)

Thus, henceforth in this section, we shall consider only those fibre bundles admitting cell decompositions of the above sort for which (2) is valid. We now write the boundary $\partial\sigma_E^{q+j}$ as the sum

$$\partial\sigma_E^{q+j} = \sigma_B^q \times (\partial\sigma_F^j) + (-1)^j(\partial\sigma_B^q) \times \sigma_F^j + \partial_2\sigma_E^{q+j} + \partial_3\sigma_E^{q+j} + \cdots,$$

where

$$\partial_k\sigma_E^{q+j} = \sum_\alpha \lambda_\alpha\sigma_{B,\alpha}^{q-k} \times \sigma_{F,\alpha}^{j+k-1}, \qquad \lambda_\alpha \in \mathbb{Z}.$$

Thus $\partial_k\sigma_E^{q+j}$ is a linear combination of the products of the $(q-k)$-dimensional cells of the base by the $(j+k-1)$-dimensional cells of the fibre. If, extending the definition of ∂_k to $k = 0, 1$, we write

$$\partial_0\sigma_E^{q+j} = \sigma_B^q \times (\partial\sigma_F^j), \qquad \partial_1\sigma_E^{q+j} = (-1)^j(\partial\sigma_B^q) \times \sigma_F^j,$$

then we shall have

$$\partial\sigma_E^{q+j} = (\partial_0 + \partial_1 + \partial_2 + \cdots)\sigma_E^{q+j}.$$

(It is important to note that although $\partial_0^2 = 0$, it may happen that $\partial_1^2 \neq 0$, essentially since the chain $(\partial\sigma_B^q) \times \sigma_F^j$, as a linear combination of cells of the form $\sigma_B^{q-1} \times \sigma_F^j$, depends on the manner in which the cell $\sigma_B^q \times \sigma_F^j$ is attached.) Since our cell decomposition of E is formally the same as that we used for the direct product $E_0 = B \times F$, we can make the identification (cf. the isomorphism at the beginning of §7)

$$C_n(E) = \sum_{q+j=n} C_q(B) \otimes C_j(F).$$

The boundary operator ∂_E then acts on an n-chain $a \otimes b$, $a \in C_q(B)$, $b \in C_j(F)$, of the cell-chain complex $C(E)$, as follows:

$$\partial_E(a \otimes b) = a \otimes \partial_F b \pm (\partial_B a) \otimes b + \partial_2(a \otimes b) + \cdots,$$

where

$$\partial_k(a \otimes b) \in C_{q-k}(B) \otimes C_{j+k-1}(F), \qquad k = 0, 1, 2, \ldots.$$

Note that the operators ∂_0 and ∂_1 have the same form as in the case of the direct product $E_0 = B \times F$ (see §7(1)), so that the operators ∂_k with $k \geq 2$ are in that case all zero. We may therefore regard the operators ∂_k, $k \geq 2$, as characterizing the degree of "skewness" or "twist" in the boundary operator ∂_E on the complex $C(E)$ as compared with ∂_{E_0}, $E_0 = B \times F$.

The investigation of the homology groups $H_k(E)$ is carried out using the method of "filtrations" or "successive approximations" (indexed by $k = 0, 1, \ldots$), based on what is called the "Leray spectral sequence". The essence of this method is contained in the following procedure:

Step 0. Since $\partial_0^2 = 0$, we can regard $\partial_0 = d_0$ as a boundary operator on n-chains in E (i.e. linear combinations over the coefficient field of n-cells of the form $\sigma_E^{q+j} = \sigma_B^q \times \sigma_F^j$) and consider it as the "zeroth approximation" to the operator σ_E. The corresponding "homology groups of the zeroth approximation" are then easily seen to be given by

$$H_n(C(E), d_0) = \sum_{q+j=n} C_q(B) \otimes H_j(F) = \sum_{q+j=n} E^{(1)}_{q,j}.$$

Thus $H_n(C(E), d_0)$ can be regarded as consisting of all q-chains in the base B with coefficients from the homology group $H_j(F)$, $q + j = n$; i.e. $E^{(1)}_{q,j} = C_q(B; H_j(F))$ (see §2).

Step 1. The operator ∂_1 induces an operator d_1 on the n-dimensional ∂_0-cycles modulo the group Im ∂_0 of n-dimensional ∂_0-boundaries (i.e. on each group $H_n(C(E), \partial_0)$, or equivalently on the q-chains in the base B with coefficients from $H_j(F)$, $q + j = n$; see above); the operator d_1 is well-defined and satisfies $d_1^2 = 0$ (see below). Thus here we take as our chain complex

$$E^{(1)} = \sum_{q,j} E^{(1)}_{q,j}, \qquad d_1: E^{(1)}_{q,j} \to E^{(1)}_{q-1,j}.$$

As is to be expected in view of the definitions of ∂_0 and ∂_1, the resulting "homology groups of the first approximation" coincide with the homology groups of the direct product:

$$H_n(E^{(1)}, d_1) = \sum_{q+j=n} H_q(B; H_j(F)) = \sum_{q+j=n} H_q(B) \otimes H_j(F) \simeq H_n(B \times F), \quad (3)$$

where the last isomorphism comes from §7(2). (Recall our blanket assumption that the coefficients are from a field.) Each element of the (q, j)th term $H_q(B; H_j(F))$ of the direct sum in (3) is of course an n-dimensional d_1-cycle $z \in E^{(1)}_{q,j} = C_q(B; H_j(F))$, determined only to within an additive n-dimensional d_1-boundary, i.e. element of Im d_1. We denote the d_1-homology groups by

$$H_n(E^{(1)}, d_1) = E^{(2)}_n = \sum_{q+j=n} E^{(2)}_{q,j} = \sum_{q+j=n} H_q(B) \otimes H_j(F) \simeq H_n(B \times F).$$

For the direct product $E_0 = B \times F$ the procedure now terminates; however, for a non-trivial skew product (i.e. fibre bundle) there will be further steps corresponding to the operators $\partial_2, \partial_3, \ldots$.

Step 2. The operator ∂_2 induces a well-defined boundary operator d_2 on the homology groups of the first approximation (i.e. on (the summands of) $E^{(2)} = \sum_{n \geq 0} E^{(2)}_n$), satisfying $d_2^2 = 0$. (We shall give the precise definition of d_2 below.) The homology groups of the complex $(E^{(2)}, d_2)$ are denoted by

$$E^{(3)}_n = H_n(E^{(2)}, d_2) = \sum_{q+j=n} H_{q,j}(E^{(2)}, d_2) = \sum_{q+j=n} E^{(3)}_{q,j},$$

$$E^{(3)} = \sum_{n \geq 0} E^{(3)}_n,$$

where the elements of the group $E^{(3)}_{q,j} = H_{q,j}(E^{(2)}, d_2)$ are d_2-cycles $z \in E^{(2)}_{q,j} = H_q(B) \otimes H_j(F)$, determined only to within addition of a d_2-boundary. From the definition of ∂_2 it follows that the boundary operator d_2 maps $E^{(2)}_{q,j}$ to $E^{(2)}_{q-2,k+1}$:

$$d_2: E^{(2)}_{q,j} \to E^{(2)}_{q-2,j+1}, \qquad E^{(2)}_{q,j} = H_q(B) \otimes H_j(F).$$

This "filtering" procedure is continued in this way, yielding at the rth step a complex (and associated boundary operator)

$$E^{(r)} = \sum E^{(r)}_{q,j}, \qquad d_r\colon E^{(r)}_{q,j} \to E^{(r)}_{q-r,j+r-1},$$

and where, furthermore,

$$E^{(r+1)} = \sum E^{(r+1)}_{q,j} = H_*(E^{(r)}, d_r).$$

It is easy to see that for all $r \geq 0$ we shall have $E^{(r)}_{q,j} = 0$ for $q < 0$ or $j < 0$ so that the operator d_r, restricted to the groups $E^{(r)}_{q,j}$ with $q < r$, is the zero operator. Hence

$$E^{(r)}_{q,j} = E^{(r+1)}_{q,j} = E^{(r+2)}_{q,j} = \cdots = E^{(\infty)}_{q,j}, \qquad q < r,$$

where $E^{(\infty)}_{q,j}$ denotes the (constant) term of this sequence.

8.1. Theorem (Leray)†

(i) *The boundary operators d_r can all be appropriately well defined so as to satisfy $d_r^2 = 0$.*

(ii) *Provided the coefficients are from a field, the* nth *homology group $H_n(E)$ of the total space E of the given fibre bundle is isomorphic to the direct sum*

$$E^{(\infty)}_n = \sum_{q+j=n} E^{(\infty)}_{q,j}.$$

(iii) *For each pair q, j the group $E^{(2)}_{q,j}$ is isomorphic to the group $H_q(B) \otimes H_j(F)$, where B and F are respectively the base and fibre of the given bundle.*

Thus we see that as the result of successive "filtrations" we ultimately obtain, to within an additive boundary, the n-cycles in E (essentially as the elements in all kernels of the homomorphisms $d_r\colon E^{(r)}_n \to E^{(r)}_{n-1}$, modulo the images under the "preceding" maps $d_r\colon E^{(r)}_{n+1} \to E^{(r)}_n$). We shall not give here the full proof of Leray's theorem; however, we shall indicate the idea of the proof below, by sketching the verification of the theorem in a special case.

8.2. Corollary. *For a fibre bundle E (of the type we are considering) the (torsion-free) ranks of the homology groups of E are less than or equal to those of the direct product (i.e. for all k the Betti numbers satisfy $b_k(E) \leq b_k(E_0)$, where $E_0 = B \times F$).*

This follows directly from the theorem since at Step 2 we have $E^{(2)}_n \simeq H_n(E_0)$, and at the next stage a "filtration" of this group takes place in the sense that only elements in the kernel of the operator d_2 on $E^{(2)}_n$ are selected,

† Of the results we have so far encountered in this book, this represents the first one of importance for which a proof without serious use of the language and machinery of homological algebra would appear to be impossible.

i.e. the d_2-cycles, from which the d_2-boundaries are then factored out to obtain $E_n^{(3)}$, which is then subjected to a further "filtration" using d_3, and so on.

We shall now define the operator d_2 on the groups $E_{q,j}^{(2)}$ precisely, and verify that $d_2^2 = 0$. Since $\partial_E = \partial_0 + \partial_1 + \partial_2 + \cdots$ and $\partial_E \partial_E = 0$, we have

$$\begin{aligned} 0 = \partial_E^2 = \partial_0^2 &+ (\partial_0\partial_1 + \partial_1\partial_0) + (\partial_1^2 + \partial_0\partial_2 + \partial_2\partial_0) \\ &+ (\partial_1\partial_2 + \partial_2\partial_1 + \partial_0\partial_3 + \partial_3\partial_0) \\ &+ (\partial_2^2 + \partial_3\partial_1 + \partial_1\partial_3 + \partial_0\partial_4 + \partial_4\partial_0) + \cdots, \end{aligned}$$

whence, applying ∂_E^2 to each summand $C_{q,j}(E)$ of $C(E)$ separately, we obtain

$$\begin{aligned} 0 &= \partial_0^2 \colon C_{q,j} \to C_{q,j-2}, \\ 0 &= \partial_0\partial_1 + \partial_1\partial_0 \colon C_{q,j} \to C_{q-1,j-1}, \\ 0 &= \partial_1^2 + \partial_0\partial_2 + \partial_2\partial_0 \colon C_{q,j} \to C_{q-2,j}, \\ 0 &= \partial_1\partial_2 + \partial_2\partial_1 + \partial_0\partial_3 + \partial_3\partial_0, \\ 0 &= \partial_2^2 + \partial_1\partial_3 + \partial_3\partial_1 + \partial_0\partial_4 + \partial_4\partial_0. \end{aligned} \tag{4}$$

(1) We first examine in greater detail the operator d_1. As noted above, the operator d_1 is required to act on d_0-cycles modulo d_0-boundaries (where $d_0 = \partial_0$), i.e. on the d_0-homology groups $E_n^{(1)} = \sum_{q+j=n} E_{q,j}^{(1)}$. Now if $\partial_0 x = 0$, i.e. if x is a d_0-cycle, then the boundary operator ∂_1, in terms of which d_1 is defined, satisfies

$$\partial_1(x + \partial_0\bar{x}) = \partial_1 x + \partial_1\partial_0\bar{x} = \partial_1 x - \partial_0(\partial_1\bar{x})$$

by (4). Hence ∂_1 does indeed induce a well-defined operator d_1 on the d_0-homology groups. From the third equation in (4) we have

$$\partial_1^2 x = -\partial_0\partial_2 x - \partial_2\partial_0 x = -\partial_0\partial_2 x$$

since $\partial_0 x = 0$. Hence $\partial_1^2 x \equiv 0 \bmod(\operatorname{Im} d_0)$, i.e. $d_1^2 = 0$ on the $E_{q,j}^{(1)}$.

(2) We now turn to the operator d_2, which is required to act on the groups $H_n(E^{(1)}, d_1) = E_n^{(2)}$. Now an n-chain x in $C(E)$ represents an element of $E_{q,j}^{(2)}$ precisely if

$$\partial_0 x = 0 \quad \text{and} \quad \partial_1 x \equiv 0 \bmod(\operatorname{Im}\partial_0), \quad \text{say } \partial_1 x = \partial_0 y. \tag{5}$$

We shall, of course, require of d_2 that if x satisfies the conditions (5) then $d_2 x$ also satisfy them. Now $\partial_2 x$ need not in fact satisfy those conditions. However, using (4) we have

$$\begin{aligned} \partial_0(\partial_2 x - \partial_1 y) &= \partial_0\partial_2 x - \partial_0\partial_1 y = -\partial_2\partial_0 x - \partial_1^2 x + \partial_1\partial_0 y = -\partial_1^2 x + \partial_1^2 x \\ &= 0, \\ \partial_1(\partial_2 x - \partial_1 y) &= -\partial_2\partial_1 x - \partial_0\partial_3 x - \partial_3\partial_0 x - \partial_1^2 y \\ &= -\partial_2\partial_0 y - \partial_0\partial_3 x + \partial_0\partial_2 y + \partial_2\partial_0 y = \partial_0(\partial_2 y - \partial_3 x) \in \operatorname{Im}\partial_0, \end{aligned}$$

whence we see that the element $\partial_2 x - \partial_1 y$ satisfies the conditions (5). We therefore define the operator d_2 by means of the formula

$$d_2 x = \partial_2 x - \partial_1 y,$$

or, in coset form, by

$$d_2 x = \partial_2 x - \partial_1 \partial_0^{-1} \partial_1 x + [\mathrm{Im}\, \partial_0 + \partial_1(\mathrm{Ker}\, \partial_0)].$$

To verify that d_2 is well-defined as an operator on the $E_n^{(2)}$, consider an arbitrary element $x + \partial_0 z + \partial_1 v = \tilde{x}$ (where $\partial_0 v = 0$) in the same coset as x. Using (4) we have

$$\begin{aligned}
\partial_2 \tilde{x} - \partial_1 \partial_0^{-1} \partial_1 \tilde{x} &= (\partial_2 x - \partial_1 \partial_0^{-1} \partial_1 x) + \partial_2 \partial_0 z + \partial_2 \partial_1 v - \partial_1 \partial_0^{-1} \partial_1 \partial_0 z \\
&\quad - \partial_1 \partial_0^{-1} \partial_1 \partial_1 v \\
&= (\partial_2 x - \partial_1 \partial_0^{-1} \partial_1 x) - \partial_0 \partial_2 z - \partial_1^2 z - \partial_1 \partial_2 v - \partial_0 \partial_3 v \\
&\quad - \partial_3 \partial_0 v + \partial_1^2 z - \partial_1 \partial_0^{-1} \partial_1 \partial_1 v \\
&= (\partial_2 x - \partial_1 \partial_0^{-1} \partial_1 x) - \partial_0(\partial_2 z + \partial_3 v) - \partial_1(\partial_2 v + \partial_0^{-1} \partial_1 \partial_1 v) \\
&\in \partial_2 x - \partial_1 \partial_0^{-1} \partial_1 x + [\mathrm{Im}\, \partial_0 + \partial_1(\mathrm{Ker}\, \partial_0)],
\end{aligned}$$

where the last step follows from

$$\partial_0(\partial_2 v + \partial_0^{-1} \partial_1 \partial_1 v) = \partial_0 \partial_2 v + \partial_1^2 v = -\partial_2 \partial_0 v = 0.$$

This establishes the correctness of the definition of d_2. It remains to verify that $d_2^2 = 0$ on $E_{q,j}^{(2)}$. With x as in (5), we have

$$\begin{aligned}
d_2^2 x &= \partial_2(\partial_2 x - \partial_1 y) - \partial_1 \partial_0^{-1} \partial_1(\partial_2 x - \partial_1 y) \\
&= -\partial_0 \partial_4 x - \partial_4 \partial_0 x - \partial_3 \partial_1 x - \partial_1 \partial_3 x + \partial_0 \partial_3 y + \partial_3 \partial_0 y + \partial_1 \partial_2 y \\
&\quad + \partial_1 \partial_0^{-1} \partial_2 \partial_1 x + \partial_1 \partial_0^{-1} \partial_0 \partial_3 x + \partial_1 \partial_0^{-1} \partial_3 \partial_0 x - \partial_1 \partial_0^{-1} \partial_0 \partial_2 y \\
&\quad - \partial_1 \partial_0^{-1} \partial_2 \partial_0 y \\
&= -\partial_0 \partial_4 x - \partial_3 \partial_0 y - \partial_1 \partial_3 x + \partial_0 \partial_3 y + \partial_3 \partial_0 y + \partial_1 \partial_2 y + \partial_1 \partial_3 x - \partial_1 \partial_2 y \\
&= -\partial_0(\partial_4 x - \partial_3 y) \in \mathrm{Im}\, \partial_0,
\end{aligned}$$

completing the verification.

(3) The boundary operator d_3 on the groups $E_{q,j}^{(3)} = H_{q,j}(E^{(2)}, d_2)$ is defined similarly, in terms of the operator ∂_3 acting on n-chains $x \in C_{q,j}(E)$ satisfying

$$\partial_0 x = 0, \qquad \partial_1 x = \partial_0 y, \qquad \partial_2 x - \partial_1 y = \partial_0 z + \partial_1 w \quad \text{where} \quad \partial_0 w = 0,$$

i.e. on the n-dimensional d_2-cycles, or rather on the cosets determined by such elements x, modulo all d_i-boundaries, $i \le 2$. One can then go on to define inductively the operators d_r for larger r by adjusting the operator $\partial_r \colon C_{q,j} \to C_{q-r,j+r-1}$, by adding to each image $\partial_r x$, $x \in E_{q,j}^{(r)}$, certain images under the maps $\partial_0, \partial_1, \ldots, \partial_{r-1}$, analogously to the definition of d_2. Then of course one needs to verify that the operator d_r: $E^{(r)}_{q,j} \to E^{(r)}_{q-r,j+r-1}$ so obtained, is well defined and satisfies $d_r^2 = 0$. We shall not give further details; for our purposes the exact form of the boundary operator d_r for $r > 2$ will not be needed.

We shall now indicate the idea of the proof of part (ii) of Leray's theorem in the particular case where, for all $i \geq 3$, the operators ∂_i are zero. Under this assumption one can establish the result by means of direct calculation without using the apparatus of homological algebra. Thus having defined d_2 and verified that $d_2^2 = 0$, we shall now show that for all n, the group $H_n(E^{(2)}, d_2) = H_n^{(3)} = E_n^{(\infty)}$ is isomorphic to the nth homology group $H_n(E)$ of the total space E of our fibre bundle (assuming the coefficients from a field).

Let x be an arbitrary non-zero element of $H_n(E)$, represented by an n-cycle $\bar{x} \in C_n(E) = \sum_{q+j=n} C_{q,j}(E)$. By the *filtration* of the element $x \in H_n(E)$, we shall mean the smallest integer q such that x is representable by a cycle $\bar{x}$ in the complete preimage $p^{-1}(B^q)$ of the q-skeleton of the base (so that x will then not be representable by a chain from $p^{-1}(B^{q-1})$):

$$\bar{x} = x_q + x_{q-1} + \cdots + x_0 = x_q + \Delta,$$

where

$$x_q \in C_{q,j}, \qquad x_{q-1} \in C_{q-1,j+1}, \ldots, x_0 \in C_{0,n}.$$

From $\partial_E \bar{x} = 0$ and our assumption that $\partial_E = \partial_0 + \partial_1 + \partial_2$, we obtain

$$\partial_E \bar{x} = \partial_0 x_q + (\partial_1 x_q + \partial_0 x_{q-1}) + (\partial_2 x_q + \partial_1 x_{q-1} + \partial_0 x_{q-2}) + (\partial_2 x_{q-1} + \partial_1 x_{q-2} + \partial_0 x_{q-3}) + \cdots = 0,$$

where we have bracketed those terms in the same group $C_{k,l}$. It follows that

$$\partial_0 x_q = 0, \qquad \partial_1 x_q = -\partial_0 x_{q-1}, \qquad \partial_2 x_q = -\partial_1 x_{q-1} - \partial_0 x_{q-2}, \ldots,$$

from which we infer that the n-chain x_q is a d_r-cycle for $r = 0, 1, 2, \ldots$, since

$$d_0 x_q = \partial_0 x_q = 0, \qquad d_1 x_q = \partial_1 x_q = -\partial_0 x_{q-1},$$

$$d_2 x_q = \partial_2 x_q - \partial_1 \partial_0^{-1} \partial_1 x_q = \partial_2 x_q + \partial_1 x_{q-1} = -\partial_0 x_{q-2}, \ldots.$$

Thus an n-chain $x_q \in C_{q,j}(E)$, corresponding as above to an n-cycle x of filtration q, represents a d_r-cycle for all $r = 0, 1, 2, \ldots$, and so remains in (i.e. represents an element of) $E^{(\infty)}_{q,j}$. (In the case we are considering $E^{(\infty)} = E^{(3)}$.)

We next show that x_q is not a d_r-boundary for any of the operators d_r ($r = 0, 1, 2$), and represents therefore a non-zero element of $E^{(\infty)}_{q,j}$. If the n-cycle x_q were a d_0-boundary, say $x_q = d_0 z = \partial_0 z$ for some $z \in C_{q,j+1}(E)$, then the filtration of the element $x = x_q + x_{q+1} + \cdots$ (omitting the bar) would be strictly less than q, since although the element $x - \partial_E z$ would represent the same element of $H_n(E)$ as x, yet we should have (using $\partial_E = \partial_0 + \partial_1 + \partial_2 + \cdots$)

$$x - \partial_E z = (x_q - \partial_0 z) + (x_{q-1} - \partial_1 z) + \cdots,$$

where $x_q - \partial_0 z = 0$. Hence x_q cannot be a d_0-boundary since by assumption the cycle x cannot be "pulled off" $p^{-1}(B^q)$. A similar argument shows that if

x_q were a d_1-boundary, say $x_q = d_1 v$ where $\partial_0 v = 0$, $v \in C_{q+1,j}$, then the cycle $x - \partial_E v$ would have filtration $< q$. Hence $x_q \notin d_1(\text{Ker } d_0)$. Finally, if we had $x_q = d_2 w = \partial_2 w - \partial_1 \partial_0^{-1} \partial_1 w$, where $\partial_0 w = 0$, $\partial_1 w = \partial_0 u$, $w \in C_{q+2,j-1}$, then the cycle $x - \partial_E w + \partial_E u$, representing the same element of $H_n(E)$ as x, would have filtration $< q$ since

$$x - \partial_E w + \partial_E u = x - \partial_1 w - \partial_2 w + \partial_0 u + \partial_1 u + \partial_2 u$$
$$= x - d_2 w + \partial_2 u = (x_q - d_2 w) + (x_{q-1} - \partial_2 u) + \cdots,$$

where $x_q - d_2 w = 0$ (and we have used $\partial_E = \partial_0 + \partial_1 + \partial_2$).

The upshot is that the n-cycle x_q, being a cycle with respect to all of the differential operators d_0, d_1, d_2, represents an element of $E_{q,j}^{(\infty)}$, which is moreover non-zero; in other words, we have an injection (in fact, a monomorphism)

$$H_n(E) \to \sum_{q+j=n} E_{q,j}^{(\infty)}. \tag{6}$$

It remains to show that this map is onto. To this end, let $x_q \in C_{q,j}(E)$ be a cycle with respect to each of the operators d_0, d_1, d_2, representing a non-zero element of $E_{q,j}^{(\infty)}$. Then

$$\partial_0 x_q = 0, \qquad \partial_1 x_q = \partial_0 y,$$
$$\partial_2 x_q - \partial_1 y = \partial_0 z + \partial_1 w, \qquad \partial_0 w = 0,$$
$$z \in C_{q-2,j+2}, \quad y, w \in C_{q-1,j+1}.$$

Setting

$$\tilde{x} = x_q + x_{q-1} + x_{q-2} + x_{q-3} + \cdots,$$

where $x_q = -(y + w)$, $x_{q-2} = -z$, we obtain

$$\partial_E \tilde{x} = \partial_0 x_q + (\partial_1 x_q - \partial_0 y - \partial_0 w) + (\partial_2 x_q - \partial_0 z - \partial_1 y - \partial_1 w) + \Delta = \Delta,$$

where $\Delta \in p^{-1}(B^{q-3})$. It can now be shown (we omit the details) that by changing the representative x_q (of the element of $E_{q,j}^{(3)} = E_{q,j}^{(\infty)}$ under consideration) and also y, w, z, choosing x_s for $s \geq q - 3$ suitably, and taking into account the assumption that $\partial_i = 0$ for $i \geq 3$, one can arrange that $\partial_E \tilde{x} = 0$. Thus a cycle $\tilde{x}$ can be constructed of filtration q, with the element of $E_{q,j}^{(\infty)}$ represented by x_q as its image under the map (6). This completes our sketch of the proof of Leray's theorem in this special case. □

In connexion with Leray's theorem we note without proof the following supplementary facts:

(1) The elements of filtration $q = 0$ are automatically d_r-cycles for all $r \geq 1$. It follows that since for each n the group $E_{0,n}^{(1)}$ is isomorphic to $H_n(F)$, the group $E_{0,n}^{(\infty)}$ is (essentially) a quotient of $H_n(F)$; the natural homomorphism $H_n(F) \to E_{0,n}^{(\infty)} \subset H_n(E)$ coincides with the homomorphism induced by an inclusion map $i: F \to E$.

(2) For $n > 0$ elements of filtration n ($j = 0$) are never boundaries; hence,

in this case, we have $E^{(2)}_{n,0} \simeq H_n(B)$ and $E^{(\infty)}_{n,0} \subset H_n(B)$. The homomorphism induced by the projection of $H_n(E)$ on its summand $E^{(\infty)}_{n,0}$:

$$\sum_{q+j=n} E^{(\infty)}_{q,j} = H_n(E) \to E^{(\infty)}_{n,0} \subset H_n(B),$$

coincides with the homomorphism induced by the projection map p of the fibre bundle:

$$p_*: H_n(E) \to H_n(B).$$

(3) The cohomological analogue of Leray's theorem is valid: There is a sequence of pairs $(E_r^{q,j}, d_r^*)$, $r = 0, 1, 2, \ldots$, of groups and homomorphisms, with the following properties:

(i) $d_r^*: E_r^{q,j} \to E_r^{q+r,j-r+1}$, $d_r^* d_r^* = 0$, $E_{r+1}^* = H^*(E_r, d_r^*)$.
(ii) $\sum_{q+j=n} E_2^{q,j} = H^n(B \times F) \simeq \sum H^q(B) \otimes H^j(F)$.
(iii) $\sum_{q,j} E_\infty^{q,j} \simeq H^*(E)$ (as groups).
(iv) For every r the complex $E_r^* = \sum E_r^{q,j}$ with operator d_r^* is dual to the corresponding pair $(E_*^{(r)}, d_r)$ appearing in the homological version of Leray's theorem. In the context of cohomology there is, however, an important additional property arising from the multiplication in cohomology algebras:
(v) Endowed with the usual multiplication in cohomology rings (see §7), each E_r^* becomes a skew-commutative algebra with identity element (and in particular $H^*(B \times F)$ and E_2^* are isomorphic as algebras): if $\alpha \in E_r^{p,i}$, $\beta \in E_r^{q,j}$, then $\alpha\beta \in E_r^{p+q,i+j}$, $\alpha\beta = (-1)^{(p+i)(q+j)}\beta\alpha$, and for each d_r^* Leibniz' formula holds (cf. §1(14), §2(15), §7, Exercise 1):

$$d_r^*(\alpha\beta) = (d_r^*\alpha)\beta \pm \alpha(d_r^*\beta). \tag{7}$$

(Note, however, that in general the algebra E_∞^* is not isomorphic to $H^*(E)$; an exception is the case where E_∞^* is a free skew-commutative algebra, when $H^*(E)$ is also such an algebra.)

Although we shall not give proofs of the above statements, we shall make use of them (especially (v)) in our calculations below.

We now apply Leray's theorem to some particular fibre bundles. (As will be evident, only the formal properties of the operators d_r, d_r^* for $r \geq 2$ will be significant, the details of their definitions playing no part in the computations.)

Examples

(a) Consider the standard fibre bundle

$$E = S^{2n+1} \to \mathbb{C}P^n = B$$

with fibre $F = S^1$ (the "general Hopf bundle"; see Part II, §24.3). By applying Leray's theorem to this bundle and using our knowledge of $H^*(S^1)$ and $H^*(S^{2n+1})$ together with the fact that $\pi_1(B) = 1$, we shall compute the algebra $H^*(\mathbb{C}P^n)$. (Recall that $H^*(\mathbb{C}P^n; \mathbb{R})$ was computed in the first example of §7.)

Here, since $\dim F = 1$, the cells σ_E^{q+j} (in the same notation as before) are defined only for $j = 0, 1$. Hence $d_i^* = 0$ for $i \geq 3$ for reasons to do with dimension, since $d_r^* E_r^{q,j} \subset E_r^{q+r,j-r+1}$. Hence $E_\infty^* = E_3^*$. We investigate the action of d_2^* on E_2^*. We have from (ii) that $E_2^* = H^*(B) \otimes H^*(F)$, where $H^*(F) = H^*(S^1) = \Lambda[u]$ say, with deg $u = 1$, $u^2 = 0$, and where the algebra $H^*(B) = H^*(\mathbb{C}P^n)$ is "unknown" (except for the datum $\pi_1 = 1$). If we regard $u \in H^1(F)$ as an element of E_2^*, namely $u = 1 \otimes u \in E_2^{0,1}$, then $d_2^* u \in E_2^{2,0} = H^2(\mathbb{C}P^n)$. In fact, $E_2^{2,0}$ is one-dimensional (as a vector space over the ground field) with $d_2^* u$ as a basis vector. To see this observe first that if we had $d_2^* u = 0$, then $1 \otimes u \in E_2^{0,1}$ would be a d_2^*-cocycle, and since no element of $E_2^{0,1}$ can be a d_2^*-coboundary (by dimensional considerations again), we should then have $E_3^1 = E_\infty^1 = H^1(E) \neq 0$, which cannot in fact be the case since $E = S^{2n+1}$. Secondly, for the one-dimensionality of $E_2^{2,0}$, note that from

$$0 = H^2(S^{2n+1}) = H^2(E) = E_\infty^2 = E_3^2,$$

and the fact that (once again for dimensional reasons) all elements of $E_2^{2,0}$ are d_2^*-cocycles, it follows that these elements must also all be d_2^*-coboundaries, i.e. $E_2^{2,0} = d_2^*(E_2^{0,1})$; since $E_2^{0,1}$ is one-dimensional (generated by $1 \otimes u$) we infer that $E_2^{2,0}$ is at most one-dimensional, as claimed. Writing $d_2^*(u) = v (\neq 0)$ (and $u \otimes v^k = uv^k$, etc.) we obtain from Leibniz's formula (7)

$$d_2^*(uv) = v^2, \qquad d_2^*(uv^k) = v^{k+1}.$$

The generalization of the preceding argument, starting from the known facts that $H^i(E) = 0$ for $1 \leq i \leq 2n$, and that all elements of $E_2^{i,0}$ are d_2^*-cocycles, yields

$$H^i(\mathbb{C}P^n) = E_2^{i,0} = d_2^*(E_2^{i-2,1}) = d_2^*(H^{i-2}(\mathbb{C}P^n) \otimes H^1(F)). \tag{8}$$

By invoking in turn the fact that $H^1(\mathbb{C}P^n) = 0$ (inferred from $\pi_1 = 1$) and the one-dimensionality of $H^2(\mathbb{C}P^n)$ (established above) it follows from (8) that $H^{2j+1}(\mathbb{C}P^n) = 0$ for all j, and that $H^{2j}(\mathbb{C}P^n)$ is one-dimensional, spanned by v^j, for $0 \leq j \leq n$. (The action of d_2^* on E_2^* is shown schematically in Figure 43, when $n = 3$.)

(b) As our second example we consider the Serre fibration $p: E \to S^n$, where $E = E(x_0)$ is the space of all paths on S^n beginning at a distinguished point x_0 (and ending anywhere on S^n), and the projection p sends such a path to its terminal point $y \in S^n$ (see Part II, §22.1). Here the fibres are all homotopically

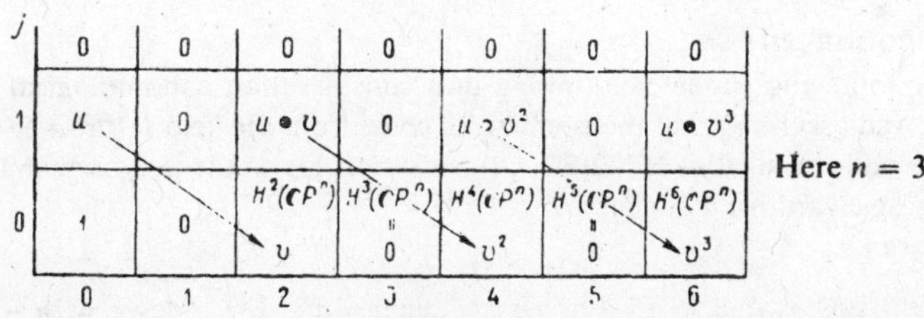

Figure 43

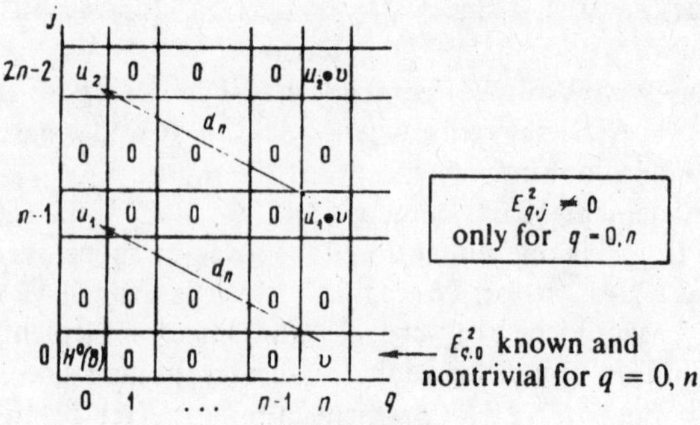

Figure 44

equivalent to the loop space $F = \Omega(S^n, x_0)$, the fibre above x_0 (see §7, Example (γ)). (In fact, the fibres of any Serre fibration (see Part II, Definition 22.1.1) are homotopically equivalent, and Leray's theorem remains valid in this more general context.) Our aim is to compute the homology group of $F = \Omega(S^n, x_0)$ using the fact that $H_*(E) = 0$ (in view of the contractibility of E; see Part II, §22.2), and our knowledge of the homology groups of the base $B = S^n$: $H_0(S^n)$ and $H_n(S^n)$ are one-dimensional, while $H_k(S^n) = 0$ for $k \neq 0$, n. Now since $E^{(2)}_{q,j} = H_q(B) \otimes H_j(F)$ it follows that $E^{(2)}_{q,j} = 0$ for $q \neq 0$, n. Hence $d_r = 0$ for $r \neq n$ and $E^{(n)}_{q,j} = E^{(2)}_{q,j}$. By using Leray's theorem (together with $H_*(E) = 0$ and our knowledge of the homology groups of the base S^n) the action of d_n on $E^{(2)}$ can without difficulty be shown to be given by (see also Figure 44):

$$d_n \colon v \to u_1,$$

$$d_n \colon v \otimes u_1 \to u_2,$$

$$\cdots\cdots\cdots\cdots\cdots$$

$$d_n \colon v \otimes u_{k-1} \to u_k,$$

whence the structure of the homology groups of $F = \Omega(S^n, x_0)$ can be inferred:

$$H_{k(n-1)}(F) \quad \text{is one-dimensional;}$$

$$H_j(F) = 0 \quad \text{for} \quad j \neq k(n-1).$$

Exercises (continued)

2. Prove the following statements (taking into consideration cohomological multiplication, and assuming that the coefficients come from the field $\mathbb{R}$ (or $\mathbb{Q}$ or $\mathbb{C}$):
   (i) If n is odd the algebra $H^*(\Omega(S^n, x_0))$ is isomorphic to the polynomial algebra in a single variable u of degree $n - 1$.
   (ii) For even n,

$$H^*(\Omega(S^n, x_0)) \simeq \mathbb{R}[v] \otimes \Lambda[u], \qquad \deg u = n - 1, \qquad \deg v = 2n - 2.$$

3. For each of the following properties, show that if any two of the spaces E, F, B (of

a fibre bundle $p: E \to B$ satisfying the hypotheses of Leray's theorem) enjoys that property, then so does the remaining one:

(i) the homology groups H_k (over a field) are all zero;

(ii) every homology group H_k has finite dimension (as a vector space);

(iii) the integral homology groups of dimension > 0 are all periodic (or, equivalently, the homology groups of dimension > 0 over any field of characteristic zero (e.g. $\mathbb{Q}$, $\mathbb{R}$, $\mathbb{C}$) are all zero);

(iv) the integral homology groups of dimension > 0 are all periodic without elements of order p, where p is a prime number (or equivalently, the homology groups of dimension > 0 with coefficients from the field $\mathbb{Z}_p$ are all zero).

4. Investigate the boundary operators d_r and d_r^* for the following fibre bundles (exploiting to this end the product operation in the cohomology algebras in Examples (ii) to (vii)):
   (i) $\mathbb{R}P^{2n+1} \to \mathbb{C}P^n$ (fibre S^1), with coefficient field $\mathbb{Z}_p$, $\mathbb{R}$ or $\mathbb{Q}$;
   (ii) $SU(n) \to S^{2n-1}$ (fibre $SU(n-1)$);
   (iii) $SO(n) \to S^{n-1}$ (fibre $SO(n-1)$);
   (iv) $S^{4n+3} \to \mathbb{H}P^n$ (fibre S^3);
   (v) $V_{n,k} \to S^{n-1}$ (fibre $V_{n-1,k-1}$); ($V_{n,k}$ is a "Stiefel manifold"; see Part II, §5.2);
   (vi) $V_{n,k}^{\mathbb{C}} \to G_{n,k}^{\mathbb{C}}$ (fibre $U(k)$); ($G_{n,k}$ is a "Grassmannian manifold"; see Part II, §5.2);
   (vii) $V_{n,k}^{\mathbb{R}} \to \hat{G}_{n,k}^{\mathbb{R}}$ (fibre $SO(k)$).

5. Prove the following assertions:
   (i) If the integral homology groups of positive dimension of a complex K are all finite (so that $H_q(K; \mathbb{R}) = 0$ for $q > 0$) then all of the homotopy groups $\pi_i(K)$, $i > 1$, of K are also finite.
   (ii) The homotopy groups $\pi_{n+i}(S^n)$ are all finite except for $i = 0$ or $n - 1$.

(*Hint*. Consider the Serre fibration $p: E \to K$, where the points of E are the paths in K beginning at a distinguished point $k_0 \in K$ (cf. Example (b) above). As before, the fibres are all homotopically equivalent to the loop space $\Omega(K, k_0)$, and E is contractible (see Part II, §22.2). Iterate this construction by taking the analogous fibration with E as base and fibre $\Omega(E, e_0)$, and so on, and consider the associated sequence of isomorphisms

$$\pi_i(K) \simeq \pi_{i-1}(\Omega(K)) \simeq \dots,$$

which continues as long as the indices exceed 1 (see Part II, Corollary 22.2.3). In conjunction with this sequence of isomorphisms use Hurewicz' theorem (Corollary 4.9): if $\pi_q(X) = 0$, $1 \le q < i$, then $\pi_i(X) \simeq H_i(X; \mathbb{Z})$. If a non-trivial π_1 is encountered, then pass to the universal covering space (which has the same higher homotopy groups), and invoke Exercise 7 below.)

There is a useful procedure for "converting a map to a fibre-space projection", defined as follows. Suppose first that the map $f: K \to L$ in question is an inclusion $K \subset L$. In this case, we take as the total space (of the intended fibre space) the space $E(K, L)$ whose points are the paths in L which begin in K (and terminate anywhere in L); it is easy to verify that $E(K, L)$ contracts to K, so that $E(K, L) \sim K$ (where $\sim$ denotes homotopy equivalence). The Serre fibration that we associate with the given inclusion $K \subset L$ is then given by the projection $p: E(K, L) \to L$ sending each path to its terminal point. (We leave to the reader the verification that this is indeed a Serre fibration; cf. Part II, Lemma 22.1.3.) In the general situation of an arbitrary map $f: K \to L$, one considers the "mapping cylinder" $C_f = (K \times [0, 1]) \cup_f L$, where each point $(x, 1)$ is identified with $f(x) \in L$. It is easy to see that $C_f \sim L$. Identifying K with $K \times 0 \subset C_f$, and applying to this inclusion the construction just described, we obtain a fibre space

$$K \sim E(K, C_f) \xrightarrow{p} C_f \sim L, \tag{9}$$

said to be *induced* from the given map f. It is not difficult to show that the map $K \to L$ determined by (9) is homotopic to the original map f.

The following exercises are intended to be solved by means of this construction.

EXERCISES (continued)

6. Prove that if a map $f: K \to L$ between simply-connected complexes induces isomorphisms $H_k(K; \mathbb{R}) \simeq H_k(L; \mathbb{R})$ between the corresponding homology groups, then it induces isomorphisms between the corresponding homotopy groups π_i, $i > 1$, tensored with $\mathbb{R}$:
$$\pi_i(K) \otimes \mathbb{R} \simeq \pi_i(L) \otimes \mathbb{R}.$$
Apply this conclusion in the particular situation where $K = S^3 \times S^5 \times \cdots \times S^{2n-1}$, $L = \mathrm{SU}(n)$, and the map $f: K \to L$ is defined in terms of the multiplication in $\mathrm{SU}(n)$.

7. Convert the natural inclusion $S^n \vee S^n \to S^n \times S^n$ into a fibre-space projection, as in the above-described construction. Find the homology groups of the fibre. Identify the (scalar-extended) homotopy groups $\pi_i(S^n \vee S^n) \otimes \mathbb{R}$.

8. Let X be a simply-connected space with free (skew-commutative) cohomology algebra $H^*(X; \mathbb{R})$. Identify the groups $\pi_i(X) \otimes \mathbb{R}$.

9. Show that if $\pi_i(X) = 0$ for $i \leq n - 1$, then $H_j(X; \mathbb{R}) \simeq \pi_j(X) \otimes \mathbb{R}$ for $j < 2n - 1$.

§9. The Extension Problem for Maps, Homotopies, and Cross-Sections. Obstruction Cohomology Classes

9.1. The Extension Problem for Maps

Consider the following problem: Let K be a cell complex and $L \subset K$ a subcomplex (for instance, $L = K^{i-1}$, the $(i - 1)$-skeleton of K), and suppose we have a map $f: L \to X$. (In order to keep the algebraic side of the exposition simple we shall assume that X is simply-connected, or at least that it is homotopically simple in the sense that $\pi_1(X)$ is abelian and acts trivially on all of the groups $\pi_i(X)$, $i > 1$ (see Part II, §22.3).) Can the map $f: L \to X$ be extended to a map $F: K \to X$?

Let σ^i be a cell of the complex K such that $\partial\sigma^i \subset L$. The restriction to $\partial\sigma^i$

of the given map $f: L \to X$, defines an element $\alpha(\sigma^i, f) \in \pi_{i-1}(X)$:

$$S^{i-1} \to \partial\sigma^i \xrightarrow{f} X.$$

It is clear that the map f can be extended to the interior of the cell σ^i if and only if $\alpha(\sigma^i, f) = 0$ in the group $\pi_{i-1}(X)$. (Hence, in particular, if $\pi_{i-1}(X) = 0$, then f can always be extended from $\partial\sigma^i \subset L$ to the whole of the cell σ^i.) If $\alpha(\sigma^i, f) \neq 0$ in $\pi_{i-1}(X)$, then, f not being extendible from $\partial\sigma^i$ to σ^i, we call the element α an "obstruction".

In attempting to extend the given map it is natural to proceed by extending it one dimension at a time, beginning with the lowest for which there are cells of that dimension not contained in L. At some stage of this procedure, when extending f to the i-skeleton say, we may encounter a non-trivial "obstruction" $\alpha(\sigma^i, f)$. It is more appropriate to regard such an "obstruction" as being determined by the cochain whose value on each i-cell is given by

$$\sigma^i \to \alpha(\sigma^i, f) \in \pi_{i-1}(X),$$

considered as an element of the group $C^i(K, L; \pi_{i-1}(X))$ of relative i-cochains with coefficients from the group $\pi_{i-1}(X)$. We denote this cochain by α_f.

9.1. Lemma. *An obstruction cochain α_f is a cocycle.*

PROOF (In the Simplicial Case Only). By definition of the coboundary operator δ, we have $\delta\alpha_f(\sigma^{i+1}) = \alpha_f(\partial\sigma^{i+1})$. Thus we need to show that α_f vanishes on $\partial\sigma^{i+1}$. Since we are assuming that K is a simplicial complex and L a subcomplex, the boundaries $\partial\sigma^{i+1}$ and $\partial\sigma^i$ are topological spheres in K, where now σ^q denotes a q-dimensional simplex. Now by definition of the cochain α_f, its value on $\partial\sigma^{i+1}$ is determined by the restriction of the given map f to the $(i-1)$-skeleton of σ^{i+1}. (Recall that for each i-simplex σ^i, $\alpha_f(\sigma^i) \in \pi_{i-1}(X)$ is determined by the map $f|_{\partial\sigma^i}: \partial\sigma^i \to X$.) Denote by α_j the element of $\pi_{i-1}(X)$ represented by the map f further restricted to the boundary of the jth i-dimensional face of σ^{i+1}. (In the notation of §3, if $\sigma^{i+1} = [0, \ldots, i+1]$, then the jth i-dimensional face of σ^{i+1} is the simplex $[0, 1, \ldots, \hat{j}, \ldots, i+1]$, where as usual the hatted symbol is understood as omitted.) Since each $(i-1)$-simplex in the $(i-1)$-skeleton of σ^{i+1} is a face of exactly two i-dimensional faces of σ^{i+1}, it follows that

$$\sum_{j=0}^{i+1} (-1)^j \alpha_j = 0 \in \pi_{i-1}(X), \tag{1}$$

whence the desired conclusion. (Note that in this argument we have made implicit use of our assumption that the action of $\pi_1(X)$ on the $\pi_i(X)$, $i \geq 0$, is trivial, in that then the choice of base point is immaterial.) □

In view of this result we call α_f an *obstruction cocycle*.

EXERCISE

1. Verify equation (1) in detail.

9.2. Lemma. *Let α_f be, as above, an obstruction cocycle. If α_f is a (relative) coboundary, i.e. if $\alpha_f = \delta\beta$ for some $\beta \in C^{i-1}(K, L; \pi_{i-1}(X))$, then the definition of (the partial extension of) f can be altered on the $(i-1)$-skeleton K^{i-1} of K, without changing its definition on the $(i-2)$-skeleton K^{i-2} or on L, in such a way that for the resulting map $\tilde{f}$ we have $\alpha_{\tilde{f}} = 0$.*

Proof. On each i-cell we have $\alpha_f(\sigma^i) = \delta\beta(\sigma^i) = \beta(\partial\sigma^i)$. We change f on each $(i-1)$-cell σ^{i-1} not contained in L, to a new map $\tilde{f}$ such that f and $\tilde{f}$ coincide on each of the boundaries $\partial\sigma^{i-1}$ and in combination yield a map $S^{i-1} \to X$ representing the element $\beta(\sigma^{i-1})$ of the group $\pi_{i-1}(X)$. In the group $C^i(K, L; \pi_{i-1}(X))$ of relative cochains, we shall then have $\alpha_{\tilde{f}} = \alpha_f - \delta\beta = 0$, whence the lemma. □

From these two lemmas we draw the following conclusion: *The "first obstruction" encountered in attempting to extend a given map $f: L \to X, L \subset K$, from the subcomplex $L \cup K^{i-1}$ to $L \cup K^i$, is determined by an element α_f of the relative cohomology group $H^i(K, L; \pi_{i-1}X)$. The vanishing of α_f in this group suffices (in view of Lemma 9.2) for the map to be extendible. Thus, in particular, the extension in question can be made if $\pi_{i-1}(X) = 0$.*

Exercise

2. Let $f: K^q \to X$ be a map from the q-skeleton of K to a $(q-1)$-connected cell complex X without cells of dimension $p, 0 < p \leq q-1$ (i.e. a "reduced" complex; cf. Theorem 4.8). Then each q-cell σ^q of the complex K determines an element $\beta(\sigma^q) \in \pi_q(X)$ via the restriction map $f|_{\sigma^q}: \sigma^q \to X$. Show that the obstruction to the extension of the map f to the $(q+1)$-skeleton K^{q+1} is the cochain $\alpha_f = \delta\beta$. (We infer that if β is a cocycle then f can be extended to K^{q+1}.)

9.2. The Extension Problem for Homotopies

We now turn to "homotopy obstructions". Here the ingredients are two maps $f, g: K \to X$, which coincide on the $(q-1)$-skeleton K^{q-1} of K. On each cell $\sigma^q \subset K^q$ the two maps f and g give rise, via their restrictions, to two maps $\sigma^q \to X$ coinciding on the boundary: $f|_{\partial\sigma^q} = g|_{\partial\sigma^q}$, and therefore yielding in combination a map $S^q \to X$ determining what is called a "distinguishing element" of $\pi_q(X)$. A distinguishing element being defined in this way for each q-cell σ^q of K, we arrive at a *difference cochain* α:

$$\alpha(\sigma^q, f, g) \in \pi_q(X).$$

Exercises (continued)

3. Show that a difference cochain α is a cocycle: $\delta\alpha = 0$. Show also that if the difference cocycle α is a coboundary, i.e. $\alpha = \delta\beta$, then f and g may be altered on K^{q-1} without changing their definition on K^{q-2}, so that they remain homotopic as maps of K^{q-1}, but now $\alpha = 0$. We conclude from this that the difference cochain may be regarded as belonging to the cohomology group $H^q(K; \pi_q(X))$.

4. Suppose we are given a pair (K, L) with $L \subset K$, and a map $f: L \to T^n$, the n-dimensional torus. An obvious necessary condition for extendibility of the map f from L to K is the following one: if $\gamma \in \pi_1(L)$ is such that under the inclusion homomorphism i_* (induced by the inclusion map $i: L \to K$) γ is mapped to the identity element: $i_*(\gamma) = 1$, then we should have $f_*(\gamma) = 1$ in the torus T^n. Show that this condition is also sufficient. Prove that the analogous homological condition, i.e. with $\gamma \in H_1(L)$, also suffices for extendibility. (A situation of this sort arises, for instance, in knot theory; see Part II, §26.)

5. Find the set $[K, T^n]$ of homotopy classes of maps $K \to T^n$ (in particular in the case $n = 1$, of the circle S^1). More generally, show that if $X = K(D, n)$ is a space such that $\pi_i(X) = 0$ for $i \neq n$, and $\pi_n(X) \simeq D$, a prescribed abelian group (an "Eilenberg–MacLane space"; see §10 below), then there is a natural one-to-one correspondence between $[K, X]$ and $H^n(K; D)$. In the case $n = 1$, verify that the elements of $H^1(K; D)$ and $[K, X]$ are determined by the homomorphisms $\pi_1(K) \to D$. (This in fact remains true for $[K, X]$ even if D is non-abelian.)

   Some examples of Eilenberg–MacLane spaces, all pertaining to the case $n = 1$, are as follows:

   $D = \mathbb{Z}, \qquad K(\mathbb{Z}, 1) \sim S^1;$

   $D = \mathbb{Z} \times \cdots \times \mathbb{Z}, \qquad K(D, 1) \sim T^n;$

   $D = \pi_1(M_g^2), \qquad K(D, 1) \sim M_g^2$ (the surface of genus $g \geq 1$);

   $D = \mathbb{Z}_m, \qquad K(D, 1) \sim S^\infty/\mathbb{Z}_m$ (in particular, $K(\mathbb{Z}_2, 1) = \mathbb{R}P^\infty = \lim\limits_{N \to \infty} \mathbb{R}P^N$);

   $D = F$ (a free group), $\qquad K(F, 1) \sim S^1 \vee \cdots \vee S^1$ (a bouquet of circles).

   (Here $\lim_{N \to \infty} \mathbb{R}P^n$ denotes the "direct limit" or union of the chain

   $$\mathbb{R}P^1 \subset \mathbb{R}P^2 \subset \cdots.)$$

   Examples of Eilenberg–MacLane spaces $X = K(D, 1)$ for various $D \simeq \pi_1(X)$, abound. However, there is only one easily constructed example of a space $K(D, n)$ with $n \geq 2$, namely (see Part II, §24.4)

   $$K(D, 2) \sim \mathbb{C}P^\infty = S^\infty/S^1, \qquad D = \mathbb{Z}.)$$

6. Let K^n be an n-dimensional complex. Find the homotopy classes of maps $K^n \to S^n$. Prove that there is a natural one-to-one correspondence between $[K^n, S^n]$ and $H^n(K^n; \mathbb{Z})$.

9.3. The Extension Problem for Cross-Sections

The problem of extending cross-sections (and homotopies between them) of a fibre bundle with projection $p: E \to B$, fibre F, and base B given as a simplicial (or cell) complex, is similar to those considered above. Once again we shall assume for the sake of simplicity that the base B is simply-connected (or at least that $\pi_1(B)$ acts trivially on the groups $\pi_i(F)$), and we shall also suppose that the fibre F is simply-connected (or at least homotopically simple).

Thus suppose $p: B^{q-1} \to E$ is a cross-section of the fibre bundle above the $(q-1)$-skeleton $B^{q-1} \subset B$; then by definition of cross-section we have $p\varphi = 1$. Let σ^q be any q-simplex of B. Above the (closed) simplex σ^q the fibre bundle is canonically identifiable with the direct product: $p^{-1}(\sigma^q) \cong \sigma^q \times F$ (see Part II, Lemma 24.4.2). On the boundary $\partial\sigma^q$ ($\cong S^{q-1}$) the cross-section $\varphi: \partial\sigma^q \to \partial\sigma^q \times F$ ($p\varphi = 1$) is, by assumption, already given. Hence via the projection map onto F we obtain a map $S^{q-1} \to F$, defining an element $\alpha(\sigma^q, \varphi) \in \pi_{q-1}(F)$ for each q-simplex $\sigma^q \subset B^q$. In this way we obtain an *obstruction cocycle* α to the attempted extension of the cross-section φ to the q-skeleton B^q.

EXERCISES (continued)

7. Prove that $\delta\alpha = 0$, i.e. that the cochain α defined above is indeed a cocycle.

8. Prove that if $\alpha = \delta\beta$ then the cross-section can be altered appropriately above B^{q-1} without changing it above B^{q-2}, in such a way that $\alpha = 0$ for the new cross-section. Consequently, an obstruction to extending a cross-section may be regarded as an element of $H^q(B; \pi_{q-1}(F))$.

9. Suppose that the fibre is the $(q-1)$-sphere S^{q-1}. Show that then the obstruction $\alpha \in H^q(B; \pi_{q-1}(F))$ is an "Euler characteristic class" of the fibre bundle (defined in Part II, §25.4, for odd $q-1$, in terms of a connexion on the bundle, with bundle group $SO(q)$, essentially as an element of $H^q(B; \mathbb{R})$).

10. Let $\varphi_1, \varphi_2: B \to E$ be two cross-sections ($p\varphi_1 = p\varphi_2 = 1$) agreeing on the $(q-1)$-skeleton $B^{q-1} \subset B$. Formulate the appropriate definition of an "obstruction to a homotopy" between the cross-sections φ_1 and φ_2:

$$\alpha(\varphi_1, \varphi_2) \in H^q(B; \pi_q(F)),$$

and investigate its properties.

Examples

(a) If the fibre is contractible, so that $\pi_i(F) = 0$ for all i, then it follows from the above exercises that cross-section always exist, and moreover that all cross-sections are homotopic. This is the situation, for instance, for the following two types of fibre bundles:

(i) For the fibre bundle over a manifold M^n whose fibre F above each point is the space of positive definite quadratic forms, it is known that cross-sections, or in other words Riemannian metrics, always exist (see Part II, §8.3), and that any two cross-sections are homotopic (i.e. any two Riemannian metrics are continuously deformable one into the other). On the other hand, these results are no longer valid for indefinite metrics (of type (p, q) say, $p + q = n$). What are the groups $\pi_i(F)$ in this case? (Note that here $F = GL(n, \mathbb{R})/O(p, q)$.)

(ii) Given a fibre bundle $E \to B$ with fibre F, we can form a new bundle over B having as fibre above each point x of B the space of all "horizontal directions" above x, i.e. subspaces of the tangent planes to E at points over

x of maximal dimension, transverse to the fibre F_x. A cross-section of this fibre bundle is then a connexion on the original bundle, shown in Part II, §24.2 to always exist.

(b) Given a vector bundle with projection $p: E \to B$, fibre $\mathbb{R}^n$, and bundle group $G = O(n)$, we can form the associated bundle $p: E_k \to B$ of orthonormal k-frames, with fibre $F_k = V_{n,k}$, the Stiefel manifold of orthonormal k-frames in $\mathbb{R}^n$ (see Part II, §5.2). (In particular, when $k = n$, we have $F_n \cong O(n)$, and when $k = 1$, $F_1 \cong S^{n-1}$.) It was (partially) shown in Part II, §24.3, that the homotopy groups of the fibre are here as follows:

$$\pi_i(V_{n,k}) = 0, \qquad i < n - k;$$

$$\pi_{n-k}(V_{n,k}) \simeq \begin{cases} \mathbb{Z}, & n - k \text{ odd, or } k = 1, \\ \mathbb{Z}_2, & n - k \text{ even.} \end{cases}$$

Hence, for each $k = 1, \ldots, n$, the obstruction cohomology class or "first obstruction" to the existence of a cross-section of the fibre bundle $E_k \to B$ will be an element

$$\alpha_k \in H^{n-k+1}(B; \pi_{n-k}(V_{n,k})).$$

9.3. Definition. The cohomology class α_k considered modulo 2 is called the kth *Stiefel–Whitney class* of the vector bundle $E \to B$, and is denoted by

$$W_q = \alpha_{n-q+1} (\text{mod } 2) \in H^q(B; \mathbb{Z}_2), \qquad q = 1, \ldots, n.$$

The polynomial

$$W(t) = W_0 + W_1(t) + \cdots + W_q t^q + \cdots + W_n t^n,$$

where we define $W_0 = 1$, is called the *Stiefel–Whitney polynomial* of the vector bundle. By the *Stiefel–Whitney classes of an n-manifold* $B = M^n$, one simply means the Stiefel–Whitney classes of the tangent bundle over M^n.

EXERCISES (continued)

11. Prove that the equation $W_1 = 0$ holds precisely if the manifold M^n is orientable. Show that W_n is the mod 2 Euler characteristic of M^n.

12. Show that for the direct product $M = M_1 \times M_2$ of manifolds (or more generally of vector bundles: $(E_1, p_1, B_1) \times (E_2, p_2, B_2) = (E_1 \times E_2, p_1 \times p_2, B_1 \times B_2)$) one has
$$W(t) = \bar{W}(t)\bar{\bar{W}}(t).$$
where $\bar{W}$ and $\bar{\bar{W}}$ are the Stiefel–Whitney polynomials of the factors M_1 and M_2, and where the coefficients of the polynomials $\bar{W}$ and $\bar{\bar{W}}$, coming respectively from $H^*(M_1; \mathbb{Z}_2)$ and $H^*(M_2; \mathbb{Z}_2)$, are multiplied using the isomorphism (4) of §7.

13. Prove that for the standard one-dimensional fibre bundle η over $\mathbb{R}P^n$ (the "generalized Möbius band"; see Part II, §24.5) we have
$$W(t) = 1 + W_1 t, \qquad W_1 \in H^1(\mathbb{R}P^n; \mathbb{Z}_2) \simeq \mathbb{Z}_2, \qquad W_1 \neq 0.$$

Compute the Stiefel–Whitney polynomial of the tangent bundle τ over $\mathbb{R}P^n$, using the following fact: $\tau \oplus 1 \cong \eta \oplus \cdots \oplus \eta$ (cf. Part II, §24.5, Exercise 1).

14. Given k vector fields $\eta_1, \ldots, \eta_k$ in general position (see Part II, §15.2) on a manifold M^n, we define the "cycle of singularities" to consist of the points of M^n where the fields take on linearly dependent values. Show that this set is indeed a cycle representing an element of $H_{k-1}(M^n; \mathbb{Z}_2)$, which is furthermore the Poincaré dual (see §6(17)) of the Stiefel–Whitney class W_{n-k+1}.

(c) As our third example, consider a complex vector bundle with projection $p: E \to B$, fibre $\mathbb{C}^n$, and group $G = U(n)$, and its associated bundle $p: E_k \to B$ of unitary complex k-frames, with fibre $F_k = V_{n,k}^{\mathbb{C}}$. Using the methods of Part II, §24.3, it can be shown that the homotopy groups of the fibre are as follows:

$$\pi_i(V_{n,k}^{\mathbb{C}}) = 0, \qquad i \leq 2(n-k);$$

$$\pi_{2(n-k)+1}(V_{n,k}^{\mathbb{C}}) \simeq \mathbb{Z},$$

so that the first obstruction to the existence of a cross-section of the bundle $E_k \to B$, is an element of the appropriate integral cohomology group:

$$c_{n-k+1} \in H^{2n-2k+2}(B; \pi_{2n-2k+1}(V_{n,k}^{\mathbb{C}})) \qquad (= H^{2n-2k+2}(B; \mathbb{Z})).$$

9.4. Definition. The obstruction cohomology classes c_q, $q = 1, \ldots, n$, are called the *Chern classes* of the given complex vector bundle $E \to B$. (The Chern classes of the tangent bundle over a complex manifold $B = M^{2n}$ are referred to as the *Chern classes of the manifold* M^{2n}.) The polynomial

$$c(t) = 1 + c_1 t + \cdots + c_q t^q + \cdots + c_n t^n$$

is called the *Chern polynomial* of the vector bundle $E \to B$.

EXERCISES (continued)

15. (Cf. Exercise 12.) Show that for a product of complex vector bundles (or complex manifolds) the formula

$$c(t) = \bar{c}(t)\bar{\bar{c}}(t)$$

is valid, where c is the Chern polynomial of the product, and $\bar{c}$ and $\bar{\bar{c}}$ are the Chern polynomials of the factors.

16. (Cf. Exercise 13.) Show that for the standard $U(1)$-bundle η over $\mathbb{C}P^n$ we have

$$c(t) = 1 + c_1 t,$$

where c_1 is a generator of $H^2(\mathbb{C}P^n; \mathbb{Z})$. Find the Chern polynomial of the tangent bundle τ over $\mathbb{C}P^n$, using the fact that $\tau \oplus 1 \cong \eta \oplus \cdots \oplus \eta$. (Verify this fact; cf. Part II, §24.5, Exercise 1.)

17. Show that for a complex manifold M^{2n} the Chern class c_n coincides with the Euler characteristic class. Investigate the Chern polynomials of Riemann surfaces.

18. Show that the structure group of a $U(n)$-bundle may be reduced to $SU(n)$ if and only if $c_1 = 0$.

19. Let $\bar{\xi}$ denote the complex conjugate of a complex vector bundle ξ over a complex manifold B (see Part II, §24.5). Prove that

$$c(t, \bar{\xi}) = c(-t, \xi).$$

that is,

$$c_{2i}(\xi) = c_{2i}(\bar{\xi}), \qquad c_{2i+1}(\xi) = -c_{2i+1}(\bar{\xi}).$$

20. Show that the Chern classes c_q, considered as elements of $H^{2q}(B; \mathbb{R})$, coincide with the characteristic classes defined in Part II, §25.4, in terms of connexions on the vector bundle. (This is particularly clear for the class c_1. Note also that this explains the fact observed in Part II, §25.5, that, given a connexion on a vector bundle, the characteristic classes defined via the curvature tensor always have, once normalized, integer-valued integrals over cycles.)
21. Prove that the Chern polynomial, considered modulo 2, of a complex vector bundle ξ, coincides with the Stiefel–Whitney polynomial of that bundle, regarded as a real vector bundle, i.e. of the bundle $r\xi$ where r is the "realization" operator (see Part II, §24.5).

(d) As our final example we consider a real vector bundle η with structure group $O(n)$. As indicated in Part II, §24.5, we can form the "complexification" of the vector bundle η, obtaining a complex vector bundle $c\eta = \xi$ with structure group $G = U(n)$, which is "self-conjugate" in the sense that the bundles ξ and $\bar{\xi}$ are isomorphic. (Verify this.) In view of this isomorphism and Exercise 19 above, the odd-indexed Chern classes satisfy

$$c_{2i+1}(\xi) = c_{2i+1}(\bar{\xi}) = -c_{2i+1}(\xi),$$

where $2c_{2i+1} = 0$. For this reason these classes are of less interest that the even-indexed ones in the present context of the complexification of a real vector bundle.

9.5. Definition. The Chern classes $(-1)^i c_{2i}$ of the complex vector bundle $\xi = c\eta$, are called (*Pontrjagin*) *characteristic classes* of the real vector bundle η, and are denoted by $p_i(\eta) \in H^{4i}(B; \mathbb{Z})$.

EXERCISES (continued)

22. Compute the classes $p_i(\mathbb{C}P^n)$.
23. Find the class $p_1(M^4_{(n)})$ of a (non-singular) manifold M^4 defined in the finite part $\mathbb{C}^3$ of $\mathbb{C}P^3$ by means of a single equation of degree n.
24. Show that the classes p_i coincide with the appropriate characteristic classes defined in Part II, §25.4, in terms of connexions on the vector bundle.
25. Prove that if an orientable manifold M^4 is the boundary of an orientable manifold: $M^4 = \partial W^5$, then $p_1(M^4) = 0$. (More generally, if $M^n = \partial W^{n+1}$, then every polynomial expression of dimension n in the classes W_q and p_s vanishes, and likewise in the non-orientable case for every polynomial of dimension n in the W_q alone.) Express the classes p_s considered modulo 2, in terms of the W_q.

§10. Homology Theory and Methods for Computing Homotopy Groups. The Cartan–Serre Theorem. Cohomology Operations. Vector Bundles

10.1. The Concept of a Cohomology Operation. Examples

The problem of computing the homotopy groups of manifolds and finite cell complexes is, generally speaking, an exceedingly difficult one. For certain non-simply-connected complexes with nontrivial action of π_1 on the π_i, this problem is in fact algorithmically insoluble in the strongest sense of mathematical logic. Even for the simplest (and most important) kinds of simply-connected complexes, for instance spheres, the determination of the structure of the homotopy groups has shown itself to be a very resistant problem, still unsolved. Isolated facts about the homotopy groups in some particular cases can be established by means of direct geometric methods (see Part II). However, there are uniform methods available for computing homotopy groups, based on the homology theory of fibre spaces in conjunction with the homotopy theory expounded in Part II. In the present section we shall first sketch a technique (foreshadowed in some of the exercises of §8 above) for obtaining information about the "rational homotopy groups" $\mathbb{Q} \otimes \pi_i(K)$ of certain simply-connected complexes K, and so about the torsion-free portions of the homotopy groups $\pi_i(K)$. The computation of the finite portions (i.e. the torsion subgroups) of the $\pi_i(K)$ requires techniques of incomparably greater complexity; we shall however indicate these techniques at least insofar as they suffice for computing the first few stable homotopy groups of spheres.

All algebraic methods of computation of the homotopy groups rely (over and above the homology theory expounded in earlier sections) on the concept of a *cohomology operation* $\theta: H^r(K, L; G_1) \to H^s(K, L; G_2)$, which has the following defining properties:

(i) the map θ should be defined for each pair K, L of complexes, with L a subcomplex of K;

(ii) the map θ should be "natural" (or in other terminology "functorial") in the sense that it commutes with (the homomorphisms induced by) continuous maps $f: (K, L) \to (K', L')$:

$$\theta f^* = f^* \theta.$$

Examples

(a) Consider the map

$$\theta: H^r(K, L; G_1) \to H^{rm}(K, L; G_1)$$

defined by $\theta(x) = x^m$; here $s = rm$, $G_1 = G_2$. When $G_1 = G_2 = \mathbb{Z}_p$ and $m = p$, where p is a prime number, then $\theta(x + y) = (x + y)^p = x^p + y^p$, and $\theta(\lambda x) = \lambda\theta(x)$ since $\lambda^p = \lambda$ for $\lambda \in \mathbb{Z}_p$, so that θ is in this case a linear transformation. However, in the case $G_1 = G_2 = \mathbb{Q}$, the rational field, the map θ will not in general be a homomorphism.

(b) Another example of a cohomology operation is provided by the (cohomological) "Bockstein homomorphism"

$$\theta = \delta_1 : H^q(K, L; \mathbb{Z}_p) \to H^{q+1}(K, L; \mathbb{Z}_p), \qquad p \text{ prime},$$

introduced (in slightly different form) in §3(11). (The definition of δ_1 is as follows: if x is any element of $H^q(K, L; \mathbb{Z}_p)$ and $\bar{x} \in C^q(K, L; \mathbb{Z})$ is an integral cochain representing x modulo p, then one sets $\delta_1(x) = ((1/p)\delta\bar{x}) \pmod p$.)

There are naturally-defined maps $\delta_2 \delta_3, \ldots$, related to the homomorphism δ_1: if $x \in \operatorname{Ker} \delta_1$, i.e. if the cocycle $(1/p)\delta\bar{x} \pmod p$ is cohomologous to zero, then $(1/p)\delta\bar{x} = py + \delta z$; hence we can define a map δ_2 of $\operatorname{Ker} \delta_1$ by

$$\delta_2(x) = y \pmod p = \frac{1}{p}\left(\frac{1}{p}\delta\bar{x} - \delta z\right). \tag{1}$$

Exercise

1. Show that δ_2 is a well-defined homomorphism

$$H^q(K, L; \mathbb{Z}_p) \supset \operatorname{Ker} \delta_1 \xrightarrow{\delta_2} H^{q+1}(K, L; \mathbb{Z}_p)/\operatorname{Im} \delta_1, \tag{2}$$

commuting with continuous maps, i.e. satisfying $f^*\delta_2 = \delta_2 f^*$ for continuous maps $f: (K, L) \to (K', L')$. Construct analogously the "higher homomorphisms"

$$\delta_k : \bigcap_{i<k} \operatorname{Ker} \delta_i \to H^{q+1} \Big/ \bigcup_{i<k} \operatorname{Im} \delta_i. \tag{3}$$

For $k \geq 2$ the maps δ_k are in general neither single-valued nor everywhere defined; for this reason they are called "higher" or "partial" cohomology operations. The significance of the "homomorphisms" δ_k is as follows: knowing the structure of the $H^q(K, L; \mathbb{Z}_p)$ and the actions of the operators δ_k, one can determine the structure of the quotients $H^q(K, L; \mathbb{Z})/\Gamma_p$ of the integral relative homology groups, where Γ_p is the subgroup of elements of finite order not divisible by p.

Exercises (continued)

2. The intersection of the kernels of the actions of all δ_k on $H^q(K; \mathbb{Z}_p)$ coincides with the group obtained by reducing the integral cohomology group $H^q(K; \mathbb{Z})$ modulo p.
3. If $x \in H^q(K; \mathbb{Z}_p)$ is an image under δ_k, i.e. $x = \delta_k y$ for some $y \in H^{q-1}(K; \mathbb{Z}_p)$, but $x \neq \delta_i z$ for $i < k$, then the element x represents a basic generator $\tilde{x}$ of $H^q(K; \mathbb{Z})$, of order p^k.

Thus knowledge of all the operators δ_k acting on the cohomology groups

$H^q(K; \mathbb{Z}_p)$ for all p (or the dual operators on the homology groups $H^q(K; \mathbb{Z}_p)$) suffices for establishing the structure of the integral cohomology (or homology) groups.

The operators δ_k have the following properties:

(i) they are defined on the groups $H^q(K, L; \mathbb{Z}_p)$ (or certain subgroups of these) and are homomorphisms;

(ii) they commute with the homomorphism

$$\delta_* \circ i^* \circ j = \hat{\delta}: H^q(K, L; \mathbb{Z}_p) \to H^{q+1}(K, L; \mathbb{Z}_p)$$

(where the maps δ_*, i^*, j are as in the (cohomology) exact sequence of the pair (K, L); cf. §5(18)), i.e. $\hat{\delta}\delta_k = \delta_k\hat{\delta}$ (verify!).

(Note that the particular map δ_1 is in fact defined on the whole of H^q, and single-valued.)

10.1. Definition. A cohomology operation with the preceding two additional properties is called *stable*.

The main fact facilitating the computation of the rational homotopy groups $\pi_i(K) \otimes \mathbb{Q}$ is the absence of non-trivial cohomology operations on the rational cohomology groups $H^q(\quad; \mathbb{Q})$, essentially different from that of raising to a power (Example (a) above). (This will be proved below.) The only stable cohomology operation on the rational (or real or complex) cohomology groups is that of multiplication by a scalar:

$$\theta: H^q \to H^q, \qquad \theta(x) = \lambda x.$$

The example of the operator δ_1 (Example (b) above) shows that there do exist non-trivial stable cohomology operations over $\mathbb{Z}_p$, i.e. on the $H^q(\quad; \mathbb{Z}_p)$. The following result (whose proof we omit) shows that in mod p cohomology there are in fact many nontrivial stable cohomology operations (see [80]).

10.2. Theorem (Steenrod)

(i) ($p = 2$) *Corresponding to each integer $i \geq 0$ there is a stable cohomology operation θ, a "Steenrod square", denoted by Sq^i, which is a homomorphism on each H^q:*

$$Sq^i: H^q(K, L; \mathbb{Z}_2) \to H^{q+i}(K, L; \mathbb{Z}_2),$$

and which has the following further properties:

(1) $Sq^i(x) = 0$ *for* $q < i$;

(2) $Sq^0 \equiv 1$;

(3) $Sq^i(x) = x^2$ *for* $q = i$;

(4) $Sq^i(xy) = \sum_{j+k=i} Sq^j(x)Sq^k(y)$;

(5) $Sq^1(x) = \delta_1 x$.

(ii) *For each odd prime p and each integer $i \geq 0$, there is a stable cohomology* *operation*

$$St_p^i\colon H^q(K, L; \mathbb{Z}_p) \to H^{q+2i(p-1)}(K, L; \mathbb{Z}_p),$$

such that

(1) $St_p^i(x) = 0$ *for* $q < 2i$;

(2) $St_p^0 \equiv 1$;

(3) $St_p^i(x) = x^p$ *for* $q = 2i$;

(4) $St_p^i(xy) = \sum_{k+j=i} St_p^j(x) St_p^k(y)$.

It turns out (this is a difficult result) that in mod p cohomology all stable cohomology operations are composites of the "Steenrod operations" figuring in this theorem, and that there are nontrivial algebraic relations amongst the latter operations. These facts together with others form, as we shall see below, the basis for a complicated procedure for computing the torsion subgroups of homotopy groups. By way of a simple illustration of the possibility of such an application, consider the 2-cell complex K_x determined to within a homotopy equivalence by an element x of $\pi_{k+q}(S^q)$:

$$K_x \sim D^{k+q+1} \cup_x S^q.$$

(Note that such complexes were considered at the conclusion of §7, in the case $q = n$, $k + q = 2n - 1$.) We know from earlier results that

$$H^q(K_x; G_1) \simeq G_1, \qquad H^{k+q+1}(K_x; G_2) \simeq G_2. \tag{4}$$

10.3. Lemma. *If there exists a non-trivial cohomology operation*

$$\theta\colon H^q(\ ; G_1) \to H^{q+k+1}(\ ; G_2)$$

such that $\theta(z) \neq 0$ *where* z *is a generator (over* G_1*) of* $H^q(K_x; G_1)$ *(see (4)), then the element* x *of* $\pi_{k+q}(S^q)$ *will be non-trivial.*

Proof. Suppose on the contrary that $x = 0$. In this case the homotopy type of the complex K_x is that of the bouquet $S^{q+k+1} \vee S^q$, so that since we are concerned only with the cohomology groups of K_x we may as well take $K_x = S^{q+k+1} \vee S^q$. Consider the map $\pi\colon K_x \to S^q$ which is the identity map on the sphere S^q of the bouquet, and identifies the sphere S^{q+k+1} with a point. Regarding the codomain S^q of the map π as the obvious subspace of K_x, we obtain thence a map $\pi\colon K_x \to K_x$ with the property that the induced map π^* is the identity map on H^q and the zero map on H^{q+k+1}:

$$\pi^* = 1\colon H^q \to H^q,$$

$$\pi^* = 0\colon H^{q+k+1} \to H^{q+k+1}.$$

Since $\theta(\pi^* z) = \pi^*\theta(z)$ by definition of a cohomology operation, it follows that

$$0 = \pi^*\theta(z) = \theta(\pi^* z) = \theta(z),$$

contradicting the hypothesis that $\theta(z) \neq 0$. This completes the proof. □

As a rather trivial example, consider the case where $k = 0$, $x \in \pi_q(S^q) \simeq \mathbb{Z}$ is the element corresponding to the integer 2^s, and $\theta = \delta_s\colon H^n(K_x; \mathbb{Z}_2) \to H^{n+1}(K_x; \mathbb{Z}_2)$.

10.2. Cohomology Operations and Eilenberg–MacLane Complexes

We have already encountered "Eilenberg–MacLane spaces" in §§8, 9; they are by definition (connected) spaces $K = K(D, n)$, D being a prescribed group, abelian if $n \geq 2$, such that

$$\pi_n(K) \simeq D, \qquad \pi_j(K) = 0 \qquad \text{for} \quad j \neq n.$$

We shall make frequent use of the fact, whose full proof we omit (see [80]), that such spaces exist for every pair D, n, and moreover as cell complexes (and their homotopic equivalents). (Note also that under fairly general conditions, which we shall assume to prevail, $K(D, n)$ is in fact determined up to a homotopy equivalence.) However, we shall outline a construction of $K(D, n)$ as a cell complex in the case where D is any finitely generated abelian group. The construction proceeds according to the following scheme:

(i) $K(D, n)$ is to have only one 0-cell σ^0;

(ii) $K(D, n)$ is to have no cells of dimension i for $1 \leq i \leq n - 1$;

(iii) The n-cells σ_j^n are chosen in one-to-one correspondence with an arbitrary finite set of generators x_j of D, and attached to σ^0 to yield the n-skeleton K^n;

(iv) The $(n + 1)$-cells σ_k^{n+1} are chosen in one-to-one correspondence with the (finitely many) defining relations

$$\gamma_k = \sum_j \lambda_{jk} x_j = 0, \qquad \lambda_{jk} \in \mathbb{Z},$$

on the x_j, and are attached to the n-skeleton K^n by means of maps $\hat{\gamma}_k\colon \partial\sigma_k^{n+1} \to K^n$ identifying $\partial\sigma_k^{n+1}$ with $\sum_j \lambda_{jk}\sigma_j^n$, yielding the $(n + 1)$-skeleton (see §4)

$$K^{n+1} = \left(\bigcup_k \sigma_k^{n+1}\right) \cup_{\hat{\gamma}_k} K^n,$$

for which we then clearly have

$$\pi_j(K^{n+1}) = 0 \qquad \text{for} \quad j < n, \qquad \pi_n(K^{n+1}) = D.$$

(v) Next, a (finite) generating set $\{\alpha_i\}$ for the group $\pi_{n+1}(K^{n+1})$ is chosen arbitrarily and corresponding $(n + 2)$-cells σ_i^{n+2} are attached to K^{n+1} by means of maps $\hat{\alpha}_i\colon \partial\sigma_i^{n+2} \to K^{n+1}$ representing the $\alpha_i \in \pi_{n+1}(K^{n+1})$, yielding the $(n + 2)$-skeleton K^{n+2}. It then follows from the proof of Theorem 4.4

that

$$\pi_j(K^{n+2}) = \pi_j(K^{n+1}) \qquad \text{for} \quad j \leq n,$$

$$0 = \pi_{n+1}(K^{n+2}) = \pi_{n+1}(K^{n+1})/(\alpha_1, \alpha_2, \ldots).$$

(vi) Iterating this construction, we next "kill" the group $\pi_{n+2}(K^{n+2})$ by suitably attaching $(n + 3)$-cells to obtain K^{n+3}, and then "kill" $\pi_{n+3}(K^{n+3})$ in the process of constructing K^{n+4}, and so on. In the limit as $n + q \to \infty$, i.e. by taking the union of the chain of $(n + q)$-skeletons so constructed, we obtain finally the desired (infinite) cell complex $K(D, n)$.

With the obvious interpretation, the following are obvious consequences of results from Part II concerning the homotopy groups:

$$K(D_1 \times D_2, n) = K(D_1, n) \times K(D_2, n),$$

$$\Omega(K(D, n)) = K(D, n - 1).$$

(The second equation, involving the loop space $\Omega(K)$ relative to some base point, is essentially a consequence of the isomorphism $\pi_j(\Omega(X)) \simeq \pi_{j+1}(X)$, established in Part II, §22.2.)

10.4. Theorem. *Given any (connected) cell complex X, each homotopy class of maps $f: X \to K(D, n)$ is fully and canonically determined by some element of the cohomology group $H^n(X; D)$, and vice versa. There is thus a natural one-to-one correspondence $[X, K(D, n)] \leftrightarrow H^n(X; D)$.*

PROOF. We first show how to canonically associate a map $f: X \to K(D, n)$, with each element $x \in H^n(X; D)$. First, we define f to be constant on the $(n - 1)$-skeleton X^{n-1} of X, i.e. to map X^{n-1} to a point, and then extend f to the n-skeleton X^n as follows. Let $\bar{x} \in C^n(X; D)$ be an n-cocycle representing x. On each n-cell $\sigma^n \subset X^n$, the functional $\bar{x}$ takes a value in D:

$$\bar{x}(\sigma^n) \in D \simeq \pi_n(K(D, n));$$

we define f on σ^n to be a map $\sigma^n \to K(D, n)$ representing the element $\bar{x}(\sigma^n)$ regarded as belonging to $\pi_n(K(D, n))$. (Note that f maps the boundary of σ^n to a point.) This determines f on each σ^n, and therefore on the whole of X^n, to within a homotopy. We can now further extend f to X^{n+1} precisely because $\bar{x}$ is a cocycle (see §9.1). Extension to skeletons X^{n+1+i} of successively higher dimensions is then always possible in light of the fact that $\pi_j(K(D, n)) = 0$ for $j \neq n$, so that the obstructions to extending f are zero *a fortiori*. Continuing in this way we ultimately obtain the desired map $f: X \to K(D, n)$, determined to within a homotopy by the given element x of $H^n(X; D)$.

For the converse, suppose we are given a map $f: X \to K(D, n)$. By Theorems 4.6 and 4.8 in combination, the map f is homotopic to one which sends the $(n - 1)$-skeleton X^{n-1} to a point. Assuming then that f maps X^{n-1} to a point, it follows that, applied to each n-cell σ^n of X, f determines an element of D, since

$$f(\sigma^n) \in \pi_n(K(D, n)) \simeq D.$$

Hence the restriction of f to the n-skeleton X^n determines an element y of $H^n(X; D)$, and this we associate with the given map f. It is not difficult to see that the map f coincides (to within a homotopy) with that constructed from y as above.

It remains to show that two maps f_1 and $f_2: X \to K(D, n)$, associated as above with cohomologous cocycles $\bar{x}_1$ and $\bar{x}_2$, are homotopic. Suppose $\bar{x}_2 = \bar{x}_1 + \delta z$. By §9, Exercise 3, f_1 can by means of a homotopy be changed on the $(n-1)$-skeleton of X to a map $\hat{f}_1$ corresponding to the cocycle $\bar{x}_1 + \delta z = \bar{x}_2$. That $\hat{f}_1$ and f_2 are homotopic then follows, as before, essentially from the fact that $\pi_j(K) = 0$ for $j \neq n$. □

10.5. Theorem. *Let M be a cell complex and L a subcomplex. The cohomology operations $\theta: H^n(M, L; D) \to H^p(M, L; G)$ are in natural one-to-one correspondence with the elements of the group $H^p(K(D, n); G)$.*

Proof. We first single out a canonically distinguished element u of $H^n(K(D, n); D)$. By Hurewicz' theorem (Corollary 4.9), $H_n(K(D, n); \mathbb{Z}) \simeq \pi_n(K(D, n)) \simeq D$. It follows that there is a canonical isomorphism

$$H^n(K; G_1) \simeq \operatorname{Hom}(D, G_1), \tag{5}$$

where G_1 is any abelian group and $\operatorname{Hom}(D, G_1)$ denotes the additive group of homomorphisms from the abelian group D to G_1. In the case $G_1 = D$, the group $\operatorname{Hom}(D, G_1) = \operatorname{Hom}(D, D)$ becomes a ring, the endomorphism ring of D, with the identity map $1: D \to D$ as multiplicative identity element. The distinguished element u that we are seeking is defined to be that element of $H^n(K; D)$ which corresponds to $1 \in \operatorname{Hom}(D, D)$ under the canonical isomorphism (5).

Examination of the proof of Theorem 10.4 shows that the one-to-one correspondence $[X, K] \leftrightarrow H^n(X; D)$ of that theorem is actually given by

$$[f] \leftrightarrow f^*(u) \in H^n(X; D), \tag{6}$$

for every map $f: X \to K$. Now let $\theta: H^n \to H^p$ be any cohomology operation. Particularizing the general pair (M, L) to $(K, *)$, and applying θ to $u \in H^n(X; D)$, we obtain the element $\theta(u) \in H^p(K; G)$, associated canonically with θ.

For the other direction, let y be an arbitrary element of $H^p(K; G)$, and let X be any complex. In view of (6), i.e. of Theorem 10.4, each element x of $H^n(X; D)$ determines, uniquely to within a homotopy, a map $f: X \to K$ such that $f^*(u) = x$. We can then define a cohomology operation θ by setting

$$\theta(x) = f^*(y) \in H^p(X; G). \tag{7}$$

Taking $K = X$, we then have in particular $\theta(u) = y$. We leave it to the reader to verify that the operation defined by (7) is indeed a cohomology operation, and to derive the theorem in the slightly more general form in which it is stated. □

10.6. Theorem. *For any finitely generated abelian group D, the cohomology algebra $H^*(K(D, n); \mathbb{Q})$ is a free skew-commutative algebra, generated by elements of the vector space $H^n(K; \mathbb{Q}) \simeq \mathrm{Hom}(D, \mathbb{Q}) = D^*$ (see (5)).*

PROOF. Being finitely generated, the abelian group D is a direct sum of cyclic groups:

$$D \simeq \mathbb{Z} \oplus \cdots \oplus \mathbb{Z} \oplus \mathbb{Z}_{m_1} \oplus \cdots \oplus \mathbb{Z}_{m_k}.$$

We shall first show that for $D = \mathbb{Z}_m$, a cyclic group of order m, we have $H^q(K; \mathbb{Q}) = 0$ for $q > 0$.

For $K = K(D, 1)$ this can be deduced from the fact noted in §9 (see also §4, Exercise 8 *et seqq.*) that

$$K(\mathbb{Z}_m, 1) \sim S^\infty/\mathbb{Z}_m = \lim_{N\to\infty} L_m^{2N-1}(1, \ldots, 1).$$

Suppose inductively that for $i < n$ we have already established that $H^q(K(\mathbb{Z}_m, i); \mathbb{Q}) = 0$ for $q > 0$, and consider the Serre fibration

$$E \to K(D, n), \qquad F \sim K(D, n-1), \tag{8}$$

where $D = \mathbb{Z}_m$, and E is the path space over $K(D, n)$ with respect to some base point $x_0 \in K(D, n)$, so that E is contractible (cf. §8, Example (b)). It follows from the formal properties of the Leray spectral sequence (see §8) for the rational cohomology of this fibre space, that since $E_2^{q,p} \simeq H^q(B) \otimes H^p(F)$, we have $E_2^{q,p} = 0$ for $p > 0$ by the inductive hypothesis, and $E_2^{q,0} \simeq H^q(B; \mathbb{Q})$ (where $B = K(D, n)$). Hence $d_r^* = 0$ for $r \geq 2$, so that

$$E_2^{q,0} = E_\infty^{q,0} (\simeq H^q(E; \mathbb{Q})) = 0,$$

whence $H^q(B; \mathbb{Q}) = H^q(K(D, n); \mathbb{Q}) = 0$ for $q > 0$. Thus for $D = \mathbb{Z}_m$, and therefore in view of the isomorphism of §7(4) also for $D = \mathbb{Z}_{m_1} \oplus \cdots \oplus \mathbb{Z}_{m_k}$, we have $H^q(K(D, n); \mathbb{Q}) = 0$ for all $q > 0$.

Suppose now that $D = \mathbb{Z}$. The desired conclusion holds for $K(\mathbb{Z}, 1)$ since $K(\mathbb{Z}, 1) \sim S^1$ (see §7, Example (α)). Suppose $n > 1$ and inductively that the conclusion of the theorem holds for $K(\mathbb{Z}, i)$ with $i < n$, so that:

(i) if n is even, $H^*(K(D, n-1); \mathbb{Q}) \simeq \Lambda[u_{n-1}]$;
(ii) if n is odd, $H^*(K(D, n-1); \mathbb{Q}) \simeq \mathbb{Q}[u_{n-1}]$.

Consider the Leray spectral sequence for the rational cohomology of the Serre fibration (8) where now, however, $D = \mathbb{Z}$.

We analyse first the case that n is even; in this case we have by the inductive hypothesis that $H^*(F) = H^*(K(D, n-1))$ is as in (i) above. By Theorem 4.8 we have $H^p(B) = H^p(K(D, n)) = 0$ for $0 < p < n$; from this, (i), and the fact that $E_2^{p,q} = H^p(B) \otimes H^q(F)$, we deduce that $E_2^{p,q} = 0$ if $0 < p < n$ or $q \neq 0$, $n-1$, and that $E_2^{0,n-1}$ is one-dimensional, generated by $u = 1 \otimes u_{n-1}$. It then follows from the properties of the cohomology spectral sequence that $d_i^* = 0$ for $i \geq 2$, $i \neq n$, whence for all p, q we have $E_2^{p,q} = \cdots = E_n^{p,q}$. From this and the fact that $H^i(E; \mathbb{Q}) = 0$ for $i > 0$, we deduce also that $E_{n+1}^{p,q} = 0$ for all p, q

whence it follows that the map

$$d_n^*\colon E_n^{p,n-1} \to E_n^{p+n,0} = E_2^{p+n,0} \simeq H^{p+n}(B; \mathbb{Q})$$

must be onto for all p. Taking $p = 0$, we infer that $E_n^{n,0} = E_2^{n,0} \simeq H^n(K(D, n); \mathbb{Q})$ is one-dimensional, generated by $d_n^*(u) = v$ say, and thence by induction that $E_n^{i,0} = 0$ unless i is a multiple of n, and that $E_n^{kn,0}$ is one-dimensional. From Leibniz' formula (see §8(7)), we have $d_n^*(uv) = v^2 \in E_n^{2n,0}$, and thence by induction that $E_2^{kn,0}$ is generated by v^k. It now follows easily that

$$H^n(K(D, n); \mathbb{Q}) \simeq \mathbb{Q}[v],$$

as required. (The generators of the (non-zero) terms $E_n^{q,p}$ of the spectral sequence are shown in Table (i).)

$d_i^* = 0, \quad i \geq 2, i \neq n;$

$d_n^*(u) = v;$

$d_n^*(v) = 0.$

$n-1$	u	0	uv	0	uv^2	
		0		0		
0	1	0	v	0	v^2	

(i)

$d_i^* = 0, \quad i \geq 2, i \neq n;$

$d_n^*(u) = v;$

$d_n^*(v) = 0.$

$2n-2$	u^2		u^2v
		0	
$n-1$	u		uv
0	1		v

(ii)

An analogous argument via the cohomology spectral sequence, again making essential use of cohomological multiplication, yields the desired result in case (ii), where n is odd; we leave the details to the reader. (See Table (ii).)

The theorem now follows from the isomorphism of §7(4) together with the fact that D decomposes as a direct product of cyclic groups. □

10.3. Computation of the Rational Homotopy Groups $\pi_i \otimes \mathbb{Q}$

The key to this computation is the following result.

10.7. Theorem (Cartan–Serre). *Let X be a simply-connected (or at least "homotopically simple") complex. Suppose that there is a dimension-preserving isomorphism from the subgroup of the rational cohomology algebra of X generated* *by the elements of dimension $< k$, to the corresponding subgroup of some free skew-commutative algebra A (whose free algebra generators of dimension $< k$ occur in dimensions α_j say). Then the following assertions are valid:*

(i) *The Hurewicz homomorphism (see the proof of Corollary 4.9)*

$$H: \pi_i(X) \otimes \mathbb{Q} \to H_i(X; \mathbb{Q})$$

has trivial kernel for all $i < k - 1$.

(ii) *The image $H(\pi_i(X) \otimes \mathbb{Q})$ consists precisely of those elements of $H_i(X; \mathbb{Q})$ which have zero scalar product with every element x of $H^*(X; \mathbb{Q})$ decomposing non-trivially as a product $x = yz$ with $\deg y > 0$, $\deg z > 0$ (i.e. with every such element x vanishing on $H_i(X; \mathbb{Q})$; see Theorem 2.9). Consequently, for $i < k - 1$ the group $\pi_i(X) \otimes \mathbb{Q}$ is isomorphic (in fact its image under H is dual) to the quotient group $H^i(X; \mathbb{Q})/\Gamma$, where Γ is generated by all elements of $H^i(X; \mathbb{Q})$ which decompose non-trivially as products.*

PROOF. In view of Theorem 10.6 the hypotheses of the theorem are fulfilled by $K(D, n)$ (with $k = \infty$) provided D is finitely generated abelian. The conclusions also hold in this case: thus (i) is immediate for $i = n$ from Hurewicz' theorem (Corollary 4.9), and for $i \neq n$ from the triviality of $\pi_i(K(D, n))$, while (ii) is easily inferred from the structure of $H^*(K(D, n); \mathbb{Q})$, as given by Theorem 10.6. The validity of the theorem more generally for products of the form

$$K = K(D_1, \alpha_1) \times K(D_2, \alpha_2) \times \cdots, \qquad \alpha_1 < \alpha_2 < \cdots \tag{9}$$

follows readily.

We shall need a certain map $f: X \to K$, where K is as in (9) with each group D_j free abelian of rank equal to the number of free generators of $H^*(X; \mathbb{Q})$ of dimension α_j (the α_j being as in the theorem); we choose f to be any map such that the induced map $f^*: H^*(K; \mathbb{Q}) \to H^*(X; \mathbb{Q})$ is an abelian-group isomorphism up to and including dimension $k - 1$. (Such a map can be constructed as the direct product of appropriate maps $X \to K(D_i, \alpha_i)$, which exist by virtue of the results of §9 (see Lemma 9.2 and Exercise 2, or the proof of Theorem 10.4).) We now convert the map f to a fibre-space projection $\tilde{f}: \tilde{X} \to \tilde{K}$, where $\tilde{X}$, $\tilde{K}$ are respectively homotopically equivalent to X, K, in accordance with the procedure described towards the end of §8; the map $\tilde{f}$ will then induce essentially the same isomorphism (up to dimension $k - 1$) as f. It follows from this isomorphism together with the (homology) spectral sequence of the fibre space $\tilde{f}: \tilde{X} \to \tilde{K}$, with fibre F say, that $H^i(F; \mathbb{Q}) = 0$ for $0 < i < k - 1$. (This can be seen as follows: From the second of the "supplementary" properties of the homology spectral sequence given in §8, we have that $E^{\infty}_{i,0} \subset H_i(B) = H_i(\tilde{K})$, and that $\tilde{f}_*: H_i(\tilde{X}) \to H_i(\tilde{K})$ coincides with the projection of $E^{(\infty)}_*$ onto $E^{(\infty)}_{n,0}$. Since $\tilde{f}_*$ is here an isomorphism for $i \leq k - 1$, we conclude that $E^{(\infty)}_{p,q} = 0$ for $0 < q$, $p + q \leq k$. It follows that for $0 < i < k - 1$, we have $0 = E^{(\infty)}_{0,i} = E^{(2)}_{0,i} \simeq H^i(F; \mathbb{Q})$, as claimed.)

Since X is by assumption simply-connected, we may take D_1 to be the first non-trivial homotopy group of X, rather than free abelian, in which case $\tilde{f}$

induces an isomorphism $\pi_2(\tilde{X}) \simeq \pi_2(\tilde{K})$. From the homotopy exact sequence of the fibre space (see Part II, §22.2) we then conclude that $\pi_1(F) = 0$.

To complete the proof we shall need the following

10.8. Lemma. *If the cohomology groups $H^i(F; \mathbb{Q})$ of a simply-connected space F are trivial for $i < k - 1$, then the groups $\pi_i(F) \otimes \mathbb{Q}$ are also trivial for $i < k - 1$.*

PROOF. It follows from the assumption that $H^i(F; \mathbb{Q}) = 0$ for $i < k - 1$, together with Hurewicz' theorem (Corollary 4.9), that the first nontrivial homotopy group $\pi_s(F)$ with $s < k - 1$ (if there is one) must be finite. Appealing again to the results of §9, we deduce the existence of a map $g: F \to K(\pi_s(F), s)$ inducing an isomorphism between the sth fundamental groups. We now convert g to a fibre space projection $\tilde{g}: \tilde{F} \to \tilde{K}(\pi_s(F), s)$ with fibre F_1 say. By exploiting in particular the fact that $\tilde{g}_*$ is an isomorphism between the fundamental groups, in conjunction with the homotopy exact sequence of the fibration (see Part II, §22.2), we obtain immediately that $\pi_s(F_1) = 0$, $\pi_i(F_1) \simeq \pi_i(F)$ for $i \neq s$. On the other hand, from the cohomology spectral sequence of the fibre space together with the assumption that $H^i(F_1; \mathbb{Q}) = 0$ for $i < k - 1$, and the fact that $H^j(K(\pi_s(F), s)) = 0$ for all $j > 0$, one deduces easily that $H^j(F_1; \mathbb{Q}) \simeq H^j(F; \mathbb{Q})$ for all j. Thus the hypothesis of the lemma certainly holds for the space F_1 whose homotopy groups, except for the sth, which is zero, are isomorphic to those of F. Hence if we repeat the foregoing argument with F_1 in place of F, and so on, inductively, we shall eventually reach the desired conclusion. □

We may now complete the proof of the Cartan–Serre theorem: From the homotopy exact sequence (tensored with $\mathbb{Q}$) of the fibration $\tilde{f}: \tilde{X} \to \tilde{K}$, with the fact just established taken into account, namely that $\pi_i(F) \otimes \mathbb{Q} = 0$ for $i < k - 1$, it is immediate that f induces an isomorphism $\pi_i(X) \otimes \mathbb{Q} \simeq \pi_i(K) \otimes \mathbb{Q}$ for $i < k - 1$. The theorem now follows by virtue of its validity for K. □

From Corollary 7.7 we deduce immediately the following

10.9. Corollary. *For any Lie group G the nontrivial rational homotopy groups $\pi_i(G) \otimes \mathbb{Q}$ correspond one-to-one with the free generators of the cohomology algebra $H^*(G; \mathbb{Q}) \simeq \Lambda[x_{i_1}, \ldots, x_{i_k}]$, of the same dimension. In particular, $\pi_i(G) \otimes \mathbb{Q} = 0$ for even i.*

We know from §8, Exercise 2, that

$$H^*(\Omega(S^n); \mathbb{Q}) \simeq \begin{cases} \mathbb{Q}[x_{n-1}], & n = 2k + 1, \\ \Lambda[x_{n-1}] \otimes \mathbb{Q}[x_{2n-2}], & n = 2k. \end{cases}$$

From this and the isomorphism $\pi_i(\Omega(S^n)) \simeq \pi_{i+1}(S^n)$ (Part II, Corollary 22.2.3) we deduce the following

10.10. Corollary. *For the sphere S^n we have:*

$$\begin{aligned} &\textit{for } n \textit{ even:} \quad \pi_n(S^n)\otimes \mathbb{Q} \simeq \mathbb{Q}, \qquad \pi_{2n-1}(S^n)\otimes\mathbb{Q}\simeq\mathbb{Q},\\ &\qquad\qquad \pi_j(S^n)\otimes\mathbb{Q}=0, \qquad j\neq n,\quad 2n-1;\\ &\textit{for } n \textit{ odd:} \quad \pi_n(S^n)\otimes\mathbb{Q}\simeq\mathbb{Q}, \qquad \pi_j(S^n)\otimes\mathbb{Q}=0, \qquad j\neq n. \end{aligned}$$

10.11. Corollary. *If X is an $(n-1)$-connected complex (i.e. $\pi_i(X)=0$ for $i<n$), then for all $q<2n-1$ the rational Hurewicz homomorphism*

$$H\colon \pi_q(X)\otimes\mathbb{Q}\to H_q(X)\otimes\mathbb{Q}$$

is an isomorphism.

(This follows from the simple observation that in the cohomology algebra of the $(n-1)$-connected complex X, nonzero elements representable as nontrivial products can occur only in dimensions $\geq 2n$ (so that, in particular, up to, but not necessarily in, dimension $2n$, $H^*(X;\mathbb{Q})$ is, as abelian group, isomorphic in the prescribed manner to a free skew-commutative algebra).)

EXERCISE

4. Compute the rational homotopy groups $\pi_i(S^k\vee S^l)\otimes\mathbb{Q}$ of the bouquet of spheres $S^k\vee S^l$.

For every simply-connected complex X the computation of the groups $\pi_i(X)\otimes\mathbb{Q}$ reduces, in view of the Cartan–Serre theorem, to the computation of the rational cohomology algebra $H^*(\Omega(X);\mathbb{Q})$, since the latter is free skew-commutative (see §7), and since $\pi_i(\Omega(X))\simeq\pi_{i+1}(X)$ (see Part II, §22.2).

10.12. Corollary. *If a simply-connected complex X has the same homotopy type as an H-space up to dimension N, then the algebra $H^*(X;\mathbb{Q})$ is free skew-commutative up to dimension $N-1$, and there is an isomorphism*

$$\left(\sum_{i\leq N-2}\pi_i(X)\otimes\mathbb{Q}\right)^*\simeq\sum_{i\leq N-2}H^i(X;\mathbb{Q})/\Gamma,$$

where Γ is generated by all elements non-trivially decomposable as products (and the asterisk indicates the dual space).

10.4. Application to Vector Bundles. Characteristic Classes

Let $G_{N,k}$ denote as usual the appropriate Grassmannian manifold (whose points are the k-dimensional linear subspaces of $\mathbb{R}^N$, $\mathbb{C}^N$ or $\mathbb{H}^N$, as the case may be; see Part II, §5.2), and consider the natural map $\psi\colon G_{N,k}\times G_{M,l}\to G_{N+M,k+l}$, determined by direct sums of subspaces. Letting $N\to\infty$, $M\to\infty$, i.e. taking direct limits, we obtain in the limit a map

$$\psi: G_{\infty,k} \times G_{\infty,l} \to G_{\infty,k+l}, \tag{10}$$

or

$$\psi: BG_k \times BG_l \to BG_{k+l},$$

where $G_{\infty,n} = BG_n$ is now being considered as the "universal classifying space" (i.e. base of the universal fibre bundle) for the group G_n, G_n being any of the groups $O(n)$, $SO(n)$, $U(n)$, $SU(n)$ or $Sp(n)$ (see Part II, §24.4; note that in the case of $SO(n)$ the points of $BG_n = BSO(n)$ are the oriented n-dimensional subspaces, and analogously for $SU(n)$).

Under each of the standard embeddings $O(n) \subset O(n+1)$, $SO(n) \subset SO(n+1)$, $U(n) \subset U(n+1)$, $SU(n) \subset SU(n+1)$, $Sp(n) \subset Sp(n+1)$, the set of homotopy classes $[X, BG]$, where X is a finite-dimensional cell complex, "stabilizes" in the sense that there is a natural one-to-one correspondence

$$[X, BG_n] \leftrightarrow [X, BG_{n+1}], \tag{11}$$

provided that: $\dim X < n - 1$ for $G = O(n)$, $SO(n)$; $\dim X < 2n - 1$ for $G = U(n)$, $SU(n)$; and $\dim X < 4n - 1$ for $G = Sp(n)$. (The analogous statement with $SO(n)$, $SO(n+1)$ in place of BG_n, BG_{n+1}, was proved in Part II, §24.3, using $SO(n+1)/SO(n) \cong S^n$; by exploiting in similar fashion the diffeomorphisms $U(n+1)/U(n) \cong S^{2n}$, $Sp(n+1)/Sp(n) \cong S^{4n}$, it may be established also for the remaining groups.) The one-to-one correspondence (11) may be established as follows. If $E_{n+1} \to BG_{n+1}$ is a universal fibre bundle for G_{n+1}, then G_n embedded in the standard way in G_{n+1} also acts on E_{n+1}, whence we obtain a universal fibre bundle $E_{n+1} \to E_{n+1}/G_n \sim BG_n$ for G_n, and thence a projection $E_{n+1}/G_n \to E_{n+1}/G_{n+1}$ with fibre G_{n+1}/G_n, an N-sphere for appropriate N. The desired conclusion now follows by applying to this fibre bundle the argument used to prove Proposition 24.3.3 of Part II.

We define (direct) limits $G_\infty = O, SO, U, SU, Sp$, by means of the following unions:

$$O = \lim_{N\to\infty} O(1) \subset O(2) \subset \cdots \subset O(N) \subset \cdots;$$

$$SO = \lim_{N\to\infty} SO(1) \subset SO(2) \subset \cdots \subset SO(N) \subset \cdots;$$

$$U = \lim_{N\to\infty} U(1) \subset U(2) \subset \cdots \subset U(N) \subset \cdots;$$

$$SU = \lim_{N\to\infty} SU(1) \subset SU(2) \subset \cdots \subset SU(N) \subset \cdots;$$

$$Sp = \lim_{N\to\infty} Sp(1) \subset Sp(2) \subset \cdots \subset Sp(N) \subset \cdots.$$

If we now equip the corresponding classifying spaces BG_∞ of the groups G_∞ with the limits of the maps given by (10):

$$\psi: BG_\infty \times BG_\infty \to BG_\infty,$$

(where $\mathbb{R}^\infty \oplus \mathbb{R}^\infty$ is suitably identified with $\mathbb{R}^\infty$, etc.), then they become H-

spaces, where we can adjoin the origin of $\mathbb{R}^\infty$, etc., as the identity element; in fact any point of BG_∞ will serve as homotopic identity element (verify!). In view of the stabilizing property (11), we may rephrase the statement that BG_∞ is a (generalized) H-space as follows: $BO(n)$ and $BSO(n)$ have the homotopy type of an H-space up to dimension $n-1$, $BU(n)$ and $BSU(n)$ up to dimension $2n-1$, and $Sp(n)$ up to dimension $4n-1$. Much later, in §25, we shall see that $BU \cong \Omega(U)$, $BSp \cong \Omega(\Omega(\Omega(SO)))$ and $BSO \cong \Omega(\Omega(\Omega(Sp)))$.

Since the groups $G_n = SO(n), U(n), SU(n), Sp(n)$ are (connected) Lie groups we know (from Corollary 10.9) the groups $\pi_i(G_n) \otimes \mathbb{Q}$, and therefore the groups $\pi_{i+1}(BG_n) \otimes \mathbb{Q}$, since $\pi_i(G_n) \simeq \pi_{i+1}(BG_n)$ (see Part II, §24.4). From Corollary 10.10 we have also that the basis elements of the dual spaces $(\pi_j(BG_n) \otimes \mathbb{Q})^*$, for j less than the appropriate dimension N, correspond one-to-one to the elements of dimension $< N$ of a multiplicative basis for the algebra $H^*(BG_n; \mathbb{Q})$, since up to dimension N the space BG_n has the homotopy type of an H-space (see above).

In fact, for any (connected) Lie group G the algebra $H^*(BG; \mathbb{Q})$ turns out to be free skew-commutative, even though BG need not be an H-space. This is a consequence of the following result of Borel:

Exercise

5. Let B be a simply-connected complex, and consider the (usual) Serre fibration $p: E \to B$ with fibre $F \sim \Omega(B)$ and E contractible. Show that if $H^*(F; \mathbb{Q})$ is a (free) exterior algebra then $H^*(B; \mathbb{Q})$ is a polynomial algebra. (*Hint*. For $H^*(F; \mathbb{Q}) = \bigwedge[x]$, imitate the treatment of case (i) in the proof of Theorem 10.6, and consider next the case $H^*(F; \mathbb{Q}) = \bigwedge[x_1, x_2]$, and then $H^*(F; \mathbb{Q}) = \bigwedge[x_1, x_2, x_3]$, and so on.)

For the classification spaces of the classical groups G we are considering, it turns out that:

$$
\begin{aligned}
H^*(BSO(2k); \mathbb{Q}) &= \mathbb{Q}[p_1, \ldots, p_{k-1}, \chi], && \deg p_i = 4i, \\
& && \deg \chi = 2k, \quad p_k = \chi^2; \\
H^*(BSO(2k+1); \mathbb{Q}) &= \mathbb{Q}[p_1, \ldots, p_k], && \deg p_i = 4i; \\
H^*(BU(k); \mathbb{Q}) &= \mathbb{Q}[c_1, \ldots, c_k], && \deg c_i = 2i; \\
H^*(BSU(k); \mathbb{Q}) &= \mathbb{Q}[c_2, \ldots, c_k], && \deg c_i = 2i; \\
H^*(BSp(k); \mathbb{Q}) &= \mathbb{Q}[\gamma_1, \ldots, \gamma_k], && \deg \gamma_i = 4i,
\end{aligned}
\tag{12}
$$

where the c_i are the Chern classes, and the remaining classes χ (the Euler class), p_i and γ_i are expressible in terms of the c_i. (This was indicated for χ and the p_i in §9.) It follows that all of these classes are integral, i.e. belong to the image under the homomorphism $H^*(BG; \mathbb{Z}) \to H^*(BG; \mathbb{Q})$.

In order to investigate the consequences for homology theory of the usual Lie-group technique involving the Cartan subgroup (i.e. maximal commutative Lie subgroup), consider now the maximal tori contained in our classical groups G:

$$T^k \overset{i}{\subset} SO(2k); \qquad T^k \overset{i}{\subset} SO(2k+1); \qquad T^k \overset{i}{\subset} U(k);$$
$$T^{k-1} \overset{i}{\subset} SU(k); \qquad T^k \overset{i}{\subset} Sp(k). \tag{13}$$

For $G = S^1$ we have $BG = BS^1 = \mathbb{C}P^\infty$ (see Part II, §24.4), whence for $G = T^n$,

$$BG = BT^n = \mathbb{C}P^\infty \times \cdots \times \mathbb{C}P^\infty.$$

It follows without difficulty that (cf. the first example of §7)

$$H^*(BT^n; \mathbb{Q}) \simeq \mathbb{Q}[t_1, \ldots, t_n], \qquad t_i \in H^2(BT^n; \mathbb{Z}). \tag{14}$$

Now it is not difficult to show (using, for example, Milnor's explicit construction of a universal fibre bundle for a topological group; see [30]) that a map $H \to G$ between Lie groups naturally induces a corresponding map $BH \to BG$ between their classifying spaces. Hence in particular the inclusions (13) induce maps i: $BT^n \to BG$, for the various n and G.

EXERCISES (continued)

6. For the above classical groups G show that the homomorphism

$$H^*(BG; \mathbb{Q}) \xrightarrow{i^*} H^*(BT^n; \mathbb{Q}),$$

has trivial kernel (i.e. is a monomorphism), and that the image $\operatorname{Im} i^*$ consists precisely of those polynomials invariant under the Weyl group (i.e. the group of those inner automorphisms of G preserving the Cartan subgroup; cf. Part II, §25.4). (It can be shown that the Weyl group of $U(n)$ can be identified with the full symmetric group on the generators t_i, while that of $SO(2n)$ contains as additional generators automorphisms inducing the maps $(t_i, t_j) \to (-t_i, -t_j)$, $t_k \to t_k$ for $k \neq i$, j, and the Weyl group of $SO(2n+1)$ (and of $Sp(n)$) contains also the n reflections $t_i \to -t_i$, $t_k \to t_k$ for $k \neq i$. The images $i^*(H^*(BG; \mathbb{Q})) \subset H^*(BT^n; \mathbb{Q})$ are thus explicitly as follows:

$$\begin{aligned}
&\text{(i) } SO(2k), \quad i^*(p_q) = \sum_{i_1 < \cdots < i_q} t_{i_1}^2 \ldots t_{i_q}^2, \quad i^*(\chi_k) = t_1 \ldots t_k;\\
&\text{(ii) } SO(2k+1), \quad i^*(p_q) = \sum_{i_1 < \cdots < i_q} t_{i_1}^2 \ldots t_{i_q}^2;\\
&\text{(iii) } U(k), \quad i^*(c_j) = \sum_{i_1 < \cdots < i_j} t_{i_1} \ldots t_{i_j}, \quad c_k = \chi;\\
&\text{(iv) } Sp(k), \quad i^*(\gamma_j) = \sum_{i_1 < \cdots < i_j} t_{i_1}^2 \ldots t_{i_j}^2.
\end{aligned} \tag{15}$$

Cf. Part II, Examples 25.4.4, where the "Newtonian polynomials" $\sum t_i^m = \tilde{c}_m$ were chosen as basis elements for the algebra of "differential-geometric characteristic classes".)

7. Derive a formula linking the classes c_i with the $\tilde{c}_m$. Find also expressions for the classes $p_j(r\xi)$ in terms of the classes $c_q(\xi)$, where $r\xi$ denotes as usual the realization of the U-bundle ξ.

8. By using the spectral sequences of the appropriate fibre bundles, verify that the structures of the cohomology algebras $H^*(BG; \mathbb{Q})$ are indeed as given by (12).

9. Show that since BG_∞ is an H-space, given a complex X of dimension $< N$, the set $[X, BG_n]$ of homotopy classes of maps (or, in view of (11) and the classification theorem for vector bundles (see Part II, §24.4), the equivalence classes of "stable" vector bundles with fibre $\mathbb{R}^n$, $\mathbb{C}^n$ or $\mathbb{H}^n$) forms an abelian group. (Here $N = n - 1$ for $O(n)$, $SO(n)$; $N = 2n - 1$ for $U(n)$, $SU(n)$; and $N = 4n - 1$ for $Sp(n)$. The addition of maps $X \to BG_n$ is defined in terms of the operation ψ given by (10), or, equivalently, via the direct sum of vector bundles (see below).) Establish the algebra isomorphism
$$[X, BG_n] \otimes \mathbb{Q} \simeq \mathrm{Hom}(H^*(BG_n; \mathbb{Q}), H^*(X; \mathbb{Q})),$$
or, in other words, show that to within a torsion element of the group $[X, BG_n]$ each vector bundle over X with group G_n is completely determined by its characteristic classes.

The *direct* or *Whitney sum* of two vector bundles over the same base B say, with structure groups $U(m)$ and $U(n)$, is defined via the embedding as diagonal $m \times m$ and $n \times n$ blocks:
$$U(m) \times U(n) \subset U(m+n)$$
(and similarly for the groups $O(k)$, $SO(k)$, $SU(k)$, $Sp(k)$; see Part II, §24.5). Under this embedding the maximal tori are clearly simply multiplied directly: $T^m \times T^n = T^{m+n}$. Now if we denote the free generators of $H^2(BT^m; \mathbb{Q})$ by $t'_1, \ldots, t'_m$, and those of $H^2(BT^n; \mathbb{Q})$ by $t''_1, \ldots, t''_n$, then as free generators of $H^2(BT^{m+n}; \mathbb{Q})$ we may in view of (14) take $t_1, \ldots, t_{m+n}$, where $t_i = t'_i$ for $i = 1, \ldots, m$, and $t_{j+m} = t''_j$ for $j = 1, \ldots, n$. In view of (15) we may identify the Chern class c_i of the universal $U(k)$-bundle with the elementary symmetric polynomial $c_i(t_1, \ldots, t_k)$ of degree i. Hence by expressing the elementary symmetric polynomial $c_i(t_1, \ldots, t_{m+n})$ in terms of $c_j(t'_1, \ldots, t'_m) = c'_j$ say, and $c_q(t''_1, \ldots, t''_n) = c''_q$, we obtain the following decomposition formula:
$$c_i = \sum_{j+q=i} c'_j c''_q,$$
or equivalently, in terms of polynomials in z,
$$c(z) = c'(z)c''(z), \tag{16}$$
where
$$c(z) = \sum c_i z^i, \qquad c'(z) = \sum c'_j z^j, \qquad c''(z) = \sum c''_q z^q, \qquad c_0 = 1.$$

Since by the classification theorem for G-bundles (see Part II, §24.4) each of the original two vector bundles over B, as also their sum, is determined up to equivalence by a map $B \to BG$, where $G = U(m)$, $U(n)$, $U(m) \times U(n)$ respectively, it can be inferred via the induced homomorphisms $H^*(BG; \mathbb{Q}) \to H^*(B; \mathbb{Q})$ (cf. Part II, Theorem 25.5.2) that the relationship (16) holds between the Chern polynomials of the original vector bundles and their Whitney sum (cf. §9, Exercise 15, where however the context was that of the "product", rather than the present "fibre product", of the vector bundles).

The *Chern character* of a $U(k)$-bundle ξ over B, is defined by

$$\operatorname{ch} \xi = \sum_{i=1}^{k} \exp(zt_i) = \sum_{i=1}^{k} \left(\sum_{n=0}^{\infty} \frac{(zt_i)^n}{n!} \right),$$

where in the manner just noted the t_i are interpreted as elements of $H^*(B; \mathbb{Q})$ via a map $B \to BG$ determining ξ. The name "character" is appropriate here in the first place in view of the fact that for the Whitney sum $\xi \oplus \eta$ of a $U(m)$- and a $U(n)$-vector bundle over B we have

$$\operatorname{ch}(\xi \oplus \eta) = \operatorname{ch} \xi + \operatorname{ch} \eta. \tag{17}$$

EXERCISE

10. Deduce (17) formally from (16) and vice versa, without having recourse to the generators t_i in the maximal torus.

The tensor product $\xi \otimes \eta$ of two vector bundles with structure groups $U(m)$ and $U(n)$ respectively, and the same base B is defined via the embedding

$$U(m) \times U(n) \to U(mn), \tag{18}$$

determined by the natural action of $U(m) \times U(n)$ on $\mathbb{C}^m \otimes \mathbb{C}^n \simeq \mathbb{C}^{mn}$ (and similarly for the groups $O(k)$, $SO(k)$, $SU(k)$ and $Sp(k)$; see Part II, §24.5). Here the relationship between the respective maximal tori is slightly more complicated than before. The embedding (18) affords a map of the maximal tori:

$$\varphi\colon T^m \times T^n \to T^{mn}$$

(yielding a map $BT^m \times BT^n \to BT^{mn}$; see the remark preceding Exercise 6). By considering the action of φ on the pairs of diagonal matrices in $U(m)$, $U(n)$ (which comprise the maximal tori in these groups) it readily follows that the induced map φ^* between the second cohomology groups is given on the free generators by

$$\varphi^*(t_{jk}) = t'_j + t''_k, \tag{19}$$

where

$$t_{jk} \in H^2(BT^{mn}; \mathbb{Q}), \qquad t'_j \in H^2(BT^m; \mathbb{Q}), \qquad t''_k(BT^n; \mathbb{Q}).$$

Since

$$\begin{aligned} \operatorname{ch} \xi \operatorname{ch} \eta &= \left(\sum_{i=1}^{m} \exp(zt'_i) \right) \left(\sum_{i=1}^{n} \exp(zt''_j) \right) = \sum_{i,j} \exp[z(t'_i + t''_j)] \\ &= \varphi^* \left(\sum_{i,j} \exp(zt_{ij}) \right), \end{aligned}$$

we infer from (19) that

$$\operatorname{ch}(\xi \otimes \eta) = \operatorname{ch} \xi \operatorname{ch} \eta. \tag{20}$$

For the n-dimensional complex projective space $\mathbb{C}P^n$ we have (see Part II, §25.4, Exercise 1)

$$\tau(\mathbb{C}P^n) \oplus 1 = \eta \oplus \cdots \oplus \eta \quad (n + 1 \text{ summands}), \tag{21}$$

where $\tau(\mathbb{C}P^n)$ is the tangent bundle, η is the general Hopf bundle over $\mathbb{C}P^n$, and 1 denotes the trivial complex line bundle over $\mathbb{C}P^n$. By Exercise 13 of §9, the Chern polynomial of η is $1 + zt$ where t is a generator of $H^2(\mathbb{C}P^n; \mathbb{Z}) \simeq \mathbb{Z}$ (see §4, Example (g)). Hence from (21) together with (16) we infer that the Chern polynomial of the vector bundle $\tau \oplus 1$ is

$$c(z) = (1 + zt)^{n+1} = 1 + c_1 z + \cdots + c_n z^n. \tag{22}$$

By Definition 9.5, the ith Pontrjagin characteristic class of a real vector bundle γ is $p_i(\gamma) = (-1)^i c_{2i}(c\gamma)$, where $c\gamma$ denotes the complexification of γ. Taking $\gamma = r\tau$, the realization of τ, and recalling that $cr\tau = \tau \oplus \bar{\tau}$ (see Part II, §25.4), we have that

$$p_i(r\tau) = (-1)^i c_{2i}(\tau \oplus \bar{\tau}).$$

Similarly, we have that

$$p_1(r\eta) = -c_2(\eta \oplus \bar{\eta}) = t^2, \qquad p_i(\eta) = 0 \qquad \text{for} \quad i > 1.$$

Hence once again from (16) we deduce that the "Pontrjagin polynomial" $p(\mathbb{C}P^n)$ of the vector bundle $r\tau$ is

$$p(z) = (1 + z^2 t^2)^{n+1} = 1 + p_1 z^2 + \cdots + p_k z^{2k} + \cdots. \tag{23}$$

EXERCISES (continued)

11. Find the Chern character of an arbitrary symmetric power $S^i\xi$, and of an exterior power $\bigwedge^i \xi$ of a $U(n)$-vector bundle ξ.
12. Find the Chern classes and character of a product of the form $\mathbb{C}P^{n_1} \times \cdots \times \mathbb{C}P^{n_s}$.
13. Investigate the class c_1 of a submanifold X_k^{n-1} of $\mathbb{C}P^n$ defined by a single (non-singular) algebraic equation of degree k. Prove that the condition $c_1(X_k^{n-1}) = 0$ is equivalent to $k = n + 1$.
14. Find the Euler characteristic $\chi(X_k^{n-1})$ and determine the integer $(c_1^{n-1}, [X_k^{n-1}])$, where $[X_k^{n-1}]$ denotes the sum of all simplexes of largest dimension in some simplicial decomposition of X_k^{n-1} (cf. proof of Proposition 3.10).
15. Examine the case $k = 4$, $n = 3$. Find the homology groups of X_4^2.
16. Prove that a hypersurface given in $\mathbb{C}P^n$ by a single non-singular equation, is simply-connected.

10.5. Classification of the Steenrod Operations in Low Dimensions

We shall now attempt to expound a method for computing the homotopy groups of spheres. The method relies first on the existence of spectral sequences with appropriate formal properties (Leray's theorem), second on the existence and formal properties of the Steenrod operations Sq^i and St_p^i (Theorem 10.2), and finally on the existence of Eilenberg–MacLane complexes $K(D, n)$ for

$D = \mathbb{Z}, \mathbb{Z}_{p^n}$ (p prime), and therefore for any finitely generated abelian group (see §10.2). As an essential initial step towards our goal we shall in this subsection find all cohomological operations modulo an arbitrary prime p.

Recall first from §9.2 (and §4, Example (g), and §7) that:

(i) $K(\mathbb{Z}, 2) \sim \mathbb{C}P^\infty$; $H^*(\mathbb{C}P^\infty; \mathbb{Z}_p) \simeq \mathbb{Z}_p[t]$, $t \in H^2(\mathbb{C}P^\infty; \mathbb{Z}_p)$ for all primes $p \geq 2$; and
(ii) $K(\mathbb{Z}_{p^h}, 1) \sim S^\infty/\mathbb{Z}_{p^h} = \lim_{N\to\infty} L_{p^h}^{2N+1}(1, 1, \ldots, 1)$.

Recall also from §4 (see Exercise 10 there) the fibre bundle

$$p\colon L_{p^h}^{2N+1}(1, 1, \ldots, 1) \to \mathbb{C}P^N, \quad \text{fibre } S^1, \tag{24}$$

obtained from the general Hopf bundle (see Part II, §24.3)

$$S^{2N+1} \to \mathbb{C}P^N, \quad \text{fibre } S^1, \tag{25}$$

by taking the orbit space of the sphere

$$S^{2N+1} = \{(z_0, \ldots, z_N) \mid \sum |z_j|^2 = 1\} \subset \mathbb{C}^{N+1},$$

under the action of the group $\mathbb{Z}_{p^h}$, given by

$$\mathbb{Z}_{p^h}\colon (z_0, \ldots, z_N) \to (\exp(2\pi i/p^h)z_0, \ldots, \exp(2\pi i/p^h)z_N).$$

With the object in view of computing $H^*(K(\mathbb{Z}_{p^h}, 1); \mathbb{Z}_p)$, consider to begin with the cohomology spectral sequence over $\mathbb{Z}$ of the general Hopf bundle (25). The argument of §8, Example (a), applies also over $\mathbb{Z}$, yielding the action of d_2^* given by Figure 43. Essentially since the boundary of a 2-cell in S^{2N+1} projects via the identification under the action of $\mathbb{Z}_{p^h}$ to a 1-chain in $L_{p^h}^{2N+1}(1, 1, \ldots, 1)$ homologous to a multiple by p^h of a 1-cycle, it follows that in the cohomology spectral sequence over $\mathbb{Z}$ of the fibre bundle (24) with total space $L_{p^h}^{2N+1}(1, 1, \ldots, 1)$, the action of d_2^* is as indicated (here u is a generator of $E_2^{0,1} \simeq \mathbb{Z}$ and v of $E_2^{2,0} \simeq \mathbb{Z}$):

1	u	0	uv	0	uv^2	0	
0	1	0	v	0	v^2	0	
	0	1	2	3	4		…

$$d_2^*\colon\ u \to p^h v, \quad v \to 0$$
$$uv \to p^h v^2,$$
$$\cdots\cdots\cdots\cdots$$
$$uv^k \to p^h v^{k+1}.$$

Since the properties of the cohomology spectral sequence of a fibre bundle listed in §8 are also valid over $\mathbb{Z}$ provided that at least one of the additive groups $H^*(B; \mathbb{Z})$, $H^*(F; \mathbb{Z})$ is torsion free (which is certainly the case here), we deduce that for

$$K = K(\mathbb{Z}_{p^h}, 1) \sim \lim_{N\to\infty} L_{p^h}^{2N+1}(1, 1, \ldots, 1),$$

we have

$$H^{2q}(K; \mathbb{Z}) \simeq \mathbb{Z}_{p^h}, \qquad q = 1, 2, 3, \ldots,$$
$$H^{2q+1}(K; \mathbb{Z}) = 0, \qquad q > 0.$$

For the cohomology spectral sequence over $\mathbb{Z}_p$ of the fibre bundle (24) it follows that

$$d_2^* u = 0, \qquad d_2^* v = 0, \qquad E_2^{p,q} = E_r^{p,q} = E_\infty^{p,q},$$

whence, in view of the fact that $E_2^{p,q} = H^p(\mathbb{C}P^n; \mathbb{Z}_p) \otimes H^q(S^1; \mathbb{Z}_p)$, we obtain

$$H^*(K(\mathbb{Z}_{p^h}, 1); \mathbb{Z}_p) \simeq \bigwedge[u] \otimes \mathbb{Z}_p[v],$$

a free skew-commutative algebra over $\mathbb{Z}_p$.

Consider now the generalized "Bockstein operators" δ_i defined in §10.1, as they apply to $H^*(K; \mathbb{Z}_p)$. From the above information about the integral cohomology of K and the properties of the operators δ_i as given in §10.1, it follows that:

$$\delta_j(uv^k) = 0, \quad j < h; \qquad \delta_q(v^k) = 0 \qquad \text{for all} \quad q;$$

$$\delta_h(u) = v; \qquad \delta_h(uv^k) = v^{k+1}.$$

(Note incidentally that the Bockstein operators obey Leibniz' formula wherever it makes sense, i.e.

$$\delta_i(\alpha\beta) = (\delta_i\alpha)\beta \pm \alpha(\delta_i\beta),$$

provided $\delta_i\alpha$ and $\delta_i\beta$ are defined. This is essentially a consequence of the fact that δ_i is defined in terms of the boundary operator on integral cochains.)

Exercise

17. Using the above information, show that if p and q are distinct primes then $H^*(K(\mathbb{Z}_{p^h}, n); \mathbb{Z}_q) = 0$. (*Hint*. Use induction in conjunction with the usual fibration $E \to K(\mathbb{Z}_{p^h}, n)$ with fibre $\Omega(K(\mathbb{Z}_{p^h}, n)) \sim K(\mathbb{Z}_{p^h}, n-1)$.)

For a given fibre bundle $p: E \to B$ with fibre F and simply-connected base B, we define a many-valued map τ called the *transgression homomorphism* by

$$\tau = (p^*)^{-1}\delta^*: H^q(F) \supset A^q \to H^{q+1}(B),$$

where $\delta^*: H^q(F) \to H^{q+1}(E, F)$ is the coboundary homomorphism (figuring in the cohomology exact sequence of the pair (E, F); see §5), p^* is the "projection homomorphism" induced by p:

$$p^*: H^q(B) \to H^q(E, F),$$

and

$$A^q = (\delta^*)^{-1}(p^* H^{q+1}(B) \cap \operatorname{Im} \delta^*)$$

is the actual domain of definition of τ as a many-valued "homomorphism" in general not everywhere defined on $H^q(F)$. The elements of A^q will be termed *transgressive*.

If $i: F \to E$ is any inclusion then the induced "pull-back homomorphism" $i^*: H^q(E) \to H^q(F)$ satisfies $\operatorname{Im} i^* = \operatorname{Ker} \delta^*$ by exactness, whence it follows easily that $A^q \supset \operatorname{Im} i^*$ and $\tau(\operatorname{Im} i^*) = 0$.

In terms of the cohomology spectral sequence of the fibre bundle, where we have $E_2^{q,0} = H^q(B)$, $E_2^{0,q} = H^q(F)$ (see §8), the transgression homomorphism

τ can be defined equivalently by:

$$A^q = \bigcap_{r \leq q} \operatorname{Ker} d_r^* = E_{q+1}^{0,q} \subset H^q(F),$$
$$\tau = d_{q+1}^* \quad \text{on the group } A^q. \tag{26}$$

Since the Steenrod operations Sq^i, St_p^i, the Bockstein operators δ_i (see §10.1) and the operation δ^*, as stable $\mathbb{Z}_p$-cohomology operations all commute with continuous maps and so in particular with p^*, and also with δ^*, it follows that they commute also with the transgression homomorphism τ. This means firstly that for every transgressive element $x \in A^q$ (i.e. for every element x at which τ is defined) and for any of the above-mentioned stable cohomology operations θ, we have $\theta x \in A^{q+j}$, i.e. the image of x is again transgressive, and secondly that

$$\theta\tau = \tau\theta, \qquad \text{for} \quad \theta = \delta, \delta_h, Sq^i, St_p^i. \tag{27}$$

We shall now compute the "stable" cohomology groups over $\mathbb{Z}_p$ of the Eilenberg–MacLane complexes $K(\mathbb{Z}, n)$ and $K(\mathbb{Z}_{p^h}, n)$ (see (i), (ii) above) in terms of the Steenrod operations.

The case $K = K(\mathbb{Z}, n)$, $p = 2$. Since we know the structure of

$$H^*(K(\mathbb{Z}, 1); \mathbb{Z}_2) \quad \text{and} \quad H^*(K(\mathbb{Z}, 2); \mathbb{Z}_2)$$

(see (i) above for the latter), we begin with the first unfamiliar case $K(\mathbb{Z}, 3)$. Consider the Serre fibration $E \to B = K(\mathbb{Z}, 3)$ with fibre $F \sim \Omega(B) \sim K(\mathbb{Z}, 2)$ above each point $y \in K(\mathbb{Z}, 3)$, consisting of all paths in $K(\mathbb{Z}, 3)$ from some prescribed point x_0 to y (see Part II, §22.1); that $F \sim K(\mathbb{Z}, 2)$ is immediate from Part II, Corollary 22.2.3, according to which $\pi_i(\Omega(B)) \simeq \pi_{i+1}(B)$. We know from §8 that

$$E_2^{p,q} = H^p(B; \mathbb{Z}_2) \otimes H^q(F; \mathbb{Z}_2)$$

in the cohomology spectral sequence over $\mathbb{Z}_2$ of this fibre space. Since by (i) $H^q(F; \mathbb{Z}_2) = 0$ for odd q, we have $E_2^{p,q} = 0$ for q odd, whence $d_2^* = 0$, and consequently $E_2^{p,q} = E_3^{p,q}$. The map $d_3^*\colon E_3^{0,2} \to E_3^{3,0}$ is bijective, since if not surjective then $E_4^{3,0} = E_\infty^{3,0} \neq 0$, and if not injective then $E_4^{0,2} = E_\infty^{0,2} \neq 0$, in either case contradicting the contractibility of the total space E of the fibre space. Let $u \in E_3^{0,2} = E_2^{0,2} = H^2(F; \mathbb{Z}_2)$ be the generator of $H^*(F; \mathbb{Z}_2) = \mathbb{Z}_2[u]$ (see (i) above). Since by (26) $\tau = d_3^*$ on $A^2 = E_3^{0,2}$, the element u is transgressive; write $v = \tau(u) = d_3^*(u)$, the generator of $E_3^{3,0} = H^3(B; \mathbb{Z}_2) \simeq \mathbb{Z}_2$. From the properties of the Steenrod operations Sq^i given in Theorem 10.2, we have

$$Sq^2(u) = u^2, \qquad Sq^4 Sq^2(u) = (u^2)^2 = u^4,$$

and by (27) these elements are transgressive. Hence again by (27)

$$\tau(u^2) = \tau Sq^2(u) = Sq^2\tau(u) = Sq^2(v) \in H^5(B; \mathbb{Z}_2),$$
$$\tau(u^4) = Sq^4 Sq^2(v) \in H^9(B; \mathbb{Z}_2),$$
$$\cdots\cdots\cdots\cdots\cdots$$

Further examination of the cohomology spectral sequence shows that in fact the algebra $H^*(B; \mathbb{Z}_2)$ is the polynomial algebra with generators $\tau(u^{2^i})$,

$i = 0, 1, \ldots$:

$$H^*(B_3; \mathbb{Z}_2) = \mathbb{Z}_2[v, Sq^2 v, Sq^4 Sq^2 v, \ldots],$$
$$B_3 = B = K(\mathbb{Z}, 3), \qquad \delta_h v = 0, \qquad Sq^3 v = v^2.$$

(To see, for instance, that $H^4(B; \mathbb{Z}_2) = 0$, one observes that $d_3^*: E_3^{1,2} \to E_3^{4,0}$ must be surjective since otherwise $E_4^{4,0} = E_\infty^{4,0} \neq 0$; thus since $E_3^{1,2} = 0$, one obtains $E_3^{4,0} = E_2^{4,0} = H^4(B) = 0$.)

Having determined the structure of $H^*(K(\mathbb{Z}, 3); \mathbb{Z}_2)$, we proceed to the next fibre space

$$E \to B_4 = K(\mathbb{Z}, 4), \qquad \text{with fibre} \quad f \sim K(\mathbb{Z}, 3).$$

A similar investigation of the cohomology spectral sequence of this fibre space yields, in addition to the former transgressive elements

$$v, Sq^2 v, \ldots, Sq^{2^i} Sq^{2^{i-1}} \ldots Sq^2 v$$

generating the non-zero $E_2^{0,q} = H^q(F; \mathbb{Z}_2)$, new ones

$$v^2 = Sq^3 v, \qquad v^4 = Sq^6 Sq^3 v, \ldots,$$

obtained by raising v to the power 2^i. From these transgressive elements the generators of $H^*(B_4; \mathbb{Z}_2)$ are obtained by applying the transgression homomorphism τ. Continuing in this way, exploiting consecutively the fibre spaces

$$E \to B_n = K(\mathbb{Z}, n), \qquad \text{with fibre} \quad F \sim K(\mathbb{Z}, n-1),$$

one obtains the following generators of the first few "stable" cohomology groups $H^{n+q}(K(\mathbb{Z}, n); \mathbb{Z}_2)$, $q < n$ (here the generator of $H^n(K(\mathbb{Z}, n); \mathbb{Z}_2)$ is denoted uniformly by u):

$q = 0$	1	2	3	4
u	0	$Sq^2 u$	$Sq^3 u$	$Sq^4 u$

$q = 5$	6	7	8	9	...
$Sq^5 u$	$Sq^6 u$ $Sq^4 Sq^2 u$	$Sq^7 u$ $Sq^5 Sq^2 u$	$Sq^8 u$ $Sq^6 Sq^2 u$	$Sq^9 u$ $Sq^7 Sq^2 u$ $Sq^6 Sq^3 u$	...

(28)

For $K = K(\mathbb{Z}_2, n)$ the argument proceeds along entirely similar lines, beginning with the fibre space

$$E \to B = K(\mathbb{Z}_2, 2), \qquad \text{with fibre} \quad F \sim K(\mathbb{Z}_2, 1) = \mathbb{R}P^\infty.$$

At this initial stage one obtains the transgressive elements

$$u \in H^1(F; \mathbb{Z}_2), \qquad Sq^1 u = \delta_1 u = u^2,$$
$$Sq^2 Sq^1 u = (u^2)^2, \ldots, Sq^{2^i} Sq^{2^{i-1}} \ldots Sq^2 Sq^1 u = (\ldots(u^2)^2 \ldots)^2.$$

By iterating the argument as before, the generators of the first few "stable" cohomology groups $H^{n+q}(K(\mathbb{Z}_2, n); \mathbb{Z}_2)$, $q < n$, turn out as follows:

$q = 0$	1	2	3	4	5	6	...
u	Sq^1u	Sq^2u	Sq^3u Sq^2Sq^1u	Sq^4u Sq^3Sq^1u	Sq^5u Sq^4Sq^1u	Sq^6u Sq^5Sq^1u Sq^4Sq^2u	...

(29)

Remark. In our ultimate application (to the computation of the first few stable homotopy groups of spheres), knowledge of the groups $H^{n+q}(K; \mathbb{Z}_2)$ and $H^{n+q}(K; \mathbb{Z}_p)$ will be required only for $q < 7$, so that the details of the verifications of the more far-reaching of our assertions will not be strictly relevant to this specific goal; for $q < 7$ they may all be verified by means of elementary arguments, of the type exemplified above, involving the appropriate spectral sequences.

The generators of the stable groups $H^{n+q}(K(\mathbb{Z}_{2^h}, n); \mathbb{Z}_2)$, $q < n$, turn out for $q \leq 4$ to be:

$q = 0$	1	2	3	4	
u	$\delta_h u$	Sq^2u	Sq^3u $Sq^2\delta_h u$	Sq^4u $Sq^3\delta_h u$	

(30)

An analogous investigation in the case of odd primes p (we omit the details) yields the following generators for the stable cohomology groups $H^{n+q}(K; \mathbb{Z}_p)$, $q < n$, of $K = K(\mathbb{Z}, n)$, $K(\mathbb{Z}_{p^h}, n)$.

For $K = K(\mathbb{Z}, n)$ the generators are given by:

$q = 0$	1	2	...	$2p - 2$	$2p - 1$	
u	0	0	0	St_p^1u	$\delta_1 St_p^1 u$	

(31)

For $K = K(\mathbb{Z}_{p^h}, n)$ the generators are:

$q = 0$	1	2	...	$2p - 2$	$2p - 1$	
u	$\delta_h u$	0	0	St_p^1u	$\delta_1 St_p^1 u$ $St_p^1\delta_h u$	

(32)

Thus we see that, at least in the dimensions indicated, all stable cohomology operations (see §10.1 for the definition)

$$\theta: H^n(K; \mathbb{Z}_p) \to H^{n+q}(K; \mathbb{Z}_p), \qquad p \geq 2,$$

reduce to successive applications of the Steenrod operations Sq^i, St_p^i, and the Bockstein operation δ_1. It follows that all stable cohomology operations

$$\theta: H^n(K; \mathbb{Z}) \to H^{n+q}(K; \mathbb{Z}_p),$$
$$\theta: H^n(K; \mathbb{Z}_{p^h}) \to H^{n+q}(K; \mathbb{Z}_p),$$

reduce (after appropriate reduction modulo p) to a succession of applications of these same operations to the generator $u \in H^n(K)$, and also, in the latter case, to $\delta_h u$.

If we define the *dimension* of a stable cohomology operation

$$\theta: H^n(X; \mathbb{Z}_p) \to H^{n+q}(X; \mathbb{Z}_p)$$

to be q, then the set of all mod p stable cohomology operations forms a graded "Steenrod algebra"

$$S_p = \sum_{q \geq 0} S^{(q)},$$

where $S^{(0)}$ consists of the scalars (i.e. the operations of scalar multiplication by the elements of $\mathbb{Z}_p$), and $S^{(q)}$ is the set of all stable operations of dimension q. From the above tables of generators of the stable cohomology groups over $\mathbb{Z}_p$ (including $\mathbb{Z}_2$) of $K(\mathbb{Z}_{p^h}, n)$, we obtain the following basis elements in low dimensions q for the algebra S_p:

$p = 2$:

$q = 0$	1	2	3	4	5	6	7	8
1	Sq^1	Sq^2	Sq^3 Sq^2Sq^1	Sq^4 Sq^3Sq^1	Sq^5 Sq^4Sq^1	Sq^6 Sq^5Sq^1 Sq^4Sq^2	Sq^7 Sq^6Sq^1 Sq^5Sq^2 $Sq^4Sq^2Sq^1$	Sq^8 Sq^7Sq^1 Sq^6Sq^2 $Sq^5Sq^2Sq^1$

(33)

$p > 2$:

$q = 0$	1	2	...	$2p - 2$	$2p - 1$	
1	δ_1	0	0	St_p^1	$\delta_1 St_p^1$ $St_p^1 \delta_1$	

(34)

There is a curious phenomenon evident from these tables of basis elements, of considerable significance, as we shall see, even in the case $p = 2$, namely that (considering the case $p = 2$) the basis of stable operations for the Steenrod algebra does not exhaust the set of all possible products (under composition) $Sq^{i_1}Sq^{i_2}\dots Sq^{i_k}$ of the Steenrod operations, so that consequently there must exist non-trivial relations amongst the Sq^i. We shall now describe a method for finding these relations. Consider the product $\mathbb{R}P_1^\infty \times \cdots \times \mathbb{R}P_n^\infty$ of n copies of $\mathbb{R}P^\infty$, and the element

$$u = t_1 \dots t_n \in H^n(\mathbb{R}P_1^\infty \times \cdots \times \mathbb{R}P_n^\infty; \mathbb{Z}_2),$$
$$0 \neq t_i \in H^1(\mathbb{R}P_i^\infty; \mathbb{Z}_2) \simeq \mathbb{Z}_2.$$

In view of the formal properties of the Steenrod operations Sq^i (see Theorem 10.2) we have $Sq^1 t_i = t_i^2$, $Sq^0 t_i = t_i$, $Sq^j t_i = 0$ for $j \neq 0, 1$, whence we can compute the result of applying any operation $Sq^{i_1}\dots Sq^{i_k}$ to u by using the property $Sq^i(xy) = \sum_{j+k=i} Sq^j(x)Sq^k(y)$. One consequence of this is that every basic operation

$$\theta\colon H^n(K(\mathbb{Z}_2^n, n); \mathbb{Z}_2) \to H^{n+q}(K(\mathbb{Z}_2^n, n); \mathbb{Z}_2), \qquad q < n,$$

acts non-trivially on u, i.e. $\theta(u) \neq 0$. We shall now verify this by direct calculation for $q \leq 8$. (Note that the basic operations for $q \leq 8$ are given in the table (33).) Denoting by σ_j the jth elementary symmetric polynomial in the t_i:

$$\sigma_j = \sum_{i_1 < \cdots < i_j} t_{i_1} \dots t_{i_j},$$

we have:

$q = 1$: $$Sq^1 u = \left(\sum_{i=1}^n t_i\right) u;$$

$q = 2$: $$Sq^2 u = \left(\sum_{i<j} t_i t_j\right) u;$$

$q = 3$: $$Sq^3 u = \left(\sum_{i<j<k} t_i t_j t_k\right) u,$$
$$Sq^2 Sq^1 u = (\sigma_1\sigma_2 + \sigma_1^3)u;$$

$q = 4$: $$Sq^4 u = \left(\sum_{i<j<k<l} t_i t_j t_k t_l\right) u = \sigma_4 u,$$
$$Sq^3 Sq^1 u = (\sigma_3\sigma_1 + \sigma_1^2\sigma_2)u;$$

$q = 5$: $$Sq^5 u = \sigma_5 u, \quad Sq^4 Sq^1 u = (\sigma_4\sigma_1 + \sigma_3\sigma_1^2)u;$$

$q = 6$: $$Sq^6 u = \sigma_6 u, \quad Sq^5 Sq^1 u = (\sigma_5\sigma_1 + \sigma_4\sigma_1^2)u;$$
$$Sq^4 Sq^2 u = (\sigma_4\sigma_2 + \sigma_3^2 + \sigma_3\sigma_2\sigma_1 + \sigma_2^3)u;$$

$q = 7$: $$Sq^7 u = \sigma_7 u, \quad Sq^6 Sq^1 u = (\sigma_6\sigma_1 + \sigma_5\sigma_1^2)u,$$
$$Sq^5 Sq^2 u = (\sigma_5\sigma_2 + \sigma_4\sigma_3 + \sigma_4\sigma_2\sigma_1 + \sigma_3\sigma_2^2)u,$$
$$Sq^4 Sq^2 Sq^1 u = (\sigma_4\sigma_2\sigma_1 + \sigma_4\sigma_1^3 + \sigma_3^2\sigma_1 + \sigma_2\sigma_1^5 + \sigma_3\sigma_2\sigma_1^2 + \sigma_2^3\sigma_1 + \sigma_2^2\sigma_1^3 + \sigma_1^7)u;$$

$$\begin{aligned} q = 8: \quad & Sq^8 u = \sigma_8 u, \quad Sq^7 Sq^1 u = (\sigma_7\sigma_1 + \sigma_6\sigma_1^2)u, \\ & Sq^6 Sq^2 u = (\sigma_6\sigma_2 + \sigma_5\sigma_3 + \sigma_5\sigma_2\sigma_1 + \sigma_4\sigma_2^2)u, \\ & Sq^5 Sq^2 Sq^1 u = (\sigma_5\sigma_2\sigma_1 + \sigma_5\sigma_1^3 + \sigma_4\sigma_3\sigma_1 + \sigma_3^2\sigma_1^2 + \sigma_3\sigma_2^2\sigma_1 + \sigma_3\sigma_1^5 \\ & \qquad + \sigma_2^3\sigma_1^2 + \sigma_2\sigma_1^6)u. \end{aligned} \tag{35}$$

It is clear from this that, at least for $q \leq 8$, the basic operations $\theta \in S^{(q)}$ act on u non-trivially, in fact linearly independently. It can be shown along these lines that, more generally, the products of the form

$$Sq^{i_k} \dots Sq^{i_1}, \qquad \text{with} \quad i_k \geq 2i_{k-1}, \quad i_{k-1} \geq 2i_{k-2}, \dots, i_2 \geq 2i_1, \tag{36}$$

constitute a full linearly independent additive basis for the Steenrod algebra S_2.

We seek the relations between the operations Sq^i in the standard form

$$Sq^i Sq^j = \sum_{a \geq 2b} \lambda_{a,b}^{i,j} Sq^a Sq^b, \qquad a \geq 2b, \quad 0 < i < 2j.$$

EXERCISE

18. Find the coefficients $\lambda_{a,b}^{i,j}$ for every q.

For $q \leq 8$ direct calculation yields

$$\lambda_{a,b}^{i,j} = \delta_{a+b}^{i+j} \binom{i-2b}{j-b-1},$$

(which is in fact valid for all q). Thus for $q \leq 8$ we have the following relations

$$\begin{aligned} & Sq^1 Sq^1 = \delta_1^2 = 0, \\ & Sq^1 Sq^2 = Sq^3, \\ & Sq^1 Sq^3 = 0, \quad Sq^1 Sq^{2q} = Sq^{2q+1}, \\ & Sq^2 Sq^2 = Sq^3 Sq^1, \quad Sq^2 Sq^3 = Sq^5 + Sq^4 Sq^1, \quad Sq^2 Sq^4 = Sq^5 Sq^1 + Sq^6, \\ & Sq^3 Sq^3 = Sq^5 Sq^1. \end{aligned} \tag{37}$$

10.6. Computation of the First Few Nontrivial Stable Homotopy Groups of Spheres

Let $f: S^n \to K = K(\mathbb{Z}, n)$ be a map representing a generator of $\pi_n(K(\mathbb{Z}, n)) \simeq \mathbb{Z}$; the induced map $f^*: H^n(K; \mathbb{Z}) \to H^n(S^n; \mathbb{Z})$ is then an isomorphism. Convert the map f to a fibre-space projection, where the total space and the base are homotopically equivalent to S^n and K respectively, in accordance with the prescription given in §8. From the exact sequence of this fibre space (see Part II, §22.2) it is easily inferred that for the fibre $F = F_1$ we have $\pi_i(F) \simeq \pi_i(S^n)$ for $i \geq n + 1$, $\pi_i(F) = 0$ for $i < n - 1$, and that $\pi_n(F) = 0$ if $\pi_{n-1}(F) = 0$. From the homology spectral sequence, whose properties, as given in §8, continue to hold over $\mathbb{Z}$ under conditions satisfied by the present fibre space, one easily verifies that $H_{n-1}(F; \mathbb{Z}) = 0$, whence by Hurewicz' theorem (Corollary 4.9) $\pi_{n-1}(F) = 0$. Thus in sum we conclude that

$$\pi_i(F) = 0 \quad \text{for } i \le n; \qquad \pi_i(F) \simeq \pi_i(S^n) \quad \text{for } i \ge n+1. \tag{38}$$

In the cohomology spectral sequence over $\mathbb{Z}_p$ of the fibre space, we have by (38)

$$E_2^m = \sum_{i+j=m} E_2^{i,j} = E_2^{m,0} + E_2^{0,m} \qquad \text{for} \quad m < 2n,$$

whence it follows, again by (38), that $E_2^{m,0} = E_{m+1}^{m,0}$, $E_2^{0,m} = E_{m+1}^{0,m}$. Hence the transgression homomorphism $\tau = d_{m+1}^*\colon E_{m+1}^{0,m} \to E_{m+1}^{m+1,0}$ is a map from $H^m(F; \mathbb{Z}_p)$ to $H^{m+1}(K; \mathbb{Z}_p)$. For $2n > m > n$ it is readily seen to be bijective using $H^m(S^n) = 0$. Hence we have the isomorphism

$$\tau\colon H^{n+q}(F; \mathbb{Z}_p) \simeq H^{n+q+1}(K; \mathbb{Z}_p), \qquad 0 < q < n,$$

which by (27) commutes with the operators Sq^i $(p = 2)$, St_p^i $(p > 2)$, δ_h. From this isomorphism and table (31) of the preceding subsection, giving the generators of the groups $H^{n+q}(K(\mathbb{Z}, n); \mathbb{Z}_p)$ for certain $q < n$, we obtain the following table of generators for the $H^{n+q}(F; \mathbb{Z}_p)$:

$p > 2$:

$q = 0$	1	...	$2p-3$	$2p-2$	
0	0	0	v	$\delta_1 v$	

Noting the easily verified triviality of $H^j(F; \mathbb{Z}_p)$ for $j \le n$ and all primes p, we now see that in fact for every prime p (including $p = 2$) we have

$$H^j(F; \mathbb{Z}_p) = 0 \qquad \text{for} \quad 0 < j < n + 2p - 3 < 2n, \quad p \ge 2,$$

whence by Theorem 2.9 the corresponding homology groups are also trivial. It now follows from the dual of Exercise 2 (§10.1) that the corresponding integral homology groups have no elements of order a power of p:

$$H_j^{(p)}(F; \mathbb{Z}) = 0 \qquad \text{for} \quad 0 < j < n + 2p - 3 < 2n, \quad p \ge 2. \tag{39}$$

On the other hand, since for $p > 2$ we have $\delta_1 v \ne 0$, the dual operator

$$\partial_1\colon H_{n+2p-2}(F; \mathbb{Z}) \to H_{n+2p-3}(F; \mathbb{Z})$$

is also non-zero, whence we infer via the statement following the proof of Proposition 3.8, that the p-component of $H_{n+2p-3}(F; \mathbb{Z})$ is $\mathbb{Z}_p$:

$$H_{n+2p-3}^{(p)}(F; \mathbb{Z}) \simeq \mathbb{Z}_p. \tag{40}$$

In view of (39) and (40), we may now conclude that for $p \ge 2$ the p-components of the homotopy groups satisfy

$$\pi_{n+q}^{(p)}(S_n) \simeq \pi_{n+q}^{(p)}(F) = 0, \qquad 0 < q < 2p - 3 < n, \quad p \ge 2, \tag{41}$$

and for $p > 2$,

$$\pi^{(p)}_{n+2p-3}(S^n) \simeq \pi^{(p)}_{n+2p-3}(F) \simeq \mathbb{Z}_p, \qquad 2p - 3 < n, \quad p > 2. \tag{42}$$

In fact the same argument serves to establish $\pi^{(2)}_{n+1}(S^n) \simeq \mathbb{Z}_2$: From the isomorphism

$$\tau\colon H^{n+q}(F; \mathbb{Z}_2) \simeq H^{n+q+1}(K(\mathbb{Z}, n); \mathbb{Z}_2)$$

and table (28), we obtain the following table of generators of the groups $H^{n+q}(F; \mathbb{Z}_2)$:

$q = 1$	2	3	4	5	6	
v	$Sq^1 v$	w $(Sq^2 v = 0)$	$Sq^1 w$ $(= Sq^2 Sq^1 v)$	$Sq^4 v$ $Sq^2 w$	$Sq^3 w$ $Sq^5 v$	

(43)

Since $Sq^1 = \delta_1$ (see Theorem 10.2), the above argument for the case $p > 2$ applies also to the present case $p = 2$ yielding $\pi^{(2)}_{n+1}(S^n) \simeq \mathbb{Z}_2$. In conjunction with (41) this then implies that

$$\pi_{n+1}(S^n) \simeq \mathbb{Z}_2. \tag{44}$$

(In table (43), the generators v and w are defined by $\tau(v) = Sq^2 u$, $\tau(w) = Sq^4 u$, where, as in table (28), u denotes the generator of $H^n(K(\mathbb{Z}, n); \mathbb{Z}_2)$. From the relation $Sq^2 Sq^2 = Sq^3 Sq^1$ and the fact that $Sq^1 u = 0$ (see table (28)), we infer that $Sq^2 v = 0$. From the relation $Sq^2 Sq^4 = Sq^5 Sq^1 + Sq^6$ it follows that $Sq^2 w = Sq^6 u$. Finally, from the relation $Sq^2 Sq^3 = Sq^5 + Sq^4 Sq^1$ we obtain $Sq^1 w = Sq^2 Sq^1 v$, as indicated in the table.)

Proceeding to the next stage, we consider a map

$$f\colon F_1 \to K(\mathbb{Z}_2, n+1) \qquad (F_1 = F),$$

with the property that the induced map $f_*\colon \pi_{n+1}(F_1) \to \pi_{n+1}(K(\mathbb{Z}_2, n+1))$ is an isomorphism. (Note the difference between the present base $K(\mathbb{Z}_2, n+1)$ and the earlier one $K(\mathbb{Z}, n)$.) Converting f to a fibre-space projection with fibre F_2 say, and arguing much as before, one finds that

$$\begin{aligned} \pi_j(F_2) &= 0 && \text{for } j \le n+1, \\ \pi_j(F_2) &= \pi_j(F_1) = \pi_j(S^n) && \text{for } j > n+1, \\ H^j(F_2; \mathbb{Z}_2) &= 0 && \text{for } 0 < j \le n+1. \end{aligned}$$

From the cohomology spectral sequence of this fibre space the following long exact sequence can be obtained in the stable range:

$$H^{n+q}(F_1; \mathbb{Z}_2) \xrightarrow{i^*} H^{n+q}(F_2; \mathbb{Z}_2) \xrightarrow{\tau} B^q \xrightarrow{f^*} H^{n+q+1}(F_1; \mathbb{Z}_2) \xrightarrow{i^*} H^{n+q+1}(F_2; \mathbb{Z}_2), \tag{45}$$

where i^* is induced from an inclusion $F_2 \subset F$, and $B^q = H^{n+q+1}(K(\mathbb{Z}_2, n+1); \mathbb{Z}_2)$

(and i^*, τ and f^* commute with the Steenrod operations). Since f^* is an isomorphism when $q = 0$, it must map the generator $\tilde{u}$ of $H^{n+1}(K(\mathbb{Z}_2, n+1); \mathbb{Z}_2)$ to the element v in table (43):

$$f^*(\tilde{u}) = v.$$

Consequently, f^*B^q consists precisely of the images of v under the Steenrod operations:

$$f^*B^q = S^{(q)}(v).$$

Since $Sq^2 v = 0$ (see (43)) and $f^*\tilde{u} = v$, we have $f^*(Sq^2\tilde{u}) = 0$, whence by exactness there is an element x of $H^{n+2}(F_2; \mathbb{Z}_2)$ such that $\tau(x) = Sq^2(\tilde{u})$. In terms of $x = \tau^{-1}(Sq^2\tilde{u})$ and $\tilde{w} = i^*(w)$, we obtain from the exactness of the sequence (45) and tables (29) and (43) the following table of generators of the $H^{n+q}(F_2; \mathbb{Z}_2)$:

$q = 2$	3	4	5	6	
	$\tilde{w}$	$(Sq^1\tilde{w} = 0)$	$Sq^2\tilde{w}$	$Sq^3\tilde{w}$	
x	$Sq^1 x$	$Sq^2 x = \delta_2 \tilde{w}$	$(Sq^3 x = 0)$	$Sq^4 x$	

(46)

Since

$$\tau(Sq^1 x) = Sq^1\tau(x) = Sq^1 Sq^2 \tilde{u} = Sq^3\tilde{u},$$

and $Sq^3\tilde{u} \neq 0$ by (29), it follows that, as indicated in the table,

$$Sq^1 x = \delta_1 x \neq 0.$$

From this we infer, via an argument similar to that used to derive (40), that in the integral homology group of F_2 of dimension $n + 2$ we have

$$H_{n+2}^{(2)}(F_2; \mathbb{Z}) \simeq \mathbb{Z}_2.$$

Taking (41) into account, we conclude that

$$\pi_{n+2}(S^n) \simeq \pi_{n+2}^{(2)}(S^n) \simeq \pi_{n+2}^{(2)}(F_2) \simeq \mathbb{Z}_2. \tag{47}$$

Before going to the next stage we show how some of the other information contained in table (46) may be verified. Observe first that since $Sq^2 Sq^2 \tilde{u} = Sq^3 Sq^1 \tilde{u} \neq 0$ (see (29)), we have

$$\tau(Sq^2 x) = Sq^2 \tau(x) = Sq^2 Sq^2 \tilde{u} \neq 0,$$

so that $Sq^2 x$ is indeed a generator of $H^{n+4}(F_2; \mathbb{Z}_2)$ as indicated in the table. To see that $Sq^1\tilde{w} = 0$, note first that $i^*v = 0$ by exactness of the sequence (45) with $q = 0$. Then since $Sq^1 w = Sq^2 Sq^1 v$ (see (43)), we deduce that

$$Sq^1\tilde{w} = i^* Sq^1 w = i^* Sq^2 Sq^1 v = Sq^2 Sq^1 (i^* v) = 0.$$

To see that $Sq^2 x = \delta_2 \tilde{w}$ and $Sq^3 x = 0$, we use a particular case of the following general lemma, which we state without proof.

10.13. Lemma. *If* $a = f^*(\bar{a}) = \delta_h w$ *and* $b = \tau^{-1}\delta_1(\bar{a})$, *then the elements* $b, \tilde{w} \in H^*(F_2; \mathbb{Z}_2)$ *satisfy* $b = \delta_{h+1}\tilde{w}$.

(The proof, which is essentially elementary, involves properties of the coboundary operator on the cochain complex $C^*(E; \mathbb{Z})$; we invite the reader to supply the details.)

Taking now $a = Sq^1 w = Sq^2 Sq^1 v = f^*(Sq^2 Sq^1 \tilde{u})$, and

$$b = Sq^2 x = \tau^{-1}(Sq^2 Sq^2 \tilde{u}) = \tau^{-1} Sq^1(Sq^2 Sq^1 \tilde{u}) = \tau^{-1}\delta_1(Sq^2 Sq^1 \tilde{u})$$

(and consequently $\bar{a} = Sq^2 Sq^1 \tilde{u}$), we infer immediately from the lemma that $Sq^2 x = \delta_2 \tilde{w}$, whence $Sq^3 x = Sq^1 Sq^2 x = \delta_1 \delta_1 \tilde{w} = 0$.

At the third (and for us final) stage one considers the fibre space

$$F_2 \xrightarrow{f} K(\mathbb{Z}_2, n+2), \qquad \text{with fibre } F_3,$$

obtained from a map f inducing an isomorphism

$$f_*: \pi_{n+2}(F_3) \simeq \pi_{n+2}(K(\mathbb{Z}_2, n+2)).$$

Once again it may be shown, much as at the first stage, that

$$\begin{aligned} \pi_j(F_3) &= 0 && \text{for} \quad j \le n+2, \\ \pi_j(F_3) &\simeq \pi_j(F_2) \simeq \pi_j(S^n) && \text{for} \quad j \ge n+3, \\ H^j(F_3; \mathbb{Z}_2) &= 0 && \text{for} \quad 0 < j \le n+2. \end{aligned}$$

In the stable range $q < n$ the spectral sequence of the fibration yields the exact sequence

$$H^{n+q}(F_2; \mathbb{Z}_2) \xrightarrow{i^*} H^{n+q}(F_3; \mathbb{Z}_2) \xrightarrow{\tau} B^{q-1} \xrightarrow{f^*} H^{n+q+1}(F_2; \mathbb{Z}_2) \xrightarrow{i^*}, \tag{48}$$

where $B^{q-1} = H^{n+q+1}(K(\mathbb{Z}_2, n+2); \mathbb{Z}_2)$. Denoting by $\hat{u}$ the generator of $H^{n+2}(K(\mathbb{Z}_2, n+2); \mathbb{Z}_2)$, it folows by the exactness of (48) with $q = 1$ that $f^*(\hat{u}) = x$, where x is the generator of $H^{n+2}(F_2; \mathbb{Z}_2)$ (see (46)). Hence

$$f^*(B^{q-1}) = S^{(q-1)}(x).$$

From the exact sequence (48), in conjunction with tables (29) and (46), we obtain the following table of generators of the first few non-trivial groups $H^{n+q}(F_3, \mathbb{Z}_2)$ (here $\hat{w} = i^*\tilde{w}$):

$q = 2$	3	4	5	
0	$\hat{w}$	$\delta_3 \hat{w}$	$Sq^2 \hat{w}$	

We now verify some of the information contained in this table. To see that $\hat{w} \ne 0$, observe that if the contrary were true, then by the exactness of (48) we should have $f^*(Sq^1 \hat{u}) = \tilde{w}$, the element $Sq^1 \hat{u}$ being the unique generator of $H^{n+3}(K(\mathbb{Z}_2, n+2); \mathbb{Z}_2)$ (see table (29)); however, then $\tilde{w} = Sq^1 f^* \hat{u} = Sq^1 x$, contradicting (46). To see that, as indicated in the table, $\delta_3\hat{w}$ generates $H^{n+4}(F_3; \mathbb{Z}_2)$, note first that from the exact sequence (48) with $q = 4$, together with information obtained earlier, one easily infers that $H^{n+4}(F_3; \mathbb{Z}_2)$ is generated by $\tau^{-1}(Sq^3\hat{u})$. Since

$$\tau^{-1}(Sq^3\hat{u}) = \tau^{-1}Sq^1Sq^2\hat{u} = \tau^{-1}\delta_1(Sq^2\hat{u}),$$

and since $f^*(Sq^2\hat{u}) = Sq^2x = \delta_2\tilde{w}$ by (46), we deduce from Lemma 10.13 with $\bar{a} = Sq^2\hat{u}$ and $\tilde{w}$, $\hat{w}$ in the roles of w, $\tilde{w}$ respectively, that $\tau^{-1}(Sq^3\hat{u}) = \delta_3\hat{w}$, whence the desired conclusion.

From this table we see that the homomorphism

$$\delta_3\colon H^{n+3}(F_3; \mathbb{Z}_2) \to H^{n+4}(F_3; \mathbb{Z}_2)$$

is non-zero, whence its dual $H_{n+4}(F_3; \mathbb{Z}_2) \to H_{n+3}(F_3; \mathbb{Z}_2)$ is also non-zero. From this we infer via Exercise 3 (dualized—see the remark following that exercise) that the 2-component of the integral homology group $H_{n+3}(F_3; \mathbb{Z})$ is cyclic of order $2^3 = 8$:

$$H^{(2)}_{n+3}(F_3; \mathbb{Z}) \simeq \mathbb{Z}_8.$$

Thus, invoking again the generalized Hurewicz theorem, we deduce that

$$\pi^{(2)}_{n+3}(S^n) \simeq \pi^{(2)}_{n+3}(F_3) \simeq \mathbb{Z}_8,$$

whence in view of (41), (42), we obtain finally

$$\pi_{n+3}(S^n) \simeq \mathbb{Z}_{24}.$$

We summarize our results as a

10.14. Theorem. *The first four stable homotopy groups $\pi_{n+q}(S^n)$, $q < n - 1$, are as follows:*

$$\pi_n(S^n) \simeq \mathbb{Z}, \qquad \pi_{n+1}(S^n) \simeq \mathbb{Z}_2, \qquad \pi_{n+2}(S^n) \simeq \mathbb{Z}_2, \qquad \pi_{n+3}(S^n) \simeq \mathbb{Z}_{24}.$$

(Note that in the cases $q = 1, 2$ these results were obtained by different methods in Part II, §23.)

EXERCISE

19. Compute the stable homotopy groups $\pi_{n+q}(S^n)$ for $q \leq 9$ (see [50]).

Remark. For $q \geq 10$ more serious difficulties arise in the computation along the above lines. These can however be overcome to some extent, so that by means of very laborious calculations the stable groups $\pi_{n+q}(S^n)$ may be determined for $q \leq 30$ or thereabouts. A general answer in reasonable form for the whole stable range of q would appear to be unattainable, although one can find in the current literature much information of a qualitative nature about the higher homotopy groups.

10.7. Stable Homotopy Classes of Maps of Cell Complexes

The following problem frequently arises: An $(n-1)$-connected cell complex K is given $(n \geq 2)$, together with a complex X of dimension $< 2n-1$, and one wishes to determine the "stable" homotopy classes of maps $X \to K$. (The complex K may be supposed without cells of dimensions $1 \leq i \leq n-1$.) A related problem (see Lemma 10.16 below) is that of determining an obstruction class $\alpha(f)$ (see §9) to the extension of a map $f: X^{n+q} \to K$ to the $(n+q+1)$-dimensional skeleton of X, where $q < n-2$.

Remark. In the preceding subsection (§10.6) we exploited instances of the following general fact: For a fibre space $E \xrightarrow{p} B$ with base B and fibre F both $(n-1)$-connected, the cohomology spectral sequence up to dimension $2n-2$ yields an exact sequence

$$H^{n+q}(E) \xrightarrow{i^*} H^{n+q}(F) \xrightarrow{\tau} H^{n+q+1}(B) \xrightarrow{i^*} H^{n+q+1}(E), \qquad q < n-1, \tag{49}$$

in view of which, as evidenced in §10.6, the complexities in the homology theory of the fibre space in stable dimensions reflect, that is, correspond in a natural way to, those of the homotopy theory of the fibration (compare (49) with the homotopy exact sequence of a fibre space given in Part II, §22.2). Observe also that in dimensions $k \leq 2n-2$ the homology theory of our fibre space is essentially the same as for a pair (E, F) with $B \sim E/F$, since in the cell decomposition of the total space E of the fibre space described in §8, non-trivial cells not contained in the base B or fibre F, occur only in dimensions $\geq 2n$ (essentially as products of cells of B and F of dimensions $\geq n$; see §8). Thus in the stable range there are several factors making for simplicity.

Returning to the problem stated initially, we begin with the following

10.15. Lemma. *Given an $(n-1)$-connected cell complex K and a complex X of dimension $< 2n-1$, the set $[X, K]$ of stable homotopy classes of maps forms an abelian group.*

Proof. Let $f, g: X \to K$ be any two maps and consider their direct product $f \times g: X \to K \times K$, $[f \times g](x) = (f(x), g(x))$. Since K is $(n-1)$-connected, we may suppose by Theorem 4.8 that K, and therefore also $K \times K$, have no cells of dimensions $1 \leq i \leq n-1$. Since by Theorem 4.6 each of the maps f, g is homotopic to a cellular map, we may assume also that f and g, and therefore their product $f \times g$, are cellular, so that in particular the image $(f \times g)(X)$ is contained in the skeleton of K of dimension $\dim X$. Since $\dim X \leq 2n-2$, it follows that $(f \times g)(X)$ lies in the bouquet $K \vee K \subset K \times K$, every cell of $K \times K$ not contained in $K \vee K$ having dimension at least $2n$. This argument applies similarly to any homotopy between cellular maps from X to $K \times K$; thus we may further assume that throughout such a homotopy the images of X are always contained in $K \vee K$.

We define the group operation on $[X, K]$ as follows. For any two classes in $[X, K]$ with representing cellular maps f and g, we define the *sum* of the classes to be the class of the map

$$f + g = \varkappa(f \times g), \tag{50}$$

where $\varkappa: K \vee K \to K$ is the "folding" map, restricting to the identity map on each member K of the bouquet. We leave to the reader the straightforward verification that this is a well-defined operation, turning $[X, K]$ into an abelian group. □

10.16. Lemma. *With X and K as in the preceding lemma, let $f: X^{n+q} \to K$ be a "stable" map (i.e. $q < n - 1$), and let*

$$\alpha(f) \in C^{n+q+1}(X^{n+q+1}; \pi_{n+q}(K))$$

be an obstruction to extending f to the $(n + q + 1)$-skeleton X^{n+q+1} of X (see §9.1). Then the obstruction depends additively on f regarded as an element of $[X^{n+q}, K]$, i.e. $\alpha(f + g) = \alpha(f) + \alpha(g)$; in particular $\alpha(\lambda f) = \lambda\alpha(f)$ for all integers λ.

PROOF. By definition the obstruction $\alpha(f)$ is zero on all but one $(n + q + 1)$-cell σ^{n+q+1}, and on that cell takes the value in $\pi_{n+q}(K)$ determined by the restriction map

$$f|_{\partial\sigma^{n+q+1}}: \partial\sigma^{n+q+1} \cong S^{n+q} \to K.$$

Given two such maps $f, g: X^{n+q} \to K$, $q < n - 1$, it is not difficult to see that under the addition of their homotopy classes defined by (50), the homotopy classes of their restrictions to $\partial\sigma^{n+q+1}$ add in accordance with the operation in $\pi_{n+q}(K)$, whence the lemma. (Note that since by hypothesis K is $(n - 1)$-connected for some $n \geq 2$, in particular $\pi_1(K)$ is trivial, so that the choice of base point for $\pi_{n+q}(K)$ is immaterial; see Part II, §22.3.) □

As in the proof of the Cartan–Serre theorem (10.7) one may construct, for appropriate free abelian groups D_j, a map

$$f: K \to \prod_{n_j \geq n}^{2n-1} K(D_j, n_j)$$

which induces, up to dimension $2n - 2$, an isomorphism between the respective rational cohomology groups (and consequently between the respective rational homotopy groups $\pi_i \otimes \mathbb{Q}$, $i \leq 2n - 2$). (Note that the rational cohomology algebra of K is "free up to dimension $2n - 2$" since by virtue of the $(n - 1)$-connectedness of K all non-trivial products in the cohomology algebra have dimension $> 2n - 2$.) We shall now indicate the construction of an "inverse" map

$$g: \left(\prod_{n_j \geq n}^{2n-1} K(D_j, n_i)\right)^{2n-2} \to K,$$

defined on the $(2n-2)$-skeleton of $\prod K(D_j, n_j)$, with the property that, up to dimension $2n-2$ the induced maps f^*, g^* of the rational cohomology groups satisfy $f^*g^* = \lambda \neq 0$ for some integer λ, i.e. $f^*g^*(x) = \lambda x$ for all $x \in H^j(K; \mathbb{Q})$ (and the induced maps f_*, g_* of the rational homotopy groups satisfy $g_*f_* = \lambda$). We construct g inductively as a cellular map, by the familiar procedure of extending its definition from each skeleton to the one of next higher dimension. Since each of the spaces K, $\prod K(D_j, n_j)$ has (or may by Theorem 4.8 be assumed to have) just one 0-cell, the initial step in the induction is determined. Suppose inductively that an appropriate inverse map g_{n+q} on the $(n+q)$-skeleton has already been defined, and make the additional assumption, to be justified below, that any obstruction to the extension of g_{n+q} to the $(n+q+1)$-skeleton has finite order (as a cohomology class). If the order of a particular such obstruction (to an appropriate extension of g_{n+q}) is μ say, then we replace g_{n+q} by μg_{n+q}, i.e. we redefine g on the $(n+q)$-skeleton to be μg_{n+q}, the sum of g_{n+q} with itself μ times in accordance with (50). By Lemma 10.16 the obstruction to the extension of μg_{n+q} will then have vanished, and μg_{n+q} can be extended. Repeating this if necessary for each of the (finitely many) $(n+q+1)$-cells in turn, we obtain a suitable "inverse" map g_{n+q+1} defined on the $(n+q+1)$-skeleton of $\prod K(D_j, n_j)$, completing the inductive step. Hence ultimately we shall have constructed the desired "inverse" map g.

It remains to justify the assumption that any obstruction class of the map g_{n+q} has finite order. To this end, observe first that the $(2n-2)$-skeleton of $\prod K(D_j, n_j)$ is in fact the bouquet (cf. proof of Lemma 10.15)

$$P = \bigvee_{n_j \geq n} (K(D_j, n_j))^{2n-2}. \tag{51}$$

Now it follows from Theorem 10.6, giving the structure of the rational cohomology algebra of an Eilenberg–MacLane complex, that in the integral cohomology algebra of each complex $(K(D_j, n_j))^{2n-1}$ elements of infinite order can occur only in the first "non-trivial" dimension n_j. Thus by carrying out the above inductive construction of g piecemeal on the individual members of the bouquet (51), we can ensure that at each stage of the induction with $n+q \geq n_j$, we encounter only obstruction classes of finite order. □

Now if X is, as before, a cell complex of dimension $\leq 2n-2$, then essentially as in the proof of the Cartan–Serre theorem we can realize any group homomorphism

$$h^*: H^*(P; \mathbb{Q}) \to H^*(X; \mathbb{Q})$$

which does not increase degree, as the homomorphism induced by a map $h: X \to P$. By composing this map with the "inverse" map g whose construction we have just sketched, we obtain the following result.

10.17. Theorem. *Given any complex X of dimension $\leq 2n-2$ and any $(n-1)$-connected complex K, the (stable) homotopy classes of maps $X \to K$ form an* *abelian group* $[X, K]$ *for which there is a natural isomorphism*

$$[X, K] \otimes \mathbb{Q} \simeq \mathrm{Hom}(H^*(K; \mathbb{Q}), H^*(X; \mathbb{Q})).$$

(Here the admissible homomorphisms in the right-hand side are arbitrary abelian group homomorphisms which do not change degree.)

Thus we see that while on the one hand each infinite-order homotopy class of maps is determined by a certain homomorphism between the rational cohomology algebras, conversely, given any purely algebraic abelian group homomorphism $a^*: H^*(K; \mathbb{Z}) \to H^*(X; \mathbb{Z})$ (or $a_*: H_*(X; \mathbb{Z}) \to H_*(K; \mathbb{Z})$) which does not increase degree, there is an integer $\lambda \neq 0$ such that the homomorphism λa^* (or λa_*) can be realized by a continuous map $f: X \to K$.

§11. Homology Theory and the Fundamental Group

Let K be a (connected) complex, cellular or even simplicial, with fundamental group $D = \pi_1(K)$, and consider the universal covering

$$p: \hat{K} \to K,$$

with D acting freely and discretely on $\hat{K}$ (see Part II, §§18.2, 18.4; the universal cover can be shown to exist for a wide class of spaces). We shall assume that the cellular (simplicial) subdivisions of K and $\hat{K}$ are such (in particular sufficiently small) that the action of D permutes the cells of $\hat{K}$, and each cell σ_γ^i of K has as complete inverse image under p a union of as many cells of $\hat{K}$ as there are elements of $D = \pi_1(K)$:

$$p^{-1}(\sigma_\gamma^i) = \sigma_{1\gamma}^i \cup \sigma_{2\gamma}^i \cup \cdots.$$

The action of D will then be such as to permute the cells $\sigma_{\alpha\gamma}^i$ comprising $p^{-1}(\sigma_\gamma^i)$.

Choosing a fixed cell $\hat{\sigma}_\gamma^i$ of $\hat{K}$ above each σ_γ^i, it follows that every cell $\sigma_{\alpha\gamma}^i$ of $\hat{K}$ is given by

$$\sigma_{\alpha\gamma}^i = g(\hat{\sigma}_\gamma^i) \qquad \text{for some} \quad g \in D = \pi_1(K),$$

and moreover that the cells $g(\hat{\sigma}_\gamma^i)$ are distinct for different $g \in D$. Hence an arbitrary integral i-chain in $\hat{K}$ has the form

$$a = \sum_{j,\gamma} \lambda_{j\gamma} g_j(\hat{\sigma}_\gamma^i), \qquad a \in C^i(\hat{K}) \quad (\lambda_{j\gamma} \in \mathbb{Z}). \tag{1}$$

Since the boundary operator ∂ on the chains in $\hat{K}$ commutes with the action of D, as well as with multiplication by integers, it is natural to introduce into the discussion the concept of the *integral group ring* $\Gamma = \mathbb{Z}[D]$, whose elements are the finite formal sums $\sum \lambda_j g_j$, $\lambda_j \in \mathbb{Z}$, $g_j \in D$, and whose multiplication is given by

$$\left(\sum_i \lambda_i g_i\right)\left(\sum_k \lambda'_k g'_k\right) = \sum_{i,k} \lambda_i \lambda'_k g_i g'_k.$$

(Clearly, Γ is a noncommutative ring if D is non-abelian.) From (1) we now see that a chain in the complex $\hat{K}$ may be regarded as a chain in $\hat{K}$ with coefficients from the group ring $\Gamma = \mathbb{Z}[D]$. Hence if Γ' is any ring and $\rho: \Gamma \to \Gamma'$ is a ring homomorphism, we can consider the chain complex with coefficients from Γ' consisting of elements of the form

$$\rho(a) = \sum_{\gamma} \rho\left(\sum_{j} \lambda_{j\gamma} g_j\right) \hat{\sigma}^i_{\gamma}, \qquad a \in C_i(\hat{K}).$$

Clearly we can multiply such chains by elements of Γ'. We next introduce a boundary operator on such chains essentially by postulating that ∂ commute with such multiplication and also with ρ. The homology groups determined by this boundary operator are called the *homology groups of K with coefficients determined by the representation* $\rho: \Gamma \to \Gamma'$, $\Gamma = \mathbb{Z}[\pi_1(K)]$. They are denoted by $H_i^{\rho}(K)$. The corresponding cohomology groups $H^i_{\rho}(K)$ are defined in the usual way in terms of the dual complex.

Examples

(a) If $\Gamma' = \Gamma$ and $\rho = 1$, then of course

$$H_i^{\rho}(K) = H_i(\hat{K}).$$

(b) If $\Gamma' = \mathbb{Z}$ and $\rho: \Gamma \to \mathbb{Z}$ is given by

$$\rho\left(\sum \lambda_j g_j\right) = \sum \lambda_j,$$

then we have (verify!)

$$H_i^{\rho}(K) = H_i(K; \mathbb{Z}).$$

(c) If $K = M^n$, a non-orientable manifold, then there is an epimorphism $\varphi: \pi_1(K) \to \mathbb{Z}_2 = \{\pm 1\}$, determined by whether or not the loops representing elements of $\pi_1(K)$ reverse the orientation of tangent frames transported around them (see Part II, §17.6). This homomorphism gives rise to a homomorphism

$$\rho: \Gamma \to \mathbb{Z}$$

defined by

$$\rho\left(\sum \lambda_i g_i\right) = \sum \lambda_i \varphi(g_i).$$

The homology groups $H_i^{\rho}(K)$ determined by this homomorphism are called "homology groups with local coefficients", and similarly for the cohomology groups $H^i_{\rho}(K)$.

Exercise

1. Prove that if M^n is a closed (non-orientable) manifold, then $H_n^{\rho}(M^n) \simeq \mathbb{Z}$.

Consider now a fibre space $E \xrightarrow{p} B$ with fibre F, where each element g of $\pi_1(B) = D$ determines (via a prescribed lift of a path representing g—see Part

II, §22.1) a continuous bijection $F \to F$, inducing an automorphism $g\colon H_q(F) \to H_q(F)$. The resulting action of D on each $H_q(F)$ extends in the obvious way to an action of the ring $\Gamma = \mathbb{Z}[D]$ on $H_q(F) = M_q$ say.

Suppose for the moment that, more generally, the ring Γ acts by means of linear transformations on a vector space M (or by means of abelian group endomorphisms on an abelian group M). Such an action will be given by a homomorphism ρ from Γ to the ring of linear transformations (endomorphisms), turning M into a "Γ-module" as they say. *Homology groups $H_i^\rho(B; M)$ with coefficients from M* are defined as follows: Taking $B = K$ in the preceding discussion, form the complex of chains over Γ in $\hat{K} = \hat{B}$, the universal cover of B. An *i-chain over M* is then defined as a finite formal sum

$$a = \sum_j m_j \hat{\sigma}_j^i, \qquad m_j \in M,$$

and an action of the ring Γ on such chains is defined by

$$\gamma(a) = \sum_j \gamma(m_j)\hat{\sigma}_j^i, \qquad g \in \Gamma,$$

where $\gamma(m)$ denotes the result of $\gamma \in \Gamma$ acting on $m \in M$. The boundary operator ∂, defined as in §4 on the cells $\hat{\sigma}_j^i$ and then extended linearly to arbitrary chains, is easily seen to commute with this action of Γ. The resulting homology groups, denoted by $H_i^\rho(B; M)$, are then the ones we sought to define. (Here ρ is the representation of Γ in the endomorphism ring of M, i.e. gives the action of Γ on the Γ-module M.) The corresponding cohomology groups $H_\rho^i(B; M)$ are defined as usual in terms of the dual complex of cochains.

Returning now to our original fibre space $E \xrightarrow{p} B$ with fibre F, where, as we saw, $H_q(F)$ has a natural Γ-module structure determined by "parallel transport", we obtain in particular the homology groups $H_i^\rho(B; H_q(F))$.

Remark. In Leray's theorem (8.1), where the base B was assumed simply-connected, we saw that $E_{q,j}^{(2)} \simeq H_q(B; H_j(F)))$. This may fail to hold if B is not simply-connected. However, it can be shown that then $E_{q,j}^{(2)} \simeq H_q^\rho(B; H_j(F))$, with the rest of the theorem remaining valid. The representation ρ can be regarded as measuring the degree of "distortion" of the operator d_1 due to the non-triviality of $\pi_1(B)$.

Example. For a covering space $E \xrightarrow{p} B$ with fibre consisting of k points, we have $H_0(F)$ free abelian of rank k, and $H_q(F) = 0$ for $q > 0$. Via the monodromy representation the group $\pi_1(B)$ permutes the points of the fibre F, and consequently acts on the group $M = H_0(F)$ by correspondingly permuting the direct summands $\mathbb{Z}$. We state some of the conclusions to be drawn in this context, in the form of exercises.

EXERCISES (continued)

2. Establish the following isomorphisms:

$$H_q^\rho(B; H_0(F)) \simeq H_q(E); \qquad H_\rho^q(B; H^0(F)) \simeq H^q(E).$$

3. Given any representation ρ of Γ by means of linear transformations of a vector space M, compute the groups $H_0^\rho(B; M)$, $H_\rho^0(B; M)$.

4. Compute the homology groups H_i^ρ of the lens space $L_m^{2n-1}(q_1, \ldots, q_{n-1})$ (see §4), where ρ is the representation of $\pi_1(L) \simeq \mathbb{Z}_m$ by the mth roots of unity acting by multiplication on $\mathbb{C} = M$. Find also linear representations
$$\rho\colon \mathbb{Z}_m \to GL(k, \mathbb{C})$$
with the property that
$$H_j^\rho(L_m^{2n-1}(q_1, \ldots, q_{n-1})) = 0 \qquad \text{for all} \quad j = 0, 1, 2, \ldots.$$
(Begin with the case $n = 2$ of 3-dimensional lens spaces.)

5. (Cf. §4, Exercise 9.) Find an explicit cell decomposition of the sphere S^{2n-1} which is invariant under the action of $\mathbb{Z}_m$ defined by the following action of a generator:
$$T\colon (z_1, \ldots, z_n) \mapsto (e^{2\pi i/m}z_1, e^{2\pi i q_1/m}z_2, \ldots, e^{2\pi i q_{n-1}/m}z_n),$$
where $|z_1|^2 + \cdots + |z_n|^2 = 1$. (For $n = 2$ there is such a cell decomposition with cells
$$T^j\sigma^0,\ T^j\sigma^1,\ T^j\sigma^2,\ T^j\sigma^3; \qquad j = 0, 1, \ldots, m-1,$$
on which the boundary operator acts as follows (cf. §4(17)):
$$\partial\sigma^0 = 0, \qquad \partial\sigma^1 = (T-1)\sigma^0,$$
$$\partial\sigma^2 = (1 + T + \cdots + T^{m-1})\sigma^1, \qquad \partial\sigma^3 = -(1 - T^q)\sigma^2.)$$

There is an interesting topological invariant of a triangulable manifold M^n, called "Reidemeister torsion", which is defined in terms of a linear representation ρ over $\mathbb{C}$ (of the group $\pi_1(M^n)$) with the property that $H_q^\rho(M^n; \mathbb{C}^n) = 0$ for $q \geq 0$ (cf. Exercise 4 above). Consider the chain complex C determined by ρ; each of the groups C_j^ρ of this complex is in fact a complex vector space with the chosen cells $\hat{\sigma}_\gamma^j$ of the universal cover of M^n as basis elements (see above; note the latitude $\hat{\sigma}_\gamma^j \to \pm g(\hat{\sigma}_\gamma^j)$ in the choice of these cells). The condition $H_q^\rho = 0$ for $q \geq 0$ is equivalent to exactness of the sequence
$$0 \to C_n^\rho \xrightarrow{\partial} C_{n-1}^\rho \xrightarrow{\partial} \cdots \xrightarrow{\partial} C_0^\rho \to 0, \tag{2}$$
where as just observed each vector space C_j^ρ comes with its distinguished basis $\{\hat{\sigma}_\gamma^j\} = e^j$ say. Consider now the following procedure: Choose a new basis $\bar{e}^{n-1}$ for C_{n-1}^ρ by extending arbitrarily the image under ∂ of the basis $e^n = \{\hat{\sigma}_\gamma^n\}$ to a full basis for C_{n-1}^ρ. Denote by $\det(e^{n-1}, \bar{e}^{n-1})$ the determinant of the matrix of this change of basis. By the exactness of (2) the basis elements in $\bar{e}^{n-1}$, representing the space $C_{n-1}^\rho/\operatorname{Im}\partial$, are mapped one-to-one by ∂ to linearly independent elements of C_{n-2}^ρ; supplement these image vectors arbitrarily to a full basis $\bar{e}^{n-2}$ of C_{n-2}^ρ. The determinant of the basis change from the old basis e^{n-2}, consisting of the chosen cells $\hat{\sigma}_\gamma^{n-2}$, to $\bar{e}^{n-2}$ is denoted by $\det(e^{n-2}, \bar{e}^{n-2})$. The process is continued in this way with C_{n-3}^ρ, and so on, yielding finally a new basis $\bar{e}^k$ for each vector space C_k^ρ, and the corresponding sequence of complex numbers $\det(e^k, \bar{e}^k)$. The *Reidemeister torsion* is then

defined to be the complex number

$$R(C, \rho) = \det(e^{n-1}, \bar{e}^{n-1}) \det(e^{n-2}, \bar{e}^{n-2})^{-1} \dots$$
$$\dots \det(e^{n-k}, \bar{e}^{n-k})^{(-1)^{k+1}} \dots \det(e^0, \bar{e}^0)^{(-1)^{n+1}}.$$

The above-noted latitude in the choice of the basic cells and their orientations implies that R is defined only to within multiplication by a constant of the form $\lambda = \pm\det \rho(g), g \in \pi_1(M^n)$.

It turns out that the Reidemeister torsion R is (up to multiplication by $\lambda \in \pm\det \rho(\pi_1)$) independent of the triangulation of M^n, and is a topological "piecewise-linear" invariant, and invariant under diffeomorphisms of M^n (see [43]). (It is not, however, a homeomorphic invariant in general—see §28.)

Exercise

6. Compute the Reidemeister torsion R for the 3-dimensional lens space $L = L_p^3(q)$ where p is prime and q is a non-zero residue modulo p, and ρ represents $\mathbb{Z}_p$ by the roots of unity acting on $M = \mathbb{C}$ via multiplication. (Recall that $L_p^3(q)$ is defined as the orbit space $S^3/\mathbb{Z}_p$ where a generator of $\mathbb{Z}_p$ acts on $S^3 = \{|z^1|^2 + |z^2|^2 = 1\} \subset \mathbb{C}^2$ according to

$$(z_1, z_2) \mapsto (e^{2\pi i/p} z_1, e^{2\pi i q/p} z_2).)$$

From Exercise 5 above (see also §4(17)) it follows easily that the mod p cohomology groups of the lens space $L_p^3(q)$, q any non-zero residue of the prime p, are given by

$$H^i(L_p^3(q); \mathbb{Z}_p) \simeq \mathbb{Z}_p, \qquad i = 0, 1, 2, 3.$$

Let u denote any generator of $H^1(L_p^3(q); \mathbb{Z}_p)$, let v denote $\delta_1 u$ (mod p), where δ_1 is the Bockstein operator (so that v generates $H^2(L; \mathbb{Z}_p)$), and let w be the generator of $H^3(L; \mathbb{Z}_p)$ obtained by reducing modulo p a generator of $H^3(L; \mathbb{Z}) \simeq \mathbb{Z}$.

Exercise

7. Prove that if u and the sign of w are chosen suitably, then the multiplication in the cohomology algebra $H^*(L; \mathbb{Z}_p)$ will be given by

$$uv = qw. \tag{3}$$

Since $\delta_1 u = v$ and w is determined up to sign, the integer factor q in (3) will, under a replacement $u \to \lambda u$, $w \to \pm w$ (with λ relatively prime to p), change to $\pm\lambda^2 q$:

$$(u \to \lambda u, w \to \pm w) \quad \Rightarrow \quad (v \to \lambda v, uv \to \pm\lambda^2 qw).$$

We conclude that: *Since the cohomology algebra of $L_p^3(q)$ and the Bockstein operator δ_1 are homotopy invariants (i.e. invariant under homotopy equivalences), all non-zero residues modulo p of the form $\bar{q} = \pm\lambda^2 q$ together comprise a complete single homotopy invariant of the lens space $L_p^3(q)$.*

Consider the first three non-trivial instances $p = 3, 5, 7$:

(a) $p = 3$. Here the non-zero residues 1, 2 form a single homotopy invariant, since $2 \equiv -1 \pmod 3$.
(b) $p = 5$. Here $q = 1, 2, 3, 4$, and $\pm\lambda^2 = 1, 4$. It follows that the pairs $\{1, 4\}$ and $\{2, 3\}$ represent distinct homotopy invariants, so that $L_5^3(1)$ and $L_5^3(2)$ are not homotopically equivalent.
(c) $p = 7$. Here $q = 1, 2, 3, 4, 5, 6$, and $\pm\lambda^2 = 1, 2, 3, 4, 5, 6$. Hence there is but one homotopy invariant of the sort in question, consisting of all non-zero residues modulo 7. (It can be shown that in fact all of the lens spaces $L_7^3(q)$ are homotopically equivalent.)

Exercise

8. Using Reidemeister torsion determine which of the lens spaces $L_7^3(q)$ are topologically inequivalent.

Remark. It is interesting to observe that here we encounter for the first time closed manifolds which have the same homotopy type but are topologically distinct. The problem of exhibiting such manifolds which are in addition simply-connected is much more difficult.

The homology and cohomology groups determined by a representation ρ of $\Gamma = \mathbb{Z}[\pi_1]$, are relevant also to the problem of extending maps $L \to X$ from a subcomplex L to a complex $K \supset L$, in the case where the fundamental group $\pi_1(X)$ acts nontrivially on the $\pi_i(X)$, and to the cognate problem of extending a cross-section of a fibre bundle. (Cf. §9 where these extension problems were examined under the assumption that X is simply-connected, or at least that $\pi_1(X)$ is abelian and acts trivially on all $\pi_i(X)$.)

Consider by way of an example the related problem of constructing a metric of given signature on an n-manifold, or, to take an interesting particular case, a metric of signature $(+---)$ on a 4-manifold M^4. Since the interior of the light cone in Minkowski space $\mathbb{R}_{1,3}$ (see Part I, §6) is contractible in canonical fashion onto the time axis, it follows that the space of possible light cones at the origin of $\mathbb{R}^4$ (or, equivalently, of quadratic forms g_{ab} of signature $(+---)$ on the vector space $\mathbb{R}^4$) is homotopically equivalent to the space $\mathbb{R}P^3$ of one-dimensional directions in $\mathbb{R}^4$ (more generally $\mathbb{R}P^{n-1}$ for $\mathbb{R}^n$). Hence our original problem is equivalent to that of constructing on M^4 a field of one-dimensional directions (i.e. a one-dimensional foliation; see Part II, §27.1), or in other words a cross-section of the bundle over M^4 of tangent lines (i.e. one-dimensional subspaces of the tangent space):

$$E \xrightarrow{p} M^4, \qquad \text{with fibre} \quad F \cong \mathbb{R}P^3. \tag{4}$$

Since the singularities of a "typical" vector field occur at isolated points of M^4 (or in other words obstructions to the problem of constructing a vector field on M^4 in general arise only in extending the field from the 3-skeleton to the 4-skeleton), such will also be the case for one-dimensional foliations. Let α be an obstruction cochain for the extension of a cross-section of the fibre bundle (4) above the 3-skeleton of some cell decomposition of M^4, to a full cross-section; as in §9.3 we have $\alpha \in C^4(M^4; \pi_3(F))$ ($= C^4(M^4; \mathbb{Z})$ since $\pi_3(\mathbb{R}P^3) \simeq \pi_3(S^3) \simeq \mathbb{Z}$). It turns out to be appropriate to regard α as a cohomology class in the group $H^4_\rho(M^4; \pi_3(F))$, where the representation ρ is determined by the action of $\pi_1(M^4)$ on $\pi_3(F)$; this is the import of the following exercise.

EXERCISE

9. Show that if $\alpha \sim 0$ in $H^4_\rho(M^4; \pi_3(F))$, then the cross-section above the 3-skeleton of the base (i.e. the direction field on the 3-skeleton) can be altered to make $\alpha = 0$, so that the altered cross-section can be extended over the whole of M^4.

If the manifold M^4 is compact and orientable, then the action of $\pi_1(M^4)$ on $\pi_3(\mathbb{R}P^3) \simeq \mathbb{Z}$ must be trivial, whence $H^4_\rho(M^4; \mathbb{Z}) = H^4(M^4; \mathbb{Z}) \simeq \mathbb{Z}$ (see e.g. Proposition 3.10 and "Poincaré duality" in §6).

EXERCISE

10. Prove that in this situation we have (as also for vector fields) $\alpha = \chi(M^4)$, the Euler characteristic.

If, on the other hand, M^4 is a (compact) non-orientable manifold, then $\pi_1(M^4)$ acts non-trivially on $\pi_3(F) \simeq \mathbb{Z}$, so that ρ is also non-trivial. It is not difficult to show that once again $H^4_\rho(M^4; \mathbb{Z}) \simeq \mathbb{Z}$.

EXERCISE

11. Show that in this case also we have $\alpha = \chi(M^4)$.

We conclude that: *In both the orientable and non-orientable cases, the precise condition for the existence of a one-dimensional foliation* (*or, equivalently, a metric of signature* $(+---)$) *on a closed manifold* M^4, *is that the Euler characteristic* $\chi(M^4) = 0$.

For non-compact manifolds M^4 it is of interest to ask whether there exists a metric g_{ab} of signature $(+---)$ which outside some compact set is "close" to the Minkowksi metric. An "open" (i.e. non-compact) manifold M^4 endowed with such a metric will admit a one-point compactification $\overline{M}^4 = M^4 \cup \{\infty\}$. It follows from the properties of the Minkowski metric that the point ∞ will necessarily be a singular point of index 2 of the "field of directors" determined by the metric of the type we are seeking (prove this!). Hence the question reduces to the constructibility on the compactified manifold $\overline{M}^4$ of a direction field with exactly one singular point, and that of index 2. It can be shown along the lines of the preceding discussion that the precise condition for the existence of such a field is that $\chi(\overline{M}^4) = 2$ (or $\chi(M^4) = 1$). (Cf. Part II, Theorem 15.2.7.)

EXERCISE

12. Prove that each homotopy class of one-dimensional distributions (i.e. direction fields), or, equivalently, metrics of signature $(1, n)$, on a manifold M^{n+1}, is determined by a homomorphism $\pi_1(M^{n+1}) \to \mathbb{Z}_2 = \{\pm 1\}$, together with a cohomology class
$$\gamma \in H^n_\rho(M^{n+1}; \pi_n(\mathbb{R}P^n)),$$
where ρ is the representation afforded by the homomorphism $\pi_1(M^{n+1}) \to \mathbb{Z}_2 \simeq \pi_1(\mathbb{R}P^n)$, the latter group acting in the usual way on $\pi_n(\mathbb{R}P^n)$. (Under the map $\pi_1(M^{n+1}) \to \mathbb{Z}_2$, closed curves, transport along which reverses orientation, are sent to -1.)

Examples

(α) Let U denote the manifold obtained by removing from $\mathbb{R}^3$ a straight line and a point off it. As noted in Part II, §17.5, the region U has the homotopy type of the bouquet $S^2 \vee S^1$. A distribution of one-dimensional directions on U will be give by a continuous map
$$f: U \to \mathbb{R}P^2$$
determined by a cross-section of the bundle analogous to (4), whose homotopy class $[f]$ is determined, according to the preceding exercise, by the induced homomorphism
$$f_*: \mathbb{Z} \simeq \pi_1(U) \to \pi_1(\mathbb{R}P^2) \simeq \mathbb{Z}_2,$$
and a cohomology class
$$\pm\gamma \in H^2_\rho(U; \pi_2(\mathbb{R}P^2)) \simeq H^2_\rho(S^2 \vee S^1; \mathbb{Z}) \simeq \mathbb{Z},$$
where the representation ρ is given by the action of $\pi_1(U) \simeq \mathbb{Z}$ on $\pi_2(\mathbb{R}P^2)$, obtained from the map f_* together with the natural action of $\pi_1(\mathbb{R}P^2)$ on $\pi_2(\mathbb{R}P^2)$. The universal cover $\hat{K}$ of $K = S^2 \vee S^1$ is as shown in Figure 45. It turns out that in the present situation ρ is non-trivial, whence all 2-cochains are cocycles none of which is cohomologous to zero, and $C^2_\rho(K) = H^2_\rho(K) \simeq \mathbb{Z}$ (verify!).

(β) We conclude this section by considering, by way of a useful application of the cohomology group H^2_ρ determined by a representation, the problem of classifying the homotopy classes of maps f from the torus T^2 to the projective plane $\mathbb{R}P^2$:
$$f: T^2 \to \mathbb{R}P^2.$$

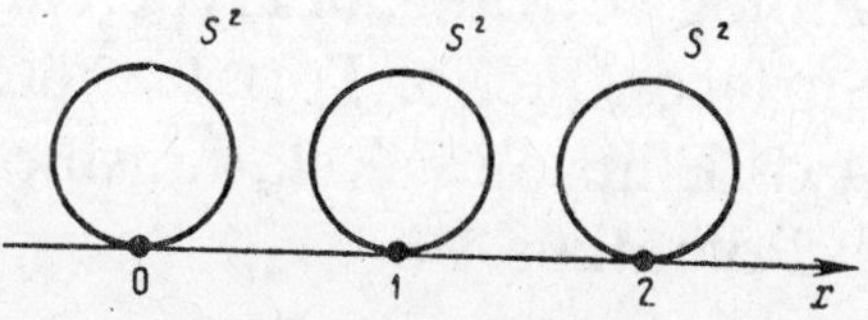

Figure 45. The action of $\pi_1(K)$ is given by: $x \mapsto x + 1$.

The simplest homotopy invariant of such a map f is the induced homomorphism between the fundamental groups

$$f_*: \mathbb{Z} \oplus \mathbb{Z} \simeq \pi_1(T^2) \to \pi_1(\mathbb{R}P^2) \simeq \mathbb{Z}_2.$$

If the homomorphism f_* is trivial, then clearly the original map f may on the one-dimensional skeleton (of the standard cell decomposition of T^2—see §4, Figure 30) be deformed to a constant map, i.e. f is homotopic to a map whose restriction to the 1-skeleton is a map to a single point of $\mathbb{R}P^2$. It follows that each homotopy class of such maps (with $f_*(\pi_1) = 0$) is naturally associated with a homotopy class of maps $S^2 \to \mathbb{R}P^2$, i.e. with an element of $\pi_2(\mathbb{R}P^2) \simeq \mathbb{Z}$.

Of greater interest is the situation where f_* is non-trivial. Without loss of generality we may suppose in this case that $f_*(a) = 1$ and $f_*(b) = 0$, where a, b are respectively a latitude and meridian on the torus. Let $f, g: T^2 \to \mathbb{R}P^2$ be two such maps with $f_* = g_*$. Decomposing the torus in the standard manner (see §4, Figure 30) into cells

$$\sigma^0, \quad \sigma_1^1 = a, \quad \sigma_2^1 = b, \quad \sigma^2,$$

we can in view of the condition $f_* = g_*$ deform f and g (i.e. apply a suitable homotopy) so that they coincide on the 1-skeleton. This assumed done, the maps f, g may be regarded as maps of σ^2 coinciding on the boundary $\partial\sigma^2$, and therefore defining a "distinguishing element" in the group $\pi_2(\mathbb{R}P^2)$, which we denote by $\alpha = \alpha(\sigma^2, f, g) \in \pi_2(\mathbb{R}P^2) \simeq \mathbb{Z}$. In fact as far as distinguishing homotopically between f and g is concerned, it suffices to regard α as an element of the cohomology group

$$H_\rho^2(T^2; \pi_2(\mathbb{R}P^2)), \tag{5}$$

where ρ is given by $f_* = g_*: \pi_1(T^2) \to \pi_1(\mathbb{R}P^2)$, taken together with the usual action of the latter group on $\pi_2(\mathbb{R}P^2)$.

EXERCISE

13. Show that if ρ is non-trivial then the group (5) is isomorphic to $\mathbb{Z}_2$.

We conclude that: *Corresponding to each (fixed) non-trivial homomorphism f_* between the fundamental groups, there are at most two distinct homotopy classes of maps $T^2 \to \mathbb{R}P^2$.*

§12. The Cohomology Groups of Hyperelliptic Riemann Surfaces. Jacobi Tori. Geodesics on Multi-Axis Ellipsoids. Relationship to Finite-Gap Potentials

A general Riemann surface R was defined in Part II, §4.2, as a non-singular surface in $\mathbb{C}^2$ given by an equation of the form $g(z, w) = 0$, where $g(z, w)$ is an analytic function of z and w. In Part II, loc. cit., and Part I, §12.3, we considered the important special case of a *hyperelliptic* Riemann surface R_g of genus g, defined by an equation

$$w^2 - P_{2g+1}(z) = 0 \qquad \text{or} \qquad w^2 - \tilde{P}_{2g+2}(z) = 0,$$

where P, $\tilde{P}$ are polynomials of degrees $2g + 1$, $2g + 2$ respectively, without multiple roots.

On an arbitrary Riemann surface one can define holomorphic differentials (i.e. holomorphic 1-forms), given in terms of local co-ordinates $z = z + iy$ by

$$\omega = f(z)\,dz,$$

where $f(z)$ is a complex-analytic function of z. The question as to which functions $f(z)$ can actually occur will be elucidated below. On a hyperelliptic Riemann surface R_g of genus $g > 0$, defined by a polynomial $P_{2g+1}(z) = \prod_{i=1}^{2g+1} (z - z_i)$, the following differentials are holomorphic:

$$\omega_k = \frac{z^{k-1}}{w}\,dz = \frac{z^{k-1}}{\sqrt{P_{2g+1}(z)}}\,dz, \qquad k = 1, 2, \ldots, g. \tag{1}$$

To see that these differentials are indeed holomorphic on R_g, observe first that away from the points $z = z_i$ (the roots of the polynomial P_{2g+1}) and the point $z = \infty$, they are obviously so. In some neighbourhood on the surface R_g of a particular root z_i, we may take as local co-ordinate $\zeta = \sqrt{z - z_i}$. Then $z = \zeta^2 + z_i$, $dz = 2\zeta\,d\zeta$, and (1) takes the form

$$\omega_k = 2\frac{(\zeta^2 + z_i)^{k-1}}{\sqrt{\prod_{j \neq i}(\zeta^2 + z_i - z_j)}}\,d\zeta, \tag{2}$$

from which the holomorphicity of the differentials ω_k at $z = z_i$ is clear. At the point at infinity $z = \infty$, as local co-ordinate $\zeta = 1/\sqrt{z}$ will serve: In terms of ζ we then have $z = 1/\zeta^2$, $dz = -2d\zeta/\zeta^3$, and consequently

$$\omega_k = -\frac{2\zeta^{2(g-k)}}{\sqrt{\prod_{i=1}^{2g+1}(1 - \zeta z_i)}}\,d\zeta, \tag{3}$$

whence it is clear that the ω_k, $k \leq g$, are holomorphic also at $z = \infty$, completing the verification.

Every holomorphic differential ω on a Riemann surface R is clearly locally exact: $\omega = f(z)\,dz = d\tilde{f}(z)$ in terms of each local co-ordinate z, where $\tilde{f}(z)$ is an anti-derivative of $f(z)$, and so also complex-analytic. Hence such a 1-form on R is closed: $d\omega = 0$. It follows, on the other hand, that a non-zero holomorphic 1-form ω on a compact Riemann surface (which includes the hyperelliptic case—see Part II, Lemma 4.2.1) cannot be globally exact since by Part II, Theorem 4.1.3, there are no non-constant holomorphic functions definable on such a surface. Under the same conditions the complex conjugate 1-form $\bar{\omega} = \overline{f(z)}\,\overline{dz}$ will then of course also be closed but not exact on R.

Returning to the hyperelliptic case, observe that the forms $\omega_1, \ldots, \omega_g$ on R_g are linearly independent over the complex numbers. Since by the preceding discussion there are no non-zero exact 1-forms on R_g, and by Part II, Theorem 4.2.2, R_g is diffeomorphic to a sphere with g handles, it follows that the forms

$$\operatorname{Re} \omega_k = \tfrac{1}{2}(\omega_k + \overline{\omega_k}), \qquad \operatorname{Im} \omega_k = \frac{1}{2i}(\omega_k - \overline{\omega_k})$$

form a basis for the real first cohomology group $H^1(R_g; \mathbb{R}) = \mathbb{R} \oplus \cdots \oplus \mathbb{R}$ ($2g$ summands).

Remark. It is known that in fact the first cohomology group $H^1(R; \mathbb{R})$ of any Riemann surface R can be analogously realized in terms of holomorphic differentials. The existence of these differentials is however difficult to establish (see [78]).

Exercise

1. Prove that any $g + 1$ holomorphic differentials on a Riemann surface of genus g are linearly dependent.

Let a_i, b_i, $i = 1, \ldots, g$, be loops on the surface $R_g \cong M_g^2$ representing the canonical generators of $\pi_1(R_g)$ (see, e.g. §4(14)) (and corresponding therefore to basic cycles for $H_1(R_g; \mathbb{Z})$ via the isomorphism of §4(13)), with pairwise intersection indices (see Part II, §15.1) given by

$$a_i \circ a_j = b_i \circ b_j = 0, \qquad a_i \circ b_j = \delta_{ij}, \qquad i, j = 1, \ldots, g. \tag{4}$$

For any two differential 1-forms ω, ω' on R_g, denote by A_i, B_i the periods of ω over the cycles a_i, b_i respectively:

$$\oint_{a_i} \omega = A_i, \qquad \oint_{b_i} \omega = B_i, \qquad i = 1, \ldots, g, \tag{5}$$

and by A_i', B_i' the corresponding periods of ω'.

12.1. Lemma. *For any two 1-forms ω, ω' on R_g with periods A_i. B_i, A_i', B_i' as above, we have*

$$\int_{R_g} \omega \wedge \omega' = \sum_{i=1}^{g} (A_i B_i' - B_i A_i'). \tag{6}$$

Proof. By cutting the surface R_g along the loops a_i, b_i, we obtain a $4g$-gon $\tilde{R}_g$ (see e.g. Theorem 3.20(ii)), on which the form ω (as also ω') will by the Poincaré lemma be exact: $\omega = df$. The exactness of ω' then entails $\omega \wedge \omega' = d(f\omega')$, whence by the general Stokes formula

$$\int_{R_g} \omega \wedge \omega' = \int_{\partial \tilde{R}_g} f\omega'. \tag{7}$$

Now let Q, Q' be points on the edges a_i, a_i^{-1} respectively, of the $4g$-gon $\tilde{R}_g$, representing the same point of the original surface R_g. It is then clear (see Figure 46) that the segment QQ' represents a cycle in R_g homologous to b_i, so

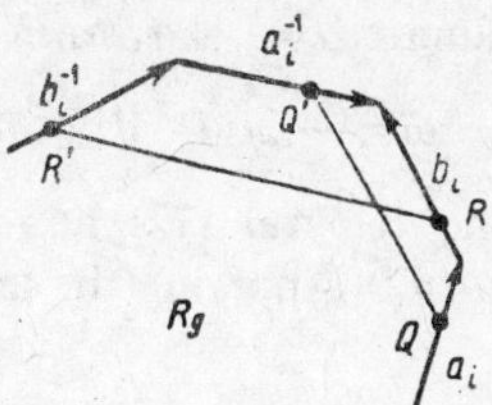

Figure 46

that, again by Stokes' theorem,

$$\int_{QQ'} \omega = f(Q') - f(Q) = \int_{b_i} \omega = B_i. \tag{8}$$

For points R, R' on the edges labelled b_i, b_i^{-1} respectively, identified with a single point of R_g, we obtain similarly

$$f(R') - f(R) = -A_i. \tag{9}$$

From (8) and (9) it follows that

$$\begin{aligned}\int_{a_i+b_i+a_i^{-1}+b_i^{-1}} f\omega' &= \int_{a_i} f\omega' + \int_{b_i} f\omega' - \int_{a_i} (f + B_i)\omega' - \int_{b_i} (f - A_i)\omega' \\ &= A_i B_i' - B_i A_i',\end{aligned}$$

whence, via (7), the lemma. □

12.2. Theorem. *The periods (A_k, B_k) and (A_k', B_k') of a pair of holomorphic differentials ω, ω' on a hyperelliptic Riemann surface R_g, satisfy the bilinear "Riemann period relation"*

$$\sum_{k=1}^{g} (A_k B_k' - B_k A_k') = 0. \tag{10}$$

Furthermore, if ω is non-zero, then the A_k, B_k satisfy the "Riemann period inequality"

$$-\frac{1}{2i} \sum_{k=1}^{g} (A_k \bar{B}_k - B_k \bar{A}_k) > 0. \tag{11}$$

Proof. Since in terms of any local co-ordinate z we have $\omega = f(z)\,dz$, $\omega' = g(z)\,dz$, for some holomorphic functions $f(z)$, $g(z)$, it follows that locally $\omega \wedge \omega' = fg\,dz \wedge dz = 0$, whence $\omega \wedge \omega' = 0$ globally. The relation (10) is now immediate from the lemma.

To establish the inequality (11), consider the integral

$$-\frac{1}{2i} \int_{R_g} \omega \wedge \bar{\omega}. \tag{12}$$

In terms of any local co-ordinate $z = x + iy$ on R_g, we have

$$\omega \wedge \bar{\omega} = -2i|f|^2 \, dx \wedge dy,$$

whence it is clear that the integral (12) has a positive value (namely $\int_{R_g} |f|^2 \, dx \wedge dy$) provided $\omega \neq 0$. Applying the lemma with $\bar{\omega}$ in the role of ω', it follows that

$$0 < -\frac{1}{2i} \int_{R_g} \omega \wedge \bar{\omega} = -\frac{1}{2i} \sum (A_k \bar{B}_k - \bar{A}_k B_k),$$

completing the proof. □

Now let $\omega_1, \ldots, \omega_g$ be members of an arbitrary basis for the vector space of holomorphic differentials on our hyperelliptic Riemann surface R_g, and write

$$A_{ij} = \oint_{a_j} \omega_i, \qquad i, j = 1, \ldots, g.$$

From the above theorem we deduce the

12.3. Corollary. *The matrix* (A_{ij}) *is non-singular.*

PROOF. The inequality (11) implies in particular that a holomorphic differential with zero A-periods must itself be zero. Now if the matrix (A_{ij}) were singular then any non-zero vector $(c_1, \ldots, c_g)$ in its kernel (i.e. mapped by it to zero) would determine a non-zero form, namely $\sum c_i \omega_i$, with zero A-periods, yielding a contradiction. □

We earlier established that the forms (1) comprise a basis for the space of holomorphic 1-forms on R_g. It follows from this corollary that we can choose a new basis

$$\varphi_k = \frac{c_{1k} z^{g-1} + \cdots + c_{gk}}{\sqrt{P_{2g+1}(z)}} \, dz = \sum_{i=1}^{g} c_{ik} \omega_{g-i+1}, \qquad k = 1, \ldots, g, \tag{13}$$

with A-periods given by

$$\oint_{a_j} \varphi_i = \delta_{ij}, \qquad i, j = 1, \ldots, g. \tag{14}$$

Write

$$B_{ij} = \oint_{b_j} \varphi_i$$

for the B-periods determined by this new basis. From Theorem 12.2 we infer also the following

12.4. Corollary. *The matrix* (B_{ij}) *is symmetric, and its imaginary part is positive definite.*

PROOF. The symmetry of (B_{ij}) follows from equation (10) with $\omega = \varphi_i, \omega' = \varphi_j$, bearing in mind (14): $A_{ij} = \delta_{ij}$.

To establish the second assertion we apply the inequality (11) to the holomorphic differential $\omega = x_1\varphi_1 + \cdots + x_g\varphi_g$, where the x_k are arbitrary real numbers. The A-periods of this differential are clearly given by $A_k = x_k$, and the B-periods by

$$B_k = x_1 B_{1k} + \cdots + x_g B_{gk},$$

so that, assuming the x_k are not all zero, the inequality (11) takes the form

$$0 < \frac{i}{2} \sum_{k=1}^{g} [x_k\overline{(x_1 B_{1k} + \cdots + x_g B_{gk})} - x_k(x_1 B_{1k} + \cdots + x_g B_{gk})]$$

$$= \sum_{k,j=1}^{g} x_j x_k \operatorname{Im} B_{jk},$$

establishing the positive definiteness of $(\operatorname{Im} B_{jk})$. □

We now introduce, in terms of the matrix (B_{ij}), a certain integral lattice Γ in the space $\mathbb{C}^g$, namely that generated by the linearly independent (over $\mathbb{R}$) vectors $e_1, \ldots, e_{2g}$ with components $(e_k)^i = \delta_{ik}, (e_{g+k})^i = B_{ik}, k = 1, \ldots, g$. The $2g$-dimensional torus $T^{2g} = \mathbb{C}^g/\Gamma$ (see Part II, §4.1) determined by this integral lattice is then called the *Jacobi torus*, or *Jacobian variety*, of the Riemann surface R_g. The preceding corollary tells us precisely (see Part II, Definition 4.1.9) that:

The Jacobi torus is abelian.

By way of an example we consider in detail the case of a surface R_1 of genus 1, an "elliptic curve":

$$w^2 = P_3(z) = (z - z_1)(z - z_2)(z - z_3).$$

Here we have just two standard cycles a_1, b_1 (indicated in Figure 47), and a single basic holomorphic differential $\varphi = c\, dz/\sqrt{P_3(x)}$, where c is chosen so that the condition $\oint_{a_1} \varphi = 1$ is satisfied. Denoting by τ the single B-period

$$\tau = B_{11} = \oint_{b_1} \varphi,$$

we have, by Corollary 12.4, $\operatorname{Im} \tau > 0$, and the vectors $1, \tau$ give rise as above to the 2-dimensional Jacobi torus T^2 of the Riemann surface R_1. Now from

Figure 47. Standard cycles on the elliptic Riemann surface R_1: $w^2 = (z - z_1) \times (z - z_2)(z - z_3)$. The dotted part of the curve b_1 lies on the second sheet.

Theorem 4.2.2 of Part II, we know that as a manifold R_1 is diffeomorphic to the torus; they are in fact biholomorphically equivalent, as we shall now show by constructing an explicit such equivalence between the Riemann surface R_1 and its Jacobi torus T^2.

To this end, we first fix on a point P_0 of R_1, and then for each point P of R_1 define the quantity $A(P)$ by

$$A(P) = \int_{P_0}^{P} \varphi = \int_{P_0}^{P} \frac{c\,dz}{\sqrt{P_3(z)}}. \tag{15}$$

Here the integral is not independent of the path of integration; however, for each path from P_0 to P its value is affected only by the addition of a cycle to that path, so that the complex number $A(P)$ is determined to within an additive integral linear combination of the A- and B-periods of the differential φ:

$$A(P) \sim A(P) + n.1 + m.\tau, \qquad n, m \in \mathbb{Z}.$$

From this it is clear that $A(P)$ is well defined as a point of the Jacobi torus T^2, or in other words that $A: R_1 \to T^2$ is a (smooth) map. The following proposition is immediate from the definition of A in (15):

12.5. Proposition. *The map $A: R_1 \to T^2$, is regular at every point P* (i.e. *has nowhere-vanishing differential—see* Part II, §10.2).

12.6. Corollary. *The map $A: R_1 \to T^2$ is a biholomorphic equivalence* (i.e. *an "isomorphism" of complex-analytic manifolds*).

PROOF. We have, from the proposition, that A is a covering map. Since, as is easily seen, the images under the map A of the generating loops a_1, b_1 of the group $\pi_1(R_1)$, are generators for $\pi_1(T^2)$, by Part II, Theorem 19.2, the covering must be trivial. □

Remark. It is known, as a result in complex-function theory, that every complex torus T^2 is the Jacobi torus of some elliptic Riemann surface.

More generally, given a Riemann surface R_g of genus $g > 0$, we can extend the definition of the map A as follows. For each set of g points $Q_1, \ldots, Q_g$ of R_g we define the vector $A(Q_1, \ldots, Q_g) = (A^1, \ldots, A^g)$ by specifying as its kth component

$$A^k(Q_1, \ldots, Q_g) = \int_{Q_0}^{Q_1} \varphi_k + \cdots + \int_{Q_0}^{Q_g} \varphi_k, \qquad k = 1, \ldots, g, \tag{16}$$

where Q_0 is any particular point of R_g, and $\varphi_1, \ldots, \varphi_g$ are as before standard basic holomorphic 1-forms on R_g (i.e. satisfying (14)). Once again the integrals depend only on the homology classes (independent of k) of the paths from Q_0 to each of the points $Q_1, \ldots, Q_g$, and the addition to a path of an integral linear combination of the basic cycles:

$$Q_0Q_k \sim Q_0Q_k + \sum_{i=1}^{g} m_i a_i + \sum_{j=1}^{g} n_j b_j,$$

leads to a corresponding change in the quantities $A^k(Q_1, \ldots, Q_g)$:

$$A^k(Q_1, \ldots, Q_g) \sim A^k(Q_1, \ldots, Q_g) + \sum_i m_i \delta_{ik} + \sum_j n_j B_{kj},$$

so that these are defined by (16) only to within an additive integral linear combination of the periods of the differentials φ_k. In terms of the vectors $e_1, \ldots, e_{2g}$ introduced above as generators of the integral lattice Γ, we have therefore

$$A(Q_1, \ldots, Q_g) \sim A(Q_1, \ldots, Q_g) + \sum_{i=1}^{g} m_i e_i + \sum_{j=1}^{g} n_j e_{g+j},$$

from which we see that $A(Q_1, \ldots, Q_g)$ is well defined as a point of the Jacobi torus $T^{2g} = \mathbb{C}^g/\Gamma$ of the given Riemann surface R_g. We shall call A, regarded as a map to the torus T^{2g}, *Abel's transformation.*

12.7. Proposition. *Abel's transformation is invertible* (*locally*) *provided that the points $Q_1, \ldots, Q_g$ are always required to be distinct.*

(In fact, it can be shown that to within a re-ordering the points $Q_1, \ldots, Q_g$ are uniquely determined by the equation $A(Q_1, \ldots, Q_g) = u$, for generic u.)

PROOF. We give the proof only under the simplifying assumption that none of the points $Q_1, \ldots, Q_g$ is a branch point of R_g. In this case the "independent" complex variable z may be used as a local co-ordinate in some neighbourhood of each point Q_k; however, since A is a function of g variables, we shall denote this local co-ordinate by $z^{(k)}$ considered as a valid local co-ordinate in the vicinity of Q_k, $k = 1, \ldots, g$. It suffices to show that the Jacobian of the map A, namely $\det(\partial A^j(Q_1, \ldots, Q_g)/\partial z^{(k)})$, is non-zero at each g-tuple $(Q_1, \ldots, Q_g)$, or, equivalently, that the same is true of the related map $\hat{A}$ obtained by replacing the φ_k in the defining formula (16) for A, by the original basis elements ω_k of (1). It is immediate that

$$\frac{\partial \hat{A}^j}{\partial z^{(k)}} = \frac{(z^{(k)})^{j-1}}{\sqrt{P_{2g+1}(z^{(k)})}}, \qquad j, k = 1, \ldots, g.$$

Hence the Jacobian of $\hat{A}$ has the form

$$\det\left[\frac{\partial \hat{A}^j}{\partial z^{(k)}}\right] = \frac{1}{\prod_{k=1}^{g} \sqrt{P_{2g+1}(z^{(k)})}} \det\begin{vmatrix} 1 & \ldots & 1 \\ \cdots & \cdots & \cdots \\ (z^{(1)})^{g-1} & \ldots & (z^{(g)})^{g-1} \end{vmatrix}$$

$$= \frac{\prod_{i<j}(z^{(i)} - z^{(j)})}{\prod_{k=1}^{g} \sqrt{P_{2g+1}(z^{(k)})}}, \tag{17}$$

from which we see that it does not vanish provided the regions of definition of the local co-ordinates $z^{(1)}, \ldots, z^{(g)}$, do not overlap. (Note that in obtaining the final expression in (17) we have used the well-known formula, familiar from algebra, for the "Vandermonde determinant".) □

Remark. The problem of inverting Abel's transformation is well known in the geometrical theory of Riemann surfaces under the name of "Jacobi's inverse problem". This problem admits an explicit solution since in fact every symmetric function in the g co-ordinates $z^{(1)}, \ldots, z^{(g)}$ of a variable n-tuple $Q_1, \ldots, Q_g$ of points of the Riemann surface R_g, can be expressed in terms of the Jacobi–Riemann θ-function defined on the (abelian) Jacobi torus T^{2g} (see Part II, §4.1). (Thus, for instance, for the elementary symmetric function $z^{(1)} + \cdots + z^{(g)}$ summing the variables, one has

$$z^{(1)} + \cdots + z^{(g)} = \frac{d^2}{du^2} \ln \theta(u_1, \ldots, u_g) + \text{const.}, \tag{18}$$

where the points $Q_1, \ldots, Q_g$ (or their co-ordinates $z^{(1)}, \ldots, z^{(g)}$) are determined up to order by the equation $A(Q_1, \ldots, Q_g) = u$, and the operator d/du is defined by

$$\frac{d}{du} = c_{11}\frac{\partial}{\partial u_1} + \cdots + c_{1g}\frac{\partial}{\partial u_g}, \tag{19}$$

the c_{1k}, $k = 1, \ldots, g$, being as in (13).)

We shall now give an application of Abel's transformation to the integration of Kovalevskaja's equations of motion of a rigid body about a fixed point (cf. Part II, §30.3). The equations in question are as follows:

$$\begin{aligned} 2\dot{p} &= qr, & \dot{\gamma}_1 &= r\gamma_2 - q\gamma_3, \\ 2\dot{q} &= -pr - \mu\gamma_3, & \dot{\gamma}_2 &= p\gamma_3 - r\gamma_1, \\ \dot{r} &= \mu\gamma_2, & \dot{\gamma}_3 &= q\gamma_1 - p\gamma_2, \end{aligned} \tag{20}$$
$$\mu = \text{const.}$$

These equations may be transposed into Hamiltonian form (which we do not give here—however, see Appendix 1). The resulting Hamiltonian system turns out to have the following integrals of motion:

$$H = 2(p^2 + q^2) + r^2 - 2\mu\gamma_1 \quad \text{(energy)}; \tag{21}$$

$$L = 2(p\gamma_1 + q\gamma_2) + r\gamma_3 \quad \text{(angular momentum)}; \tag{22}$$

$$K = (p^2 - q^2 + \mu\gamma_1)^2 + (2pq + \mu\gamma_2)^2 \quad \text{(Kovalevskaja's integral)}, \tag{23}$$

and there is in addition the constraint $\gamma_1^2 + \gamma_2^2 + \gamma_3^2 = 1$ to be taken into account.

Consider the (joint) level surface determined by these three integrals, i.e. given say by equations $H = 6h$, $L = 2l$, $K = k^2$, where h, l, k are constants. These equations, together with the constraint $\gamma_1^2 + \gamma_2^2 + \gamma_3^2 = 1$, define a 2-

dimensional surface (an "invariant manifold" of the dynamical system (20)). On this surface we introduce co-ordinates s_1, s_2 (the "Kovalevskaja variables") defined by

$$s_{1,2} = 3h + \frac{R(x_1, x_2) \mp \sqrt{R(x_1)R(x_2)}}{(x_1 - x_2)^2},$$

where

$$x_{1,2} = p \pm iq, \qquad R(z) = -z^4 + 6hz^2 + 4\mu lz + \mu^2 - k^2,$$

$$R(x_1, x_2) = -x_1^2 x_2^2 + 6hx_1 x_2 + 2\mu l(x_1 + x_2) + \mu^2 - k^2.$$

EXERCISE

2. Show that in terms of the variables s_1, s_2 the system (20), restricted to the level surface we are considering, takes the form

$$\dot{s}_1 = \pm\frac{i\sqrt{\Phi(s_1)}}{2(s_1 - s_2)}, \qquad \dot{s}_2 = \mp\frac{i\sqrt{\Phi(s_2)}}{2(s_1 - s_2)}, \tag{24}$$

where $\Phi(z)$ is the following fifth-degree polynomial:

$$\Phi(z) = \{z[(z - 3h)^2 + \mu^2 - k^2] - 2\mu^2 l^2\}(z - 3h - k)(z - 3h + k). \tag{25}$$

Remark. The system (24) has essentially the same form as that arising from the "commutativity equation" on the level surface determined by two integrals (see Part II, Proposition 30.3.1, and below).

The single-valued functions on the right-hand sides of equations (24) can be considered as defined on the hyperelliptic Riemann surface of genus 2 determined by the equation $w^2 = \Phi(z)$. We may therefore regard (24) as representing the equations of motion of a pair of points P_1, P_2 on this Riemann surface.

Suppose, for instance, that the roots of the polynomial $\Phi(z)$ are real and distinct, denoted in order by $a_0 < a_1 < a_2 < a_3 < a_4$. If we assume also that the initial data $s_1(0), s_2(0)$ (i.e. the initial z-co-ordinates of the points P_1, P_2 on the Riemann surface) are real and satisfy

$$a_1 \leq s_1(0) \leq a_2, \qquad a_3 \leq s_2(0) \leq a_4,$$

then at all future times t the co-ordinates $s_1(t)$, $s_2(t)$ will remain real and in the same intervals. (Since $\Phi(z)$ has degree 5 and positive leading coefficient, we must have $\Phi(z)$ negative for z in the intervals (a_1, a_2) or (a_3, a_4), so that under the present assumptions the system (24) becomes real.) It follows that the points $P_1 = P_1(t)$, $P_2 = P_2(t)$ move on the Riemann surface in cycles projecting onto the line-segments $[a_1, a_2]$ and $[a_3, a_4]$ in the z-plane (as shown schematically in Figure 48). The cycle above the segment $[a_1, a_2]$ is obtained by joining together at the ends of that segment two copies $[a_1, a_2]^+$, $[a_1, a_2]^-$ in the Riemann surface, and similarly for the other cycle. The "phase point" therefore moves on a torus of two real dimensions in the Riemann surface corresponding to the equation $w^2 = \Phi(z)$. To integrate the system (24)

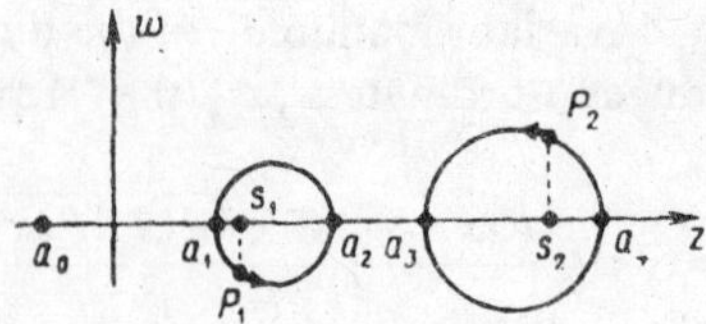

Figure 48

we now apply to it (in essence) Abel's transformation as determined by this Riemann surface. Since here the genus $g = 2$, there are just two basic holomorphic differentials, namely $dz/\sqrt{\Phi(z)}$ and $z\,dz/\sqrt{\Phi(z)}$. Hence we have, in our previous notation,

$$\begin{aligned} \hat{A}^1(P_1, P_2) &= \int_{P_0}^{P_1} \frac{dz}{\sqrt{\Phi(z)}} + \int_{P_0}^{P_2} \frac{dz}{\sqrt{\Phi(z)}}, \\ \hat{A}^2(P_1, P_2) &= \int_{P_0}^{P_1} \frac{z\,dz}{\sqrt{\Phi(z)}} + \int_{P_0}^{P_2} \frac{z\,dz}{\sqrt{\Phi(z)}}, \end{aligned} \tag{26}$$

where P_0 is an arbitrary fixed point of the Riemann surface.

12.8. Proposition. *Under the transformation* (26) *the Kovalevskaja equations* (24) *are transformed into the following linear system with constant coefficients*:

$$\begin{aligned} \frac{d}{dt}\hat{A}^1(P_1(t), P_2(t)) &= 0, \\ \frac{d}{dt}\hat{A}^2(P_1(t), P_2(t)) &= \pm\frac{i}{2}. \end{aligned} \tag{27}$$

Proof. Away from the branch points $a_0, \ldots, a_4$, we may use the variables s_1, s_2 as co-ordinates in neighbourhoods of $P_1(t)$, $P_2(t)$ respectively. From (26) and (24) in turn we shall then have

$$\frac{d}{dt}\hat{A}^1(P_1(t), P_2(t)) = \frac{\dot{s}_1}{\sqrt{\Phi(s_1)}} + \frac{\dot{s}_2}{\sqrt{\Phi(s_2)}} = \pm\frac{i}{2(s_1 - s_2)} \mp \frac{i}{2(s_1 - s_2)} = 0,$$

$$\frac{d}{dt}\hat{A}^2(P_1(t), P_2(t)) = \frac{\dot{s}_1 s_1}{\sqrt{\Phi(s_1)}} + \frac{\dot{s}_2 s_2}{\sqrt{\Phi(s_2)}} = \frac{\pm is_1 \mp is_2}{2(s_1 - s_2)} = \pm\frac{i}{2},$$

as claimed. □

Exercise

3. Show that the system

$$\frac{ds_1}{dt} = \frac{s_2\sqrt{\Phi(s_1)}}{s_1 - s_2}, \qquad \frac{ds_2}{dt} = \frac{s_1\sqrt{\Phi(s_2)}}{s_2 - s_1}, \tag{28}$$

also goes under the transformation (26) into a linear system with constant coefficients.

From Proposition 12.8 we deduce immediately that

$$\hat{A}^1(P_1(t), P_2(t)) = \hat{A}^1(P_1(t_0), P_2(t_0)),$$
$$\hat{A}^2(P_1(t), P_2(t)) = \hat{A}^2(P_1(t_0), P_2(t_0)) \pm \frac{i}{2}(t - t_0). \tag{29}$$

Thus, once mapped onto the Jacobi torus, the Kovalevskaja equations become easy to solve. To obtain the explicit dependence of the variables s_1 and s_2 on the time t, it then only remains to invert the transformation (26), i.e. to solve Jacobi's inverse problem for that transformation. We conclude that:

After extension to the complex domain, the invariant manifold $\{H = 6h, L = 2l, K = k^2, \gamma^2 = 1\}$ *of the Kovalevskaja system* (21) *becomes identifiable with the Jacobi torus* T^4 *of the Riemann surface determined by the equation* $w^2 = \Phi(z)$.

We shall now give two further examples of Hamiltonian systems integrable in a similar manner by means of the appropriate Abel transformations, i.e. systems whose invariant manifolds, extended to the complex domain, are essentially Jacobi tori of certain Riemann surfaces.

Examples

(a) Recall (from Part II, §30.2) that the "commutativity equation"

$$[\mathscr{L}, A_2 + c_1 A_1 + c_2 A_0] = 0, \tag{30}$$

where $\mathscr{L}$ is the Sturm–Liouville operator, $\mathscr{L} = -d^2/dx^2 + u(x)$, A_0, A_1, A_2 are certain differential operators with respect to x of orders 1, 3, 5 respectively, and c_1, c_2 are constants, can be written in the equivalent Lagrangian form

$$\frac{\delta L}{\delta u(x)} = 0, \tag{31}$$

with Lagrangian L of the form (cf. Part II, §30.2(23))

$$L = L(u, u', u'') = \tfrac{1}{2}(u'')^2 - \tfrac{5}{2}u''u^2 + \tfrac{5}{2}u^4 + c_1(\tfrac{1}{2}(u')^2 + u^3) + c_2u^2 + c_3u, \tag{32}$$

where c_3 is another constant. As mentioned towards the end of §30.3 of Part II, the solutions of the system (31) turn out to be just the finite-gap (here 2-gap) periodic (and almost periodic) potentials of the operator $\mathscr{L}$. The corresponding Hamiltonian system with two degrees of freedom has two independent integrals J_1, J_2 which are "in involution" (meaning that their Poisson bracket is zero), as a consequence of which the system is completely integrable (see Part II, §28.4). As noted in Part II, §30.3, if, assuming $c_1 = 0$, one introduces the co-ordinates γ_1, γ_2 defined on the level surfaces of these two integrals by

$$u = -2(\gamma_1 + \gamma_2),$$
$$\tfrac{1}{8}(3u^2 - u'') = \gamma_1\gamma_2 - \tfrac{1}{2}\sum_{i<j} \lambda_i\lambda_j, \tag{33}$$

where $\lambda_0, \ldots, \lambda_4$ are the roots of a certain polynomial $P_5(\lambda)$ of degree 5 given in terms of c_2, c_3 and the integrals of motion J_1, J_2 essentially by

$$P_5(\lambda) = \lambda^5 + \frac{c_2}{4}\lambda^3 + \frac{c_3}{16}\lambda^2 + \left(\frac{J_1}{32} + \frac{c_2^2}{4}\right)\lambda + \frac{1}{2^7}\left(\frac{J_2}{2} - c_2 c_3\right), \tag{34}$$

then in terms of these co-ordinates equation (31) takes on the same form as the system (24) with s_i replaced by γ_i and t by x (see Part II, Proposition 30.3.1). Hence the system (31) of the present example can also be solved by means of Abel's transformation. (Here the pertinent Riemann surface of genus 2 is that determined by the polynomial $P_5(\lambda)$.)

Note, incidentally, that the system (28) of Exercise 3, with s_i replaced by γ_i, describes the time-dependence of certain solutions $u(x, t)$ of the Korteweg–de Vries equation (see Part II, §30.4). (Verify!)

Remark. The commutativity equations of higher order can also be integrated by means of Abel transformations, having as they do invariant manifolds which, extended to the complex domain, become Jacobi tori of hyperelliptic Riemann surfaces of higher genera.

(b) In Neumann's problem concerning the motion of a particle on the 2-dimensional unit sphere

$$|x|^2 \equiv \sum_{i=0}^{2} (x^i)^2 = 1,$$

under the action of a quadratic potential of the form

$$U(x) = \frac{1}{2}\sum_{i=0}^{2} a_i (x^i)^2, \qquad a_i = \text{const.},$$

the equations of motion turn out to be

$$\ddot{x}^i = -a_i x^i + \lambda(t) x^i, \qquad i = 0, 1, 2, \tag{35}$$

$$|x|^2 \equiv \sum_{i=0}^{2} (x^i)^2 = 1, \tag{36}$$

where $\lambda(t)$ is the Lagrange multiplier arising from the constraint (36). The system defined by equations (35) and (36) can be obtained as the restriction to the sphere $|x|^2 = 1$ of the Hamiltonian flow in $\mathbb{R}^6$ with Hamiltonian

$$H = \frac{1}{2}\sum_{i=0}^{2} a_i (x^i)^2 + \tfrac{1}{2}(|x|^2 |y|^2 - (xy)^2). \tag{37}$$

Exercise

4. Prove that the functions

$$F_k(x, y) = (x^k)^2 + \sum_{i \neq k} \frac{(x^k y^i - x^i y^k)^2}{(a_i - a_k)}, \qquad k = 0, 1, 2, \tag{38}$$

together form a full set of independent integrals "in involution" (see the preceding example) for the system with Hamiltonian (37).

It is readily verified that in terms of these integrals the Hamiltonian H is itself given by

$$H = \frac{1}{2}\sum_{i=0}^{2} a_i F_i. \tag{39}$$

Exercise

5. Verify that the change of variables

$$x' = y, \qquad y' = -x, \qquad H' = \sum_{i=0}^{2} a_i^{-1} F_i,$$

transforms the above Hamiltonian flow on the sphere into a geodesic flow (see Part II, §28.3) on the 3-axis ellipsoid

$$\sum_{i=0}^{2} \frac{(x^i)^2}{a_i} = 1,$$

(and so transforms Neumann's problem into the problem, solved by Jacobi, of finding the geodesics on such an ellipsoid).

We shall show that Neumann's problem, and therefore also Jacobi's problem, can be integrated by means of an appropriate Abel transformation. In fact, following recent work in this area, we shall indicate how Neumann's problem can be reduced to the type of problem considered in Example (a), i.e. to that of solving a "commutativity equation" of the form (30) (or (31)) with "2-gap potentials" as solutions.

To this end, denote by ψ_0, ψ_1, ψ_2 eigenfunctions of the operator $\mathscr{L} = -d^2/dx^2 + u(x)$, corresponding to the eigenvalues a_0, a_1, a_2 respectively, i.e. ψ_0, ψ_1, ψ_2 are non-trivial solutions of the differential equations

$$\mathscr{L}\psi_i = a_i\psi_i, \qquad i = 0, 1, 2,$$

or, more explicitly,

$$\psi_i'' = -a_i\psi_i + u(x)\psi_i, \qquad i = 0, 1, 2. \tag{40}$$

These equations obviously have the same form as the equations (35) arising in the Neumann problem: under the change $x \to t$, $\psi_i \to x_i$, $u(x) \to \lambda(t)$ (the Lagrange multiplier), (40) is transformed into (35). However, there remains the constraint $\sum (x^i)^2 = 1$ still to be satisfied; to achieve this one chooses the potential to be "2-gap", and moreover such that the roots $\lambda_0, \ldots, \lambda_4$ of the corresponding polynomial $P_5(\lambda)$ (see (34)) satisfy

$$\begin{gathered} \lambda_0 = a_0 < \lambda_1 < \lambda_2 = a_1 < \lambda_3 < \lambda_4 = a_2, \\ P_5(\lambda) = \prod_{i=0}^{4} (\lambda - \lambda_i). \end{gathered} \tag{41}$$

(As noted in §30.3, the roots $\lambda_0, \ldots, \lambda_4$ are the right-hand end-points of the "gaps" in the spectrum of the operator $\mathscr{L}$.) It turns out that the solutions of (40) that we need have simple expressions in terms of the variables γ_1, γ_2 defined (essentially) as in (33):

EXERCISES

6. Show that the functions

$$\psi_i(x) = \alpha_i\sqrt{(a_i - \gamma_1(x))(a_i - \gamma_2(x))}, \qquad i = 0, 1, 2, \tag{42}$$

where the α_i are constants, satisfy the respective equations (40) if

$$u = -2(\gamma_1 + \gamma_2) + \sum_{i=0}^{4} \lambda_i,$$

$$\gamma_1' = \frac{2i\sqrt{P_5(\gamma_1)}}{\gamma_1 - \gamma_2}, \qquad \gamma_2' = \frac{2i\sqrt{P_5(\gamma_2)}}{\gamma_2 - \gamma_1}.$$

7. Show that if in the preceding exercise the constants $\alpha_0, \alpha_1, \alpha_2$ are chosen in the form

$$\alpha_i = \left[\prod_{j \neq i} (a_i - a_j)\right]^{-1/2}, \qquad i = 0, 1, 2, \tag{43}$$

then the functions ψ_i defined in (42) satisfy the constraint

$$\psi_0^2 + \psi_1^2 + \psi_2^2 \equiv 1.$$

Thus by means of the formulae (42) and (43) Neumann's problem (and consequently also Jacobi's problem of finding the geodesics on an ellipsoid) is reduced to Jacobi's inverse problem for the Riemann surface of genus 2 with branch points given by (41).

It is noteworthy that although the invariant tori and flows defined on them, via the appropriate Abel transformations, by the equation (31) for the 2-gap potentials, and by the systems arising in the Neumann and Jacobi problems, all coincide (even in the complex domain!), the original three Hamiltonian systems are not canonically equivalent (see Part I, Definition 34.3.3). (Verify!)

The higher-dimensional analogues of the systems with 2 degrees of freedom occurring in the Neumann and Jacobi problems, can be integrated in essentially the same manner, by means of a similar reduction using finite-gap potentials.

§13. The Simplest Properties of Kähler Manifolds. Abelian Tori

We begin by recalling the definition of a Kähler (or Kählerian) manifold.

13.1. Definition (cf. Part I, §27.2). A complex manifold M^{2n} (of n complex dimensions) endowed with a Hermitian metric $ds^2 = g_{\alpha\beta}\, dz^\alpha\, d\bar{z}^\beta$, where $g_{\alpha\beta} =$

$g_{\beta\alpha}$ (cf. loc. cit.) is called a *Kähler manifold* if the associated real 2-form

$$\Omega = \frac{i}{2} \sum_{\alpha,\beta} g_{\alpha\beta}\, dz^\alpha \wedge d\bar{z}^\beta \tag{1}$$

is closed: $d\Omega = 0$.

Recall that, as mentioned in Part II, Example 28.2.2(a), for a Kähler metric the form $\Omega^n = \Omega \wedge \cdots \wedge \Omega$ (n factors) is a non-zero multiple of the volume element:

$$\Omega^n = c\, dV = c\sqrt{\det(g_{\alpha\beta})}\, dz^1 \wedge \cdots \wedge dz^n \wedge d\bar{z}^1 \wedge \cdots \wedge d\bar{z}^n, \qquad c \neq 0. \tag{2}$$

From this we deduce the following

13.2. Corollary. *The forms Ω^i, $i = 1, \ldots, n$, on a compact, Kähler manifold M^{2n}, are not cohomologous to zero. It follows that the cohomology groups $H^{2i}(M^{2n}; \mathbb{R})$ are non-zero for $i = 1, \ldots, n$.*

PROOF. If the form Ω^i were exact for some i, $1 \leq i \leq n$, say $\Omega^i = d\omega$, then the form Ω^n would also be exact, since we should then have

$$\Omega^n = \Omega^i \wedge \Omega^{n-i} = d(\omega \wedge \Omega^{n-i}),$$

and from the general Stokes formula (and the compactness of M^{2n}) it would follow, since M^{2n} is without boundary, that $\int_{M^{2n}} \Omega^n = 0$. However, since M^{2n} is oriented (see Part II, §4.1), we have

$$\int_{M^{2n}} \Omega^n = c \int_{M^{2n}} dV \neq 0,$$

and we have reached a contradiction. Hence Ω^i cannot be exact. □

Examples

(a) By dimensional considerations every Riemann surface (with the induced metric on it) is a Kähler manifold.

(b) We shall define a natural Hermitian metric on complex projective space $\mathbb{C}P^n$, with respect to which it is a Kähler manifold. To this end, consider the 2-form on $\mathbb{C}^{n+1}$ given by

$$ds^2 = \sum_{k=0}^{n} dz^k\, d\bar{z}^k - \left(\sum_{k=0}^{n} z^k\, d\bar{z}^k\right)\left(\sum_{j=0}^{n} \bar{z}^j\, dz^j\right), \tag{3}$$

or rather the restriction of this form to the sphere

$$S^{2n+1}: |z^0|^2 + \cdots + |z^n|^2 = 1.$$

Now $\mathbb{C}P^n$ is obtained from $\mathbb{C}^{n+1} \setminus \{0\}$ (or, equivalently, from S^{2n+1}) by identifying vectors which are scalar multiples of each other (see Part II, §2.2). Hence the form (3) will define a (clearly Hermitian) metric on $\mathbb{C}P^n$ provided it is invariant under transformations of the form

$$z^k \to e^{i\varphi} z^k, \qquad \bar{z}^k \to e^{-i\varphi} \bar{z}^k.$$

To verify this, note first that under such a transformation we have

$$dz^k \to e^{i\varphi}(dz^k + iz^k\, d\varphi), \qquad d\bar{z}^k \to e^{-i\varphi}(d\bar{z}^k - i\bar{z}^k)\, d\varphi.$$

Hence, bearing in mind that $\sum z^k \bar{z}^k = 1$ on the sphere S^{2n+1}, we obtain

$$\sum_k dz^k\, d\bar{z}^k \to \sum_k dz^k\, d\bar{z}^k + i\left[\sum_k (z^k\, d\bar{z}^k - \bar{z}^k\, dz^k)\right] d\varphi + (d\varphi)^2,$$

$$\sum_k z^k\, d\bar{z}^k \to \sum_k z^k\, d\bar{z}^k - i\, d\varphi, \qquad \sum_j \bar{z}^j\, dz^j \to \sum_j \bar{z}^j\, dz^j + i\, d\varphi,$$

from which the invariance of ds^2 follows:

$$\sum_k dz^k\, d\bar{z}^k - \left(\sum_k z^k\, d\bar{z}^k\right)\left(\sum_j \bar{z}^j\, dz^j\right) \to \sum_k dz^k\, d\bar{z}^k - \left(\sum_k z^k\, d\bar{z}^k\right)\left(\sum_j \bar{z}^j\, dz^j\right).$$

Hence the form ds^2 may be regarded as a Hermitian metric on $\mathbb{C}P^n$. The real form (1) determined by this metric is then

$$\Omega = \frac{i}{2}\sum dz^k \wedge d\bar{z}^k - \frac{i}{2}\left(\sum_k z^k\, d\bar{z}^k\right) \wedge \left(\sum_j \bar{z}^j\, dz^j\right). \tag{4}$$

Since $\sum z^k \bar{z}^k = 1$ on the sphere S^{2n+1}, it follows that

$$\sum z^k\, d\bar{z}^k + \sum \bar{z}^k\, dz^k = 0,$$

so that the second term on the right-hand side of equation (4) vanishes, leaving

$$\Omega = \frac{i}{2}\sum dz^k \wedge d\bar{z}^k. \tag{5}$$

The closure of this form was established in §1 in the course of computing the cohomology algebra of $\mathbb{C}P^n$. Hence the space $\mathbb{C}P^n$, endowed with the metric (3), is a Kähler manifold.

(c) We conclude our examples with that of a compact, complex manifold, called the *Hopf manifold* (of n complex dimensions), which does not admit a Kähler structure. The manifold in question is the orbit space $(\mathbb{C}^n \setminus \{0\})/\Gamma$ of the manifold $\mathbb{C}^n \setminus \{0\}$, under the action of the group Γ generated by the transformation $z \mapsto 2z$. It is easy to see that this orbit space is indeed a compact, complex manifold, which is, moreover, homeomorphic to the direct product $S^1 \times S^{2n-1}$. Since $H^2(S^1 \times S^{2n-1}; \mathbb{R}) = 0$ (see §7(4)), it follows from Corollary 13.2 that this manifold is non-Kähler for $n > 1$.

Given a Kähler manifold M^{2n} one may, as in §1 (see especially Theorem 1.13), consider the *periods* of the associated form Ω, i.e. the integrals of Ω over the 2-cycles of M^{2n}. The Kähler manifold M^{2n} is called a *Hodge manifold* if these periods are all integer-valued, or at least become so after multiplication by a fixed number, i.e. after renormalization: $\Omega \to \lambda\Omega$.

For example, since for the manifold $\mathbb{C}P^n$ we know that $H_2(\mathbb{C}P^n; \mathbb{Z}) \simeq \mathbb{Z}$, we can, by multiplying the metric ds^2 defined by (3) by a suitable constant factor, arrange that the period of the associated form Ω over a generating 2-cycle (and therefore over every 2-cycle) be integer-valued.

EXERCISE

1. As generator of the group $H_2(\mathbb{C}P^n; \mathbb{Z})$ one may take the submanifold $\mathbb{C}P^1$ defined in $\mathbb{C}P^n$ by the equations $z^2 = \cdots = z^n = 0$, in terms of the usual homogeneous co-ordinates $z^0, z^1, \ldots, z^n$ (see §4, Example (g)). Calculate the period of the form $\Omega = (i/2)\sum_{k=0}^{n} dz^k \wedge d\bar{z}^k$ over this cycle. (Its reciprocal will then give the normalizing factor.)

13.3. Proposition. *A complex-analytic submanifold N^{2m} of a Kähler manifold M^{2n} is again Kähler. Furthermore if M^{2n} is Hodge, so is N^{2m}.*

PROOF. Let $f: N^{2m} \to M^{2n}$ be the inclusion map, ds^2 the given Hermitian metric on M^{2n} endowing it with its Kählerian structure, and Ω the associated closed 2-form. The metric induced on N^{2m} from ds^2 is then also Hermitian, and has the pullback $f^*\Omega$ as its associated 2-form. Since $df^*\Omega = f^*\, d\Omega = 0$, the latter form is closed, so that N^{2m} is indeed a Kähler manifold.

For the second statement, note first that for any 2-cycle c in the manifold N^{2m}, we have

$$\int_c f^*\Omega = \int_{f_*c} \Omega.$$

If M^{2n} is a Hodge manifold, then the right-hand side, the period of Ω over the cycle f_*c, is integral. Hence the left-hand side, the period of $f^*\Omega$ over the cycle c, is integral, so that N^{2m} is a Hodge manifold, as claimed. □

It can be shown (cf. Part II, §24.5) that every complex-analytic submanifold of $\mathbb{C}P^n$ is a projective algebraic variety (or union of such). (A *projective variety* is defined as the set of simultaneous solutions of a finite set of equations

$$\begin{gathered} F_1(z^0, \ldots, z^n) = 0, \\ \cdots\cdots\cdots\cdots\cdots\cdots \\ F_k(z^0, \ldots, z^n) = 0, \end{gathered} \tag{6}$$

where $F_1, \ldots, F_k$ are homogeneous polynomials:

$$F_i(cz^0, \ldots, cz^n) = c^\alpha F_i(z^0, \ldots, z^n);$$

it is usually further required that this set be irreducible with respect to unions of such solution sets. The term "complete intersection" is also used if the variety has dimension $n - k$ and each of the equations (6) defines a hypersurface in $\mathbb{C}P^n$.) For such submanifolds we deduce the following

13.4. Corollary. *Every non-singular complex submanifold of $\mathbb{C}P^n$ is a Hodge manifold.*

Remark. Every such submanifold N^{2m} is homotopic to a cycle in some triangulation of $\mathbb{C}P^n$ (recall that, as noted in §3, every differentiable manifold, hence in particular N^{2m}, is triangulable, and apply Theorem 4.6). For compact, complex submanifolds N^{2m} this cycle is never homologous to zero. This can be seen as follows: Denote by $f: N^{2m} \to \mathbb{C}P^n$ the inclusion map, and by Ω the standard form (5) on $\mathbb{C}P^n$; then $f^*\Omega$ is the 2-form on N^{2m} arising in the usual way from the induced Kähler metric on N^{2m}. Hence by the result mentioned earlier (see (2)) the form $(f^*\Omega)^m$ is a non-zero constant multiple of the volume-element on N^{2m}. Since N^{2m} is orientable (by virtue of being complex-analytic) and compact, and so has finite, non-zero volume, it follows that

$$\int_{N^{2m}} (f^*\Omega)^m \neq 0, \qquad \text{whence} \quad \int_{f_*N^{2m}} \Omega^m \neq 0. \tag{7}$$

Hence $f_* N^{2m}$ cannot be (homotopic to) a boundary in $\mathbb{C}P^n$ since Stokes' theorem together with the closure of Ω^m would otherwise give $\int_{f_*N^{2m}} \Omega^m = 0$, contradicting (7).

We next turn to the question of which complex tori T^{2n} can be endowed with a Hodge structure. Since a complex torus is obtained as a quotient space $\mathbb{C}^n/\Gamma$ where the integer lattice Γ is generated by $2n$ vectors $e_1, \ldots, e_{2n}$ linearly independent over $\mathbb{R}$, it follows that any Hermitian metric on $\mathbb{C}^n$ with constant components will determine a Kähler metric on T^{2n}. Conversely, given any Kähler metric on T^{2n}, then by averaging this metric (essentially by taking the $2n$-dimensional multiple integral of each of its components over the whole torus), we obtain an Hermitian metric with constant components.

EXERCISE

2. Prove that if the initial metric on T^{2n} is Hodge then the metric resulting from taking its average in this way will again be Hodge, moreover with the same periods (assuming the torus to have unit volume).

In view of this we may without loss of generality restrict attention to Hermitian metrics on T^{2n} with constant components. Each such metric on T^{2n} clearly determines an Hermitian scalar product on $\mathbb{C}^n = \mathbb{R}^{2n}$ given by

$$H(x, y) = \sum_{\alpha,\beta=1}^{n} h_{\alpha\beta} z_1^\alpha \bar{z}_2^\beta, \qquad h_{\beta\alpha} = \bar{h}_{\alpha\beta}, \quad x = (z_1^\alpha), \quad y = (z_2^\beta). \tag{8}$$

We shall regard $H(x, y)$ as a complex-valued bilinear form on $\mathbb{R}^{2n} \times \mathbb{R}^{2n}$. We clearly have

$$H(y, x) = \overline{H(x, y)}, \qquad H(ix, y) = iH(x, y). \tag{9}$$

Writing $H(x, y) = F(x, y) + iG(x, y)$, where $F(x, y)$ and $G(x, y)$ are real-valued, we obtain from (9) that

$$G(x, y) = -G(y, x), \qquad F(x, y) = G(ix, y),$$

whence we see that $G(x, y)$ is an alternating (i.e. skew-symmetric) real-valued form, and that the original form $H(x, y)$ can be recovered from its imaginary part $G(x, y)$ alone. Since by definition a Hermitian metric is supposed to determine a Riemannian metric (in terms of the real co-ordinates obtained from the given complex ones in the usual way), it follows also that $F(x, y)$ is an ordinary real, positive definite, bilinear form.

13.5. Proposition. *A complex torus $T^{2n} = \mathbb{C}^n/\Gamma$ can be endowed with the structure of a Hodge manifold if and only if there exists a real, skew-symmetric, bilinear form $G(x, y) = -G(y, x)$ on $\mathbb{C}^n \times \mathbb{C}^n = \mathbb{R}^{2n} \times \mathbb{R}^{2n}$, with the following two properties (the "Frobenius relations"):*

(i) *the form $F(x, y) = G(ix, y)$ is symmetric and positive definite;*
(ii) *$G(e_\alpha, e_\beta)$ is an integer for every pair e_α, e_β of vectors from a generating set $\{e_1, \ldots, e_{2n}\}$ for the integral lattice Γ.*

PROOF. We know from the preceding discussion that condition (i) is equivalent to the existence of a Kähler metric on T^{2n} (namely, that given by the Hermitian form $H(x, y) = G(ix, y) + iG(x, y)$). Hence it suffices to show that condition (ii) is equivalent to T^{2n} being a Hodge manifold with respect to this metric. It is not difficult to see, by generalizing Example (b) of §4, that as basic 2-cycles in T^{2n} the following will serve:

$$c_{\alpha\beta} = \{\lambda e_\alpha + \mu e_\beta | 0 \leq \lambda, \mu \leq 1\}, \qquad \alpha < \beta,$$

whence one easily calculates the rank of $H_2(T^{2n}; \mathbb{Z})$ to be $\binom{2n}{2} = n(2n-1)$ (cf. §1, Example (a)). Thus the proof reduces to showing that T^{2n}, endowed with the metric defined by $H(x, y)$, is Hodge if and only if integration of the imaginary part $G(x, y)$ of $H(x, y)$ over each basic 2-cycle $c_{\alpha\beta}$, yields an integer. This is however almost immediate since in the interior of each 2-cycle $c_{\alpha\beta}$, this form can be expressed as $G(e_\alpha, e_\beta)\, d\lambda \wedge d\mu$, so that its integral over $c_{\alpha\beta}$ is just $G(e_\alpha, e_\beta)$. □

In Part II, §4.1, we defined a complex torus $T^{2n} = \mathbb{C}^n/\Gamma$ to be *abelian* if the associated matrix (B_{kj}) is symmetic with positive definite imaginary part. (Here the complex entries B_{kj} are defined by

$$e_{n+k} = \sum_{j=1}^{n} B_{kj} e_j, \qquad 1 \leq k \leq n,$$

where $e_1, \ldots, e_{2n}$ are generators for the integral lattice Γ, so arranged that the first n are linearly independent over $\mathbb{C}$.) It was shown in the preceding section that the Jacobi tori of hyperelliptic Riemann surfaces are abelian. We shall now prove that

13.6. Proposition. *Abelian tori are Hodge manifolds.*

PROOF. Let $T^{2n} = \mathbb{C}^n/\Gamma$ be an abelian torus, and denote by $H(x, y)$ the Hermitian form on $\mathbb{C}^n \times \mathbb{C}^n$ given by

$$H(x, y) = \beta_{kj} z_1^k \bar{z}_2^j, \qquad x = (z_1^1, \ldots, z_1^n), \quad y = (z_2^1, \ldots, z_2^n),$$

where the indicated components of the vectors x, y are relative to the basis $\{e_1, \ldots, e_n\}$ for $\mathbb{C}^n$, and where the matrix (β_{kj}) of coefficients is the inverse of the positive definite matrix $\operatorname{Im}(B_{kj})$. The imaginary part of the form $H(x, y)$ is then

$$G(x, y) = \operatorname{Im} H(x, y) = \frac{1}{2i} \beta_{kj}(z_1^k \bar{z}_2^j - z_2^j \bar{z}_1^k). \tag{10}$$

We shall verify that the alternating form $G(x, y)$ is integer-valued on pairs of generators $e_1, \ldots, e_{2n}$ of the integral lattice Γ. For $m, l \leq n$, we have from (10):

$$G(e_m, e_l) = \frac{1}{2i} \beta_{kj}(\delta_m^k \delta_l^j - \delta_l^j \delta_m^k) \equiv 0;$$

$$G(e_m, e_{n+l}) = \sum_j \frac{1}{2i} \beta_{kj}(\delta_m^k \bar{B}_{lj} - B_{lj}\delta_m^k) = -\delta_m^k \sum_j \beta_{kj}(\operatorname{Im} B)_{jl} = -\delta_m^k \delta_{kl}$$
$$= -\delta_{ml} = -G(e_{n+l}, e_m);$$

$$G(e_{n+m}, e_{n+l}) = \sum_{k,j} \frac{1}{2i} \beta_{kj}(B_{mk}\bar{B}_{lj} - B_{lj}\bar{B}_{mk}) = \sum_{k,j} \beta_{kj}(b''_{mk} b'_{lj} - b'_{mk} b''_{lj}) = b'_{lm} - b'_{ml}$$
$$= 0,$$

where in the last calculation we have introduced the notation $b'_{jk} = \operatorname{Re} B_{jk}$, $b''_{jk} = \operatorname{Im} B_{jk}$. This completes the proof of the proposition. □

EXERCISE

3. Prove the converse assertion: Every Hodge torus is abelian.

We conclude by remarking that the importance of the class of Hodge (or, equivalently, abelian) tori derives from the fact that any such torus can, by means of θ-functions, be explicitly realized as a non-singular algebraic subvariety of a complex projective space (Lefschetz). (This, in fact, holds true more generally for arbitrary Hodge manifolds (Kodaira; see [15]).)

§14. Sheaf Cohomology

It is appropriate to describe one further type of (co)homology theory which is of considerable significance in various areas of mathematics (not, however, falling within the ambit of the present book).

In what follows, coverings of a set by a collection of subsets will be assumed *locally finite*, i.e. to have the property that only finite subcollections can have non-empty intersection.

14.1. Definition. A *presheaf* F consists of a topological space X, with each open set U of which there is associated an abelian group (or ring or field) F_U; it is further required that corresponding to each inclusion $U \subset V$ of open sets of X there is a "restriction homomorphism" or "pullback"

$$i_{UV}: F_V \to F_U, \tag{1}$$

obeying the natural composition rule: if $U \subset V \subset W$, then $i_{UW} = i_{UV} i_{VW}$. (Note that a presheaf F restricts to a presheaf $F|_U$ on each open set U of X.)

14.2. Definition. A presheaf F is called a *sheaf* on X if it has the following three additional properties:

(i) For every open set U of X and covering of U by open sets of X: $U = \bigcup_\alpha U_\alpha$, if $f \in F_U$ satisfies $i_{U_\alpha U}(f) = 0$ for all α, then $f = 0$.

(ii) Corresponding to each point of X there is an open neighbourhood U which is "sufficiently small" for it to have the property that for every cover of U: $U = \bigcup_\alpha U_\alpha$ by open sets of X, every collection of "compatible" or "matching" elements $f_\alpha \in F_{U_\alpha}$ (think of them as maps defined on U_α) can be obtained by restricting a single element $f \in F_U$ to the various U_α; thus here if

$$U = \bigcup_\alpha U_\alpha, \qquad U_{\alpha\beta} = U_\alpha \cap U_\beta = U_{\beta\alpha},$$

and

$$i_{U_{\alpha\beta}U_\beta} f_\beta = i_{U_{\beta\alpha}U_\alpha} f_\alpha \quad \text{(compatibility condition)},$$

then

$$i_{U_\alpha U} f = f_\alpha.$$

(iii) The object $F_\varnothing$ (abelian group or ring or field) associated with the empty set is the zero object: $F_\varnothing = \{0\}$.

With each covering $\{U_\alpha\}$ of a set by its subsets one can associate a certain simplicial complex $N\{U_\alpha\}$, the *nerve of the covering*, in the following way:

(i) the vertices (0-simplexes) σ_α^0 are taken to be just the U_α;

(ii) the edges (1-simplexes) $\sigma_{\alpha\beta}^1$ are to be in one-to-one correspondence with the pairs (U_α, U_β) where $U_\alpha \neq U_\beta$ and $U_\alpha \cap U_\beta \neq \varnothing$;

(iii) the triangles (2-simplexes) $\sigma_{\alpha\beta\gamma}^2$ are to correspond one-to-one to the triples $(U_\alpha, U_\beta, U_\gamma)$ of distinct subsets such that $U_\alpha \cap U_\beta \cap U_\alpha \neq \varnothing$;

(iv) in general the k-simplexes are given by the $(k+1)$-tuples $(U_{\alpha_0}, \ldots, U_{\alpha_k})$ of distinct subsets in the covering, with non-empty intersection: $U_{\alpha_0} \cap \cdots \cap U_{\alpha_k} \neq \varnothing$.

Given a presheaf F on a topological space X and an open cover $\{U_\alpha\}$ of X, one can then define the "cohomology groups of the covering with coefficient presheaf F": The *k-dimensional cochains* are the linear functionals (or homomorphisms) on the set of integral k-chains $\sum_i n_i \sigma_{\alpha_{0i}\ldots\alpha_{ki}}^k$ in the nerve $N\{U_\alpha\}$ of the covering, with values in the direct limit (see below) of the corresponding F_{U_i} with the homomorphisms (1) between them, where $U_i = U_{\alpha_{0i}} \cap \cdots \cap U_{\alpha_{ki}}$; furthermore, on each simplex $\sigma^k_{\alpha_{0i} \ldots \alpha_{ki}}$ the functionals are to take their values in F_{U_i}. The *coboundary* δc^k of each such cochain c^k is defined by

$$(\delta c^k, \sigma^{k+1}_{\alpha_0 \ldots \alpha_{k+1}}) = \sum_{q=0}^{k+1} (-1)^q i_{UU_q}(c^k, \sigma^k_{\alpha_0 \ldots \hat{\alpha}_q \ldots \alpha_{k+1}}), \tag{2}$$

where as usual hatted symbols are to be regarded as omitted, and

$$(c^k, \sigma^k_{\alpha_0 \ldots \hat{\alpha}_q \ldots \alpha_{k+1}}) \in F_{U_q},$$

$$U = U_{\alpha_0} \cap \cdots \cap U_{\alpha_{k+1}},$$

$$U \subset U_q = U_{\alpha_0} \cap \cdots \cap \hat{U}_{\alpha_q} \cap \cdots \cap U_{\alpha_{k+1}}.$$

The kth *cohomology group of the covering with coefficient presheaf F* is then the quotient of the group of k-cocycles by the k-coboundaries:

$$H^k(N\{U_\alpha\}; F) = \operatorname{Ker} \delta / \operatorname{Im} \delta = Z^k/B^k. \tag{3}$$

Now let $\{V_\beta\}$ be a refinement of the covering $\{U_\alpha\}$ (i.e. another covering of X such that each V is contained in some U) with the property that whenever $V_\beta \cap U_\alpha$ is non-empty then $V_\beta \subset U_\alpha$. Any assignment to the V_β of containing U_α clearly extends to a unique simplicial map between the nerves of the respective coverings:

$$\varphi_{UV}: N\{V_\beta\} \to N\{U_\alpha\},$$

which induces, via the appropriate restriction homomorphisms (1), a map between the corresponding k-dimensional cochain complexes, and thence between the cohomology groups (cf. §5):

$$\varphi^*_{UV}: H^k(N\{U_\alpha\}; F) \to H^k(N\{V_\beta\}; F).$$

The cohomology groups (3) of all possible open coverings of X, together with the homomorphisms φ^*_{UV} between them (when defined), form a "direct system"; the kth *cohomology group with coefficient sheaf F* (assuming now that F is a sheaf), denoted by $H^k(X; F)$, is then defined to be the "direct limit" of this system (obtained as the quotient of the direct sum of the $H^k(N\{U_\alpha\}; F)$ over all open coverings $\{U_\alpha\}$, by the subgroup generated by all differences of the form $x_U - \varphi^*_{UV}(x_U)$, $x_U \in H^k(N\{U_\alpha\}; F)$). Thus elements $x_U \in H^k(N\{U_\alpha\}; F)$, $x_W \in H^k(N\{W_\beta\}; F)$ represent the same element of $H^k(X; F)$ precisely if for some common refinement $\{V_\gamma\}$ of both $\{U_\alpha\}$ and $\{W_\beta\}$ and appropriate φ_{UV}, φ_{WV}, we have

$$\varphi^*_{UV} x_U = \varphi^*_{WV} x_W = x_V \in H^k(N\{V_\gamma\}; F).$$

Examples

(a) *A constant sheaf* on a topological space X is one where F_U is one and the same abelian group G for all $U \neq \varnothing$, and the maps i_{UV}, $\varnothing \neq U \subset V$, are all the identity map: $i_{UV} = 1_G$.

In particular, if $X = M^n$, a manifold, and $\{U_\alpha\}$ is an open covering of M^n with the property that for all k, all intersections of the form $U_{\alpha_0} \cap \cdots \cap U_{\alpha_k}$ are contractible (which will be the case if, for instance, the U_α are sufficiently small convex subsets (relative to some metric on M^n)), then for all k

$$H^k(N\{U_\alpha\}; G) \simeq H^k(M^n; G).$$

EXERCISE

1. Prove that the nerve of such a covering of M^n is a complex "homologically equivalent" to M^n.

(b) Let Y denote the reals or complexes (or some other topological abelian group). The *sheaf F of germs of continuous functions* (or, more briefly, *continuous sheaf*) with values in Y, is obtained by taking F_U to be the additive group (or ring or vector space) of all continuous functions $U \to Y$. Sheaves of germs of functions in more restrictive classes (smooth, holomorphic, algebraic, etc.) are defined analogously (with the appropriate Y).

EXERCISE

2. Show that if F is a sheaf (on $X = M^n$) of germs of functions in some class or other, then $H^0(X; F)$ consists essentially of the functions in that class that are defined on the whole manifold M^n. Deduce that $H^0(U; F|_U)$ is identifiable with F_U. (This leads to the following general definition: Given a presheaf F the *sheaf $\tilde{F}$, determined by the presheaf F*, is defined by specifying $\tilde{F}_U = H^0(U; F|_U)$, together with the obvious restriction homomorphisms.)

The group $H^1(X; F)$, with F a sheaf of germs of functions, arises for instance in connexion with the following problem. Let $X = M^{2n}$ be a complex manifold, and let $\{U_\alpha\}$ be a covering of X by open sets U_α, on each of which there is given a "principal part" f_α of an (unknown) function f from X to $\mathbb{C}$, say. (These "principal parts" might, for example, be given by the various Laurent series of the unknown function f in (punctured) neighbourhoods of its poles.) Given such "principal parts" f_α (perhaps without prior knowledge of the existence of an original function f), the problem is then to find a corresponding globally defined meromorphic function f such that each difference $f - f_\alpha$ is holomorphic in the region U_α. Clearly the appropriate "compatibility condition" must hold: on each region of intersection $U_{\alpha\beta} = U_\alpha \cap U_\beta$, the difference $f_\alpha - f_\beta = g_{\alpha\beta}$ must be holomorphic.

EXERCISE

3. Relate this problem to the first cohomology group $H^1(X; F)$ with coefficient sheaf F defined by taking each F_U to be the vector space of holomorphic functions on U (whence $F_U = H^0(U; F|_U)$). Show that if $H^1(X; F) = 0$ then there is a meromorphic function f solving the problem.

(c) From a continuous map $f\colon Y \to X$ of topological spaces, one can construct sheaves F^j on X, $j = 0, 1, 2, \ldots$, as follows. Each open set $U \subset X$ uniquely determines, of course, its complete inverse image $f^{-1}(U)$, open in Y; for each $j \geq 0$ set

$$F_U^j = H^j(f^{-1}(U))$$

(the jth integral cohomology group of the space $f^{-1}(U)$); taken together with the appropriate restriction homomorphisms, this defines the sheaf F^j. The resulting cohomology groups $H^q(X; F^j)$, $j \geq 0$, $q \geq 0$, are of interest in particular for the role they play in the most general (cohomological) version of Leray's theorem (see §8): with the $E_2^{q,j}$ replaced by the correspondingly indexed $H^q(X; F^j)$, that theorem, as given in §8, generalizes essentially to the present context.

EXERCISE

4. Show that if $p: Y \to X$ is a fibre bundle projection, where the base X is a simply-connected cell complex, then
$$H^q(X; F^j) \simeq H^q(X; H^j(\hat{F})),$$
where $\hat{F}$ is the fibre: $\hat{F} = p^{-1}(x)$.

(d) Sheaves (in general of non-commutative groups) provide an alternative approach to the problem of classifying fibre bundles with given base X and structure group G. (This problem was considered from a different point of view in Part II, §24.4.) Let $(E, X, p: E \to X, \hat{F}, G)$ be a fibre bundle over X, with fibre $\hat{F}$ and structure group G. Recall (from Part II, §24.1) that the structure of the fibre bundle is give by an open covering $\{U_\alpha\}$ of X with the property that $p^{-1}(U_\alpha) \cong U_\alpha \times \hat{F}$, together with "glueing functions", i.e. transition functions, determined by continuous maps on the regions of overlap:

$$T_{\alpha\beta}: U_\alpha \cap U_\beta \to G, \qquad (T_{\beta\alpha}(x))^{-1} = T_{\alpha\beta}(x),$$

with the property that on intersections of the form $U_\alpha \cap U_\beta \cap U_\gamma$,

$$T_{\alpha\beta}(x) T_{\beta\gamma}(x) T_{\gamma\alpha}(x) = 1, \qquad \text{for every} \quad x \in U_\alpha \cap U_\beta \cap U_\gamma. \tag{4}$$

From (2) (with $k = 1$ and $\sigma^2_{\alpha_0\alpha_1\alpha_2} = (U_\alpha, U_\beta, U_\gamma)$) we see that (4) is precisely the condition for the triple $(T_{\alpha\beta}, T_{\beta\gamma}, T_{\gamma\alpha})$ to be a "generalized" 1-cocycle in the nerve of the covering $\{U_\alpha\}$, taking its values in the set of continuous maps $U_\alpha \to G$, i.e. with coefficient sheaf F where F_U is the set of continuous maps $U \to G$. (Here a "generalized 1-cocycle" is defined by specifying the images in the appropriate F_U (in the direct limit) of distinct 1-simplexes, and extending linearly; if G is non-abelian then the set of such 1-cocycles will not form a group.) If the bundle is trivial: $E = X \times \hat{F}$, then there will exist, at least for some refinement of a given covering, maps $\varphi_\alpha: U_\alpha \to G$, such that $\varphi_\alpha(x)^{-1}\varphi_\beta(x) = T_{\alpha\beta}(x)$ for all x in $U_\alpha \cap U_\beta$, so that the maps $T_{\alpha\beta}$ are 1-coboundaries and the trivial bundle corresponds to the "zero" of the first cohomology group of the covering. It is not difficult to see that, more generally, the equivalence classes of G-bundles over X correspond one-to-one to the elements of $H^1(X; F)$ (not a group if G is non-abelian), with F as above.

EXERCISES

5. Calculate $H^1(X; F)$ for abelian G.

6. Prove that for the sheaf F on a manifold M^n, defined by taking F_U to be the vector space of smooth functions (e.g. real-valued) on some closed region $\bar{U} \supset U$, the cohomology groups (other than the zeroth) of M^n with coefficient sheaf F, are all trivial;
$$H^q(M^n; F) = 0, \quad q > 0; \qquad H^0(M^n; F) \simeq C^*(M^n),$$
where $C^*(M^n)$ denotes the algebra of smooth functions on M^n (of the appropriate class). (The analogous statement for holomorphic sheaves over complex manifolds is false in general.) (*Hint*. Use the fact, which can be established along the lines of §8 of Part II, that a smooth function defined on a closed region $\bar{U}$ can be extended to a smooth functon on the whole manifold M^n.)

7. Similarly, given a vector bundle with base $B = M^n$, and defining a sheaf F on M^n by taking F_U to be the vector space of smooth cross-sections of the (restricted) bundle over U (more precisely over a closed region $\bar{U} \supset U$) we have
$$H^q(M^n; F) = 0, \qquad q > 0,$$
with $H^0(M^n; F)$ the vector space of all smooth cross-sections of the vector bundle (and again the holomorphic version of this result is invalid).

Let $F^{(0)}, F^{(1)}, F^{(2)}$ be sheaves on $X = M^n$ such that for every sufficiently small open ball U in M^n there is an exact sequence of groups
$$0 \to F_U^{(0)} \xrightarrow{\alpha_U} F_U^{(1)} \xrightarrow{\beta_U} F_U^{(2)} \to 0.$$
where the maps α_U, β_U commute (appropriately) with the restriction homomorphisms $i_{VU}: F_U^{(k)} \to F_V^{(k)}$, $U \supset V$, $k = 0, 1, 2$. We leave it as an exercise for the reader to construct the corresponding "cohomology exact sequence of the triple of sheaves":
$$\begin{aligned} 0 \to H^0(M^n; F^{(0)}) &\xrightarrow{\alpha} H^0(M^n; F^{(1)}) \xrightarrow{\beta} H^0(M^n; F^{(2)}) \xrightarrow{\delta} H^1(M^n; F^{(0)}) \\ &\xrightarrow{\alpha} H^1(M^n; F^{(1)}) \xrightarrow{\beta} H^1(M^n; F^{(2)}) \xrightarrow{\delta} H^2(M^n; F^{(0)}) \xrightarrow{\alpha} \cdots. \end{aligned} \tag{5}$$

Examples

(α) Let $F^{(0)}$ be the constant sheaf on M^n with $G = \mathbb{Z}$, $F^{(1)}$ the sheaf of germs of smooth real-valued functions on M^n, and $F^{(2)}$ the sheaf of germs of smooth functions with values in $G = S^1 \cong \mathbb{R}/\mathbb{Z}$. As an exercise calculate $H^1(M^n; F^{(2)})$ using the exact sequence (5). (In view of Example (d), as a result of this calculation one obtains a classification of smooth fibre bundles over M^n with group $G = S^1$.)

(β) Let M^{2n} be a complex manifold. Let $F^{(0)}$ be the constant sheaf on M^{2n} with $G = \mathbb{Z}$, let $F^{(1)}$ be the sheaf on M^{2n} defined by taking $F_U^{(1)}$, for each complex chart U, to be the vector space of holomorphic functions on U, and finally let $F^{(2)}$ be defined by taking as $F_U^{(2)}$ the multiplicative group of non-vanishing holomorphic functions on U. We then define each map $\alpha_U: \mathbb{Z} \to F_U^{(1)}$ by setting $\alpha_U(1)$ equal to any non-zero constant function on U, and define $\beta_U: F_U^{(1)} \to F_U^{(2)}$ by $\beta_U(f) = \exp(2\pi i f)$.

EXERCISE

8. Show that the group $H^1(M^{2n}; F^{(2)})$ affords a classification of the holomorphic, complex line bundles over M^{2n} (cf. Part II, §25.3). Establish a relationship between the group $H^1(M^{2n}; F^{(1)})$ and the problem of classifying the holomorphic, complex line bundles over M^{2n} which are topologically (not necessarily as complex manifolds) trivial, i.e. topological direct products.

(γ) As observed in Part II, §24.5, a tensor field on a manifold M^n may be regarded as a cross-section of the tensor product of the appropriate tensor powers of the tangent and cotangent bundles over M^n. In the context of tensors obvious sheaves suggest themselves: here the F_U will be the vector spaces of smooth tensor fields (of the appropriate rank) on the regions $U \subset M^n$. In particular, for each integer $i \geq 0$ we obtain a sheaf F^i by taking F_U^i to be the vector space of smooth skew-symmetric tensors of rank i, or, in other words, of differential forms of rank i, on each $U \subset M^n$. The differential operator d then determines an "exact sequence of sheaves" (actually, an exact sequence of groups $F_U^{(i)}$, for each U of an appropriate covering of M^n; see below):

$$0 \to R \to F^0 \xrightarrow{d} F^1 \xrightarrow{d} \cdots \xrightarrow{d} F^n \to 0; \tag{6}$$

here R denotes the constant sheaf with $G = \mathbb{R}$, sent via the second map in (6) to the constant functions on M^n (closed forms of rank 0); thereafter the differential operator d determines the homomorphism, sending each form of rank i on U to its differential, a form on U of rank $i + 1$. Assuming that the regions U covering M^n are chosen to be sufficiently small n-dimensional balls, the exactness of the sequence (6) is immediate from the Poincaré lemma (see Corollary 1.6 *et seqq.*), which asserts that any closed form ω of rank > 0 on a manifold, is locally exact, i.e. for each point of the manifold there is an open neighbourhood U and a form ω' on U such that $\omega = d\omega'$ on U.

If we distinguish the closed 1-forms from among the 1-forms defined on each $U \subset M^n$, we obtain the sheaf Z^1 of germs of closed 1-forms on M^n, where $Z_U^1 \subset F_U^1$ is given by $Z_U^1 = \operatorname{Ker} d$, $d\colon F_U^1 \to F_U^2$, and the following exact sequence of sheaves (actually of groups):

$$0 \to R \to F^0 \to Z^1 \to 0. \tag{7}$$

The corresponding cohomology exact sequence (5) is then:

$$0 \to H^0(M^n; R) \to H^0(M^n; F^0) \xrightarrow{d} H^0(M^n; Z^1) \xrightarrow{\delta} H^1(M^n; R) \to H^1(M^n; F^0).$$

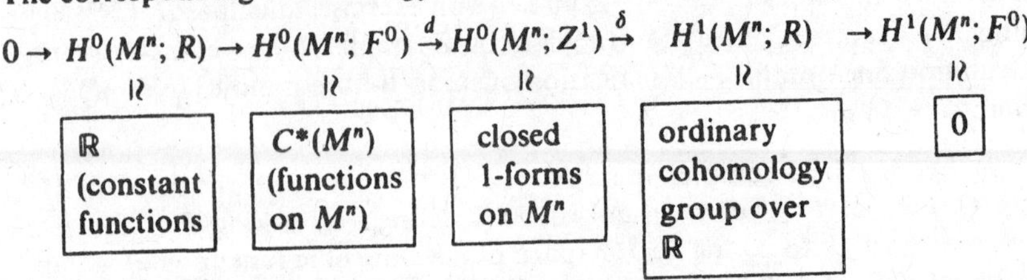

Here the indicated isomorphisms are either easy to see or have been noted earlier; thus for instance the last isomorphism, $H^1(M^n; F^0) = 0$, comes from the following

EXERCISE

9. From the exactness it follows that the homomorphism

$$\delta\colon H^0(M^n; Z^1) \to H^1(M^n; R)$$

is onto (i.e. an epimorphism). Since the kernel consists of the differentials df of smooth functions f on M^n; we conclude that

$$H^1(M^n; R) \simeq \operatorname{Ker} d/\operatorname{Im} d = H^1(M^n; Z^1)/\{df\}.$$

(Cf. the definition given in §1 of the first cohomology group of M^n as consisting of the classes of closed 1-forms modulo the exact ones.)

By elaborating this argument one can establish "De Rham's theorem", mentioned in §§3, 6: *The cohomology groups of a manifold M^n defined* (as in §1) *in terms of differential forms, are* (*naturally*) *isomorphic to the corresponding simplicial cohomology groups $H^q(M^n; \mathbb{R})$ for all q.* We conclude this section by indicating the argument in the case $q = 2$ (having given it above for $q = 1$).

Consider the sheaves $\bar{F}^0$ and Z^2 defined respectively by

$$\bar{F}^0_U = F^0_U/R$$

(where R is identified with its image under the second map in (7)), and

$$Z^2_U = \operatorname{Ker} d; \qquad d\colon F^2_U \to F^3_U.$$

(Thus for each U, Z^2_U is the vector space of closed 2-forms on U.) We infer the corresponding "exact sequences of sheaves"

$$0 \to R \to F^0 \to F^0/R \to 0, \tag{8}$$

and (from (6))

$$0 \to F^0/R \xrightarrow{d} F^1 \to Z^2 \to 0. \tag{9}$$

From the cohomology exact sequence (see (5)) corresponding to the exact sequence (8) one easily infers (again using Exercise 6) that

$$H^1(M^n; F^0/R) \simeq H^2(M^n; R),$$

and from the cohomology exact sequence arising from (9), using the fact that $H^1(M^n; F^1) = 0$ (which follows from Exercise 7), that

$$H^0(M^n; Z^2)/\{d\omega\} \simeq H^1(M^n; F^0/R),$$

where $\{d\omega\}$ denotes the vector space of all exact 2-forms on M^n (identified with the image under the map $d\colon H^0(M^n; F^1) \to H^0(M^n; Z^2)$). Since $H^0(M^n; Z^2)$ can be identified with the vector space of closed 2-forms on M^n, we finally obtain the desired isomorphism:

$$Z^2(M^n)/\{d\omega\} \simeq H^2(M^n; R).$$

CHAPTER 2

Critical Points of Smooth Functions and Homology Theory

§15. Morse Functions and Cell Complexes

Let M be an n-dimensional, compact, smooth manifold on which a Morse function f is given. (Recall from Part II, §10.4, that a *Morse function* is by definition a smooth function all of whose critical (i.e. stationary) points are non-degenerate, i.e. the Hessian $(\partial^2 f/\partial x^i\,\partial x^j)$ is non-singular at those points.) We wish to investigate the structure of the level surfaces $f_c = \{x \mid f(x) = c\}$, and the associated regions $M_c = \{x \mid f(x) \le c\}$ determined by restricting the range of f. We begin with the following

15.1. Lemma (M. Morse). *Let M be a smooth manifold, f a smooth (real-valued) function on M, and x_0 a non-degenerate critical point of f. There exist local co-ordinates $y^1, \ldots, y^n$ in some neighbourhood of x_0, in terms of which the function f is given in that neighbourhood by*

$$f(y^1, \ldots, y^n) = -(y^1)^2 - \cdots - (y^\lambda)^2 + (y^{\lambda+1})^2 + \cdots + (y^n)^2.$$

(The integer λ is called the *index* of the critical point; cf. Part II, Definition 10.4.2.)

PROOF. By way of preparing for the (very similar) proof for general n, we consider first the simplest case $n = 2$.

In view of the local character of the assertion of the lemma, we may restrict attention to an open disc $D_\varepsilon^2(0)$ of radius $\varepsilon > 0$ and with centre the critical point x_0, co-ordinatized by co-ordinates x^1, x^2 in such a way that $x_0 = (0, 0) = 0$, the origin of these co-ordinates. We may clearly also assume without loss of generality that $f(0) = 0$. We first show that there are smooth functions g_1, g_2 such that $f = x^1 g_1 + x^2 g_2$, $g_i(0) = \partial f(0)/\partial x^i$. To see this, observe first that for each fixed $x = (x^1, x^2)$,

$$\int_0^1 \frac{d}{dt} f(tx)\, dt = f(1 \cdot x) - f(0 \cdot x) = f(x).$$

Since

$$\frac{d}{dt} f(tx) = \frac{\partial f(tx)}{\partial x^\alpha} \frac{d(tx^\alpha)}{dt} = \frac{\partial f(tx)}{\partial x^\alpha} x^\alpha,$$

(where as usual summation over the repeated index α is understood), it follows that

$$f(x) = \int_0^1 \frac{\partial f(tx)}{\partial x^\alpha} x^\alpha\, dt = x^\alpha \int_0^1 \frac{\partial f(tx)}{\partial x^\alpha}\, dt = x^\alpha g_\alpha(x),$$

where

$$g_\alpha(x) = \int_0^1 \frac{\partial f(tx)}{\partial x^\alpha}\, dt. \tag{1}$$

Since grad $f(0) = 0$, we have $g_\alpha(0) = 0$. Hence the same argument can be applied to each g_α to yield smooth functions $h_{\alpha\beta}$ such that $g_\alpha(x) = x^\beta h_{\alpha\beta}(x)$. Thus

$$f(x) = x^\alpha x^\beta h_{\alpha\beta}(x),$$

where by symmetrizing we may suppose $h_{\alpha\beta}(x) = h_{\beta\alpha}(x)$. We now show that

$$h(0) = (h_{\alpha\beta}(0)) = \left(\frac{\partial^2 f(0)}{\partial x^\alpha\, \partial x^\beta}\right).$$

Observe first that since $(d/dt)g_\alpha(tx) = x^\beta(\partial g_\alpha(tx)/\partial x^\beta)$, we have

$$g_\alpha(x) = \int_0^1 \frac{d}{dt} g_\alpha(tx)\, dt = x^\beta \int_0^1 \frac{\partial g_\alpha(tx)}{\partial x^\beta}\, dt,$$

whence by (1)

$$g_\alpha(x) = x^\beta \int_0^1 \frac{\partial}{\partial x^\beta}\left(\int_0^1 \frac{\partial f(t\tau x)}{\partial x^\alpha}\, d\tau\right) dt = x^\beta \int_0^1 \int_0^1 \frac{\partial^2 f(t\tau x)}{\partial x^\alpha \partial x^\beta}\, d\tau\, dt.$$

Comparison of this expression for $g_\alpha(x)$ with $g_\alpha(x) = x^\beta h_{\alpha\beta}(x)$, then readily yields

$$h_{\alpha\beta}(0) = \frac{\partial^2 f(0)}{\partial x^\alpha \partial x^\beta}, \tag{2}$$

as claimed.

Thus we now have our function f given in the disc $D_\varepsilon^2(0)$ by

$$f = (x^1)^2 h_{11} + 2x^1 x^2 h_{12} + (x^2)^2 h_{22}, \tag{3}$$

where the matrix $(h_{\alpha\beta}(0))$ is symmetric and, in view of (2), non-singular. Since every such matrix is similar (via a real linear transformation of co-ordinates) to a diagonal matrix with non-zero diagonal entries, we may assume that

$h_{11}(0) \neq 0$ by applying an appropriate linear change of the co-ordinates if necessary. This done, we shall in fact have $h_{11}(x) \neq 0$ in some neighbourhood of the origin (the critical point x_0), in view of the continuity of the $h_{\alpha\beta}$. In that neighbourhood the expression (3) for f is equivalent to the following one:

$$\begin{aligned} f(x) &= h_{11}\left((x^1)^2 + 2x^1x^2\frac{h_{12}}{h_{11}} + (x^2)^2\frac{h_{12}^2}{h_{11}^2}\right) + (x^2)^2\left(h_{22} - \frac{h_{12}^2}{h_{11}}\right) \\ &= h_{11}\left(x^1 + \frac{h_{12}}{h_{11}}x^2\right)^2 + (x^2)^2\left(h_{22} - \frac{h_{12}^2}{h_{11}}\right). \end{aligned}$$

Since the matrix $(h_{\alpha\beta}(0))$ is non-singular, we shall also have $h_{11}h_{22} - h_{12}^2 \neq 0$ in some neighbourhood of the origin. The change to co-ordinates y^1, y^2 given by

$$y^1 = \sqrt{|h_{11}|}\left(x^1 + \frac{h_{12}}{h_{11}}x^2\right), \qquad y^2 = \sqrt{\left|h_{22} - \frac{h_{12}^2}{h_{11}}\right|}\, x^2,$$

is therefore invertible (since it has non-zero Jacobian at 0) in some neighbourhood of the origin, and in terms of these co-ordinates f has the form

$$\tilde{f}(y^1, y^2) = \pm(y^1)^2 \pm (y^2)^2.$$

This completes the proof of the lemma in the case $n = 2$.

We now prove the Morse lemma for general n. Initially, we proceed as in the case $n = 2$ above, obtaining functions $h_{\alpha\beta}$, $\alpha, \beta = 1, \dots, n$, $h_{\alpha\beta} = h_{\beta\alpha}$, with the properties that $\det(h_{\alpha\beta}(0) \neq 0$, and in terms of certain co-ordinates $x^1, \dots, x^n$ in some neighbourhood of the given critical point $x_0 = 0$, we have

$$f(x) = x^\alpha x^\beta h_{\alpha\beta}. \tag{4}$$

Suppose inductively that we have found co-ordinates $y^1, \dots, y^n$ in some neighbourhood of $x_0 = 0$ such that

$$f(y) = \pm(y^1)^2 \pm \cdots \pm (y^{k-1})^2 + \sum_{\alpha,\beta \geq k} y^\alpha y^\beta P_{\alpha\beta}(y),$$

where $P_{\alpha\beta}(y) = P_{\beta\alpha}(y)$ and the (symmetric) matrix $(P_{\alpha\beta}(0))$ is non-singular. (The initial case of the induction follows from (4), with the $h_{\alpha\beta}$ in the role of the $P_{\alpha\beta}$.)

Since the matrix $P_{\alpha\beta}(0)$ is symmetric and non-singular and of the form shown in Figure 49, it can be brought into diagonal form by means of a real linear change of the last $n - k + 1$ co-ordinates $y^k, \dots, y^n$ (not involving $y^1, \dots, y^{k-1}$). In particular, therefore, we may suppose the co-ordinates $y^k, \dots, y^n$ so chosen beforehand that $P_{kk}(0) \neq 0$, and consequently $P_{kk}(y) \neq 0$ in some neighbourhood of the origin. Write $q(y) = \sqrt{|P_{kk}(y)|}$, and define a co-ordinate change $(y^i) \to (z^i)$ on that neighbourhood by

$$z^i = y^i \quad \text{for} \quad 1 \leq i \leq k-1 \text{ and } k+1 \leq i \leq n,$$

$$z^k = q(y)\left(y^k + \sum_{i>k} y^i \frac{P_{ik}(y)}{P_{kk}(y)}\right).$$

$$(P_{\alpha\beta}) = \left(\begin{array}{ccc|ccc} \pm 1 & & 0 & & & \\ & \ddots & & & 0 & \\ 0 & & \pm 1 & & & \\ \hline & & & P_{kk} & & * \\ & 0 & & & * & \\ & & & & & \ddots \\ & & & * & & * \end{array}\right)$$

Figure 49. $P_{\alpha\alpha} = \pm 1$ for $1 \le \alpha \le k-1$.

It is clear that at the origin the Jacobian of this transformation is

$$\left.\frac{\partial z^k}{\partial y^\gamma}\right|_0 = q(0) = \sqrt{|P_{kk}(0)|} \neq 0,$$

i.e. $\det I(y \to z) \neq 0$ (see Figure 50). Hence by the Inverse Function Theorem the functions $z^1, \ldots, z^n$ will serve as local co-ordinates in some neighbourhood of 0. In terms of these co-ordinates the function f has the following form:

$$\begin{aligned} f(z) = \sum_{i \le k-1} \pm (z^i)^2 + P_{kk}\frac{(z^k)^2}{q^2(y)} - 2P_{kk}\frac{z^k}{q(y)}\sum_{i>k} y^i \frac{P_{ik}}{P_{kk}} + P_{kk}\left(\sum_{i>k} y^i \frac{P_{ik}}{P_{kk}}\right)^2 \\ + 2\left(\frac{z^k}{q(y)} - \sum_{i>k} y^i \frac{P_{ik}}{P_{kk}}\right)\sum_{i>k} y^i P_{ik} + \sum_{\alpha,\beta \ge k+1} y^\alpha y^\beta P_{\alpha\beta} \\ = \pm (z^1)^2 \pm \cdots \pm (z^k)^2 + \sum_{\alpha,\beta \ge k+1} z^\alpha z^\beta \tilde{P}_{\alpha\beta}. \end{aligned}$$

This completes the inductive step, and thence the proof of the theorem. □

Remark. This lemma is not of especial significance for investigating the behaviour of a level surface of the function f near non-degenerate critical points, since the topology of such a surface near these points is essentially already determined (via Taylor's theorem) by the form d^2f.

15.2. Lemma. *Let $f(x)$ be a smooth function on a closed (i.e. compact and boundary-less) manifold M, and let $[a, b]$, $a < b$, be a closed interval containing*

$$I(y \to z)|_0 = \begin{pmatrix} 1 & & & & & & \\ & \ddots & & & & 0 & \\ & & 1 & & & & \\ & & & q(0) & & & \\ & & & & 1 & & \\ & 0 & & & & \ddots & \\ & & & & & & 1 \end{pmatrix}$$

Figure 50

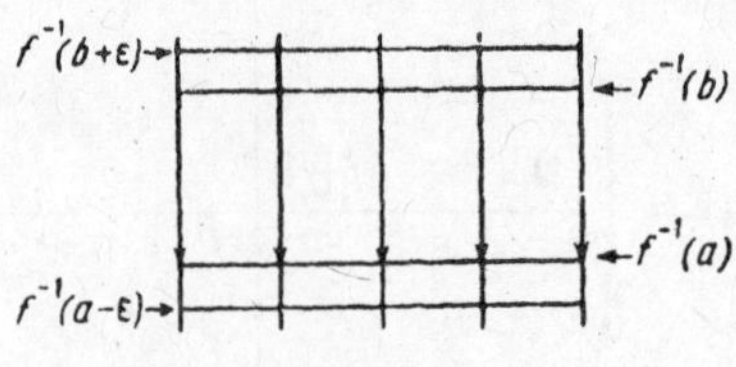

Figure 51

no critical values of f (i.e. with the property that $f^{-1}[a, b]$ contains no critical points). Then the manifold $f_a = \{x \mid f(x) = a\}$ is diffeomorphic to f_b, and the manifold-with-boundary $M_a = \{x \mid f(x) \leq a\}$ is diffeomorphic to M_b.

PROOF. In view of the compactness of M (and its consequent completeness relative to an appropriate metric) there exists a number $\varepsilon > 0$ such that the larger interval $[a - \varepsilon, b + \varepsilon]$ contains no critical value of f (since otherwise either a or b would be a critical value). Endow M with a Riemannian metric (see Part II, §8.2), and consider the vector field $v(x)$ on M obtained from the covector field grad $f(x)$ (in local co-ordinates) by using the metric to "raise the index" (see Part I, §19.1). This vector field has no singularities in the interior of the manifold-with-boundary $f^{-1}[a - \varepsilon, b + \varepsilon]$, and it is orthogonal to each level surface $f^{-1}(\alpha)$, $a \leq \alpha \leq b$. Consider the integral trajectories of the field $v(x)$ beginning at the points of $f^{-1}(b)$ and ending at $f^{-1}(a)$ (see Figure 51). It is not difficult to see that since M is compact there is a smooth deformation of the surface $f^{-1}(b)$ onto the surface $f^{-1}(a)$ moving the points of $f^{-1}(b)$ along these integral trajectories, which realizes a diffeomorphism between $f^{-1}(b)$ and $f^{-1}(a)$.

The desired diffeomorphism between M_a and M_b follows essentially from the fact just established that $f^{-1}[a, b]$ is diffeomorphic to $f_a \times I$, where I is a (closed) interval. The details are left to the reader. □

From the Morse lemma one can readily infer the behaviour of a level surface of a smooth function f on an n-manifold M^n, in the vicinity of the non-degenerate critical points. Let $x_0 \in M^n$ be such a point where (we may assume) $f(x_0) = 0$. By the Morse lemma (15.1), in some sufficiently small neighbourhood $U(x_0)$ of x_0 we may introduce curvilinear co-ordinates $x^1, \ldots, x^n$ in terms of which f has the form

$$f(x) = -(x^1)^2 - \cdots - (x^\lambda)^2 + (x^{\lambda+1})^2 + \cdots + (x^n)^2 \tag{5}$$

(where λ is the index of the critical point). We saw in the course of proving the Morse lemma that we may also suppose that x_0 is the origin of this co-ordinate system: $x_0 = 0$. Hence for sufficiently small $\varepsilon > 0$ the three hypersurfaces $f_0, f_\varepsilon, f_{-\varepsilon}$ are given, in the vicinity of $x_0 = 0$, by the following three equations:

$$-(x^1)^2 - \cdots - (x^\lambda)^2 + (x^{\lambda+1})^2 + \cdots + (x^n)^2 = \begin{cases} 0 \\ \varepsilon . \\ -\varepsilon \end{cases}$$

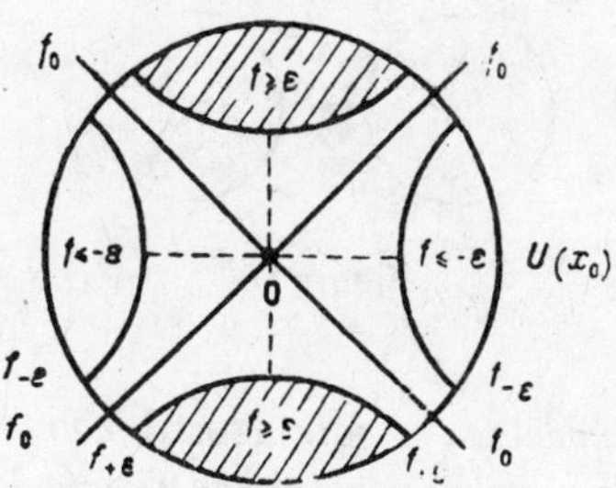

Figure 52

Thus (provided $\lambda \neq 0, n$) near the point x_0 the surface f_0 has the form of a cone with apex at x_0, while the surfaces $f_{\pm\varepsilon}$ are hyperboloids (see Figure 52).

15.3. Lemma. *Let M, f, λ and ε be as above, with $\lambda \neq 0, n$. If $f^{-1}[-\varepsilon, \varepsilon]$ contains exactly one critical point, then the manifold M_ε has the homotopy type of the space obtained by attaching a single λ-dimensional cell σ^λ to $M_{-\varepsilon}$ at its boundary $f_{-\varepsilon}$ ($\partial M_{-\varepsilon} = f_{-\varepsilon}$).*

PROOF. It suffices to produce a homotopy $\varphi_t: M_{+\varepsilon} \to M_{+\varepsilon}$, with $\varphi_0 = 1$, and φ_1 a map $M_{+\varepsilon} \to M_{-\varepsilon} \cup \sigma^\lambda$ which is the identity on $M_{-\varepsilon}$. Let $v(x)$ be the vector field on M determined by the covector field $-\text{grad}\, f$ (cf. the preceding proof), and as above let U be a neighbourhood of the critical point in which there exist co-ordinates satisfying (5), with the critical point at the origin. Outside U and $M_{-\varepsilon}$, let φ_t be the deformation of $M_{+\varepsilon}$ along the integral trajectories of the field $v(x)$, and inside U define φ_t as indicated schematically in Figure 53 (and take φ_t to be the identity map on $M_{-\varepsilon}$). Thus inside U parts of the various level surfaces f_c, $-\varepsilon < c < \varepsilon$, are "flattened" by φ_t onto a λ-dimensional disc $D^\lambda(x^1, \ldots, x^\lambda)$ (represented schematically in Figure 53 by the line segment AB) with boundary ($\cong S^{\lambda-1}$) smoothly embedded in the boundary $f_{-\varepsilon}$ of the region $M_{-\varepsilon}$. The end result of the deformation is shown in Figure 54. (Note that in these figures $\lambda = 1$, and the sphere S^0 is the pair of points A, B.) □

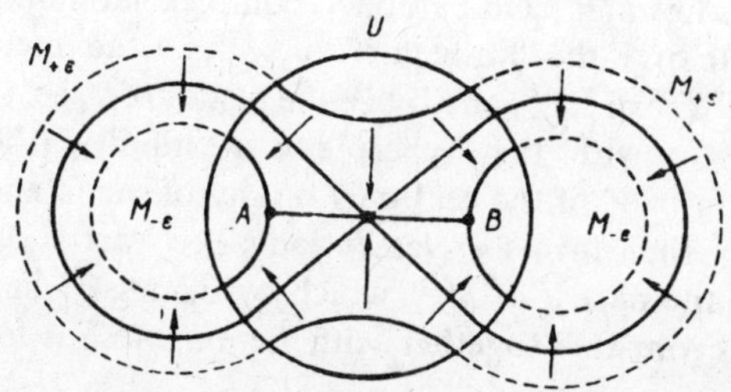

Figure 53

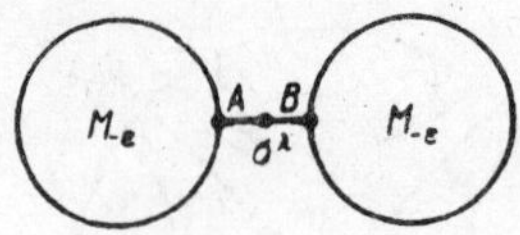

Figure 54

15.4. Theorem. *Any (connected) smooth, closed manifold M is homotopically equivalent to a finite cell complex constructed from a Morse function on M in such a way that each critical point P_λ of index λ gives rise, as in the preceding lemma, to a λ-dimensional cell, thus determining a one-to-one correspondence between the critical points and the cells of the complex.*

PROOF. By Theorem 10.4.3 of Part II, there certainly exists a Morse function on M with the property that on each level surface there is at most one critical point, i.e. on each "critical surface" there is exactly one critical point. It now follows from the preceding two lemmas, once the latter of these (15.3) has been extended appropriately to cover the cases $\lambda = 0, n$, that M is homotopically equivalent to a cell space of the appropriate sort; we leave the (straightforward) details to the reader. An appeal to Theorem 4.4 (or rather its proof), according to which a cell space has the homotopy type of a cell complex, then completes the proof. □

The imposition on f of certain conditions related to analyticity leads to restrictions on the possible indices of the critical points.

EXERCISES

1. If $f = \text{Re } F(z^1, \ldots, z^n)$ is the real part of a complex-analytic function F on $\mathbb{C}^n$, then every non-degenerate critical point of f has index n.

2. If f is a harmonic function on $\mathbb{R}^n$ then the non-degenerate critical points of f cannot have index 0 or n. (This is essentially the "maximum priciple" for harmonic functions.)

(Note, however, that on a compact (complex) manifold no non-constant complex-analytic or harmonic functions exist (cf. Part II, §4.1).)

We conclude this section by sketching a topological application of the result of Exercise 1. Let M^{2n} be a compact, complex submanifold of $\mathbb{C}P^N = \mathbb{C}^N \cup \mathbb{C}P_\infty^{N-1}$. Denote by V the "finite part" of M^{2n}, i.e. the intersection of M^{2n} with the "finite part" $\mathbb{C}^N$ of $\mathbb{C}P^N$, and by W the rest of M^{2n}, i.e. the intersection $M^{2n} \cap \mathbb{C}P^{N-1}$ of M^{2n} with the hyperplane at infinity (the "hyperplane section"). Restriction to V of the real part of any of the (standard) complex co-ordinates for $\mathbb{C}^N$ then furnishes (essentially; see Part II, §11.2) a Morse function f on the finite part V of M^{2n}, which by Exercise 1 has all its critical points of index n. From this together with Lemma 15.3 it follows that the manifold is homotopically equivalent to a cell complex of the form

$$[W \cup \sigma_1^n \cup \cdots \cup \sigma_k^n] \cup \sigma^{2n},$$

where k is the number of critical points of f. (Give a precise proof of this; use the argument of Lemma 15.3, and note that V is non-compact.) From this decomposition of M^{2n} we readily infer the following isomorphisms:

$$\pi_i(W) \simeq \pi_i(M^{2n}), \qquad i < n - 1;$$

$$H_i(W) \simeq H_i(M^{2n}), \qquad i < n - 1 \text{ or } n < i < 2n;$$

and also that the homomorphism $H_{n-1}(W) \to H_{n-1}(M^{2n})$ induced by the inclusion is onto, i.e. an epimorphism.

§16. The Morse Inequalities

There is a close connexion between the (number of) critical points of a function f on a smooth closed manifold M^n and certain topological invariants of the manifold, in particular its homology groups and Euler characteristic. For instance, the number $\sum_{\lambda \geq 0} (-1)^\lambda \mu_\lambda(f)$, where $\mu_\lambda(f)$ is the number of critical points of index λ of the Morse function f, is actually independent of that function, coinciding as it does with the Euler characteristic of M^n (see Part II, Corollary 15.2.5, for the orientable case of this). Another link between the ranks of the homology groups and the number of critical points of each index of a Morse function, is exhibited in the following

16.1. Theorem. *Let M^n be a smooth, closed manifold, and let $b_k(M^n)$ denote the dimension of kth homology group of M^n over any field (i.e. dimension as a vector space). Then for any Morse function f on M^n the following "Morse inequalities" hold:*

$$\mu_\lambda(f) \geq b_\lambda(M^n), \qquad \lambda = 0, 1, \ldots, n, \tag{1}$$

where $\mu_\lambda(f)$ is the number of critical points of f of index λ.

PROOF. By Theorem 15.4 the manifold M^n is homotopically equivalent to a cell complex K in which the number of cells of each dimension λ coincides with the number $\mu_\lambda(f)$ of critical points of index λ. Since (by Corollary 5.4) homotopically equivalent spaces have isomorphic homology groups, and since the rank of the λth homology group $H_\lambda(K)$ is at most the number of cells of dimension λ, the theorem follows. □

This result does not however account for all correlations between the Betti numbers $b_\lambda(M^n)$ (= rank $H_\lambda(M^n)$) on the one hand, and on the other hand the numbers $\mu_\lambda(f)$ (which may by the above be regarded simply as the numbers of λ-dimensional cells in the complex $K \sim M^n$). For instance, by Proposition

2.8 we know also that

$$\sum_{\lambda \geq 0} (-1)^\lambda b_\lambda = \sum_{\lambda \geq 0} (-1)^\lambda \mu_\lambda(f). \tag{2}$$

(Cf. the above-mentioned result that the right-hand side is equal to the Euler characteristic of M^n.) The totality of such relations is most conveniently arrived at via a certain property of the following two polynomials (or "generating functions"). These are the *Poincaré polynomial of the manifold*, denoted by $P(M^n, t)$ and defined by $P(M^n, t) = \sum b_\lambda t^\lambda$, and the *Poincaré polynomical of the function f on* M^n, defined by $Q(M^n, f, t) = \sum \mu_\lambda(f)t^\lambda$. (These polynomials may in fact be defined for every finite cell complex, in the latter case simply by taking μ_λ to be the number of cells of dimension λ.) The relation (2) is then equivalent to the divisibility of the polynomial $(Q - P)$ by $(1 + t)$. It turns out that the (integer) polynomial $(Q - P)/(1 + t)$ has all its coefficients non-negative. We shall prove a stronger version of this below, where the function f is permitted to have (finitely many) degenerate critical points. From this we shall derive a version of the Morse inequalities, applying to a somewhat more general class of functions than Morse functions.

We first make the following definitions. In these definitions f is assumed to be an infinitely differentiable function on a manifold M^n, i.e. of class C^∞.

16.2. Definition. A point $x_0 \in M^n$ is *topologically regular* for the function f if there is an open neighbourhood $U = U(x_0)$ homeomorphic to a direct product of the form $(f^{-1}(a) \cap U) \times I(-\varepsilon, \varepsilon)$, where $a = f(x_0)$ (i.e. to the direct product of that part of the level surface through x_0 lying in U, with an (open) interval (see Figure 55)), and moreover by means of a homeomorphism which "respects fibres", i.e. which identifies each level surface $f^{-1}(a + t)$ (or rather that part of it in U) with the cross-section $(f^{-1}(a) \cap U, t)$ of the product. A point x_0 of M^n which is not topologically regular for f is called a *bifurcation point* of f.

If $x_0 \in M^n$ is a non-degenerate critical point of a smooth function f on M^n, then x_0 is a bifurcation point; this follows essentially from the Morse lemma—see Figure 56 for the case where the index $\lambda \neq 0, n$. However, a degenerate critical point need not be a bifurcation point, as the following simple example shows.

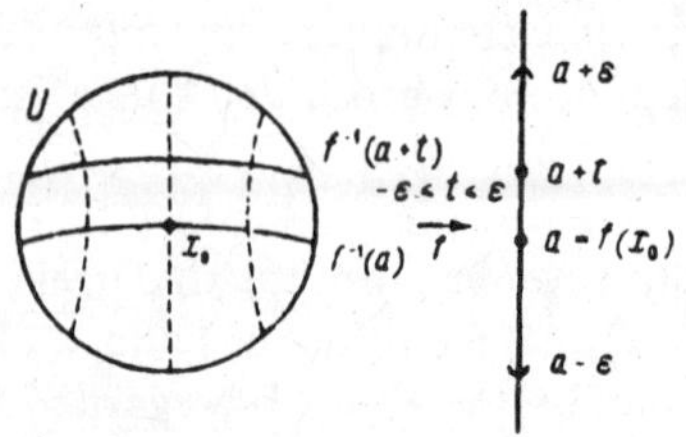

Figure 55

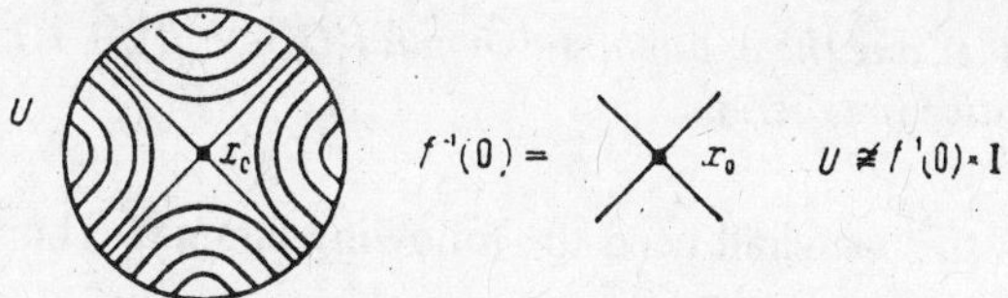

Figure 56

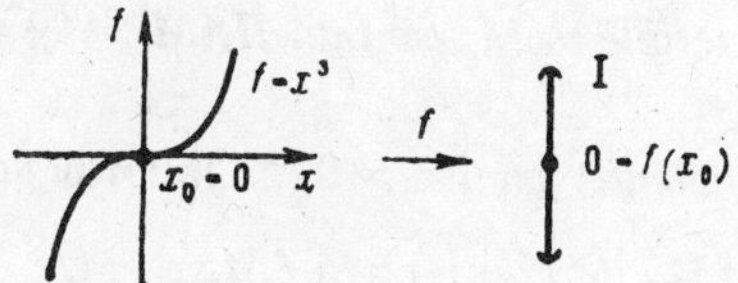

Figure 57

Example. Consider $M = \mathbb{R}^1(x)$, $f(x) = x^3$, $x_0 = 0 \in \mathbb{R}^1$. Although x_0 is a degenerate critical point of f, it is nonetheless clearly toplogically regular (i.e. not a bifurcation point) for f (see Figure 57).

In what follows we shall assume that our smooth manifold M^n is closed, and restrict attention to those smooth functions on it that have only finitely many bifurcation points (e.g. Morse functions), i.e. these will be the "admissible" functions. Let $c_1, \ldots, c_N$ $(N < \infty)$ be those values (necessarily critical) of such a function f on such a manifold M^n whose complete inverse images $f^{-1}(c_\alpha)$ contain bifurcation points (i.e. each $f^{-1}(c_\alpha)$ should contain at least one bifurcation point). Note that the bifurcation points are isolated since by assumption there are only finitely many of them. Write $\{x\}_\alpha$ for the set of bifurcation points on the level surface $\{x \mid f(x) = c_\alpha\}$, and, as usual, $M_{c_\alpha} = \{x \mid f(x) \leq c_\alpha\}$, $\alpha = 1, \ldots, N$. The relative homology groups $H_k(M_{c_\alpha}, M_{c_\alpha} \backslash \{x\}_\alpha)$ (see §5) turn out to provide the appropriate means for generalizing the polynomial $Q(M^n, f, t)$ introduced above. (Note that in view of the fact that the bifurcation points are isolated, the groups $H_k(M_{c_\alpha}, M_{c_\alpha} \backslash \{x\}_\alpha)$ may, if one prefers, be replaced by the groups $H_k(M_{c_\alpha}, M_{c_\alpha} \backslash \{U(x)\}_\alpha)$, where $\{U(x)\}_\alpha$ is a collection of suitably small open neighbourhoods of the points in $\{x\}_\alpha$.)

16.3. Definition. The (*generalized*) *Poincaré polynomial of the function* $f\colon M^n \to \mathbb{R}$, is defined by

$$Q(M^n, f, t) = \sum_{k=0}^{n} \sum_{\alpha=1}^{N} b_k(M_{c_\alpha}, M_{c_\alpha} \backslash \{x\}_\alpha) t^k,$$

where $b_k(X, Y) = \dim(H_k(X, Y)$ (over any field).

16.4. Theorem. *Let* $P(M^n, t)$ *(as before) and* $Q(M^n, f, t)$ *(as just defined) be the Poincaré ploynomials of* M^n *and of* f. *The difference* $(Q - P)$ *is divisible by the* *polynomial* $(1 + t)$, *and the quotient polynomial* $(Q - P)/(1 + t)$ *has all its coefficients non-negative (integers).*

To establish this we shall need the following lemmas. The first of these is clear from the proof of Lemma 15.2.

16.5. Lemma. *Let* $a < b$ *be two numbers in the range of* $f: M^n \to \mathbb{R}$, *with the property that the interval* $[a, b]$ *contains none of the critical values* $c_1, \ldots, c_N$ *of* f. *Then* M_b *is contractible to* M_a, *so that* $H_k(M_b, M_a) = 0$ *for all k.*

16.6. Lemma. *For sufficiently small* $\varepsilon > 0$ *the following equality holds:*

$$b_k(M_{c_\alpha}, M_{c_\alpha}\backslash\{x\}_\alpha) = b_k(M_{c_\alpha+\varepsilon}, M_{c_\alpha-\varepsilon}).$$

PROOF. By considering the restriction of f to $M_{c_\alpha}\backslash\{x\}_\alpha$ we infer from the preceding lemma that for sufficiently small ε the manifold $M_{c_\alpha}\backslash\{x\}_\alpha$ (possibly with boundary) contracts to $M_{c_\alpha-\varepsilon}$. The preceding lemma (or rather its proof) also implies that for sufficiently small ε the manifold (possibly with boundary) $M_{c_\alpha+\varepsilon}\backslash\{x\}_\alpha$ can be contracted onto $M_{c_\alpha}\backslash\{x\}_\alpha$ by means of a deformation throughout which the latters space remains pointwise fixed. This clearly extends to a contraction of $M_{c_\alpha+\varepsilon}$ onto M_{c_α} by letting the points of $\{x\}_\alpha$ also remain fixed. It readily follows from the existence of these two contractions ($M_{c_\alpha+\varepsilon}$ onto M_{c_α}, and $M_{c_\alpha}\backslash\{x\}_\alpha$ onto $M_{c_\alpha-\varepsilon}$) and the definition of the relative homology groups, that

$$H_k(M_{c_\alpha}, M_{c_\alpha}\backslash\{x\}_\alpha) \simeq H_k(M_{c_\alpha+\varepsilon}, M_{c_\alpha-\varepsilon}),$$

whence the lemma. □

The next (and final) lemma concerns the following three polynomials of "Poincaré type", defined for each pair of reals a, b, $a < b$:

$$P(M_a) = \sum_k b_k(M_a)t^k;$$

$$P(M_b, M_a) = \sum_k b_k(M_b, M_a)t^k \quad (M_b \supset M_a);$$

$$P(\text{Im } \partial) = \sum_k \dim(\text{Im } \partial_{k+1})t^k,$$

where $\partial_{k+1}: H_{k+1}(M_b, M_a) \to H_k(M_a)$ is the boundary operator figuring in the homology exact sequence of the pair (M_b, M_a) (see Theorem 5.6).

16.7 Lemma. *For each* a, b, $a < b$, *these three polynomials are related by the equation*

$$P(M_b, M_a) - \{P(M_b) - P(M_a)\} = (1 + t)P(\text{Im } \partial).$$

PROOF. Consider the following segment of the homology exact sequence of the pair (M_b, M_a) (see §5(18)):

$$H_{k+1}(M_b, M_a) \xrightarrow{\partial_{k+1}} H_k(M_a) \xrightarrow{i_*} H_k(M_b) \xrightarrow{j} H_k(M_b, M_a) \xrightarrow{\partial_k} H_{k-1}(M_a).$$

The exactness yields the following chain of equalities:

$$b_k(M_b, M_a) = \dim(\operatorname{Im} j) + \dim(\operatorname{Im} \partial_k);$$

$$\begin{aligned}\dim(\operatorname{Im} j) &= b_k(M_b) - \dim(\operatorname{Im} i)\\ &= b_k(M_b) - \{b_k(M_a) - \dim(\operatorname{Im} \partial_{k+1})\}\\ &= \{b_k(M_b) - b_k(M_a)\} + \dim(\operatorname{Im} \partial_{k+1}).\end{aligned}$$

It follows that

$$\begin{aligned}b_k(M_b, M_a) - \dim(\operatorname{Im} j) &= b_k(M_b, M_a) - \{b_k(M_b) - b_k(M_a)\} - \dim(\operatorname{Im} \partial_{k+1})\\ &= R_k - \dim(\operatorname{Im} \partial_{k+1}) = \dim(\operatorname{Im} \partial_k),\end{aligned}$$

where $R_k = b_k(M_b, M_a) - \{b_k(M_b) - b_k(M_a)\}$. From

$$R_k = \dim(\operatorname{Im} \partial_{k+1}) + \dim(\operatorname{Im} \partial_k),$$

we obtain

$$t^k R_k = t^k \dim(\operatorname{Im} \partial_{k+1}) + t(t^{k-1} \dim(\operatorname{Im} \partial_k)),$$

whence

$$\sum_k t^k R_k = (1 + t)P(\operatorname{Im} \partial),$$

completing the proof. □

Proof of Theorem 16.4. As before denote by $c_1, \ldots, c_N$ ($N < \infty$) the critical values, in increasing order, of f such that each complete inverse image $f^{-1}(c_i)$ contains at least one bifurcation point. Let $a_0, \ldots, a_N$ be real numbers (non-critical for f) separating the c_i, i.e. satisfying

$$a_0 < c_1 < a_1 < c_2 < \cdots < a_{N-1} < c_N < a_N,$$

as shown in Figure 58. By Lemma 16.7 for each $i = 0, \ldots, N - 1$, we have

$$P(M_{a_{i+1}}, M_{a_i}) - \{P(M_{a_{i+1}}) - P(M_{a_i})\} = (1 + t)P(\operatorname{Im} \partial)_i.$$

Summing these equations over i from 0 to $N - 1$, we obtain

$$\left\{\sum_{i=0}^{N-1} P(M_{a_{i+1}}, M_{a_i})\right\} - P(M_{a_N}) + P(M_{a_0}) = (1 + t)K(t), \qquad (3)$$

where $K(t)$ is a polynomial with non-negative integer coefficients (since the $P(\operatorname{Im} \partial)_i$ have this property). By Lemma 16.6

$$P(M_{a_{i+1}}, M_{a_i}) = P(M_{c_{i+1}}, M_{c_{i+1}} \setminus \{x\}_{i+1}),$$

so that (3) can be rewritten as

$$Q(M, f, t) - P(M_{a_N}) + P(M_{a_0}) = (1 + t)K(t). \qquad (4)$$

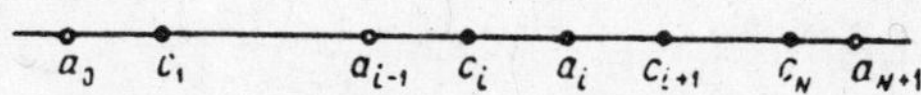

Figure 58

Now since M is compact, $f(M)$ is also compact, i.e. a closed and bounded subset of $\mathbb{R}$, so that we may choose $a_N > \max_{x\in M} f(x)$, $a_0 < \min_{x\in M} f(x)$. We shall then have $M_{a_N} = M^n$, whence $P(M_{a_N}) = P(M^n, t)$, and $M_{a_0} = \varnothing$, whence $P(M_{a_0}) = 0$, and (4) becomes

$$Q(M, f, t) - P(M, t) = (1 + t)K(t),$$

completing the proof. □

We turn now to the consequences of this theorem. We shall assume that the coefficients are from the real field $\mathbb{R}$. As in the theorem let f be any admissible smooth function on a closed manifold M^n, with relevant critical values $c_1, \ldots, c_N$. Recall that the Poincaré polynomial of the manifold is (suppressing the t) $P(M) = \sum_{k\geq 0} b_k t^k$ where the b_k ($=\dim H_k$) are the Betti numbers of the manifold M^n, and that the Poincaré polynomial of the function f is given by $Q(M, f) = \sum_{k\geq 0} \mu_k t^k$, where by Definition 16.3,

$$\mu_k = \sum_{\alpha=1}^{N} \dim H_k(M_{c_\alpha}, M_{c_\alpha}\backslash\{x\}_\alpha). \tag{5}$$

We shall call the numbers μ_k the *Morse numbers* of the function f. (For Morse functions these numbers have a clear geometrical significance; see below and cf. Theorem 16.1.) By Theorem 16.4 we have

$$Q(M, f) - P(M) = \sum_k (\mu_k - b_k)t^k = (1 + t)K(t), \tag{6}$$

where $K(t)$ is a polynomial with non-negative integer coefficients. It follows that the polynomial $\sum(\mu_k - b_k)t^k$ also has non-negative coefficients, whence $\mu_k \geq b_k$, i.e. *the kth Betti number b_k of a closed manifold M^n does not exceed the kth Morse number of any admissible function on M^n*. (Cf. the Morse inequalities (1).) Further, putting $t = -1$ in (6) we obtain (cf. (2))

$$\sum_k (-1)^k \mu_k = \sum_k (-1)^k b_k.$$

(Note that the right-hand side of this equation is the Euler characteristic $\chi(M^n)$ of the manifold M^n, defined in §2 as the alternating sum of the Betti numbers (see Definition 2.7).) We conclude that: *the alternating sum of the Morse numbers of an arbitrary admissible function f on a closed manifold M^n is a homotopy invariant of the manifold* (and so, in particular, is the same for every such function f).

Finally, from the power series expansion of $(1 + t)^{-1}$:

$$(1 + t)^{-1} = \sum_{\alpha=0}^{\infty} (-1)^\alpha t^\alpha,$$

we obtain

$$K(t) = \left(\sum_k (\mu_k - b_k)t^k\right) \sum_{\alpha=0}^{\infty} (-1)^\alpha t^\alpha,$$

so that in the power series represented by the right-hand side (after multiplying out and gathering like powers of t) the coefficients must be non-negative. Thus for each fixed non-negative integer λ we have

$$(\mu_0-b_0)(-1)^\lambda+(\mu_1-b_1)(-1)^{\lambda-1}+(\mu_2-b_2)(-1)^{\lambda-2}+\cdots+(\mu_\lambda-b_\lambda)\geq 0,$$

or, equivalently,

$$\mu_\lambda-\mu_{\lambda-1}+\mu_{\lambda-2}-\cdots\pm\mu_0\geq b_\lambda-b_{\lambda-1}+b_{\lambda-2}-\cdots\pm b_0.$$

We now fulfil our promise to show that when the admissible function is Morse, the Morse numbers μ_k have a particularly transparent geometric interpretation (namely as the number of critical points of index k, so that in this case the definition of μ_k via (5) reduces to the original definition—for Morse functions—as the number of critical points of index k). Thus let f be a Morse function on a closed manifold M^n, let x_0 be a critical (and therefore a bifurcation) point of f, of index λ say, and let c be the corresponding critical value: $f(x_0)=c$. We shall assume that x_0 is the only critical point on the critical surface $f^{-1}(c)$. (By Theorem 10.4.3 of Part II, Morse functions having this property for every critical value, are everywhere dense in the set of all smooth functions on M^n.) We shall show that the dimension of the (real) vector space $H_k(M_c, M_c\backslash\{x_0\})$ is 1 or 0 according as $\lambda=k$ or $\lambda\neq k$. By Theorem 5.9, if X is a cell complex and Y a subcomplex of X, then $H_k(X, Y)\simeq H_k(X/Y)$. Applying this to the pair $(M_{c+\varepsilon}, M_{c-\varepsilon})$ (homotopically equivalent to such a pair (X, Y) essentially by Lemma 15.3 or Theorem 15.4), we obtain the isomorphism

$$H_k(M_{c+\varepsilon}, M_{c-\varepsilon})\simeq H_k(M_{c+\varepsilon}/M_{c-\varepsilon}). \tag{7}$$

Now by Lemma 15.3 again, $M_{c+\varepsilon}$ is homotopically equivalent to $M_{c-\varepsilon}\cup\sigma^\lambda$, i.e. to $M_{c-\varepsilon}$ with a λ-dimensional cell σ^λ attached, so that

$$H_k(M_{c+\varepsilon}/M_{c-\varepsilon})\simeq H_k(\sigma^\lambda/\partial\sigma^\lambda)\simeq H_k(S^\lambda), \tag{8}$$

where S^λ is as usual the λ-dimensional sphere. The isomorphisms (7) an (8) together give the desired result:

$$H_k(M_c, M_c\backslash\{x_0\})\simeq H_k(S^\lambda)\simeq\begin{cases}\mathbb{R} & \text{if } k=\lambda, 0,\\ 0 & \text{if } k\neq\lambda, 0.\end{cases}$$

We conclude the section with two instructive examples involving degenerate critical points.

Examples

(a) The function $f(x, y)=\operatorname{Re}(z^n)$, where $z=x+iy$, has the origin $(0, 0)$ as a degenerate critical point. The behaviour of the level surfaces near the origin is indicated (in the case $n=3$) in Figure 59 (where $c=0$). It is not difficult to see that in this case (namely $n=3$) we have $M_{c+\varepsilon}/M_{c-\varepsilon}\cong S^1\vee S^1$. Examination of the proof of Theorem 10.4.3 of Part II, reveals that in applying a (linear) perturbation to a smooth function f (with the object of obtaining a Morse function), each degenerate critical point may give rise to several non-

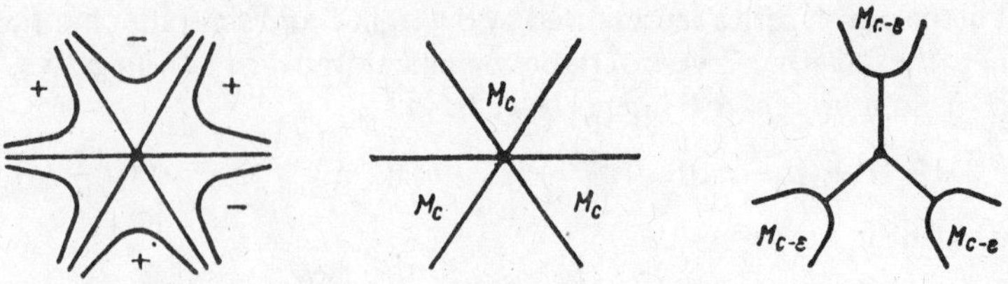

Figure 59

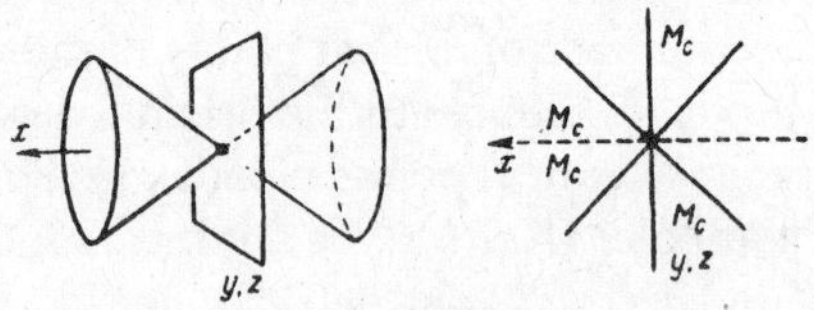

Figure 60

degenerate critical points. (Consider for example $f: \mathbb{R} \to \mathbb{R}$ defined by $f(x) = x^3$, and its perturbation to $g(x) = x^3 - ax$, a small and positive.) In the present example such a perturbation yields $(n-1)$ non-degenerate critical points in place of the single degenerate critical point $(0, 0)$. This is a reflection of the general fact that the polynomial $Q(M^n, f, t)$ is unchanged by sufficiently small perturbations of f (since $Q(M^n, f, t)$ is defined (see Definition 16.3 and Lemma 16.6) in terms of the relative homology groups $H_k(M_{c+\varepsilon}, M_{c-\varepsilon})$ whose structure is obviously unaffected by such perturbations of f). Thus, for a given admissible function f, one can infer from the form of the polynomial $Q(M^n, f, t)$ the number of non-degenerate critical points of each index λ arising from the degenerate critical points of f when the latter is perturbed (by a sufficiently small amount) to yield a Morse function.

(b) Finally, consider the function

$$f(x, y, z) = x^3 - 3x(y^2 + x^2),$$

for which the origin is, once again, a degenerate critical point. We leave it to the reader to convince himself, using Figure 60, that $M_{c+\varepsilon}/M_{c-\varepsilon} \cong S^1 \vee S^2$, and thence to compute the relative homology groups $H_k(M_{c+\varepsilon}, M_{c-\varepsilon})$.

§17. Morse–Smale Functions. Handles. Surfaces

We shall now show that on any closed manifold M^n there exists a Morse function whose critical values have the same ordering us the indices of the corresponding critical points, i.e. $f(x_\lambda) = f(x_\mu)$ whenever the respective indices λ, μ are equal, $\lambda = \mu$, and $f(x_\lambda) > f(x_\mu)$ if $\lambda > \mu$. These are called *Smale functions* (or *nice functions*). (Note that unlike the general Morse functions,

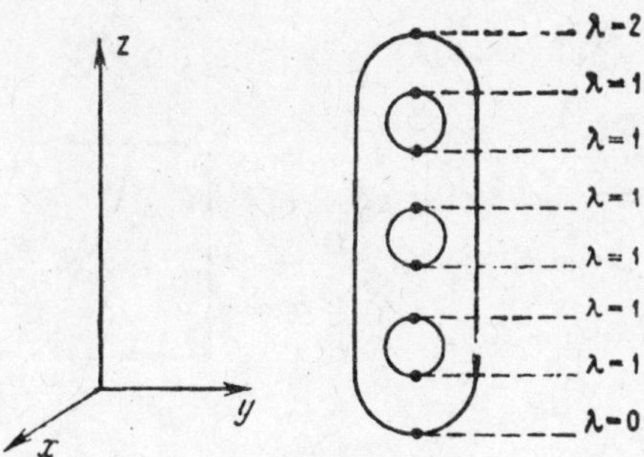

Figure 61

the set of Smale functions is not always everywhere dense in the space of all smooth functions on M^n.) In fact, there exist Smale functions with the further property of having only one point whose value is a local maximum and one point whose value is a local minimum. For the 2-dimensional orientable surfaces M_g^2 for instance, the appropriate height functions (see Part II, §11.2) are Smale functions with this additional property, as is clear from Figure 61.

17.1. Theorem. *On any closed, smooth manifold M^n, there exist Smale functions with exactly one local-maximum point (of index $n = \dim M^n$) and one local-minimum point (of index 0).*

Remark. If we use an appropriate such Smale function to build, as in Theorem 15.4, a cell complex homotopically equivalent to the given manifold M^n, then during this building process each new cell will be attached to cells of lower dimension.

Before embarking on the proof of the theorem we introduce the following auxiliary concept.

17.2. Definition. Given a Morse function f on M^n, a smooth vector field ξ on M^n is said to be *gradient-like* for f if the following two conditions are fulfilled:

(i) the Lie (i.e. directional) derivative $\xi(f)$ of f in the direction of the field ξ should be non-zero away from the critical points $x_1, \ldots, x_N$ of f, i.e. $\xi(f) \neq 0$ on $M \setminus \{x_1, \ldots, x_N\}$;

(ii) there is a neighbourhood $U(x_i)$ of each critical point x_i, $i = 1, \ldots, N$, such that in some co-ordinates on $U(x_i)$ in terms of which f has the form guaranteed by the Morse lemma:

$$f(x)|_{U(x_i)} = f(x_i) - \sum_{k=1}^{\lambda} (x^k)^2 + \sum_{k=\lambda+1}^{n} (x^k)^2, \tag{1}$$

the field ξ has essentially the form

$$\xi(x) = (-x^1, \ldots, -x^\lambda, x^{\lambda+1}, \ldots, x^n).$$

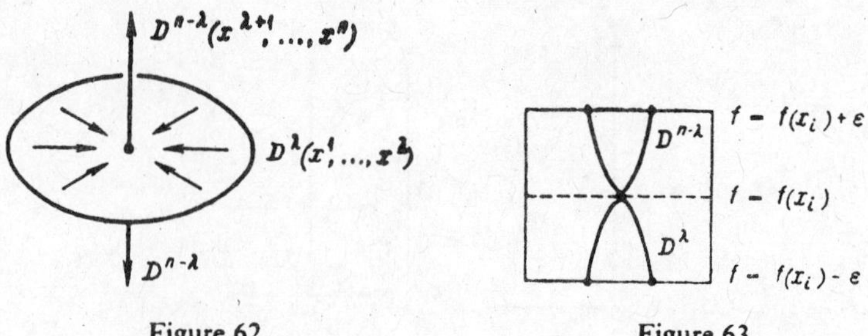

Figure 62 Figure 63

For every Morse function f on M^n there is clearly at least one such vector field, namely that obtained from the covector field grad f by using a metric on M^n to convert to a vector field (cf. proof of Lemma 15.2).

Proof of Theorem 17.1. Let f be a Morse function on M^n, let x_0 be any one of its critical points, of index λ say, and let ξ be a gradient-like vector field for f. Consider the so-called "separatrix diagram" of the point x_0, consisting of all the integral trajectories of the field ξ entering or leaving the point x_0. It is easy to see (from Definition 17.2) that, in an appropriate closed neighbourhood $U(x_0)$ of x_0, those trajectories flowing into x_0 together comprise a (closed) disc $D^\lambda = D^\lambda(x^1, \ldots, x^\lambda)$, and those emanating from x_0 fill a disc $D^{n-\lambda} = D^{n-\lambda}(x^{\lambda+1}, \ldots, x^n)$ (as shown in Figure 62). (Here the co-ordinates $x^1, \ldots, x^n$ in $U(x_0)$ are as in Definition 17.2.) Consider, for sufficiently small ε, the two spheres

$$\begin{aligned} S^{\lambda-1} &= D^\lambda \cap \{x \mid f(x) = f(x_0) - \varepsilon\}, \\ S^{n-\lambda-1} &= D^{n-\lambda} \cap \{x \mid f(x) = f(x_0) + \varepsilon\}. \end{aligned} \tag{2}$$

In view of the form (1) taken by f in the neighbourhood $U(x_0)$, we may suppose that in fact $S^{\lambda-1} = \partial D^\lambda$ and $S^{n-\lambda-1} = \partial D^{n-\lambda}$ (as shown schematically in Figure 63). If we now imagine the discs D^λ, $D^{n-\lambda}$ "blown up", i.e. expanded outwards from x_0, along the respective integral trajectories, then their boundaries, the spheres $S^{\lambda-1}$ and $S^{n-\lambda-1}$, will also expand (as their points move outwards along the integral trajectories) without self-intersection until such time as another critical point is encountered. (This is clear in view of the fact that integral trajectories can intersect only at critical points; cf. Part I, §23.1.) It is convenient to break up the proof from this point on into two lemmas.

17.3. Lemma. *Suppose that there are exactly two critical points x_0 and y_0 of the Morse function f in the layer $f^{-1}(a') \cup (M_{b'} \setminus M_{a'}) = f^{-1}[a', b']$, and that these satisfy*

$$a' < a = f(x_0) < f(y_0) = b \le b', \tag{3}$$

and suppose further that in this layer we also have

$$D^{n-\lambda}(x_0) \cap D^{\lambda'}(y_0) = \varnothing, \qquad \lambda = \operatorname{ind}(x_0), \quad \lambda' = \operatorname{ind}(y_0), \tag{4}$$

(where now $D^{n-\lambda}(x_0)$ denotes the disc determined by those portions of the integral trajectories leaving x_0 which lie in the region $f^{-1}[a', b']$, and similarly for the other discs $D^{\lambda'}(y_0)$, etc.). Then there exists a new Morse function g on the manifold M with the following properties: $g(x) = f(x)$ outside $f^{-1}[a', b']$; g has the same critical points as f; any gradient-like field for f is also gradient-like for g; $g(x_0) > g(y_0)$; and g and f differ by a constant in the neighbourhoods $U(x_0)$ and $U(y_0)$ of x_0 and y_0.

PROOF. It follows from the hypotheses of the lemma, in particular assumptions (3) and (4), that in the layer $f^{-1}[a', b']$ the separatrix diagrams of the points x_0 and y_0 do not intersect (as shown schematically in Figure 64), i.e.

$$[D^{n-\lambda}(x_0) \cup D^{\lambda}(x_0)] \cap [D^{n-\lambda'}(y_0) \cup D^{\lambda'}(y_0)] = \varnothing. \tag{5}$$

Writing

$$W = f^{-1}[a', b'], \qquad A = D^{n-\lambda}(x_0) \cup D^{\lambda}(x_0), \qquad B = D^{n-\lambda'}(y_0) \cup D^{\lambda'}(y_0),$$

it is not difficult to see that

$$\begin{aligned} W \backslash (A \cup B) &\cong \{f^{-1}(b') \backslash [(A \cup B) \cap f^{-1}(b')]\} \times I[a', b'] \\ &\cong \{f^{-1}(a') \backslash [(A \cup B) \cap f^{-1}(a')]\} \times I[a', b']; \end{aligned}$$

this follows by means of an argument similar to that used in the proof of Lemma 15.2: a diffeomorphism between the manifolds

$$f^{-1}(b') \backslash [S^{n-\lambda'-1}(y_0) \cup S^{n-\lambda-1}(x_0)] \text{ and } f^{-1}(a') \backslash [S^{\lambda'-1}(y_0) \cup S^{\lambda-1}(x_0)] \tag{6}$$

can be realized by moving the points of one of them along the integral trajectories of the vector field in the region $f^{-1}[a', b']$ until they reach the other. It is then clear that each of the cylinders

$$\{f^{-1}(b') \backslash [S^{n-\lambda'-1}(y_0) \cup S^{n-\lambda-1}(x_0)]\} \times I[a', b'],$$
$$\{f^{-1}(a') \backslash [S^{\lambda'-1}(y_0) \cup S^{\lambda-1}(x_0)]\} \times I[a', b'],$$

is diffeomorphic to $W \backslash (A \cup B)$ by means of a diffeomorphism identifying in each case the lid and base with the manifolds (6), and the vertical fibres with the segments of the appropriate integral trajectories of ξ (one of which is shown in Figure 64).

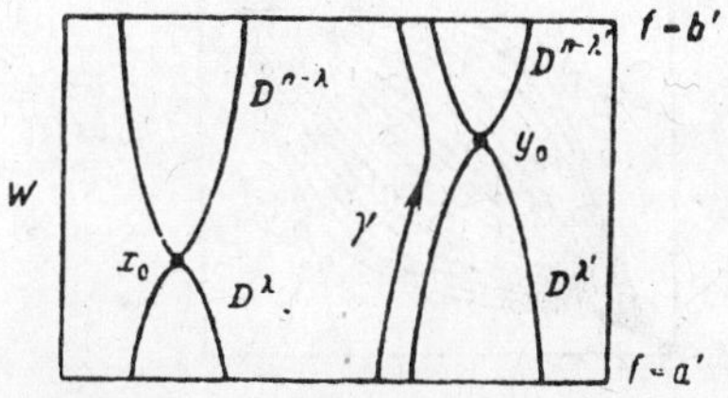

Figure 64

Let $\alpha(x)$ be a smooth function on $f^{-1}(a')$ such that $\alpha(x) \equiv 0$ on a sufficiently small neighbourhood of $A \cap f^{-1}(a')$, and $\alpha(x) \equiv 1$ on a sufficiently small neighbourhood of $B \cap f^{-1}(a')$. (Such a function exists in view of the fact that $A \cap B = \varnothing$; see (5).) We extend this function to the whole of W by defining it to be constant on the segments of the integral trajectories of ξ contained in W. Since these trajectories do not intersect in $W \backslash (A \cup B)$, we shall in this way obtain a smooth function $\alpha(x)$ on W, constant on the (segments of) integral trajectories in W, vanishing on some neighbourhood $U(A)$ of A, and equal to 1 on some neighbourhood $U(B)$ of B.

The second ingredient for concocting the desired function $g(x)$ is a smooth function $\bar{z} = \rho(\bar{x}, \bar{y})$ of two variables, of the form shown in Figure 65. The projections onto the $(\bar{z}, \bar{x})$-plane of the sections of the graph of $\bar{z} = \rho(\bar{x}, \bar{y})$ by the planes $\bar{y} = t$ (const.), t varying from 0 to 1, are to be as shown in Figure 66. Further conditions that the function ρ is required to satisfy, are as follows:

(i) $(\partial/\partial\bar{x})(\rho(\bar{x}, \bar{y})) > 0$ for all $(\bar{x}, \bar{y})$, and for each fixed $\bar{y}$ $(0 \leq \bar{y} \leq 1)$, $\rho(\bar{x}, \bar{y})$ increases from 0 to 1 as $\bar{x}$ increases from 0 to 1;

(ii) $\rho(a, 0) = b$; $\rho(b, 1) = a$;

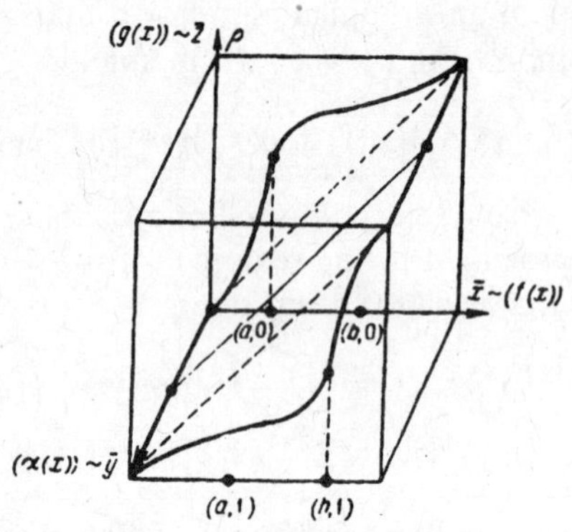

Figure 65

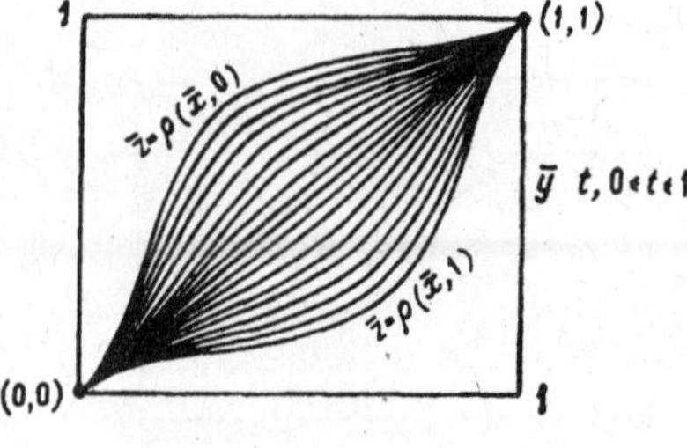

Figure 66

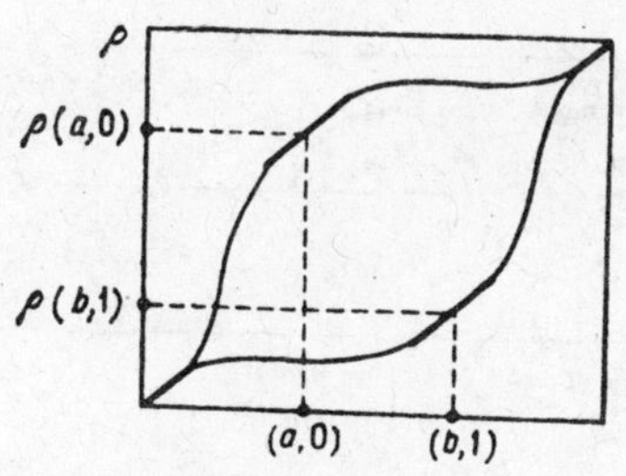

Figure 67

(iii) $(\partial/\partial\bar{x})(\rho(\bar{x}, 0) \equiv 1$ for all $\bar{x}$ in some neighbourhood of a, and $(\partial/\partial\bar{x})(\rho(\bar{x}, 1)) \equiv 1$ for all $\bar{x}$ in some neighbourhood of b (see Figure 67).

(Note that by composing f with a suitable function on $\mathbb{R}$ we may suppose $a' = 0$, $b' = 1$.)

The function g we are seeking is now defined in W by $g(x) = \rho(f(x), \alpha(x))$, $x \in W$. We then have

$$g(x_0) = \rho(f(x_0), \alpha(x_0)) = \rho(a, 0) > \rho(b, 1) = \rho(f(y_0), \alpha(y_0)) = g(y_0),$$

so that $g(x_0) > g(y_0)$. Note also that if x is such that $f(x) = a'(=0)$, then $g(x) = \rho(0, \alpha(x)) = 0$ (from Figure 65) $= a'$, and similarly $f(x) = b'$ implies $g(x) = 1(=b')$. Furthermore, it can be arranged that at such $x = (x^i)$, $\partial g/\partial x^i = \partial f/\partial x^i$ for all i, by defining ρ appropriately (verify this!). Hence the function $g(x)$ defined as above on W and set equal to $f(x)$ outside W, is smooth. The other properties required of g are readily verified from the conditions (i), (iii) imposed on ρ. □

17.4. Lemma. *Let $f: M^n \to \mathbb{R}$, $W = f^{-1}[a', b']$, $x_0, y_0 \in W$, $f(x_0) < f(y_0)$, all be as in the preceding lemma. Suppose that*

$$\lambda = \lambda(x_0) \geq \lambda(y_0) = \lambda', \tag{7}$$

where $\lambda = \lambda(x_0)$ and $\lambda' = \lambda(y_0)$ are the indices of the critical points x_0, y_0. Then there exists a Morse function g on M^n having all of the properties stated in the conclusion of the previous lemma; in particular, g has the same critical points as f, and $g(x_0) > g(y_0)$.

Proof. Let A, B be as in the proof of the preceding lemma (relative to some gradient-like field ξ on M^n). If $A \cap B = \varnothing$, then that lemma applies to give the desired conclusion. We shall show that if $A \cap B \neq \varnothing$, then the additional assumption (7) allows us to deform the field by an arbitrarily small amount so as to obtain the situation $A \cap B = \varnothing$ already dealt with.

We may assume without loss of generality (as noted in the previous proof) that $a' = 0$, $b' = 1$, and furthermore that

$$a' = 0 < f(x_0) < \tfrac{1}{2} < f(y_0) < 1 = b'.$$

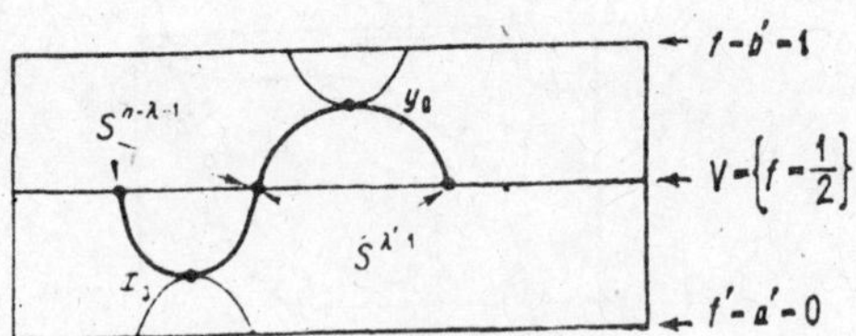

Figure 68

Consider the surface $V = \{x \mid f(x) = \frac{1}{2}\}$. If $A \cap B \neq \varnothing$ then since $f(x_0) < f(y_0)$, the spheres

$$S^{n-\lambda-1}(x_0) = V \cap D^{n-\lambda}(x_0), \qquad S^{\lambda'-1}(y_0) = V \cap D^{\lambda'}(y_0),$$

must intersect: $S^{n-\lambda-1}(x_0) \cap S^{\lambda'-1}(y_0) \neq \varnothing$ (see Figure 68). (Here, as before, $D^{n-\lambda}(x_0)$ denotes the totality of points on (segments of) integral trajectories emanating from x_0, contained in W, and analogously for $D^{\lambda'}(y_0)$.) Since $\frac{1}{2}$ is not a critical value of f, its complete inverse image V is (by the Implicit Function Theorem; cf. Part II, §10.2) an $(n-1)$-dimensional smooth manifold with $S^{n-\lambda-1}(x_0)$ and $S^{\lambda'-1}(y_0)$ as smooth submanifolds. Since

$$\begin{aligned}\dim S^{n-\lambda-1}(x_0) + \dim S^{\lambda'-1}(y_0) &= n - \lambda - 1 + \lambda' - 1 \\ &= n - (\lambda - \lambda') - 2 < n - 1,\end{aligned}$$

where in the final inequality we have used (7): $\lambda - \lambda' \geq 0$, it follows from the general theorem on t-regularity (Theorem 10.3.2 of Part II; see especially the remark ending §10.3 of Part II), that there exists an arbitrarily small isotopy of the given inclusion $i: S^{\lambda'-1} \to V$, into a nearby embedding of $S^{\lambda'-1}$ in V, such that the image under this embedding does not intersect the submanifold $S^{n-\lambda-1}(x_0) \subset V$. Clearly, such an isotopy may be extended to a small neighbourhood of $V \subset W$, and then to the whole of W by taking it to be the identity deformation outside that neighbourhood. If the gradient-like field ξ is subjected to this deformation, then provided the deformation is sufficiently close to the identity, the resulting field will still be gradient-like and will now have the desired property, namely that the separatrix diagrams A and B arising from it (in W) are disjoint (see Figure 69). Hence the preceding lemma may be applied, completing the proof. □

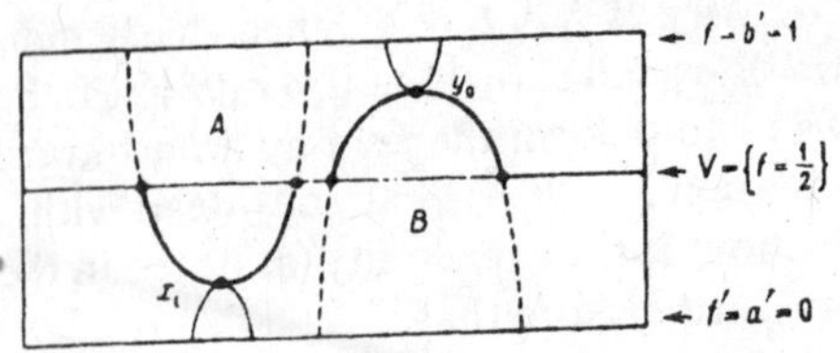

Figure 69

The existence of Smale functions on M^n is now almost immediate from this lemma. The second statement of Theorem 17.1 (asserting the existence of Smale functions with exactly one maximum point and one minimum point) we leave to the reader as an exercise (and a relatively easy one, at least in the 2-dimensional case). □

We now examine more minutely the process described in the proof of Lemma 15.3, of attaching cells σ^λ to the submanifold $M_{-\varepsilon}$. To be more specific, we shall investigate the changes in the submanifold $M_{-\varepsilon}$ occasioned by "passage through the critical point x_λ", from the point of view of differentiabililty, i.e. how the manifold $M_{-\varepsilon}$ is altered by applying to it the operation of "attaching handles", as it is called.

We begin by defining an *n-dimensional handle of index* λ: this is just the direct product $H_\lambda^n = D^\lambda \times D^{n-\lambda}$, where D^q denotes as usual the (closed) disc, or ball, of dimension q. Thus the handle H_λ^n is a manifold-with-boundary, and its boundary is, of course,

$$\partial H_\lambda^n = [(\partial D^\lambda) \times D^{n-\lambda}] \cup [D^\lambda \times (\partial D^{n-\lambda})] = (S^{\lambda-1} \times D^{n-\lambda}) \cup (D^\lambda \times S^{n-\lambda-1}). \tag{8}$$

The operation of *attaching a handle* H_λ^n to a manifold-with-boundary K^n is then defined as follows: Write $V^{n-1} = \partial K^n$, and let $S^{\lambda-1} \subset V^{n-1}$ be a sphere smoothly embedded in V^{n-1} in such a way that a sufficiently small tubular neighbourhood $T_\varepsilon(S^{\lambda-1})$ of radius ε (relative to some Riemannian metric on K^n) in V^{n-1}, is diffeomorphic to the product $S^{\lambda-1} \times D^{n-\lambda}$,

$$T_\varepsilon(S^{\lambda-1}) \cong S^{\lambda-1} \times D^{n-\lambda},$$

via a diffeomorphism identifying the $(n - \lambda)$-dimensional disc through each point s of the sphere and normal to it, with the fibre $s \times D^{n-\lambda}$ (see Figure 70; also Part II, Corollary 11.3.3). The handle H_λ^n is now attached to K^n by identifying the component $S^{\lambda-1} \times D^{n-\lambda}$ of the boundary of the handle H_λ^n (see (8)) with this tubular neighbourhood, by means of the obvious diffeomorphism

$$S^{\lambda-1} \times D^{n-\lambda} \to T_\varepsilon(S^{\lambda-1}) \cong S^{\lambda-1} \times D^{n-\lambda}.$$

(See Figures 71 and 72 for the cases $n = 2, 3$ respectively, $\lambda = 1$, and Figure 73 for the case $n = 3$, $\lambda = 2$.) Since the manifold-with-boundary K^n and the handle H_λ^n come with prescribed smooth structures on them, we need to "smooth the corners" arising at the points of $\partial T_\varepsilon(S^{\lambda-1}) = S^{\lambda-1} \times S^{n-\lambda-1}$ (as

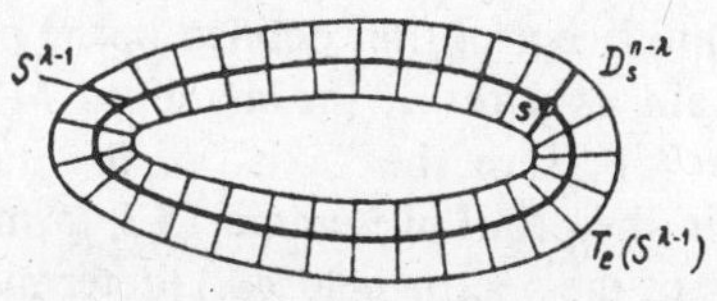

Figure 70

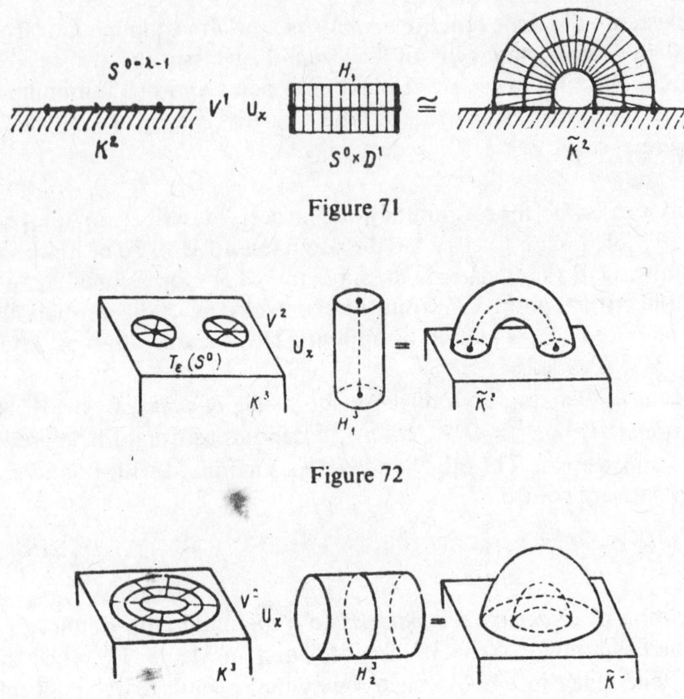

Figure 71

Figure 72

Figure 73

indicated in the final illustration of Figure 71), to obtain finally a smooth manifold $\tilde{K}^n$ with smooth boundary $\tilde{V}^{n-1}$.

17.5. Theorem. *Every connected, closed (i.e. compact and without boundary) smooth manifold M^n is diffeomorphic to a union of finitely many handles H_λ^n (λ variable), where the handles H_λ^n are in one-to-one correspondence with the critical points P_λ (of index λ) of some Morse function f on M^n.*

PROOF. Since by Lemma 15.2 the absence of critical points in an interval $[a, b]$ implies that the manifolds M_a, M_b are diffeomorphic, it suffices to determine how $M_{-\varepsilon}$ changes in "passing through a critical point" P_λ of index λ, i.e. in passing from $M_{-\varepsilon}$ to the manifold M_ε (assuming $f(P_\lambda) = 0$). (Note that as in the proof of Theorem 15.4 we may suppose that there is just the critical point corresponding to the critical value 0.) By applying to M_ε the deformation shown in Figure 74 (reminiscent of that defined in the proof of Lemma 15.3; see Figure 53), we obtain as a result the manifold $M_{-\varepsilon}$ with a handle H_λ^n attached (as in Figure 75). Here the "axis" of the handle H_λ^n is the disc $D^\lambda(x^1, \ldots, x^\lambda)$ figuring in the proof of Lemma 15.3, formed from segments of the integral trajectories of the vector field $v(x)$ (determined as before by the covector field $-\operatorname{grad} f(x)$ on M_ε), emanating from the singular point P_λ (cf.

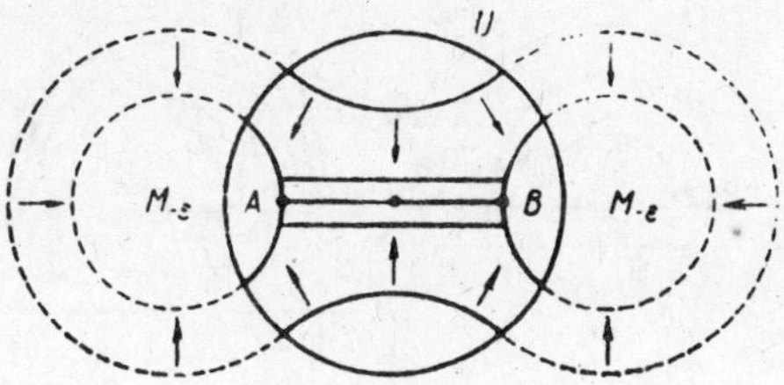

Figure 74

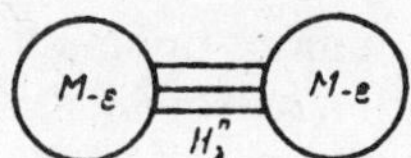

Figure 75

also Figure 62). With this we end the proof, leaving it to the reader to convince himself that this deformation does indeed define a diffeomorphism between M_ε and the manifold $M_{-\varepsilon}$ with the handle H_λ^n attached. □

The following converse of this theorem holds: *Given a decomposition of a manifold M as a "sum" of handles, there exists a Morse function f on M which gives rise in the manner just described to this handle decomposition of M.* We sketch a proof of this using induction on the index λ of the handles in the given decomposition. Thus we begin by defining our Morse function f on each handle H_0^n (identified beforehand in the given handle decomposition with a disc D^n) by taking its centre C to be (non-degenerate) critical, and the concentric $(n-1)$-spheres with centre C (and union D^n) to be (components of) level surfaces f_c of f, with the value c decreasing as the radius of the spheres decreases, until it reaches a local minimum at the centre C. This leaves some latitude in the definition of f on the H_0^n; in particular, we may clearly arrange that f take on the same constant value on the boundaries of all the handles H_0^n. Suppose inductively that f is already defined on the smooth manifold-with-boundary M_a made up of all handles of index less than λ, and further that $f(x) \le a$ for all x in M_a, and $V^{n-1} = \{x \mid f(x) = a\}$ forms its boundary. Let H_λ^n be a handle of index λ (i.e. identified with $D^\lambda \times D^{n-\lambda}$ in the given handle decomposition) attached to M_a at its boundary V^{n-1}. The way in which f is now defined on $M_a \cup H_\lambda^n$ is shown in the right-hand diagram of Figure 76, where the new level surfaces are indicated, together with a new critical point of index λ. (Note that slight changes in the definition of f on M_a are involved in extending f smoothly to the handle H_λ^n.) It is not difficult to see that this extension can be carried out so as to yield a smooth (Morse) function which is again constant on the boundary of $M_a \cup H_\lambda^n$, so that the process can be continued. □

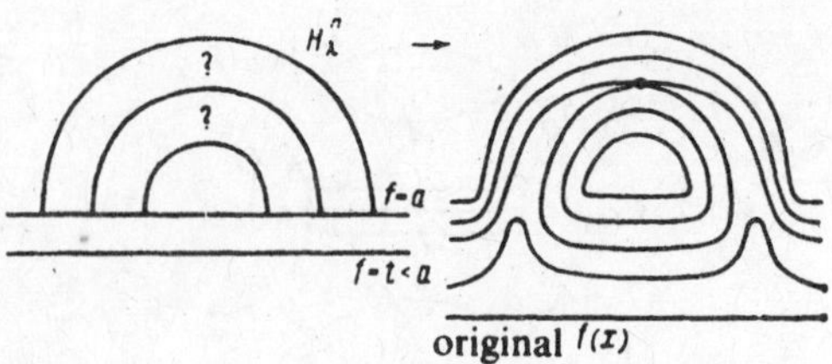

Figure 76

We now consider in greater detail the particular case of the representation of 2-dimensional manifolds as unions of handles H_λ^2 in accordance with the above theorem. Our investigation will incidentally yield another proof of the classification theorem for closed 2-manifolds (see Theorem 3.20 *et seqq.*).

Thus let M^2 be a closed 2-manifold, and let f be a Smale function with a single local-minimum point x_0 (i.e. critical point of index 0), critical points $x_1, \ldots, x_N$ of index 1, and a single local-maximum point x_{N+1} (i.e. critical point of index 2), satisfying

$$f(x_i) < f(x_{i+1}), \qquad 0 \le i \le N. \tag{9}$$

(We are guaranteed the existence of such functions by Theorem 17.1, or, more precisely, in view of the strictness of the inequalities (9), by Lemma 17.4 in conjunction with Theorem 10.4.3 (iv) of Part II, asserting the existence of Morse functions with at most one critical point on each level surface.) In fact, by composing with a suitable smooth function from $\mathbb{R}$ to $\mathbb{R}$, we may suppose that $0 \le f(x) \le N + 1$, and $f(x_i) = i$, $i = 0, \ldots, N + 1$.

We now carry out, in the present context of our closed 2-manifold M^2 with this special Smale function given on it, the construction described in the preceding theorem for representing M^2 as a union of handles. To begin with, it is clear from various earlier discussions that any submanifold of the form $\{x \mid 0 \le f(x) \le \eta < 1\}$ may be considered a handle $H_0^2 \cong D^2$ (homotopically equivalent to a point, i.e. 0-dimensional cell σ^0). Passage through the critical point x_1 then involves (as in the proof of Theorem 17.5) attaching a handle H_1^2 (see Figure 77). This can be carried out in essentially only two ways, as shown in Figure 78; these two ways of attaching H_1^2 to H_0^2 yield manifolds-with-boundary which are homotpically equivalent (to the circle S^1), but diffeomorphically distinct, since if the attaching is performed as in Figure 78(I), we obtain $H_0^2 \cup_I H_1^2 \cong S^1 \times D^1$, a cylinder, which is orientable, whereas the attachment as in Figure 78(II) yields a Möbius band (which is of course non-orientable).

Iterating this procedure as we pass through the critical points $x_2, \ldots, x_N$ in turn, the operation of attaching a handle H_1^2 at each stage will, as in the first instance, amount essentially to attaching either a cylinder $S^1 \times D^1$ or a Möbius band, or, up to a homotopy equivalence, attaching a 1-cell σ_i^1, $1 \le i \le N$, Thus after passing through x_N ($f(x_N) = N$) we shall have con-

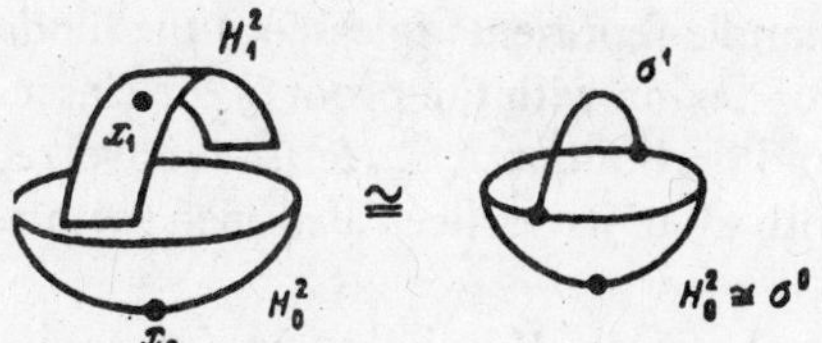

Figure 77

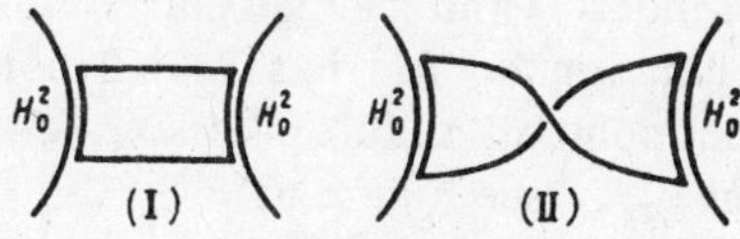

Figure 78

structed a 2-manifold homotopically equivalent to a bouquet of circles $\bigvee_{i=1}^N S_i^1$, with one circle $S_i^1 = \sigma_i^1 \cup \sigma^0$ for each critical point x_i of index 1. As the last step in the construction we attach a handle $H_2^2 \cong D^2$, i.e. a 2-cell σ^2. The final outcome of the construction is therefore a manifold diffeomorphic to a union of handles of the form

$$M^2 \cong H_0^2 \cup \underbrace{H_1^2 \cup \cdots \cup H_1^2}_{N} \cup H_2^2,$$

and homotopically equivalent to a cell complex of the form

$$M^2 \sim \sigma^0 \cup \sigma_1^1 \cup \cdots \cup \sigma_N^1 \cup \sigma^2.$$

Note that at the final, $(N+1)$ st, step, the handle $H_2^2 = D^2$ can be attached to the manifold-with-boundary K^2 constructed up to that point, in essentially only one way, namely by means of a diffeomorphism between the boundaries $\partial D^2 \cong S^1$ and $\partial K^2 \cong S^1$.

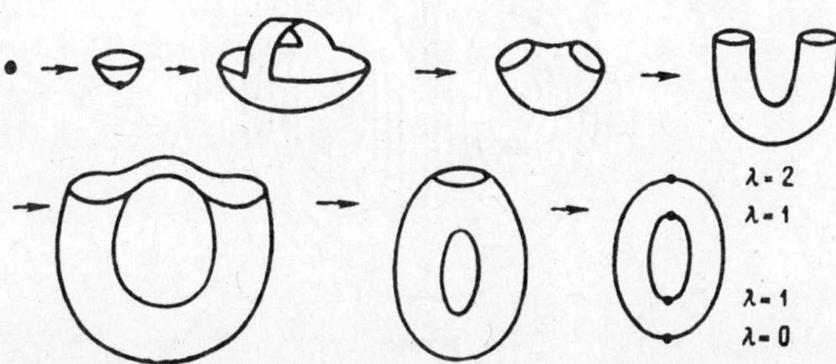

Figure 79

Remark. This final handle represents in essence the fundamental polygon W considered in §3 in connexion with the proof of the classification theorem for closed 2-manifolds, and the bouquet $\bigvee_{i=1}^{N} S_i^1$ (see above) represents the boundary of W, however with all of its vertices identified (with a single point).

The above-described process of building up the manifold M^2 by a succession of handle adjunctions is illustrated in Figure 79 in the case $M^2 = M_{g=1}^2 = T^2$, the torus, embedded in a standard way in $\mathbb{R}^3$ so that the height function given by $f(P) = z$ is a Smale function with a single minimum point x_0, two saddle points x_1, x_2 (of index 1) and a single maximum point x_3. For $g > 1$ the analogous height function on M_g^2 has $2g + 2$ non-degenerate critical points: a single minimum point x_0, saddle points $x_1, \ldots, x_{2g}$ (of index 1), and a single maximum point x_{2g+1} (see Figure 61).

On any of the orientable surfaces M_g^2 one may, on the other hand, define a smooth height function $f: M_g^2 \to \mathbb{R}$ with just four critical points, namely a minimum point, a maximum point, and two saddle points, degenerate for $g > 1$, by embedding M_g^2 suitably in $\mathbb{R}^3$; an appropriate embedding in the case $g = 2$ is shown in Figure 80.

We leave it to the reader as an exercise to show that there is no embedding of $M_{g>0}^2$ into $\mathbb{R}^3$ determining a smooth height function with fewer than four

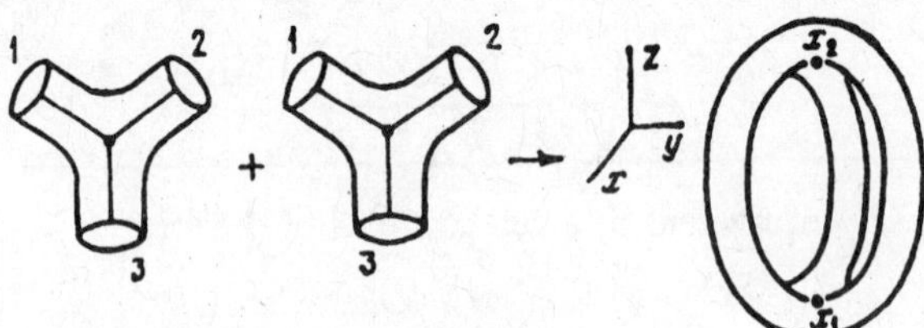

Figure 80

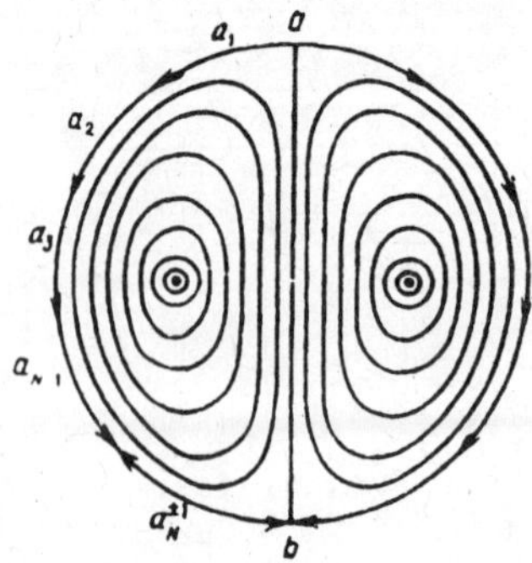

Figure 81

critical points. However, there nonetheless exists for every closed 2-manifold $M^2_{g>0}$ or M^2_μ (the non-orientable case) a smooth function f (not realizable as a height function) with just three critical points. To see this recall first that each such manifold M^2 can be retrieved from a word of the canonical form (see the second of the remarks concluding §3)

$$w = a_1 \dots a_N \cdot a_1^{-1} \dots a_{N-1}^{-1} a_N^{\pm 1},$$

by identifying edges carrying the same label a_i in the corresponding fundamental polygon (and identifying all vertices with a single vertex). The desired function f may then be taken to be one with level curves as shown in Figure 81, where the critical point to the left of the segment ab is a minimum point say, and that to the right a maximum point, and where the unique (degenerate) saddle point is the point in M^2 represented in the figure by the various vertices of the fundamental polygon. Clearly this still leaves some latitude in defining f, so that in some sufficiently small neighbourhood of the saddle point it may be taken to have the form $f(x, y) = \operatorname{Re}[(x + iy)^k]$. (As an exercise, express k as a function of g or μ.)

By Part II, Theorem 10.4.3, one may obtain a Morse function from f by applying an arbitrarily small perturbation; as a result of such a perturbation (e.g. $f(x, y) \to \operatorname{Re} \sum_{\alpha=1}^{k} (z - \varepsilon_\alpha)$ in some neighbourhood of the saddle point,

Figure 82

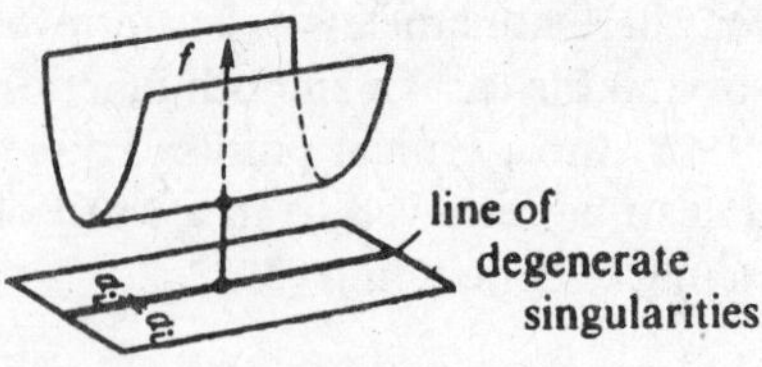

Figure 83

where $z = x + iy$ and $\varepsilon_i \neq \varepsilon_j$ for $i \neq j$) the degenerate saddle point must give rise to a certain minimum number of non-degenerate critical points ($2g$ if $M^2 = M_g^2$). (The effect of such a perturbation on the integral trajectories of the vector field $v(x, y)$ obtained (as before) from grad $f(x, y) =$ grad $\operatorname{Re}[(x + iy)^4]$, in a neighbourhood of the saddle point, is indicated in Figure 82; there the degenerate saddle point gives rise to three non-degenerate critical points, or, equivalently, three non-degenerate singularities of the vector field v. Note that such a saddle point will not in fact occur in the case of a function defined on a 2-manifold M^2 via Figure 81; verify that for $M^2 = M_{g=2}^2$ (whence $N = 4$) the appropriate value of k is 5.

Remark. Note that in Figure 81, on neither side of the segment ab are there edges carrying the same label a_i. If the segment ab had not been so chosen, then in M^2 such an edge a_i would have consisted entirely of (degenerate) critical points, i.e. we should have obtained instead a function with a continuum of critical points (see Figure 83).

§18. Poincaré Duality

In topology, algebraic geometry, and homological algebra, there fall under the single heading "Poincaré duality" several assertions to the effect that under various hypotheses the homology and cohomology groups of "complementary dimensions" are isomorphic. The simplest result of this type, due essentially to Poincaré, states that for a closed, connected, orientable, smooth manifold M one has the following isomorphisms among the real homology groups $H_i(M; \mathbb{R})$:

$$H_k(M; \mathbb{R}) \simeq H_{n-k}(M; \mathbb{R}), \qquad n = \dim M, \quad k = 0, \ldots, n,$$

or, in other words, the corresponding Betti numbers of M are equal: $b_k(M) = b_{n-k}(M)$. If the manifold M is non-orientable then, as shown in §6, we have the following version of "Poincaré duality", now over the integers modulo 2:

$$H_k(M; \mathbb{Z}_2) \simeq H_{n-k}(M; \mathbb{Z}_2), \qquad n = \dim M.$$

We shall, however, confine ourselves here largely to the orientable case; the exposition for non-orientable manifolds is analogous.

Let M^n be an orientable, closed manifold. The following construction of mutually "dual" cell complexes K and $\tilde{K}$, homotopically equivalent to M^n, underlies all manifestations of Poincaré duality (cf. §6). We first define the sense in which the complexes K and $\tilde{K}$ are *dual* as follows: To each cell σ^i of K there should correspond in a one-to-one manner a cell $D(\sigma^i) = \tilde{\sigma}^{n-i}$ of $\tilde{K}$ of complementary dimension; the one-to-one correspondence ("duality operator") D between the cells of K and the cells of $\tilde{K}$ is required to have the further "incidence-preserving" property, that for each pair σ^i, σ^{i-1} of cells of K, the incidence number $[\sigma^i : \sigma^{i-1}]$ (see §4) coincides, up to difference in sign (which in any case depends only on the prescribed orientations of the cells of the complex), with the incidence number $[\tilde{\sigma}^{n-i+1} : \tilde{\sigma}^{n-i}]$ of the images of the pair of cells σ^{i-1}, σ^i under the duality operator D, i.e.

$$[\sigma^i : \sigma^{i-1}] = \pm[\tilde{\sigma}^{n-i+1} : \tilde{\sigma}^{n-i}]. \tag{1}$$

(In the non-orientable case the incidence numbers are considered modulo 2, so that it is instead required only that $[\sigma^i : \sigma^{i-1}] \equiv [\tilde{\sigma}^{n-i+1} : \tilde{\sigma}^{n-i}] \bmod 2$.)

The construction of the complexes K and $\tilde{K}$ utilizes a Smale function f, i.e. a Morse function whose critical values are ordered according to their indices: $f(x_i) > f(x_j)$ if $\lambda(x_i) > \lambda(x_j)$, for every pair of critical points x_i, x_j. (The existence of Smale functions was established in the preceding section; see Theorem 17.1) As noted earlier we may assume also that there is at most one critical point on each level surface of f. Clearly the function $g = -f$ has the same critical points x_i as f, and if x_i has index λ_i as a critical point of f, then it has index $n - \lambda_i$ as a critical point of g. As the cell decomposition K of the manifold M^n we take that afforded by the Smale function f in accordance with the construction described in Lemma 15.3 and Theorem 15.4, and as $\tilde{K}$ the cell complex constructed in the same manner using $g = -f$. In order to define the duality operator D we need to examine more closely the way in which the complexes K, $\tilde{K}$ are obtained from f, g: Recall that, as shown in the proof of Lemma 15.3, if x_i is a critical point of f of index λ_i, with corresponding critical value $f(x_i) = a_i$, then the manifold $M_{a_i+\varepsilon}$ is homotopically equivalent to the manifold $M_{a_i-\varepsilon}$ with a cell $\sigma_i^{\lambda_i}$ (or D^{λ_i}) attached at its boundary (as in the left-hand diagram in Figure 84). Similarly, using $g = -f$ instead of f, we have

$$M^{(-f)}_{-a_i+\varepsilon} \sim M^{(-f)}_{-a_i-\varepsilon} \cup \tilde{\sigma}_i^{n-\lambda_i},$$

(where $M^{(-f)}_{-a_i+\varepsilon} = \{x \mid -f(x) \leq -a_i + \varepsilon\}$), as indicated in the right-hand diagram of Figure 84; here passage through the critical points x_i proceeds in the opposite order, i.e. in the direction of increasing $-a_i$ (and $n - \lambda_i$). The correspondence D is then defined by $D(\sigma_i^{\lambda_i}) = \tilde{\sigma}_i^{n-\lambda_i}$, for each pair of cells $\sigma_i^{\lambda_i}$, $\tilde{\sigma}_i^{n-\lambda_i}$, one such pair arising from each critical point x_i (see Figure 85).

It remains to verify that the operator D has the requisite property (1). Consider a pair of cells σ_i^λ, $\sigma_j^{\lambda-1}$ of the complex K; by definition (see §4) the incidence number $[\sigma_i^\lambda : \sigma_j^{\lambda-1}]$ is the degree of the map p_{ij}^λ: $S_i^{\lambda-1} \to S_j^{\lambda-1}$, where $S_i^{\lambda-1} = \partial(\sigma_i^\lambda)$, the boundary of the cell σ_i^λ, and p_{ij}^λ is the composite of the attaching map $\partial\sigma_i^\lambda \to K^{\lambda-1}$ of the cell σ_i^λ to the $(\lambda - 1)$-skeleton of K, followed by the natural projection $K^{\lambda-1} \to K^{\lambda-1}/K^{\lambda-2}$, and finally the projection of the

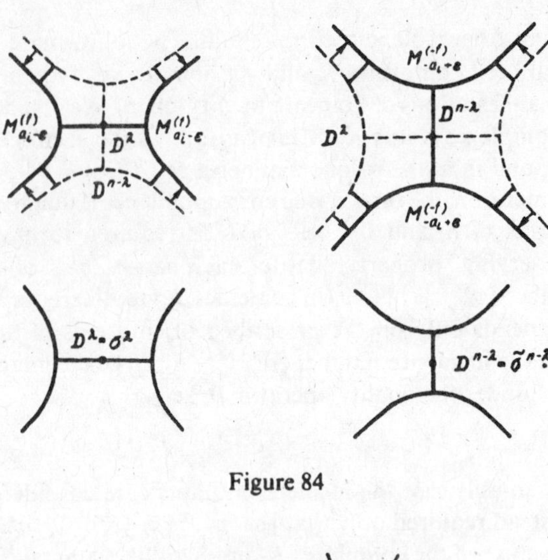

Figure 84

σ^λ D $\tilde{\sigma}^{n-\lambda}$

Figure 85

quotient $K^{\lambda-1}/K^{\lambda-2} \cong \bigvee S_k^{\lambda-1}$ onto its jth component $S_j^{\lambda-1}$ (see Figure 86). Now it is not difficult to see that, up to sign, the degree of the map p_{ij}^λ coincides with the intersection index (Part II, Definition 15.1.1) of $\partial\sigma_i^\lambda$ (after attachment to $K^{\lambda-1}$) and $\tilde{\sigma}_j^{n-\lambda+1} = D(\sigma_j^{\lambda-1})$ (appropriately immersed in M^n). (This is a consequence of the fact that $\tilde{\sigma}_j^{n-\lambda+1}$ is in essence transverse to $\sigma^{\lambda-1}$ in M^n; consider for instance the case $n = 2$, $\lambda = 2$, $\deg p_{ij}^2 = 2$: here the circle $\partial\sigma_i^2$ is wrapped twice round $S_j^{\lambda-1}$ by the map p_{ij}^2, and so meets the transverse cell $\tilde{\sigma}_j^1$ twice, in the same sense each time. See also Figure 87, where $\lambda = 2$, $n = 3$.) Now the intersection index of $\partial\sigma_i^\lambda$ and $\tilde{\sigma}_j^{n-\lambda+1}$ in M^n, is by definition (see the footnote to Exercise 10 of §7) the linking coefficient $\{\partial\sigma_i^\lambda, \partial\tilde{\sigma}_j^{n-\lambda+1}\}$. The analogous argument shows that the incidence number $[\tilde{\sigma}_j^{n-\lambda+1} : \tilde{\sigma}_i^{n-\lambda}]$ is equal (up to sign) to the linking coefficient $\{\partial\tilde{\sigma}_j^{n-\lambda+1}, \partial\sigma_i^\lambda\}$ in M^n. Since the linking coefficient of two spaces is essentially independent of the order in which they are taken (verify this!), it follows that

$$[\sigma_i^\lambda : \sigma_j^{\lambda-1}] = [\tilde{\sigma}_j^{n-\lambda+1} : \tilde{\sigma}_i^{n-\lambda}] = [D(\sigma_j^{\lambda-1}) : D(\sigma_i^\lambda)],$$

so that the complexes K and $\tilde{K}$ are dual under the operator D, as claimed.

As constructed above (see Figure 85) the duality operator D has the property that each cell σ_j^λ (appropriately immersed in M^n) of the complex K intersects the image cell $D(\sigma_j^\lambda) = \tilde{\sigma}_j^{n-\lambda}$ of $\tilde{K}$, transversely in exactly one (internal) point (consider, for instance, the case $n = 2$, $\lambda = 1$). (For orientable M^n, the cells of K and $\tilde{K}$ may be oriented so that the intersection index

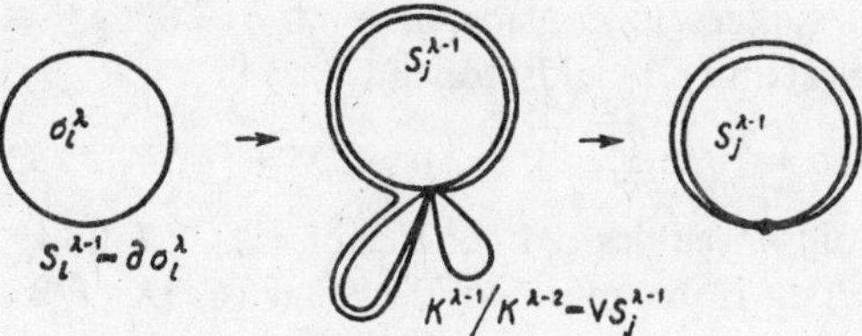

Figure 86

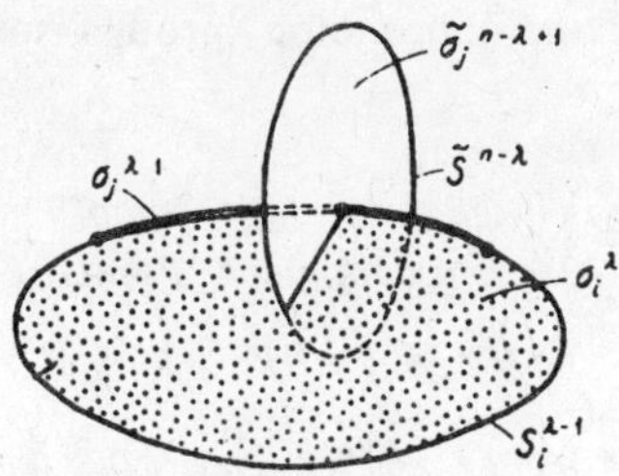

Figure 87

$\sigma_j^\lambda \circ \tilde{\sigma}_j^{n-\lambda}$ is always $+1$. For non-orientable M^n one considers this intersection index modulo 2.) Every other pair of cells σ^λ, $\tilde{\sigma}^\mu$ $(D(\sigma^\lambda) \neq \tilde{\sigma}^\mu)$ has empty intersection. Since the respective cells of K and $\tilde{K}$ form bases for the integral (among other) chain complexes

$$C(K) = \sum_{\lambda=0}^{n} C_\lambda(K) \qquad \text{and} \qquad C(\tilde{K}) = \sum_{\lambda=0}^{n} C_{n-\lambda}(\tilde{K}),$$

the intersection index of pairs of cells σ_i^λ, $\tilde{\sigma}_q^{n-\lambda} = D(\sigma_q^\lambda)$, which we have just observed to be given by

$$\sigma_i^\lambda \circ \tilde{\sigma}_q^{n-\lambda} = \sigma_i^\lambda \circ D(\sigma_q^\lambda) = \delta_{iq}, \tag{2}$$

extends, for each λ, to a non-degenerate bilinear map $C_\lambda(K) \times C_{n-\lambda}(\tilde{K}) \to \mathbb{Z}$ (or $\mathbb{R}$, etc.), given by

$$a \circ b = \sum_{i,j} a_i b_j (\sigma_i^\lambda \circ \tilde{\sigma}_j^{n-\lambda}) \qquad \text{where} \quad a = \sum_i a_i \sigma_i^\lambda, \quad b = \sum_j b_j \tilde{\sigma}_j^{n-\lambda}.$$

From the definition (see §4(13)) of the boundary operator ∂ on $C(K)$:

$$\partial(\sigma_i^\lambda) = \sum_k [\sigma_i^\lambda : \sigma_k^{\lambda-1}] \sigma_k^{\lambda-1}, \tag{3}$$

and on $C(\tilde{K})$, and the fact that D satisfies

$$[\sigma_i^\lambda : \sigma_j^{\lambda-1}] = [D\sigma_j^{\lambda-1} : D\sigma_i^\lambda], \tag{4}$$

it follows that for $a \in C_\lambda(K)$, $b \in C_{n-\lambda+1}(\tilde{K})$, we have

$$(\partial a) \circ b = a \circ (\partial b). \tag{5}$$

(To verify this it suffices to consider $a = \sigma_i^\lambda$, $b = \tilde{\sigma}_q^{n-\lambda+1} = D(\sigma_q^{\lambda-1})$; direct calculation using (2) and (3) yields, for this a and b,

$$(\partial a) \circ b = [\sigma_i^\lambda : \sigma_q^{\lambda-1}], \qquad a \circ (\partial b) = [\tilde{\sigma}_q^{n-\lambda+1} : \tilde{\sigma}_i^{n-\lambda}],$$

whence, in view of (4), the desired conclusion (5).) It follows readily that the complex $(C(\tilde{K}), \partial)$ is isomorphic to the dual $(C^*(K), \partial^*)$ of the complex $(C(K), \partial)$, under the map given by $\tilde{\sigma}_i^{n-\lambda} \mapsto (\sigma_i^\lambda)^*$, where $(\sigma_i^\lambda)^*$ is the linear functional on $C_\lambda(K)$ taking the value 1 on σ_i^λ and 0 on all other λ-dimensional cells in K (see §2, in particular equation (13) there). Since by Theorem 15.4 both K and $\tilde{K}$ are homotopically equivalent to M^n, and therefore by Corollary 5.4 have their homology and cohomology groups isomorphic to those of M^n, we infer the following

18.1. Theorem. *For every closed, orientable, smooth manifold M^n of dimension n, and each $k = 0, \dots n$, there is a canonical "Poincaré-duality isomorphism"*

$$H_k(M^n) \simeq H^{n-k}(M^n) \tag{6}$$

(where the coefficients are from $\mathbb{Z}$ or $\mathbb{R}$, etc.), whence it follows in particular that the Betti numbers satisfy

$$b_k = b_{n-k}, \qquad k = 0, \dots, n.$$

Furthermore, there exists a non-degenerate, bilinear map

$$H_n \times H_{n-k} \to \mathbb{Z} \qquad (\text{or } \mathbb{R}, \text{etc.})$$

(which in the case where the coefficients are from $\mathbb{Z}$ is unimodular), called the "intersection index of cycles". Consequently, for $k = n/2$ (n even), there exists a non-degenerate bilinear form (i.e. scalar product) on $H_k(M^n)$, satisfying $a \circ b = (-1)^k b \circ a$. (In the non-orientable case the analogous conclusions hold over $\mathbb{Z}_2$.)

Examples

(a) Recall (essentially from Proposition 3.10) that for a closed, connected, orientable manifold M^n we have $H_0(M^n) \simeq \mathbb{Z} \simeq H_n(M^n)$. For non-orientable, closed, connected M^n, on the other hand, we have (by Proposition 3.11)

$$H_0(M^n; \mathbb{Z}) \simeq \mathbb{Z}, \qquad H_n(M^n; \mathbb{Z}) = 0.$$

while modulo 2,

$$H_0(M^n; \mathbb{Z}_2) \simeq \mathbb{Z}_2 \simeq H_n(M^n; \mathbb{Z}_2),$$

as claimed in the theorem.

(b) In the case of an orientable, closed, 2-dimensional surface M_g^2, the above theorem guarantees the existence of a non-degenerate, alternating, bilinear form (afforded by the intersection index of cycles) on $H_1(M^2; \mathbb{Z}) \simeq \mathbb{Z} \oplus \cdots \oplus \mathbb{Z}$ ($2g$ summands). This has in essence been noted earlier in several places: denoting as usual the canonical basic cycles for this free abelian group by

$a_1, \ldots, a_g, b_1, \ldots, b_g$, we have (cf. §12(4))

$$a_i \circ a_j = b_i \circ b_j = 0, \qquad a_i \circ b_j = \delta_{ij}.$$

(c) The intersection index of cycles likewise furnishes the first homology group $H_1(\mathbb{R}P^2; \mathbb{Z}_2)(\simeq \mathbb{Z}_2)$ of the projective plane $\mathbb{R}P^2$ (of course non-orientable) with a non-degenerate bilinear form $(a \circ b)$ mod 2. From the non-degeneracy it follows that the generator x of $H_1(\mathbb{R}P^2; \mathbb{Z}_2)$ (represented by the projective line $\mathbb{R}P^1 \subset \mathbb{R}P^2$) satisfies

$$x \circ x = 1 \pmod 2.$$

(d) Again let M^n be closed, connected and orientable. As observed earlier (cf. Proposition 3.10) the n-chain $[M^n] = \sum_i \sigma_i^n$, where the summation is over all n-cells of any fixed complex homotopically equivalent to M^n, is a cycle (in fact, a generator of $H_n \simeq \mathbb{Z}$), so that the "diagonal" $\Delta = \sum_i (\sigma_i^n \times \sigma_i^n)$ is an n-cycle of $M^n \times M^n$ (see §7(1)). It follows from Part II, §15.2 (see especially Theorem 15.2.3) that the intersection index $\Delta \circ \Delta$ (after the two copies of Δ are brought into general position with respect to each other in $K \times K \sim M^n \times M^n$) is equal to the total index of any smooth vector field on M^n, which in turn coincides with the Euler characteristic of M^n (see loc. cit.). For general k, $0 \le k \le 2n$, the bilinear form $a \circ b$ (given by the intersection index of cycles) on the group $H_k(M^n \times M^n; \mathbb{R})$ is determined by the corresponding forms on the $H_j(M^n; \mathbb{R})$ as follows: In view of the isomorphism (see §7(2))

$$H_k(M^n \times M^n; \mathbb{R}) \simeq \sum_{q+l=k} H_q(M^n; \mathbb{R}) \otimes H_l(M^n; \mathbb{R}),$$

bases $\{z_i^{(j)}\}$ for the vector spaces $H_j(M^n; \mathbb{R})$ yield the basis

$$\{z_i^{(q)} \otimes z_j^{(l)} | q + l = k\}$$

for $H_k(M^n \times M^n; \mathbb{R})$, and the intersection index of an arbitrary pair of such basis elements satisfies

$$(z_i \otimes z_j) \circ (z_k' \otimes z_l') = (-1)^{\dim z_j + \dim z_k'}(z_i \circ z_k')(z_j \circ z_l').$$

(Verify this!) It follows, in particular, that for an intersection index of this form to be non-zero, it it necessary that

$$\dim z_i + \dim z_k' = n = \dim z_j + \dim z_l'.$$

Exercises

1. Let $f: M^n \to M^n$ be a given smooth self-map of an orientable, closed manifold M^n, whose induced homomorphisms (actually linear transformations) $f_{k,*}: H_k(M^n; \mathbb{R}) \to H_k(M^n; \mathbb{R})$ are known. Show that the intersection index $\Delta \circ \Delta_f$, where Δ is (essentially as above) the diagonal of $M^n \times M^n$, and $\Delta_f = \{x, f(x)\}$, the graph of f, is given by

$$\Delta \circ \Delta_f = \sum_{k=0}^{n} (-1)^k \operatorname{tr} f_{k,*}, \tag{7}$$

(valid also for non-orientable M^n with $\mathbb{Z}_2$ replacing $\mathbb{R}$). (Here tr $f_{k,*}$ denotes the trace of the linear transformation $f_{k,*}$.) (Hint. Consider first the simplest cases: $M^n = S^n, T^n, \mathbb{R}P^2, M^2 = M_g^2$.)

Remark. By Theorem 15.3.2 of Part II, the intersection index $\Delta \circ \Delta_f$ is equal to the (signed) number of fixed points (assumed "non-degenerate") of f, i.e. the "Lefschetz number" of f. If, in particular, f is homotopic to a constant map, i.e. to a map to a single point, then $f_{k,*} = 0$ for $k > 0$ and $f_{0,*}$ is the identity linear transformation, whence by (7),

$$\Delta \circ \Delta_f = \operatorname{tr} f_{0,*} = 1,$$

which is essentially Corollary 15.3.3 of Part II.

2. Since by Theorem 2.9, $H_k(M^n; \mathbf{k}) \simeq H^k(M^n; \mathbf{k})$ for any field $\mathbf{k}$, one obtains from (6) a Poincaré-duality isomorphism between the cohomology groups of complementary dimensions. Prove that this isomorphism can be retrieved from cohomological multiplication (introduced in §7) by showing that the bilinear form $\langle a, b\rangle$ defined for each pair of elements $a \in H^q(M^n; \mathbf{k})$, $b \in H^{n-q}(M^n; \mathbf{k})$ by
$$\langle a, b\rangle = (ab, [M^n]), \tag{8}$$
is non-degenerate. (Recall that the right-hand side of (8) denotes the value of the product $ab \in H^n(M^n; \mathbf{k})$ at the cycle $[M^n]$, defined for instance as in Example (d) above.) This convenient alternative approach to the duality isomorphism may in fact be extended to the case where the coefficients are from $\mathbb{Z}$, when torsion may be present, by means of the following formula involving the "cap product" (see §7(7)):
$$Da = a \frown [M^n], \tag{9}$$
where $a \in H^k(M^n; \mathbb{Z})$, and $a \frown [M^n] \in H_{n-k}(M^n; \mathbb{Z})$ (cf. §7, Exercise 6). Note that in view of the formula (see §7(7))
$$((a \frown [M^n]), b) = (ab, [M^n]),$$
where a, b are from the appropriate cohomology groups, when the coefficients are from a field the formula (9) reduces to (8) and so represents nothing new.

3. Let M be a finite cell complex and $K \subset M$ a subcomplex such that $M \backslash K$ is an orientable smooth manifold. Establish the following "Lefschetz-duality" isomorphisms:
$$H_i(M, K; \mathbb{Z}) \simeq H_i(M/K; \mathbb{Z}) \simeq H^{n-i}(M \backslash K; \mathbb{Z}), \qquad i > 0;$$
$$H^i(M, K; \mathbb{Z}) \simeq H^i(M/K; \mathbb{Z}) \simeq H_{n-i}(M \backslash K; \mathbb{Z}), \qquad i > 0.$$
Investigate the case $i = 0$.

4. Let K^m be a finite cell complex embedded in the n-sphere S^n, $m < n$. Establish the following "Alexander-duality" isomorphisms:
$$H_i(K^m; \mathbb{Z}) \simeq H^{n-i-1}(S^n \backslash K^m; \mathbb{Z}), \qquad 0 < i < n-1;$$
$$H^i(K^m; \mathbb{Z}) \simeq H_{n-i-1}(S^n \backslash K^m; \mathbb{Z}), \qquad 0 < i < n-1.$$
Examine the special cases $i = 0, n-1$.

5. Let M^n be a smooth, closed, orientable manifold, and let $H_k(M^n; \mathbb{Z}) = R_k \oplus T_k$ be a decomposition of H_k as a direct sum of a free abelian group R_k and a finite abelian group T_k (the torsion subgroup). Show that $R_k \simeq R_{n-k}$, $T_k \simeq T_{n-k-1}$. (*Hint.* Isomorphisms $R_k \simeq R^k$, $T_k \simeq T^{k+1}$ are valid for every finite cell complex.)

We conclude this section with the following remark. Recall that the Euler characteristic $\chi(M^n)$ may be defined as the alternating sum of the Betti numbers $b_i = \dim H_i(M^n; \mathbb{R})$ (or $\dim H_i(M^n; \mathbb{Z}_2)$ if M^n is non-orientable). If M^n is closed then Poincaré duality gives $b_i = b_{n-i}$, whence for all odd-dimensional such $M^n = M^{2k+1}$, orientable or not, one has

$$\chi(M^{2k+1}) = \sum(-1)^i b_i = 0.$$

§19. Critical Points of Smooth Functions and the Lyusternik–Shnirelman Category of a Manifold

We saw in §16 that for a Morse function on a manifold M, i.e. a function whose critical points ae all non-degenerate, the number μ_k of critical points of index k is bounded below by the corresponding Betti number $b_k = \dim H_k(M; G)$, where $G = \mathbb{R}$ (or $\mathbb{Z}_2$ or indeed $\mathbb{Z}_p$, p prime): $b_k \leq \mu_k$. It follows that a Morse function on, for instance, the 2-dimensional surface M_g^2 must have at least $(2g + 2)$ critical points all told. However, the problem of estimating the number of critical points of *arbitrary* smooth functions (not necessarily Morse) on M is considerably more complicated. Simple examples show that the number of critical points may for certain smooth functions be much smaller then for Morse functions; as observed earlier (in §§16, 17) as a result of an arbitrarily small perturbation of a Morse function, several non-degenerate critical points may merge together into a single degenerate critical point, thereby reducing the number of critical points. (At the conclusion of §17 we saw that while, as noted above, a Morse function on the surface M_g^2 must have at least $(2g + 2)$ critical points, there exists a smooth function f on M_g^2 with just three critical points, one of which is a maximum point, one a minimum point, and the third a degenerate critical point yielding $2g$ non-degenerate critical points upon application of a suitable arbitrarily small homotopy of f.) Nonetheless there are, as we have seen in §16, inequalities of the type $\sum_k \pm \mu_k \geq \sum_k \pm b_k$ still valid for arbitrary, not necessarily Morse, functions on M; however, for arbitrary functions the numbers μ_k no longer have the simple interpretation they have in the case of Morse functions (namely as the number of critical points of index k), but provide rather a measure of the "degree of complexity" of the critical points, no longer directly related to their number. Moreover, since the numbers μ_k are defined for general smooth functions f in terms of the bifurcation points of f (see Definition 16.2), and, as noted in §16, a degenerate critical point need not always be a bifurcation point, it follows that in the inequalities of the type $\sum_k \pm \mu_k \geq \sum_k \pm b_k$ obtained in §16, some of the critical points may fail to be taken into account, so that in general those inequalities cannot be used to estimate the number of critical points. It turns out that there is a certain numerical topological invariant of M, called the "Lyusternik-Shnirelman category" of M, which does afford a useful lower bound for the number of critical points of an arbitrary smooth function on M. We begin by defining a relative version of this invariant for a closed subset A of an arbitrary (Hausdorff) topological space X.

19.1. Definition. Let X be a Hausdorff topological space and $A \subset X$ a closed set of X. The closed set A is said to be of *category* $k = \operatorname{cat}_X(A)$ *with respect to the topological space* X if k is the least number for which A can be written as a union

$$A = A_1 \cup \cdots \cup A_k$$

of k closed sets $A_1, \ldots, A_k$, each of which is contractible in X (to a point). (Thus $\operatorname{cat}_X A$ may take the values $1, 2, 3, \ldots$, and in general also infinite values.)

Remark. We shall, in what follows, for simplicity assume that X is connected (although A and the A_i need not be). In the case $A = X$ we write $\operatorname{cat}_X(X) = \operatorname{cat}(X)$, the "Lyusternik-Shnirelman category of the space X".

In the following few lemmas we establish the basic properties of $\operatorname{cat}_X(A)$.

19.2. Lemma. *If $A \subset B$ where A, B are closed subsets of X, then $\operatorname{cat}_X(A) \le \operatorname{cat}_X(B)$.*

Proof. Writing $q = \operatorname{cat}_X(B)$, we have $B = B_1 \cup \cdots \cup B_q$ where the B_i are closed sets contractible in X (to a point). It follows that each set $A_i = A \cap B_i$ is closed and contractible in X to a point, and, since $A \subset B$, we have $A = A_1 \cup \cdots \cup A_q$. Hence $\operatorname{cat}_X(A) \le q$, as claimed. □

19.3. Lemma. *If A, B are any closed sets of the topological space X, then*

$$\operatorname{cat}_X(A \cup B) \le \operatorname{cat}_X(A) + \operatorname{cat}_X(B).$$

Proof. If

$$A = A_1 \cup \cdots \cup A_k, \qquad B = B_1 \cup \cdots \cup B_l$$

where the A_i and B_j are closed and contractible in X, then

$$A \cup B = (A_1 \cup \cdots \cup A_k) \cup (B_1 \cup \cdots \cup B_l),$$

whence it is immediate that $\operatorname{cat}_X(A \cup B) \le k + l$. □

19.4. Corollary. *If A, B are closed subsets of X with $A \subset B$, then*

$$\operatorname{cat}_X(\overline{B \backslash A}) \ge \operatorname{cat}_X(B) - \operatorname{cat}_X(A),$$

where $\overline{B \backslash A}$ denotes the closure of the set $B \backslash A$ in X.

Proof. Since $B = A \cup (\overline{B \backslash A})$, we have from the preceding lemma

$$\mathrm{cat}_X(B) \le \mathrm{cat}_X(A) + \mathrm{cat}_X(\overline{B \backslash A}),$$

whence the corollary. □

19.5. Lemma. *Let A, B be closed sets of X, and suppose that B is continuously deformable in X to a subset of A (i.e. there is a homotopy φ_t between the inclusion $i: B \to X$ and some map $\varphi_1: B \to X$ satisfying $\varphi_1(B) \subset A$; note here that of course $\varphi_1(B)$ need not be homeomorphic to B). Then $\mathrm{cat}_X(A) \ge \mathrm{cat}_X(B)$.*

Proof. Write $k = \mathrm{cat}_X(A)$, and let $A = A_1 \cup \cdots \cup A_k$, where the A_i are closed and contractible in X. Write $R_i = \varphi_1(B) \cap A_i$; then $\bigcup_{i=1}^k R_i = \varphi_1(B)$ since $\varphi_1(B) \subset A$. From the homotopy

$$\varphi: B \times [0, 1] \to X, \qquad \varphi_0 = \varphi(\ ,0) = i, \qquad \varphi(\ ,1) = \varphi_1,$$

we obtain a continuous map $\alpha: B \to \varphi_1(B)$ (essentially just φ_1), defined by $\alpha(b) = \varphi(b, 1) = \varphi_1(b)$. Writing $B_i = \alpha^{-1}(R_i)$, $i = 1, \ldots, k$, we have

$$B = B_1 \cup \cdots \cup B_k, \tag{1}$$

since $\bigcup_{i=1}^k R_i = \varphi_1(B)$. Now application of the homotopy φ_t to each B_i deforms it in X to $\varphi_1(B_i) = R_i$, which is in turn deformable in X to a point (i.e. contractible) since $R_i \subset A_i$ and the A_i are contractible. Hence (1) expresses B as a union of closed sets contractible in X, so that $\mathrm{cat}_X(B) \le k$ as claimed. (See Figure 88.) □

19.6. Lemma. *Let A be a compact, closed subset of a manifold M. There exists $\varepsilon > 0$ (depending on A) such that $\mathrm{cat}_M(U_\varepsilon(A)) = \mathrm{cat}_M(A)$, where $U_\varepsilon(A)$ denotes the closed ε-neighbourhood (with respect to some Riemannian metric on M or region thereof) of the subset $A \subset M$.*

Proof. Since $A \subset U_\varepsilon(A)$, Lemma 19.2 implies that $\mathrm{cat}_M(A) \le \mathrm{cat}_M(U_\varepsilon(A))$. For the reverse inequality note first that $\mathrm{cat}_M(A)$ is finite in view of the compactness of A (verify this!). Write $k = \mathrm{cat}_M(A)$ and $A = \bigcup_{i=1}^k A_i$ where each A_i is closed and contractible in M. Since M is a manifold and k is finite, there exists $\varepsilon > 0$ such that for all $i = 1, \ldots, k$, $U_\varepsilon(A_i)$, the closed ε-neighbourhood of A_i, is also contractible in M. Since $U_\varepsilon(A) = \bigcup_{i=1}^k U_\varepsilon(A_i)$, it follows that $\mathrm{cat}_M(U_\varepsilon(A)) \le k$, as required. □

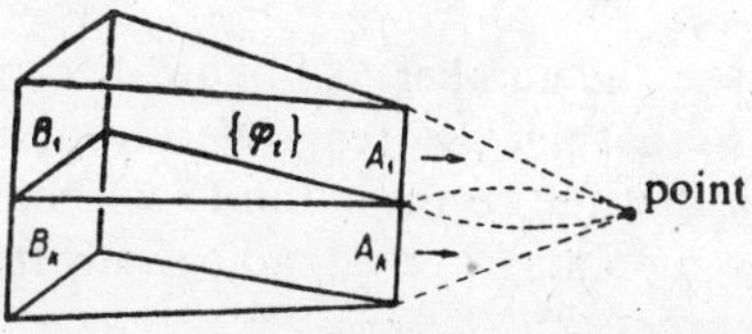

Figure 88

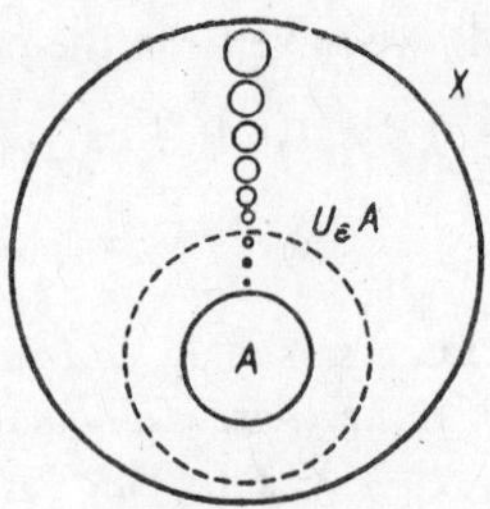

Figure 89

Remark. This lemma need not hold if M is not a manifold. A counterexample is suggested by Figure 89, where although A is contractible in X, every ε-neighbourhood of A has infinitely many "holes" in it, represented by the converging sequence of small discs (not forming part of X).

19.7. Lemma. *Let M be a manifold and let A, B_n, $n = 1, 2, \ldots$, be closed subsets in M such that $A = \lim_{n\to\infty} B_n$ (in the sense that $\rho(A, B_n) \to 0$ as $n \to \infty$, where for arbitrary subsets $C, D \subset M$, $\rho(C, D)$ is defined by*

$$\rho(C, D) = \sup_{x\in C}\left(\inf_{y\in D} \rho(x, y)\right) + \sup_{x\in D}\left(\inf_{y\in C} \rho(x, y)\right),$$

$\rho(x, y)$ being the distance function on M arising from some Riemannian metric). If $\operatorname{cat}_M(B_n) \geq k$ for all n, then $\operatorname{cat}_M(A) \geq k$.

PROOF. By Lemma 19.6 there exists $\varepsilon > 0$ such that $\operatorname{cat}_M(U_\varepsilon(A)) = \operatorname{cat}_M(A)$. Since $\rho(A, B_n) \to 0$ it is easy to see that there exists some number N such that $B_n \subset U_\varepsilon(A)$ for all $n > N$. The lemma now follows since for these n

$$k \leq \operatorname{cat}_M(B_n) \leq \operatorname{cat}_M(U_\varepsilon(A)) = \operatorname{cat}_M(A),$$

where for the second inequality we are invoking Lemma 19.2. □

19.8. Theorem. *Let M^n be a smooth, connected, closed manifold and $f(x)$ a smooth real-valued function on M^n. The number k (which may be infinite) of distinct critical points of f is bounded below by $\operatorname{cat}(M^n)$:*

$$k \geq \operatorname{cat}(M^n).$$

In fact, as we shall see, the number p of bifurcation points of f is at least $\operatorname{cat}(M^n)$: $p \geq \operatorname{cat}(M^n)$, so that this theorem suffers from the same "drawback" as was noted above, in that the contribution to k from those critical points which are not bifurcation points (i.e. are topologically regular; see Definition 16.2), is not taken into account.

As an illustrative preliminary to proving the theorem, we consider a relationship subsisting between the set of critical points of a function f on a certain manifold, and the eigenvalues of a certain bilinear form on the manifold. As manifold we take the sphere S^{n-1} embedded in the standard manner in $\mathbb{R}^n(x^1, \ldots, x^n)$:

$$S^{n-1} = \{x \in \mathbb{R}^n | (x^1)^2 + \cdots + (x^n)^2 = |x|^2 = 1\}.$$

as bilinear form any real symmetric bilinear form $B(x, y)$ ($= g_{ij}x^i y^j$ where the g_{ij} are constants) on $\mathbb{R}^n$, and we take $f(x) = B(x, x)$, $|x| = 1$, as the function f on S^{n-1}. We shall now find the critical points of f. To this end, let $x_0 \in S^{n-1}$, and $\bar{a} \in T_{x_0}(S^{n-1})$, the tangent space to S^{n-1} at x_0; the directional derivative $df/d\bar{a}|_{x_0}$ of f at x_0 in the direction $\bar{a}$ can be calculated as follows: if $x(t)$ is any smooth curve on S^{n-1} satisfying $x(0) = x_0$, $\dot{x}(0) = \bar{a}$, then since for every x, y in $\mathbb{R}^n$, $\langle Bx, y\rangle = \langle x, By\rangle$, where B is the linear transformation $B: \mathbb{R}^n \to \mathbb{R}^n$ given by $B(x^i) = (g_{si}x^s)$ (from which it is clear that $(d/dt)Bx(t) = B\dot{x}(t)$ since the g_{si} are constants), and $\langle\ ,\ \rangle$ is the usual Euclidean scalar product on $\mathbb{R}^n$, it follows that

$$\left.\frac{df}{d\bar{a}}\right|_{x_0} = \left.\frac{d}{dt}f(x(t))\right|_{t=0} = \left.\frac{d}{dt}B(x(t), x(t))\right|_{t=0}$$

$$= \left.\frac{d}{dt}\langle Bx(t), x(t)\rangle\right|_{t=0} = (\langle B\dot{x}, x\rangle + \langle Bx, \dot{x}\rangle)|_{t=0} = 2\langle Bx_0, \bar{a}\rangle.$$

Hence, a point x_0 of S^{n-1} is a critical point of f if and only if $\langle Bx_0, \bar{a}\rangle = 0$ for every vector $\bar{a}$ in $T_{x_0}(S^{n-1})$, or, equivalently, if and only if the vector Bx_0 is orthogonal in $\mathbb{R}^n$ to the tangent hyperplane $T_{x_0}(S^{n-1})$, i.e., since the radius vector x_0 is orthogonal to this tangent plane (S^{n-1} being embedded in the standard way in $\mathbb{R}^n$), $Bx_0 = \lambda x_0$ for some real number λ.

Thus a point x_0 of S^{n-1} is a critical point of f precisely if the radius vector is an eigenvector of the linear transformation B. Let $\lambda_0 \leq \lambda_1 \leq \cdots \leq \lambda_{n-1}$ be the eigenvalues of B (real since B is symmetric, though not necessarily all distinct), and let $v_0, \ldots, v_{n-1}$ be corresponding unit eigenvectors. We may in fact suppose that $v_0, \ldots, v_{n-1}$ form an orthonormal system, since, again by the symmetry of the linear transformation B, a pair of eigenvectors corresponding to distinct eigenvalues will automatically be orthogonal, while each set of r eigenvectors corresponding to a single eigenvalue of multiplicity r may be chosen pairwise orthogonal. We shall then have

$$f(v_i) = \langle Bv_i, v_i\rangle = \langle \lambda_i v_i, v_i\rangle = \lambda_i,$$

so that the eigenvalues λ_i are the critical values of $f: S^{n-1} \to \mathbb{R}$. For each $i = 0, 1, \ldots, n-1$, let M_i denote the collection $\{S^i\}$ of all i-dimensional "equators" of S^{n-1}, i.e. sections of S^{n-1} by $(i+1)$-dimensional planes in $\mathbb{R}^n$ through the origin. It is a known fact of the theory of quadratic forms that for each $i = 0, 1, \ldots, n-1$,

$$\lambda_i = \inf_{M_i}\left(\max_{x \in S^i} f(x)\right). \tag{2}$$

(We suggest the reader attempt to prove this formula for himself; note in particular its consistency with the assumed ordering $\lambda_0 \leq \cdots \leq \lambda_{n-1}$ of the eigenvalues.) Clearly the group $SO(n)$ acts transitively on each set M_i, since each i-dimensional equator S^i can be obtained from a fixed one S_0^i by means of some "rotation", i.e. element of $SO(n)$.

19.9. Proposition. *The number of distinct critical points of the function $f(x) = \langle Bx, x \rangle$ on the sphere $S^{n-1} \subset \mathbb{R}^n$, is at least twice the number of sets M_i, i.e.*

$$\textit{number of critical points of } f \geq 2n. \tag{3}$$

PROOF. If the eigenvalues λ_i of B are all distinct then the critical points of f are precisely the points of S^{n-1} at the tips of the eigenvectors $\pm v_i$, so that in this case they number exactly $2n$. If, on the other hand, there is a pair $i, j, i < j$, such that $\lambda_i = \lambda_j$, then there will be a $(j - i)$-dimensional equator S^{j-i} the radius vectors of whose points are all eigenvectors of the particular eigenvalue $\lambda_i = \lambda_j$, so that all the points of S^{j-i} will be (degenerate) critical points of f. Hence in this case there is a continuum of critical points, whence (3) certainly holds. □

Remark. Since the function $f(x) = \langle Bx, x \rangle$ is obviously invariant under the reflection $x \to -x$, it follows that it determines a function $\bar{f}$ on the projective space $\mathbb{R}P^{n-1}$. Reformulated in terms of this function, the above proposition states that: *the function $\bar{f}$ on $\mathbb{R}P^{n-1}$ defined by $\langle Bx, x \rangle$ on $S^{n-1} \subset \mathbb{R}^n$, has at least n critical points, i.e. at least as many critical points as there are sets M_i.*

We now seek to generalize this line of argument from the rather special situation considered above to that of an arbitrary smooth function f on an arbitrary closed, smooth manifold $M = M^n$. We first need to introduce appropriate analogues of the various entities featuring in the above preliminary version. Thus in place of the sphere S^{n-1} we take the manifold M^n, in place of the function $f(x) = B(x, x)$ we take our arbitrary smooth function f on M^n, and in place of the group $SO(n)$ of rotations (preserving each set M_i of i-dimensional equators) a set of homotopies preserving certain collections of closed subsets of M^n, these collections representing the analogues of the M_i; finally, in the role of the eigenvalues λ_i of the linear transformation B we consider certain analogues defined in terms of the aforementioned collections of closed sets in M^n.

For each $i \geq 0$, the appropriate analogue of M_i is, to be specific, the set (also denoted by M_i) of all closed subsets $X \subset M^n$ satisfying $\operatorname{cat}_M(X) \geq i$:

$$M_i = \{X \subset M^n | X \text{ closed}, \operatorname{cat}_M(X) \geq i\}.$$

Clearly $M_{i+1} \subset M_i$. We turn the set $\theta(M^n)$ of all closed subsets of M^n into a metric space by defining the distance $\rho(X, Y)$ between two closed subsets X, Y to be

$$\rho(X, Y) = \sup_{x \in X} \left(\inf_{y \in Y} \rho(x, y) \right) + \sup_{y \in Y} \left(\inf_{x \in X} \rho(x, y) \right),$$

and we write (cf. Lemma 19.7)

$$Y = \lim_{p\to\infty} X_p \qquad \text{if} \quad \lim_{p\to\infty} \rho(Y, X_p) = 0, \quad Y, X_p \in \theta(M^n).$$

19.10. Lemma. *Each collection $M_i \subset \theta(M^n)$ is preserved by the above limit operation (i.e. is a closed set in the metric space $(\theta(M^n), \rho)$), and also by homotopies of the inclusions of subsets in M (i.e. under continuous deformations of subsets of M).*

PROOF. Let $X = \lim_{p\to\infty} X_p$, where $X_p \in M_i$, $p = 1, 2, \ldots$. By definition of M_i, we have $\text{cat}_M(X_p) \geq i$, whence by Lemma 19.7 $\text{cat}_M(X) \geq i$, or, equivalently, $X \in M_i$, proving that M_i is closed in the metric space $\theta(M^n)$. For the second statement of the lemma, let $B \in M_i$, and let $A = \varphi_1(B) \subset M^n$ be the set obtained from B by means of a homotopy $\varphi_t \colon B \to M^n$, with φ_0 the inclusion. Then by Lemma 19.5, $\text{cat}_M(A) \geq \text{cat}_M(B)$, whence $A \in M_i$, completing the proof. □

By analogy with (2) we now define, in the present general context,

$$\lambda_i = \inf_{X \in M_i} \left(\max_{x \in X} f(x) \right). \tag{4}$$

Since M^n is by assumption compact, we have $\text{cat}(M^n) < \infty$. Write $N = \text{cat}(M^n)$. It follows easily from the definitions of the collections M_i that

$$\theta(M^n) = M_0 = M_1 \supset M_2 \supset \cdots \supset M_N = M_{N+1} = \cdots; \tag{5}$$

thus $\lambda_0 = \lambda_1$, and the chain of M_i becomes constant at M_N. Now any smooth function f on M^n determines, via the sequence (5), functions $f_0, f_1, \ldots, f_N$, defined on $M_0, M_1, \ldots, M_N$ respectively, given by

$$f_i(X) = \max_{x \in X} f(x), \qquad \text{for each} \quad X \in M_i.$$

It is immediate that

$$\lambda_i = \inf_{X \in M_i} f_i(X). \tag{6}$$

Since $M_{i+1} \subset M_i$, it is clear that λ_i increases with i:

$$\lambda_0 = \lambda_1 \leq \lambda_2 \leq \cdots \leq \lambda_N, \qquad N = \text{cat}(M^n).$$

Since by Lemma 19.10 each M_i is closed in the metric space $\theta(M^n)$, there must, in view of (6), exist in M_i and element X_i^0 at which f_i attains the value λ_i, i.e. there is in each M_i a set X_i^0 such that $\lambda_i = \max_{x \in X_i^0} f(x)$.

19.11. Lemma. *On each level surface*

$$f_{\lambda_i} = \{x \in M^n \mid f(x) = \lambda_i\},$$

there is at least one critical point of f.

PROOF. Suppose that on the contrary a level surface f_{λ_i} contains no critical points. For this i let $X_i^0 \in M_i$ be the element of M_i such that $f_i(X_i^0) = \lambda_i$, that is (see above)

$$\lambda_i = \max_{x \in X_i^0} f(x). \tag{7}$$

Since X_i^0 is a closed subset of M^n, it contains a point x_i^0 say, satisfying $f(x_i^0) = \lambda_i$ (we were assuming this implicitly in using "max" earlier, rather than "sup"). Thus $x_i^0 \in X_i^0 \cap f_{\lambda_i}$. The assumption that grad $f(x) \neq 0$ for all $x \in f_{\lambda_i}$ together with the transversality of grad f to the level surface at each point of it and the compactness of M^n, imply that the level surface f_{λ_i} can be deformed (diffeomorphically) along the integral trajectories of the vector field $v(x)$ determined, via a Riemannian metric on M^n, by the covector field $-$grad $f(x)$, onto some nearby level surface $f_{\lambda_i - \varepsilon}$ (see Figure 90). It follows, once again exploiting the compactness of M^n, that there is a smooth isotopy of the identity map $M^n \to M^n$ which remains the identity outside some small neighbourhood of the region $\{x \in M^n | \lambda_i - \varepsilon \leq f(x) \leq \lambda_i\}$, and deforms the region $\{x \in M^n | f(x) \leq \lambda_i\}$ onto $\{x \in M^n | f(x) \leq \lambda_i - \varepsilon\}$, and its boundary f_{λ_i} onto $f_{\lambda_i - \varepsilon}$ (see Figure 90). If $\tilde{X}_i^0$ is the end result of the restriction of this homotopy to X_i^0, then in view of Lemma 19.5, $\operatorname{cat}_M(\tilde{X}_i^0) \geq \operatorname{cat}_M(X_i^0)$ (in fact we have equality here), whence $\operatorname{cat}_M(\tilde{X}_i^0) \geq i$, i.e. $\tilde{X}_i^0 \in M_i$. However, since

$$\tilde{X}_i^0 \subset \{x \in M^n | f(x) \leq \lambda_i - \varepsilon\},$$

and f_{λ_i} is deformed onto $f_{\lambda_i - \varepsilon}$, it follows from (7) that

$$\max_{x \in \tilde{X}_i^0} f(x) = \lambda_i - \varepsilon,$$

whence

$$\inf_{X \in M_i} \left(\max_{x \in X} f(x) \right) \leq \lambda_i - \varepsilon < \lambda_i,$$

contradicting (4). □

19.12. Lemma. *Suppose $\lambda_i = \lambda_{i+p}$ for some $p > 0$. Then the set S of critical points of the function f on the level surface f_{λ_i} satisfies* $\operatorname{cat}_M(S) \geq p + 1$.

(Contrast this result with the special case considered earlier where $f(x)$ was given in the form $\langle Bx, x \rangle$ on S^{n-1}: there the λ_j were the eigenvalues of the linear transformation B, and the equality of $\lambda_i, \ldots, \lambda_{i+p}$ implies (as was noted

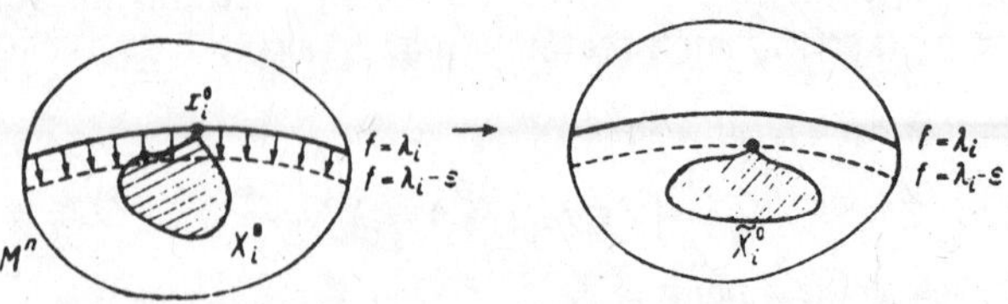

Figure 90

in the proof of Proposition 19.9) that the section S^p of S^{n-1} by the plane spanned by the corresponding eigenvectors $v_i, \ldots, v_{i+p}$, consists entirely of critical points; however, since $\mathrm{cat}_{S^{n-1}}(S^p) \leq 2$, the parallel between the earlier subsets M_i, $i = 0, \ldots, n-1$, and the present M_i (of which there are just three, namely M_0, M_1, M_2, in the case of S^{n-1}) is not exact.)

PROOF OF LEMMA 19.12. Suppose that, on the contrary,

$$\mathrm{cat}_M(S) \leq p. \tag{8}$$

Since S is closed, by Lemma 19.6 there exists $\varepsilon > 0$ such that $\mathrm{cat}_M(S) = \mathrm{cat}_M(U_\varepsilon(S))$. Let $X^0_{i+p} \in M_{i+p}$ be such that

$$\max_{x \in X^0_{i+p}} f(x) = \lambda_{i+p};$$

the existence of such a set was established earlier. Consider the closed set (indicated in Figure 91)

$$X^0 = \overline{X^0_{i+p} \backslash (X^0_{i+p} \cap U_\varepsilon(S))}.$$

For this set we have (using Lemmas 19.4, 19.2, together with (8))

$$\begin{aligned} \mathrm{cat}_{M^n}(X^0) &\geq \mathrm{cat}_{M^n}(X^0_{i+p}) - \mathrm{cat}_{M^n}(X^0_{i+p} \cap U_\varepsilon S) \\ &\geq \mathrm{cat}_{M^n}(X^0_{i+p}) - \mathrm{cat}_{M^n}(U_\varepsilon S) \\ &= \mathrm{cat}_{M^n}(X^0_{i+p}) - \mathrm{cat}_{M^n}(S) \geq i + p - p = i, \end{aligned}$$

whence $X^0 \in M_i$. Now

$$\lambda_{i+p} = \max_{x \in X^0_{i+p}} (f(x)) \geq \max_{x \in X^0} (f(x)) \geq \lambda_i,$$

where the first inequality follows from $X^0 \subset X^0_{i+p}$, and the second from the definition of λ_i and the fact that $X^0 \in M_i$. Since $\lambda_i = \lambda_{i+p}$ we infer that

$$\max_{x \in X^0} f(x) = \lambda_i.$$

Now by arguing as in the proof of Lemma 19.11, with a manifold of the form $P = M^n \backslash U_\varepsilon(S)$ in place of M^n, i.e. with sufficiently small closed neighbourhoods of the (isolated) critical points of f removed from M^n, and with f replaced by its restriction to P (and using the obvious fact that $\mathrm{cat}_P(A) = \mathrm{cat}_M(A)$ for all closed sets $A \subset P$), one can show that in fact any set $X^0_i \in M_i$ satisfying

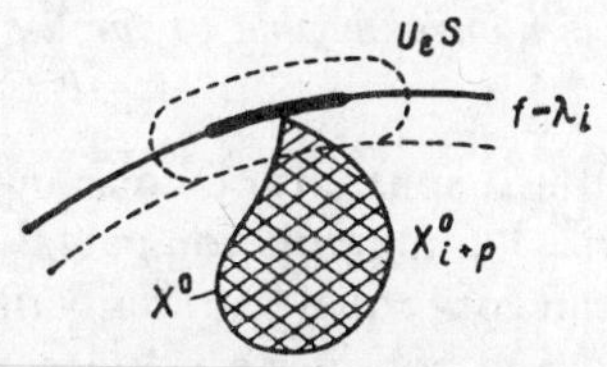

Figure 91

$\max_{x \in X_i^0} f(x) = \lambda_i$ must contain a critical point of f. Hence, in particular, X^0 contains a critical point. However, since $X^0 \cap S = \varnothing$ by construction, we have a contradiction. □

PROOF OF THEOREM 19.8. We wish to show that the number of critical points of an arbitrary smooth function f on a compact smooth manifold M^n is at least $\operatorname{cat}(M^n)$. Consider the chain of sets

$$\theta(M^n) = M_0 = M_1 \supset M_2 \supset \cdots \supset M_N,$$

where $N = \operatorname{cat}(M^n)$. If the corresponding numbers λ_j are all distinct (except for λ_0, λ_1 which are necessarily equal):

$$\lambda_0 = \lambda_1 < \lambda_2 < \cdots < \lambda_N,$$

then the N level surfaces f_{λ_j}, $j = 1, \ldots, N$, are pairwise disjoint, and since by Lemma 19.11 there is at least one critical point on each of them, there must be at least $N = \operatorname{cat}(M^n)$ critical points altogether.

If, on the other hand, $\lambda_i = \lambda_{i+p}$ occurs with "multiplicity" $p + 1, p > 0$, then by Lemma 19.12 the set S of critical points on the corresponding level surface f_{λ_i} satisfies $\operatorname{cat}_M(S) \geq p + 1$, so that there must be at least $p + 1$ distinct points in S (otherwise S would be representable as a union of fewer than $p + 1$ closed sets contractible in M). Hence each level surface f_{λ_j} contains at least as many critical points of f as there are occurrences of λ_j in the sequence $\lambda_1 \leq \lambda_2 \leq \cdots \leq \lambda_N$, whence again the total number of critical points is at least $N = \operatorname{cat}(M^n)$, completing the proof. □

Remark. A close examination of the proof reveals that, as noted earlier, the lower bound $\operatorname{cat}(M^n)$ is actually valid for the size of the smaller set of bifurcation points of f. We leave to the reader the details of the verification of this.

Having shown that the number of critical points of a smooth function on a compact manifold M^n is at least $\operatorname{cat}(M^n)$, it is natural to ask if this lower bound is best possible, generally speaking, i.e. if there is a plentiful supply of non-trivial examples of functions on manifolds M^n with exactly $\operatorname{cat}(M^n)$ critical points. We shall now show that in fact for the compact, connected 2-dimensional manifolds M^2 the bound is attained. We begin by computing $\operatorname{cat}(M^2)$ for these manifolds.

19.13. Proposition. *For a closed, connected 2-dimensional manifold M^2, we have* $\operatorname{cat}(M^2) = 2$ *if M^2 is homeomorphic to the 2-sphere S^2, and otherwise* $\operatorname{cat}(M^2) = 3$.

PROOF. The assertion for the 2-sphere is obvious; in fact, it is not difficult to show that S^2 is characterized among the compact 2-manifolds by having its Lyusternik–Shnirelman category equal to 2 (verify this!). Thus suppose M^2 is not homeomorphic to S^2, and consider a cell decomposition of M^2 of the (standard) form

$$\sigma^0 \cup \left(\bigcup_{\alpha=1}^{q} \sigma_\alpha^1 \right) \cup \sigma^2,$$

i.e. a bouquet of circles $\bigvee_{\alpha=1}^{q} S_\alpha^1$, to which a single 2-cell σ^2 is (suitably) attached (see §4, in particular Figures 29, 30 there). Let $U_\varepsilon(\bigvee_{\alpha=1}^{q} S_\alpha^1)$ be an open neighbourhood in M^2 of the 1-skeleton of this cell decomposition, sufficiently small for the complement

$$\bar{D}^2 = M^2 \setminus U_\varepsilon\left(\bigvee_{\alpha=1}^{q} S_\alpha^1 \right)$$

to be a closed disc (essentially just the 2-cell σ^2 with an open neighbourhood of its boundary removed; see Figure 92). We now represent M^2 as the union of three closed sets (see Figure 93): $M = A_1 \cup A_2 \cup A_3$, where $A_1 = \bar{D}^2$,

$$A_2 = U_\varepsilon(\textstyle\bigvee_\alpha S_\alpha^1) \cap W_\eta(\sigma^0) \qquad \text{and} \qquad A_3 = \overline{U_\varepsilon(\textstyle\bigvee_\alpha S_\alpha^1) \setminus A_2};$$

here $W_\eta(\sigma^0)$ is a suitably small neighbourhood of σ^0. Now $A_1 = \bar{D}^2$ and A_2 are both obviously contractible (in fact over themselves), and A_3 is contractible over itself to q points, and is therefore contractible to a single point in the connected manifold M^2. Hence $\text{cat}(M^2) \leq 3$. The proposition now follows from the remark made earlier in the proof that $\text{cat}(M^2) \neq 2$ if M^2 is not homeomorphic to S^2. □

Turning now to smooth functions on the compact manifolds M^2, we observe first that the standard height function on the sphere S^2 has exactly two

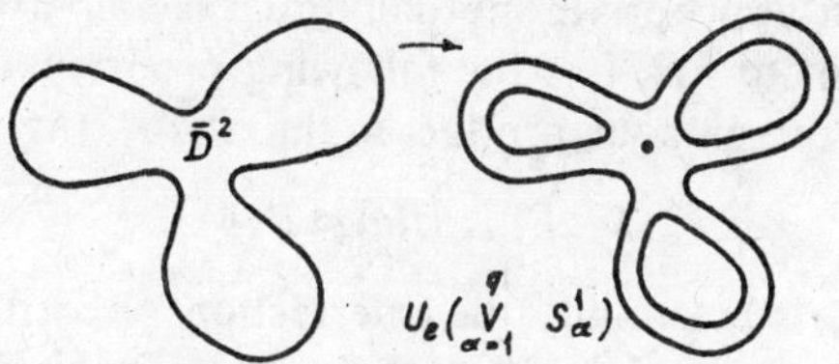

Figure 92

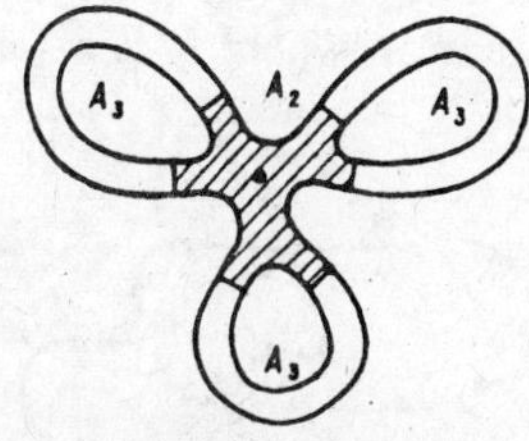

Figure 93

critical points, so that the lower bound $\operatorname{cat}(S^2)$ for the number of critical points of any smooth function on S^2 is achieved. On compact smooth M^2 other than S^2, it was indicated in §17 (see in particular Figure 81) how to construct a smooth function with exactly three critical points. Hence, in view of the preceding proposition, the lower bound $\operatorname{cat}(M^2)$ for the number of critical points is indeed attained for all compact, smooth 2-manifolds M^2.

Computation of $\operatorname{cat}(M^n)$ presents in general a non-trivial problem, the precise value being ascertained only with great difficulty. Since for particular concrete manifolds M^n upper bounds for $\operatorname{cat}(M^n)$ can be obtained by constructing coverings $\bigcup_{i=1}^N A_i = M^n$ by closed sets A_i contractible in M^n, the main difficulty lies in finding reasonable lower bounds for $\operatorname{cat}(M^n)$. There is, however, such a lower bound deriving from the structure of the integral cohomology ring $H^*(M^n; \mathbb{Z})$ (or the rings $H^*(M^n; \mathbb{Z}_p)$, p prime) in the orientable case, and the ring $H^*(M^n; \mathbb{Z}_2)$ in the non-orientable case.

The lower bound in question is the *cohomological length* of the manifold M^n, defined to be the largest positive integer k for which there exist non-zero products $\alpha_1 \ldots \alpha_k$ of k elements $\alpha_1, \ldots, \alpha_k$ of positive degree in $H^*(M^n; \mathbb{Z})$ ($H^*(M^n; \mathbb{Z}_2)$ if M^n is non-orientable). (Here the product is the multiplication in the ring $H^*(M^n; \mathbb{Z})$ (or $H^*(M^n; \mathbb{Z}_2)$) introduced in §7.)

19.14. Proposition. *If M^n is a compact smooth manifold of cohomological length k, then*

$$\operatorname{cat}(M^n) \geq k + 1.$$

PROOF. It can be shown that the Poincaré-duality isomorphism $D: H^k(M^n; \mathbb{Z}) \to H_{n-k}(M^n; \mathbb{Z})$ (constructed as in §18 from the duality operator $D: K \to \tilde{K}$ between appropriate mutually dual cell complexes K, $\tilde{K}$ homotopically equivalent to M^n) has the following property: if α, β are any two cocycles and $\alpha\beta$ represents the product in the ring $H^*(M^n; \mathbb{Z})$ then

$$D(\alpha\beta) = D(\alpha) \cap D(\beta), \tag{9}$$

where $D(\alpha) \cap D(\beta)$ is essentially the intersection of corresponding singular cycles in M^n. (In the case where $D(\alpha)$, $D(\beta)$ are realizable as submanifolds γ_1, γ_2, possibly with singularities, of M^n, the appropriate singular cycle $\gamma_1 \cap \gamma_2$ can in fact be obtained as the intersection of γ_1 and γ_2 after these have been brought into general position in M^n (see Figure 94).) (The relation (9) is most easily seen in the simple situation where K is a simplicial complex and $\tilde{K}$ is

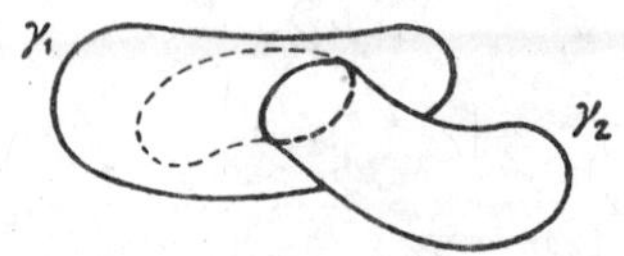

Figure 94

the dual complex defined (as in §6) in terms of the barycentric subdivision of K. Consider a $(k+l)$-simplex $\sigma^{k+l} = [\alpha_0, \alpha_1, \ldots, \alpha_{k+l}]$ of K, and let α and β be the basic cocycles taking the value 1 on $\sigma_1^k = [\alpha_0, \alpha_1, \ldots, \alpha_k]$ and $\sigma_2^l = [\alpha_k, \alpha_{k+1}, \ldots, \alpha_{k+l}]$ respectively, and 0 on all other simplexes. Then as defined in §7 the (cup) product $\alpha\beta = \alpha \smile \beta$ takes the value

$$(\alpha \smile \beta, \sigma^{k+l}) = (\alpha, \sigma_1^k)(\beta, \sigma_2^l)$$

on σ^{k+l} and 0 on every other $(k+l)$-simplex. It is not difficult to see geometrically that indeed $D(\sigma^{k+l}) = D(\sigma^k) \cap D(\sigma^l)$ where $D\colon K \to \tilde{K}$ is the canonical dual simplicial operator (verify this for low values of k and l!), whence (9) in this case.)

Noting (to accommodate the non-orientable case) that the analogue of (9) holds over $\mathbb{Z}_2$, let $\alpha_1, \ldots, \alpha_k \neq 0$ be a non-zero product of length k in $H^*(M^n; \mathbb{Z})$ (or $H^*(M^n; \mathbb{Z}_2)$ as the case may be), with $\deg \alpha_i > 0$, and write $\gamma_i = D(\alpha_i)$, $i = 1, \ldots, k$. Then by (9)

$$D(\alpha_1 \alpha_2 \ldots \alpha_k) = \gamma_1 \cap \gamma_2 \cap \cdots \cap \gamma_k = \gamma,$$

where the cycle γ is not homologous to zero since D is an isomorphism. Suppose now that $\operatorname{cat}(M^n) \le k$, i.e. that there exist closed subsets $A_1, \ldots, A_s$, $s \le k$, of M^n, with $M^n = \bigcup_{i=1}^s A_i$ and each A_i contractible in M^n to a point. We may assume $s = k$ by taking, if necessary, $A_{s+1}, \ldots, A_k$ to each consist of a single point of M^n. For each $i = 1, \ldots, k$, since A_i is contractible in M^n, it follows that for every $k > 0$ the kth homology group $H_k(M; \mathbb{Z})$ embeds naturally in the corresponding relative homology group $H_k(M, A_i; \mathbb{Z})$ (and this of course remains true over $\mathbb{Z}_2$), whence each cycle γ_i is homologous to a cycle $\tilde{\gamma}_i$ in $M^n \backslash A_i$, i.e. γ_i may be "pulled away" from A_i. Hence the cycle $\gamma = \bigcap_{i=1}^k \gamma_i$ is homologous to

$$\bigcap_{i=1}^k \tilde{\gamma}_i \subset \bigcap_{i=1}^k (M \backslash A_i) = M \backslash \bigcup_{i=1}^k A_i = \varnothing,$$

i.e. γ is null-homologous, and we have reached a contradiction. We must therefore have $\operatorname{cat}(M^n) > k$, and the proposition is proved. □

We now use this proposition to calculate $\operatorname{cat}(M^n)$ for some particular manifolds M^n.

Examples

(a) Any closed 2-dimensional manifold M^2 other than the sphere S^2 has cohomological length 2. (This can be verified directly starting from the standard cell decomposition of M^2; see §4.) Hence, by the proposition, $\operatorname{cat}(M^2) \ge 3$, confirming our earlier claim.

(b) We next show that $\operatorname{cat}(\mathbb{R}P^n) = n + 1$. We begin by establishing the inequality

$$\operatorname{cat}(\mathbb{R}P^n) \le n + 1 \tag{10}$$

directly. Consider the (standard) decomposition $\mathbb{R}P^n = \bigcup_{i=1}^{n+1} A_i$, where A_i is the open n-dimensional disc consisting of all lines

$$\{\lambda(x^1, \ldots, x^i, \ldots, x^{n+1}) | x^i \neq 0\}$$

in $\mathbb{R}^{n+1}$, i.e. points of $\mathbb{R}P^n$ with ith homogeneous co-ordinate non-zero (see Part II, §2.2). Since the A_i together form a finite open covering of $\mathbb{R}P^n$, there is contained in each A_i a closed disc A_i' differing from A_i by sufficiently little for $A_i', \ldots, A_{n+1}'$ to still cover $\mathbb{R}P^n$. Since each A_i' is contractible (over itself in fact), the desired inequality (10) follows immediately from the definition of $\operatorname{cat}(\mathbb{R}P^n)$.

For the reverse inequality, $\operatorname{cat}(\mathbb{R}P^n) \geq n + 1$, it suffices, in view of Proposition 19.14, to show that the (mod 2) cohomological length of $\mathbb{R}P^n$ is at least n. Now the cohomology $\mathbb{Z}_2$-algebra of $\mathbb{R}P^n$ can be shown to be isomorphic to the algebra of "truncated" polynomials of degree $\leq n$ over $\mathbb{Z}_2$ in a single variable x_1 (cf. §4(12) and the first example of §7):

$$H^*(\mathbb{R}P^n; \mathbb{Z}_2) \simeq \mathbb{Z}_2[x_1]/(x_1^{n+1}),$$

where, in the usual notation, (x_1^{n+1}) denotes the ideal generated by x_1^{n+1}. Since x_1^n is an appropriate non-zero product of length n in this algebra the desired result is immediate. This completes the proof that $\operatorname{cat}(\mathbb{R}P^n) = n + 1$.

(c) Finally, we show that for the n-dimensional torus T^n, we have $\operatorname{cat}(T^n) = n + 1$. Since T^n is a Lie group it follows essentially from Corollary 7.7 that its integral cohomology ring $H^*(T^n; \mathbb{Z})$ is isomorphic to the free exterior algebra $\Lambda[x_1, \ldots, x_n]$ on n generators. Since the product $x_1 \ldots x_n$ is non-zero in this algebra we infer, via Proposition 19.14, that $\operatorname{cat}(T^n) \geq n + 1$.

For the reverse inequality observe first that since $T^n = S^1 \times T^{n-1}$, we may represent T^n as the bouquet $S^1 \vee T^{n-1}$ with an n-cell σ^n attached:

$$T^n = (S^1 \vee T^{n-1}) \cup \sigma^n.$$

The desired inequality $\operatorname{cat}(T^n) \leq n + 1$ is then an easy consequence (via Lemma 19.3 and an induction) of the following

19.15. Lemma. *For the bouquet $X \vee S^n$ of the n-sphere S^n with an arbitrary path-connected cell complex X for which* $\operatorname{cat}(X) \geq 2$, *we have*

$$\operatorname{cat}(X \vee S^n) = \operatorname{cat} X.$$

PROOF. Write $\operatorname{cat}(X) = k$, and let $A_1, \ldots, A_k$ be k closed subsets of X covering X and each contractible in X. Let x_0 be the point where the bouquet is "gathered": $X \vee S^n = X \vee_{x_0} S^n$. Represent S^n as the union of two closed discs D_1^n and D_2^n such that $x_0 \in D_1^n$, $x_0 \notin D_2^n$, let A_{i_0} be one of the A_i's containing x_0, and let j_0 be any index other than i_0: $1 \leq j_0 \leq k$, $j_0 \neq i_0$. Set $B_i = A_i$ for $i \neq i_0, j_0$, $B_{i_0} = A_{i_0} \cup D_1^n$, and $B_{j_0} = A_{j_0} \cup D_2^n$ (see Figure 95). We then have $X \vee S^n = \bigcup_{i=1}^k B_i$, where each B_i is closed and contractible in $X \vee S^n$. (The path-connectedness of X and S^n is needed here, in particular, since $A_{j_0} \cap D_2^n = \varnothing$.) Hence $\operatorname{cat}(X \vee S^n) = \operatorname{cat} X$, as claimed. □

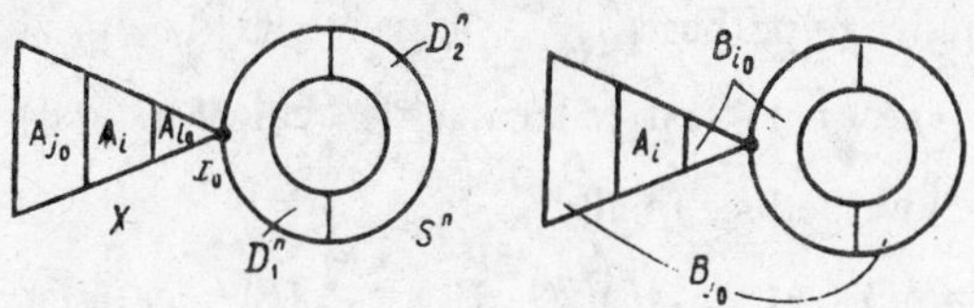

Figure 95

EXERCISE

Show that, more generally, for any path-connected topological spaces X and Y,

$$\operatorname{cat}(X \vee Y) = \max\{\operatorname{cat}(X), \operatorname{cat}(Y)\}.$$

We conclude the section with the following result for Serre fibrations, i.e. covering maps $p: E \to B$ with the "covering homotopy property" (see Part II, §22.1). (Recall that the fibre F of such a covering map is uniquely defined to within a homotopy equivalence.)

19.16. Proposition. *Let $p: E \to B$ be a Serre fibration with fibre F. Then*

$$\operatorname{cat}(E) \le \operatorname{cat}_E(F)\operatorname{cat}(B). \tag{11}$$

PROOF. We shall actually prove the following more general version of this inequality: *for every closed subset Y of the base space B, one has*

$$\operatorname{cat}_E(p^{-1}(Y)) \le \operatorname{cat}_E(F)\operatorname{cat}_B(Y), \tag{12}$$

where $p^{-1}(Y)$ is the complete inverse image of Y; clearly, setting $Y = B$ in (12) yields the desired inequality (11).

To begin with, consider the case $\operatorname{cat}_B(Y) = 1$ (the first step of an induction), where (12) reduces to

$$\operatorname{cat}_E(p^{-1}(Y)) \le \operatorname{cat}_E(F). \tag{13}$$

In this case, in view of the covering homotopy property, any homotopy $\varphi_t: Y \to B$ contracting Y to a point x_0 in the base B, can be "covered" by a homotopy $\tilde{\varphi}_t: p^{-1}(Y) \to E$, contracting $p^{-1}(Y) \subset E$ into the fibre F above the point x_0. It is then immediate from Lemma 19.5 that $\operatorname{cat}_E(F) \ge \operatorname{cat}_E p^{-1}(Y)$, as required.

Proceeding to the inductive step, suppose $\operatorname{cat}_B(Y) = k > 0$, and let $A_1, \ldots, A_k$ be k closed subsets of B, each contractible in B, such that $Y = \bigcup_{i=1}^k A_i$. Write $\tilde{Y} = \bigcup_{i=1}^{k-1} A_i$, $A = A_k$; then $Y = \tilde{Y} \cup A$, where $\operatorname{cat}_B(\tilde{Y}) = k - 1$, $\operatorname{cat}_B(A) = 1$. Now by Lemma 19.3

$$\begin{aligned}\operatorname{cat}_E(p^{-1}(Y)) &= \operatorname{cat}_E(p^{-1}(\tilde{Y} \cup A)) = \operatorname{cat}_E(p^{-1}(\tilde{Y}) \cup p^{-1}(A)) \\ &\le \operatorname{cat}_E(p^{-1}(\tilde{Y})) + \operatorname{cat}_E(p^{-1}(A)).\end{aligned} \tag{14}$$

Since A is contractible in B, we have from (13), with A in the role of Y, that $\operatorname{cat}_E(p^{-1}(A)) \le \operatorname{cat}_E(F)$, whence by (14)

$$\operatorname{cat}_E(p^{-1}(Y)) \le \operatorname{cat}_E(p^{-1}(\tilde{Y})) + \operatorname{cat}_E(F). \tag{15}$$

Now by the inductive hypothesis

$$\operatorname{cat}_E(p^{-1}(\tilde{Y})) \le \operatorname{cat}_E(F)\cdot \operatorname{cat}_B(\tilde{Y}) = \operatorname{cat}_E(F)(k-1).$$

From this and (15) we deduce finally

$$\operatorname{cat}_E(p^{-1}(Y)) \le \operatorname{cat}_E(F)(k-1) + \operatorname{cat}_E(F) = \operatorname{cat}_E(F)\cdot k,$$

completing the inductive step, and thereby the proof. □

There exist fibre spaces for which equality is achieved in (11). For instance, for the Hopf fibration $p: S^3 \to S^2$ with fibre S^1 (see Part II, §24.3, Example (a)), we have $\operatorname{cat}(E) = \operatorname{cat}(S^3) = 2$, $\operatorname{cat}(B) = \operatorname{cat}(S^2) = 2$, and $\operatorname{cat}_E(F) = \operatorname{cat}_{S^3}(S^1) = 1$ (since S^1 is contractible in S^3), so that indeed, in this case, $\operatorname{cat}(E) = \operatorname{cat}_E(F)\operatorname{cat}(B)$.

§20. Critical Manifolds and the Morse Inequalities. Functions with Symmetry

An important situation occurs when the degenerate critical points of a smooth function f on a manifold M^n form a so-called "non-degenerate critical manifold". The precise conditions of interest are as follows:

(i) the equation grad $f = 0$ should define a collection of smooth submanifolds $W_k \subset M^n$ (of dimensions α_k say);
(ii) the Hessian d^2f should at each point of each manifold W_k define a quadratic form of rank $n - \alpha_k$, in the sense that at each such point d^2f should determine a non-degenerate bilinear form on the subspace of dimension $n - \alpha_k$ of the tangent space to M^n orthogonal to W_k (as usual with respect to some Riemannian metric on M^n).

Under these conditions the W_k are called *critical manifolds* of the function f.

Functions of this type arise naturally in the context of a manifold on which a Lie group of transformations acts, namely as functions invariant under the action of the Lie group. They may also be constructed from maps $\psi: M^n \to M^{n-q}$ to manifolds of lower dimension than n, in the form $f(x) = g(\psi(x))$ where ψ has rank $n - q$ and g is a Morse function on M^{n-q}.

20.1. Definition (cf. Definition 10.4.2 of Part II, and Lemma 15.1 above). The *index* of a connected critical manifold $W_k \subset M^n$ of a function f on M^n, is the number λ of negative squares in the form d^2f as a quadratic form on the (above-mentioned) subspace of the tangent space of M^n orthogonal to W_k at each point of W_k, after it has been brought into canonical diagonal form. (Note that the index λ is independent of the particular point of W_k used to define it, in view of the non-degeneracy of d^2f as a bilinear form on the aformentioned orthogonal subspace, and the connectedness of W_k.)

Throughout this section we shall assume the manifold M^n to be compact, and the given smooth function f defined on it to have finitely many critical values $a_1, \ldots, a_N$. We shall further suppose that each critical level surface f_{a_j} contains at most one critical manifold W_j. Under these assumptions the basic invariants of a critical manifold of f turn out to be, somewhat analogously to the case of Morse functions as considered in §16, the "local Betti numbers", i.e. the ranks of the appropriate relative homology groups (over $\mathbb{Z}_2$ or $\mathbb{R}$):

$$b_k(M_{a_j}, M_{a_j} \setminus W_j) = \operatorname{rank} H_k(M_{a_j}, M_{a_j} \setminus W_j),$$

where as before M_a denotes the set $\{x \mid f(x) \leq a\}$. It can be shown much as in the proof of the analogous Lemma 16.6 that, provided there are no critical values of f other than a_j in the interval $[a_j - \varepsilon, a_j + \varepsilon]$ and the points of W_j account for all critical points corresponding to the critical value a_j, then

$$b_k(M_{a_j+\varepsilon}, M_{a_j-\varepsilon}) = b_k(M_{a_j}, M_{a_j} \setminus W_j). \tag{1}$$

Now as a consequence of Theorem 16.4, we deduced in §16 the following generalization of the Morse inequalities for a function on M^n with only finitely many bifurcation points:

$$\mu_k = \sum_j b_k(M_{a_j+\varepsilon}, M_{a_j-\varepsilon}) \geq b_k(M^n).$$

The analogue of this in the present context, where the degenerate critical points actually comprise whole submanifolds (of positive dimension) of M^n, turns out to be an equality involving topological invariants (the Betti numbers) of the critical manifolds, and their indices:

20.2. Theorem

(i) *Let M^n, f, W_j, a_j, λ_j, ε be as above (with, in particular, W_j comprising all critical points on the level surface f_{a_j}). Then for each j*

$$b_k(W_j) = b_{k+\lambda_j}(M_{a_j+\varepsilon}, M_{a_j-\varepsilon}), \tag{2}$$

where b_k, $b_{k+\lambda_j}$ are the appropriate Betti numbers of mod 2 *homology.*

(ii) *If the manifold M^n is orientable then for each simply-connected critical manifold W_j, equation* (2) *holds between the appropriate Betti numbers* $b_r = \operatorname{rank} H_r(\ ;\mathbb{R})$, *i.e. with coefficients from* $\mathbb{R}$.

As a preliminary to the proof we need to obtain a more detailed topological picture of the way the critical manifold W_j is situated in M^n. To this end, we endow M^n with a Riemannian metric and consider a closed ε-neighbourhood $U(W_j)$ of W_j in M^n with ε sufficiently small for the interior of $U(W_j)$ to be diffeomorphic to the normal bundle over W_j (see Part II, §7.2 and Corollary 11.3.3), with projection the restriction (to the interior of $U(W_j)$) of

$$p: U(W_j) \to W_j,$$

and fibre the interior of the closed disc $D^{n-\alpha_j}$ of radius ε, identifiable with the normal plane $\mathbb{R}^{n-\alpha_j}$ to W_j at each point. From the second of our basic assumptions about critical manifolds (see above), it follows that as a quadratic form at each point x of W_j, $d^2f|_x$ determines a direct decomposition

$$\mathbb{R}_x^{n-\alpha_j} = R_x^+ \oplus R_x^-, \qquad \dim R_x^- = \lambda_j,$$

of the fibre $\mathbb{R}_x^{n-\alpha_j}$ through x, where $d^2f|_x$ is positive definite on R_x^+ and negative definite on R_x^-. This induces a direct decomposition of the bundle $U(W_j)$ having factors $U^+(W_j)$ (with fibre R_x^+ above each $x \in W_j$) and $U^-(W_j)$ (with fibres R_x^-), and we have

$$U^+(W_j),\ U^-(W_j) \subset U(W_j) \subset M^n.$$

If we identify W_j with the 0th cross-section of $U(W_j)$ (i.e. the set of points $(x, 0)$, $x \in W_j$, of the normal bundle), then it is easy to see by means of the usual argument (essentially amounting to the "second-derivative test") that the restriction of the function f to $U^-(W_j)$ takes on its largest value (namely a_j) on $W_j \subset U^-(W_j)$.

20.3. Lemma (cf. Lemma 15.3). *Let M^n, f, a_j, ε be as above (so that in particular the critical points in $f^{-1}[a_j - \varepsilon, a_j + \varepsilon]$ are exactly the points of the critical manifold W_j). Then the manifold $M_{a_j+\varepsilon}$ has the homotopy type of the space obtained by attaching the manifold $U^-(W_j)$ to $M_{a_j-\varepsilon}$ by means of a map $\varphi\colon \partial U^-(W_j) \to \partial M_{a_j-\varepsilon}$.*

The proof of this lemma essentially imitates that of the analogous Lemma 15.3: instead of attaching a single cell σ^{λ_j} corresponding to an isolated non-degenerate critical point x_j of index λ_j, in the present situation one in effect attaches a smooth family of cells $\sigma_x^{\lambda_j}$, one for each point x of W_j (where now λ_j is the index of the critical manifold W_j), together forming the manifold U^-W_j. As noted already in the remark in §15, the full strength of the Morse lemma (15.1), according to which a Morse function f can be put into canonical quadratic form in the vicinity of a non-degenerate critical point, is not strictly speaking essential for the later results of §15; it is at least intuitively clear that the crucial condition is rather the non-degeneracy of the form d^2f (or, in the present context, of its restriction to the normal plane to W_j at each point), which determines the topological structure of the level surfaces of f in a neighbourhood of the critical point (or here the critical manifold W_j). Thus in the situation of the present lemma the non-degeneracy of the bilinear form $d^2f|_x$, restricted to the normal plane to W_j at each point $x \in W_j$, determines the topological structure of the level surfaces neighbouring on f_{a_j}, i.e. contained in $U(W_j)$, in a completely analogous manner. □

The boundary $\partial U^-(W_j)$ may clearly be regarded as a fibre bundle over W_j with fibre $S_x^{\lambda_j-1}(=\partial R_x^-)$ above each point $x \in W_j$, i.e. with fibres the boundaries $\partial\sigma_x^{\lambda_j}$ of the cells $\sigma_x^{\lambda_j}(\cong D^{\lambda_j})$ parametrized by the points x of W_j (see above).

(Note that the bundles U^- and ∂U^-, with respective fibres D^{λ_j} and S^{λ_j-1}, need not in general be trivial.) If the base W_j is simply-connected (as assumed in part (ii) of the theorem) then both fibre bundles U^- and ∂U^- will be orientable (as also W_j). Since the proof of part (ii) exploits only the orientability of these spaces the hypothesis that W_j be simply-connected may be replaced by the weaker requirement that both W_j and $U^-(W_j)$ be orientable.

20.4. Lemma. *Let $U^-(W_j)$ be the fibre bundle defined above, with base W_j and fibre D^{λ_j}. Then in* mod 2 *homology there exist isomorphisms*

$$\begin{aligned} H^{\lambda_j+q}(U^-, \partial U^-) &\simeq H^q(W_j), \\ H_{\lambda_j+q}(U^-, \partial U^-) &\simeq H_q(W_j). \end{aligned} \tag{3}$$

If W_j and $U^-(W_j)$ are orientable then there are analogous isomorphisms also in the case where the coefficients are from $\mathbb{Z}$ or $\mathbb{R}$.

PROOF. By Exercise 3 of §18 (on "Lefschetz duality"), also valid over $\mathbb{R}$, and over $\mathbb{Z}_2$ in the non-orientable case, we have, noting that U^- has dimension $\alpha_j + \lambda_j$,

$$H^p(\text{Int } U^-) \simeq H^p(U^- \backslash \partial U^-) \simeq H_{\alpha_j+\lambda_j-p}(U^-, \partial U^-), \tag{4}$$

provided $\alpha_j + \lambda_j - p > 0$. (Here Int U^- denotes the interior of U^-.) By ordinary Poincaré duality we have (recalling that dim $W_j = \alpha_j$):

$$H^p(W_j) \simeq H_{\alpha_j-p}(W_j). \tag{5}$$

Since Int U^- is homotopically equivalent to W_j, the left-hand sides of (4) and (5) are isomorphic, whence also the right-hand sides. Setting $q = \alpha_j - p$ then yields the first isomorphism in (3); the second is obtained similarly. □

PROOF OF THEOREM 20.2. From Lemma 20.3, the Excision Theorem (5.9), and Lemma 20.4 in turn, we have

$$\begin{aligned} H_{k+\lambda_j}(M_{a_j+\varepsilon}, M_{a_j-\varepsilon}) &\simeq H_{k+\lambda_j}(M_{a_j-\varepsilon} \cup_\varphi U^-, M_{a_j-\varepsilon}) \\ &\simeq H_{k+\lambda_j}(U^-, \partial U^-) \simeq H_k(W_j), \end{aligned}$$

whence the theorem. □

Examples

(a) Let M^2 be a surface of revolution about the z-axis in $\mathbb{R}^3$, and let f be the height function on M^2 whose value at each point of M^2 is the z-co-ordinate of that point. The critical manifolds W_j of f (of positive dimension) are then circles S^1, so that $\alpha_j = 1$. The index λ_j will be either 0 (in the case where the points of W_j all correspond to a local minimum value of f), or 1 (when the points of W_j are all local-maximum points). There may occur also isolated critical points (corresponding to a local maximum or minimum) on the z-axis itself.

(b) The following fibre bundles over the sphere S^k were introduced in Part II, §24.3, Example (c):

(i) $SO(n+1) \xrightarrow{p} S^n$ (with fibre $SO(n)$);
(ii) $U(n) \xrightarrow{p} S^{2n-1}$ (with fibre $U(n-1)$);
(iii) $Sp(n) \xrightarrow{p} S^{4n-1}$ (with fibre $Sp(n-1)$).

On each of the spheres S^n, S^{2n-1}, S^{4n-1} take a function $g(x)$ satisfying $0 \le g(x) \le 1$, and having exactly one minimum point x_0 (with $g(x_0) = 0$) and one maximum point x_1 (with $g(x_1) = 1$). By composing with the appropriate projection p we shall then obtain a function

$$f(x) = g(p(x))$$

on each of the corresponding total spaces $SO(n+1)$, $U(n)$, $Sp(n)$ with, in each case, exactly two critical manifolds $W_0 = p^{-1}(x_0)$ and $W_1 = p^{-1}(x_1)$ having respective indices $\lambda_0 = 0$, corresponding to the absolute minimum value 0, and $\lambda_1 = n$, $2n-1$ or $4n-1$ (according to the bundle (i), (ii) or (iii) in question), corresponding to the absolute maximum 1. Now from Theorem 20.2 and the definition of the relative homology groups, we have, noting that $W_0 \cong SO(n) \cong W_1$ and taking $M = SO(n+1)$,

$$b_{j-n}(SO(n)) = b_j(M_1, M_{1/2}) \ge b_j(M_1) - b_j(M_{1/2}). \tag{6}$$

Since $M_1 = M = SO(n+1)$, and by Lemmas 20.3, 20.4,

$$b_j(M_{1/2}) = b_j(U^-(W_0), \partial U^-(W_0)) = b_j(W_0) = b_j(SO(n)),$$

we infer from (6) that

$$b_{j-n}(SO(n)) \ge b_j(SO(n+1)) - b_j(SO(n)),$$

or

$$b_j(SO(n+1)) \le b_{j-n}(SO(n)) + b_j(SO(n)). \tag{7}$$

One shows similarly that

$$b_j(U(n)) \le b_{j-(2n-1)}(U(n-1)) + b_j(U(n-1)), \tag{8}$$

$$b_j(Sp(n)) \le b_{j-(4n-1)}(Sp(n-1)) + b_j(Sp(n-1)). \tag{9}$$

Since in each case we have the appropriate orientability (verify this!) it follows, as noted above, that the three inequalities (7), (8), (9) are valid not only over $\mathbb{Z}_2$ but also over $\mathbb{Z}$ and $\mathbb{R}$.

EXERCISE

1. Prove that one actually has equality in (7), (8), (9) when $j < n$ in the case of $SO(n+1)$, $j < 2n-1$ for $U(n)$, and $j < 4n-1$ for $Sp(n)$.

From the detailed knowledge of the structure of the ring $H^*(U(n); \mathbb{R})$ obtained in §7 (see Example (β)) it follows that in fact equality holds in (8) for all j, and a similar approach shows that this is the case also for $Sp(n)$, i.e. (9)

holds for all j. It is a more difficult problem to show that for $SO(n)$ one always has equality in (7) when the coefficients are from $\mathbb{Z}_2$, and, for odd n, over $\mathbb{R}$.

(c) Let M be a Riemannian manifold realized as a homogeneous space of a Lie group G of isometries of M, with stationary group H: $M \cong G/H$ (see §1 above, or Part II, §5.1). Consider the function f on M given by $f(x) = \rho^2(x, x_0)$, where $\rho(x, y)$ is the distance function on M determined by the given Riemannian metric, and x_0 is a point fixed by H, i.e. $Hx_0 = x_0$ (recall that H is defined to be the set of all $g \in G$ fixing some point $x_0 \in M$). Clearly the function f is H-invariant: $f(Hx) = f(x)$ for all $x \in M$.

Exercise

2. Investigate the critical manifolds of the function $f(x) = \rho^2(x, 1)$ in the case $M = SO(n)$ (and also $U(n)$, $Sp(n)$) where the Lie group in question is $G = SO(n) \times SO(n)$ acting on $SO(n)$, endowed with a two-sided invariant metric, as follows:
$$G\colon x \mapsto g_1 x g_2^{-1}, \qquad x \in SO(n), \qquad (g_1, g_2) \in G.$$
Here the stationary subgroup H is the diagonal subgroup:
$$H = \{(g, g) | g \in SO(n)\} \simeq SO(n),$$
which leaves f invariant since $g1g^{-1} = 1$. (Note that here H-invariance of f is equivalent to invariance under the inner automorphisms of $SO(n)$: $f(gxg^{-1}) = f(x)$.)

(d) Let G be a Lie group and $T\colon G \to GL(n, \mathbb{R})$ a real matrix representation. The *character* of the representation T is then by definition the function f given by
$$f(x) = \chi_T(x) = \operatorname{tr}(Tx), \qquad x \in G.$$
The character $f = \chi_T$ affords a further example of a function invariant under the inner automorphisms of G, since
$$\operatorname{tr}(Tx) = \operatorname{tr}(T(gxg^{-1})).$$

Exercise

3. Investigate the critical manifolds of the character $f = \chi_T$ in the case where $G = SO(n)$ (and also $U(n)$, $SU(n)$, $Sp(n)$) and T is an irreducible representation, paying particular attention to the cases $G = SO(3)$, $SO(4)$, $SU(3)$. (Note that the non-trivial, real, irreducible representations of $SO(2)$ are all of degree 2 of the form
$$T_n(\varphi) = \begin{pmatrix} \cos(n\varphi) & \sin(n\varphi) \\ -\sin(n\varphi) & \cos(n\varphi) \end{pmatrix},$$
with corresponding character
$$f_n(\varphi) = \chi_T(\varphi) = 2 \cos n\varphi.)$$

In attempting Exercises 2 and 3 the appropriate first step is to characterize the orbits of the respective groups of inner automorphisms (of $SO(n)$, $U(n)$,

$SU(n)$, $Sp(n)$). For the groups $SO(3)$ and $SU(2) \simeq Sp(1)$ we have the following simple situation:

EXERCISE

4. Prove that every non-central orbit of $SU(2) \simeq Sp(1)$, under the action of its group of inner automorphisms, is diffeomorphic to the sphere S^2. (The centre of $SU(2) \simeq Sp(1)$ is $\{1, -1\}$; the orbits of these elements are of course singletons.)

Consider the case $G = U(n)$. It is easy to verify (using the Jordan canonical form of a matrix) that every matrix A in $U(n)$ is conjugate to a diagonal matrix; in fact, this conjugation can be realized using a matrix in $U(n)$ itself: $A \to gAg^{-1}$ where gAg^{-1} is diagonal, for some $g \in U(n)$. Now the properties of a diagonal matrix A are of course determined by its diagonal entries, i.e. eigenvalues, taken according to their multiplicities. We may group these eigenvalues into blocks as shown, where each λ_j has modulus 1, $\lambda_j = \exp(2\pi i \varphi_j)$, and has multiplicity l_j (whence $l_1 + \cdots + l_k = n$):

$$A = \left[\begin{array}{cccc} \begin{array}{|ccc|}\hline \lambda_1 & & \\ & \ddots & \\ & & \lambda_1 \\ \hline\end{array} & 0 & & \\ 0 & \begin{array}{|ccc|}\hline \lambda_2 & & \\ & \ddots & \\ & & \lambda_2 \\ \hline\end{array} & & 0 \\ & & \ddots & \\ 0 & & & \begin{array}{|ccc|}\hline \lambda_k & & \\ & \ddots & \\ & & \lambda_k \\ \hline\end{array} \end{array}\right] \tag{10}$$

EXERCISE

5. Prove that the orbit $\{gAg^{-1} | g \in G\}$ of a matrix of the form (10) is diffeomorphic to $U(n)/(U(l_1) \times \cdots \times U(l_k))$, where $U(l_1), \ldots, U(l_k)$ are the obvious subgroups of $U(n)$.

An orbit containing a matrix A of the form (10) with all of its n eigenvalues distinct, may be thought of as being "in general position" among all orbits. In this case $U(l_j) \cong U(1) \cong S^1$, and by Exercise 5 the orbit is diffeomorphic to

$$U(n)/(U(1) \times \cdots \times U(1)) = U(n)/T^n,$$

where the $U(1)$'s are the appropriate subgroups of $U(n)$.

In each case ($G = SO(n)$, $U(n)$, $SU(n)$, $Sp(n)$) we obtain from a representation $T: G \to GL(m, \mathbb{R})$ a function $f = \chi_T$ on the manifold $M = G$ invariant under the action of G by its inner automorphisms. However, essentially since there will in general be orbits of this action which are not diffeomorphic to one another, the orbit space M/G, with points the conjugacy classes of G, will not be a manifold. Thus although $f(x)$ can be expressed in the form $\varphi(p(x))$ where φ is some function on M/G (and $p: M \to M/G$ is the natural projection), Morse theory, as expounded in the preceding sections, cannot in general be applied to study M/G.

(e) The "crystallographic groups" (see Part I, §20) afford an interesting class of examples of the type just mentioned, however with the group acting instead discretely on the manifold. Let G_n denote the group of all proper isometries of n-dimensional Euclidean space $\mathbb{R}^n$. It is easy to verify that G_n is a semi-direct product of the translation subgroup (which we may identify with $\mathbb{R}^n$) and the rotation subgroup $SO(n) \simeq G_n/\mathbb{R}^n$ (cf. Part I, §4). Let K be any discrete subgroup of G_n (thus for $n = 3$, K is precisely an (orientation-preserving) crystallographic group, by definition). According to a well-known theorem (cf. the first remark in §20 of Part I) the translations contained in K form a normal subgroup $N = K \cap \mathbb{R}^n$ with finite quotient group $D_K = K/N$, whose elements are in natural one-to-one correspondence with the rotations ρ (about the origin) with the property that $\rho\tau \in K$ for some translation $\tau \in \mathbb{R}^n$. As a discrete subgroup of $\mathbb{R}^n$, the group N must be free abelian of rank $\leq n$, and if rank $N = n$, then the quotient group $\mathbb{R}^n/N$ is identifiable with the n-dimensional torus T^n on which the finite group D_K acts by conjugation:

$$g(x) = gxg^{-1}, \qquad x \in T^n = \mathbb{R}^n/N, \quad g \in D_K = K/N. \tag{11}$$

Hence any smooth function $\tilde{f}$ on $\mathbb{R}^n$ invariant under the action of such a "crystallographic" group K on $\mathbb{R}^n$ (which action may be identified with conjugation of the translation subgroup $\mathbb{R}^n$) gives rise to a smooth function f on the torus $T^n = \mathbb{R}^n/N$, invariant under the action (11) of the finite group D_K on T^n. If f is a Morse function (or at least, as in §16, has only finitely many bifurcation points) then one can, in this sort of situation, often obtain an improvement of the Morse inequalities (see §16) involving submanifolds made up of fixed points under the action of individual elements of D_K. The key to this improvement, in the more general setting of a finite group D acting on a compact Riemannian manifold M^n, is provided by the following

20.5. Lemma. *Let D be a finite group of isometries of a compact Riemannian manifold M^n, let $d \in D$, $d \neq 1$, be a particular non-trivial element of D, with the property that one of the connected components of the set of fixed points of M^n under the isometry d is a submanifold $W_d \subset M^n$, and let f be a D-invariant smooth function on M^n. Then any point $x_0 \in W_d$ which is a critical point of the restriction of f to W_d, will also be a critical point of the unrestricted function f on M^n.*

PROOF. As before, by using the metric on M^n to "raise the index" (see Part I, §19.1), we obtain from the covector field grad $f(x)$ a vector field $\xi(x)$ on M^n. The invariance of the function f and the metric under the transformation $d \in D$ implies that for each $x \in M^n$, the induced map between the tangent spaces at

x and $d(x)$, sends $\xi(x)$ to $\xi(d(x))$, where $\xi(d(x))$ is the vector determined by grad $f(d(x))$. Let x now be a point of W_d and resolve $\xi = \xi(x)$ into a sum:

$$\xi = \xi_1 + \xi_2,$$

where ξ_1 is tangential to W_d, and ξ_2 normal to W_d. Since $d(x) = x$, we have $\xi \mapsto \xi$ under the self-map of the tangent space at x induced by d, and since W_d consists entirely of fixed points of d, clearly also $\xi_1 \mapsto \xi_1$, whence also $\xi_2 \mapsto \xi_2$. Now if $\xi_2 \neq 0$, then there will be in some neighbourhood of x a unique geodesic through x with tangent vector ξ_2 at x (see Part I, §29.2), which since x and ξ_2 are preserved by d, and d is an isometry, will likewise be pointwise preserved by d, contradicting the hypothesis that W_d represent a full connected component of the set of fixed points of M^n under d. Hence $\xi_2 = 0$, and the field $\xi(x)$ is tangential to W_d at each of its points. The lemma now follows easily. □

From this lemma it follows in particular that an isolated fixed point of an isometry $d \in D$ of M^n, will be a critical point of every D-invariant smooth function on M^n.

Given a specific manifold M^n and finite group D of isometries, exploitation of the above lemma to obtain improvements of the Morse inequalities for appropriate D-invariant functions f on M^n, requires knowledge of the fixed manifolds W_d of the elements $d \in D$, of their relative situation (for various d) in M^n, and of the inclusion homomorphisms from their homology groups to the corresponding homology groups of M^n. From the Morse inequalities (see Theorem 16.1), as applied to the restriction $f|_{W_d}$ of a Morse function f on M^n, we have

$$\mu_k(f|_{W_d}) \geq b_k(W_d).$$

Note, however, that a critical point on W_d may have different indices according as it is regarded as a critical point of W_d or M^n. Even in the case where D is cyclic (of order m say) with generator d, the Morse inequalities for a D-invariant Morse function f on M^n may be improved upon, provided one has some knowledge of the inclusions

$$W_d \subset W_{d^2} \subset \cdots \subset M^n = W_{d^m}, \qquad d^m = 1.$$

Consider, by way of illustration, the special case where $m = 2$, i.e. $D \simeq \mathbb{Z}_2$, and the manifold W_d has dimension $n - 1$ and separates the manifold M^n into two diffeomorphic parts:

$$M^n = M_1 \cup M_2, \qquad \partial M_1 = \partial M_2 = W_d = W,$$

with d acting as follows:

$$d\colon M_1 \to M_2,\ M_2 \to M_1; \qquad d|_{\partial M_1} = 1 = d|_{\partial M_2}.$$

In terms of the exact homology sequence of the pair (M_1, W) (see Theorem 5.6):

$$\xrightarrow{j} H_{q+1}(M_1, W) \xrightarrow{\partial_*} H_q(W) \xrightarrow{i_*} H_q(M_1) \xrightarrow{j} H_q(M_1, W) \xrightarrow{\partial_*},$$

we define

$$\hat{b}_k(M_1, W) = b_k(M_1) + \operatorname{rank}(H_k(M_1, W)/\operatorname{Im} j).$$

EXERCISE

6. Given a Morse function f on M^n, prove that the number of critical points of index k of $f|_{M_1}$ is at least $\hat{b}_k(M_1, W)$.

§21. Critical Points of Functionals and the Topology of the Path Space $\Omega(M)$

For "infinite-dimensional smooth manifolds" there are natural analogues of the Morse and Lyusternik–Shnirelman theories expounded above. On such a "manifold" M^∞ one can, as in the finite-dimensional case, consider functions $F(\gamma)$, $\gamma \in M^\infty$ (usually, however, called "functionals" in the infinite-dimensional context). The concept of a "critical point" extends naturally and readily to such functionals $F(\gamma)$ on M^∞; however, the appropriate extension of the notion of the "index" of a critical point requires some preliminary groundwork. An example of an infinite-dimensional manifold is afforded by the space $\Omega(M, p, q)$ of (parametrizations of) piecewise-smooth paths in a finite-dimensional manifold $M = M^n$, joining the point p to the point q; and in fact rather than develop here the theory of infinite-dimensional manifolds, we shall confine our attention to this particular example.

Thus here p and q are of course particular fixed points of M, and a (parametrized) piecewise-smooth path (i.e. a typical element of $\Omega(M, p, q)$) is a continuous map γ: $[0, 1] \to M$ such that $\gamma(0) = p$, $\gamma(1) = q$, and for some finite subdivision

$$0 = t_0 < t_1 < \cdots < t_k = 1$$

of the interval $[0, 1]$, the restriction of γ to each closed interval $[t_i, t_{i+1}]$ is a smooth map $[t_i, t_{i+1}] \to M$ (of whatever smoothness degree needed in the sequel), $i = 0, \ldots, k - 1$. (The piecewise-smoothness condition, rather than simply smoothness, turns out, as we shall see (in §22), to be the technically convenient hypothesis under which to establish the homotopic equivalence of Ω to a union of cells, analogously to the result in the finite-dimensional case (Theorem 15.4) to the effect that a finite-dimensional, compact, smooth manifold is homotopically equivalent to a finite cell complex.) One may take the "compact-open" topology (see Example (γ) of §7) as the topology on $\Omega(M, p, q)$, or make it into a metric space by defining a natural distance function between paths in terms of some Riemannian metric on M (see §22 below); however it is the additional structure, by virtue of which the space Ω qualifies as a smooth infinite-dimensional manifold, that will be exploited (implicitly) in what follows. With each point $\gamma \in \Omega$ we associate in the following way a certain infinite-dimensional vector space $T_\gamma\Omega$, the natural candidate for the title of "tangent space" to Ω at the "point" γ:

21.1. Definition. *The tangent space* to the path space $\Omega = \Omega(M, p, q)$ at a point γ, is the vector space of all piecewise-smooth vector fields v along the path γ in M, satisfying $v(0) = 0$, $v(1) = 0$. A *variation with parameter* u, $-\varepsilon \le u \le \varepsilon$ (where $\varepsilon > 0$ is small), *of a path* $\gamma \in \Omega$, leaving the end-points p and q of γ fixed, is a map $\tilde{\alpha}$: $[-\varepsilon, \varepsilon] \to \Omega$, such that $\tilde{\alpha}(0) = \gamma$, and for some subdivision

$$0 = t_0 < t_1 < \cdots < t_k = 1$$

of $[0, 1]$, the map $\alpha(u, t)$ defined by $\alpha(u, t) = \tilde{\alpha}(u)(t)$, is a smooth map of each rectangle $-\varepsilon \le u \le \varepsilon, t_i \le t \le t_{i+1}$, to M (and satisfies $\alpha(u, 0) = p, \alpha(u, 1) = q$); it is also required that $\alpha(u, t)$ be continuous on the whole rectangle $-\varepsilon \le u \le \varepsilon$, $0 \le t \le 1$, and that for each fixed u the partial derivative $(\partial\alpha/\partial u)(u, t)$ be piecewise-smooth on $[0, 1]$ (see Figure 96).

Since a variation $\tilde{\alpha}$, as just defined, associates with each u, $-\varepsilon \le u \le \varepsilon$, a point of Ω (i.e. a piecewise-smooth path in M), namely $\tilde{\alpha}(u)$, we may clearly consider $\tilde{\alpha}$ as a path, or trajectory, in the space Ω itself, parametrized by u (see Figure 97). The *velocity* or *tangent vector to the trajectory* $\tilde{\alpha}(u)$, $-\varepsilon \le u \le \varepsilon$, at the point $\gamma = \tilde{\alpha}(0)$, is then by definition the vector field

$$v = \frac{\partial\alpha}{\partial u}(0, t), \qquad (\alpha(u, t) = \tilde{\alpha}(u)(t), \tilde{\alpha}(0) = \gamma), \tag{1}$$

along $\gamma = \gamma(t)$, which by definition of a variation (see above) is piecewise-smooth and therefore belongs to the tangent space $T_\gamma\Omega$. It is not difficult to verify the converse of this: given an arbitrary piecewise-smooth vector field $v \in T_\gamma\Omega$ (i.e. a field $v(t)$ along $\gamma(t)$) vanishing at $t = 0, 1$, there exists a trajectory $\tilde{\alpha}$ in Ω such that $(\partial/\partial u)\alpha(0, t) = v(t)$, $0 \le t \le 1$. In the conventional notation of the calculus of variations, one writes $\delta\gamma$ for the field $v(t)$ given by (1).

Now let F be a real-valued function on Ω, and as above let $\gamma \in \Omega$, let $\tilde{\alpha}$ be a variation of γ, and write $v = \delta\gamma$. Consider the partial derivative $(\partial/\partial u)F(\tilde{\alpha}(u))|_{u=0}$, assuming it exists. (For the specific examples of functionals that we shall be dealing with below, the existence of this derivative will be obvious.) Observe that the derivative $(\partial/\partial u)F(\tilde{\alpha}(u))$ is closely analogous to the directional derivative of a smooth function f on a finite-dimensional manifold

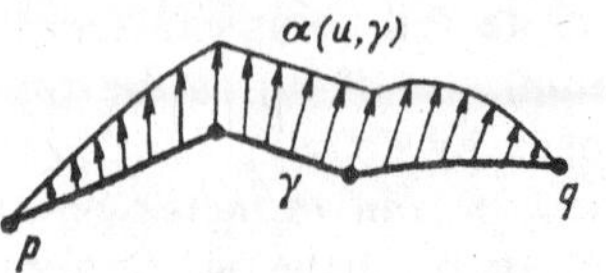

Figure 96

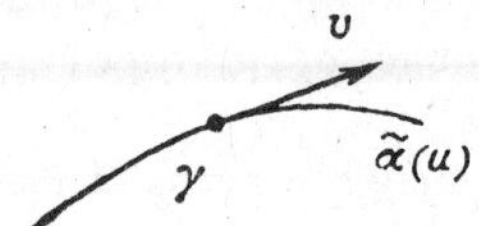

Figure 97

N in the direction of the tangent vector to a curve in N. Pursuing this analogy further, we define a path $\gamma_0 \in \Omega$ to be *critical* for $F = F(\gamma)$, if

$$\frac{\partial}{\partial u} F(\tilde{\alpha}(u))|_{u=0} \equiv 0$$

for every variation $\tilde{\alpha}$ of the path γ_0 (or, as one might say, if the "variational derivative $\delta F/\delta \gamma$" vanishes; cf. for instance, Part I, §31.1).

Our primary interest is in the following specific functionals on Ω (cf. Part I, §31.1, Examples (a) and (b)):

$$\text{the "action" of a path:}\quad E(\gamma) = \int_0^1 \left|\frac{d\gamma}{dt}\right|^2 dt;$$

$$\text{the length of a path:}\quad L(\gamma) = \int_0^1 \left|\frac{d\gamma}{dt}\right| dt.$$

(Here the norm $|\ |$ is that determined by some Riemannian metric on M.) The functionals L and E enjoy the obvious relationship $L^2 \le E$, with equality occurring precisely if $|\dot{\gamma}| = \text{const.}$, i.e. if the parameter t (parametrizing $\gamma = \gamma(t)$) is proportional to arc-length, i.e. is "natural" (cf. Part I, §31.2(c). (Note that in general, although the functional L is parameter-independent, the value of E at γ depends on the parametrization of γ.)

We now recall from Part I various facts we shall need concerning the variational derivative of the functional $E(\gamma)$ on Ω. As above we shall denote by $\tilde{\alpha} = \tilde{\alpha}(u)$ a variation of a path γ, and by $v = \delta\gamma$ the vector field on γ representing the velocity vector of the trajectory $\tilde{\alpha}(u)$ in Ω; recall also the notation $\dot{\gamma}(t)$ for the tangent vector to γ at $\gamma(t)$, $\nabla_{\dot{\gamma}}(\dot{\gamma})$ for the directional covariant derivative of $\dot{\gamma}$ in the direction of the path γ (see Part I, §29.1), and $\Delta\dot{\gamma}(t) = \dot{\gamma}(t^+) - \dot{\gamma}(t^-)$ for the "jump" in the velocity vector $\dot{\gamma}$ at time t. It follows by direct calculation, essentially as in Part I, §31.2, Example (b), that at those t where $\gamma(t)$ is smooth

$$\frac{1}{2}\frac{\delta E}{\delta \gamma^l} = -(\ddot{\gamma}^k + \Gamma^k_{ij}x^i x^j)g_{kl} = -\nabla_{\dot{\gamma}}(\dot{\gamma}^k)g_{kl},$$

where (Γ^k_{ij}) is the connexion compatible with the Riemannian metric (g_{kl}) on M (see Part I, §29.3). Hence essentially from the "formula for the first variation" (see formula (11) of Part I, §31.1), taking account of the finitely many points t_i where $\gamma(t)$ is not smooth, we deduce the following

21.2. Theorem. *For the functional E above and any variation $\tilde{\alpha} = \tilde{\alpha}(u)$ of $\gamma \in \Omega$, we have*

$$\frac{1}{2}\frac{\partial}{\partial u} E(\tilde{\alpha}(u))|_{u=0} = -\sum_i \langle v(t_i), \Delta\dot{\gamma}(t)\rangle - \int_0^1 \langle v(t), a(t)\rangle\, dt, \tag{2}$$

where $v(t)$ is as in (1), $a(t) = \nabla_{\dot{\gamma}}\dot{\gamma}$, the directional covariant derivative, the summation is over those t_i at which $\gamma(t)$ fails to be smooth (there being only finitely *many such t_i in view of the piecewise-smoothness of γ), and $\langle\ ,\ \rangle$ denotes the scalar product determined by the metric on M.*

As a corollary we have the following

21.3. Theorem (cf. Theorem 31.2.1 of Part I). *A path $\gamma_0 \in \Omega$ is a critical point of the functional $E(\gamma)$ precisely if γ_0 is a smooth geodesic (parametrized by the natural parameter $t = (\text{arc-length})/L(\gamma_0)$).*

PROOF. If $\gamma_0(t)$ is a geodesic then, essentially by definition (see Part I, §29.2), we have $\Delta\dot{\gamma}(t) \equiv 0$, $a(t) \equiv 0$, whence by (2)

$$\frac{\partial}{\partial u} E(\tilde{\alpha}(u))|_{u=0} \equiv 0. \tag{3}$$

Suppose, on the other hand, that (3) holds for every variation $\tilde{\alpha}(u)$ of the path $\gamma_0(t)$. Let $g(t)$ be any (smooth) function such that $g(t) \geq 0$ and $g(t) = 0$ precisely at those points $t_i \in [0, 1]$ where $\Delta\dot{\gamma}(t_i) \neq 0$, and consider the vector field $v(t) = g(t)a(t)$. As noted above there is a variation $\tilde{\alpha}(u)$ with $(\partial/\partial u)\alpha(0, t) = v(t)$, and for this variation we shall then have

$$0 = \frac{\partial}{\partial u} E(\tilde{\alpha}(u))|_{u=0} = -\int_0^1 \langle a(t), a(t)\rangle g(t)\, dt,$$

whence $a(t) = \nabla_{\dot{\gamma}_0}(\dot{\gamma}_0) \equiv 0$ on each smooth segment of the path $\gamma_0(t)$, i.e. each smooth segment of γ_0 is a geodesic arc (parametrized by a natural parameter; cf. Part I, §31.2).

Suppose now that γ_0 fails to be smooth at points t_i. Choosing a variation $\tilde{\alpha}(u)$ such that $v(t_i) = \Delta\dot{\gamma}_0(t_i)$, we shall then have

$$0 = \frac{1}{2}\frac{\partial}{\partial u} E(\tilde{\alpha}(u))|_{u=0} = -\sum_i \langle \Delta\dot{\gamma}(t_i), \Delta\dot{\gamma}(t_i)\rangle,$$

whence $\Delta\dot{\gamma}_0(t_i) = 0$ for each i. Hence $\gamma_0(t)$ is smooth for all $t \in [0, 1]$, and the proof is complete. □

We now recall from Part I the "formula for the second variation" as it applies to the functional E. We first define a *two-parameter variation* of a path $\gamma \in \Omega$: this is a map $\alpha\colon U \times [0, 1] \to M$ (satisfying conditions similar to those required of a one-parameter variation), where U is an open neighbourhood of the origin $(0, 0)$ of the parameter space $\mathbb{R}^2(u_1, u_2)$, and $\alpha(0, 0, t) = \gamma(t)$, $0 \leq t \leq 1$. Analogously to the one-parameter case we write $\tilde{\alpha}(u_1, u_2)(t) = \alpha(u_1, u_2, t)$; this notation makes it evident that the two-parameter variation $\tilde{\alpha}$ is a parametrized piece of surface in the space Ω. Under the appropriate conditions the two vector fields along γ defined by

$$v_1(t) = \frac{\partial\alpha}{\partial u_1}(0, 0, t), \qquad v_2(t) = \frac{\partial\alpha}{\partial u_2}(0, 0, t), \tag{4}$$

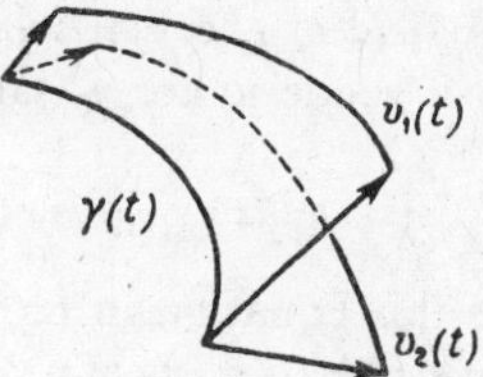

Figure 98

are piecewise-smooth and therefore belong to the tangent space $T_\gamma\Omega$; and it is not difficult to verify that conversely given any two piecewise-smooth vector fields v_1 and v_2 along γ, there exists a two-parameter variation α of γ satisfying (4) (see Figure 98).

The *Hessian of a functional* F on Ω, relative to a pair v_1, v_2 of piecewise-smooth vector fields on γ, is defined by

$$d^2F(v_1, v_2) = \left.\frac{\partial^2 F(\tilde{\alpha}(u_1, u_2))}{\partial u_1 \partial u_2}\right|_{u_1=u_2=0}, \tag{5}$$

where $\tilde{\alpha}(u_1, u_2)$ is a two-parameter variation of γ satisfying (4). The "formula for the second variation" of the functional E, established in Part I (see the theorem and exercise of §36.1 of Part I), expresses the Hessian in terms of v_1, v_2, certain directional covariant derivatives along γ, and the Riemann curvature tensor of M, in the case where $\gamma = \gamma_0$ is a smooth geodesic:

21.4. Theorem (cf. Part I, §36.1). *Let the path $\gamma_0 \in \Omega$ be a critical point of the functional $E(\gamma)$ (i.e. a smooth geodesic) and $\tilde{\alpha}(u_1, u_2)$ a two-parameter variation of γ_0; write $v_i = (\partial\tilde{\alpha}/\partial u_i)(0, 0)$, $i = 1, 2$. Then*

$$\begin{aligned}\frac{1}{2}\frac{\partial^2 E(\tilde{\alpha})}{\partial u_1 \partial u_2}(0, 0) = &-\sum_i \langle v_2(t_i), \Delta(\nabla_{\dot{\gamma}_0} v_1(t_i))\rangle \\ &- \int_0^1 \langle v_2(t), \nabla_{\dot{\gamma}_0}\nabla_{\dot{\gamma}_0} v_1(t) + R(\dot{\gamma}_0, v_1)\dot{\gamma}_0\rangle\, dt,\end{aligned} \tag{6}$$

where the $\Delta(\nabla_{\dot{\gamma}_0} v_1(t_i)) = \nabla_{\dot{\gamma}_0} v_1(t_i^+) - \nabla_{\dot{\gamma}_0} v_1(t_i^-)$ are the "jumps" in the directional covariant derivative $\nabla_{\dot{\gamma}_0} v_1(t)$ at its points of discontinuity t_i, $R = (R^i_{jkl})$ is the Riemann curvature tensor of M (see Part I, §30.1), and $\langle\ ,\ \rangle$ denotes the scalar product determined by the metric on M.

Since by Theorem 21.3 the path γ_0 is smooth, we may if need be confine ourselves to 2-parameter variations $\tilde{\alpha}$ of γ_0 for which the vector fields $v_1(t)$ and $v_2(t)$ are smooth (i.e. "unbroken"). With this restriction on $\tilde{\alpha}$ the first term in the formula (6) drops out leaving

$$\frac{1}{2}\frac{\partial^2 E(\tilde{\alpha})}{\partial u_1 \partial u_2}(0, 0) = -\int_0^1 \langle v_2, \nabla_{\dot{\gamma}_0}\nabla_{\dot{\gamma}_0} v_1 + R(\dot{\gamma}_0, v_1)\dot{\gamma}_0\rangle\, dt. \tag{7}$$

We next define a *Jacobi field of the functional* $E(\gamma)$ (cf. Part I, §36.2) to be a vector field $v(t)$ along a geodesic arc γ_0 satisfying Jacobi's differential equation

$$(\nabla_{\dot{\gamma}_0})^2 v + R(\dot{\gamma}_0, v)\dot{\gamma}_0 = 0. \tag{8}$$

It is convenient to re-write this equation in terms of the co-ordinates in a neighbourhood of γ_0 determined by a basis of n vector fields $e_1(t), \ldots, e_n(t)$ on γ_0, parallel along γ_0 (i.e. satisfying $\nabla_{\dot{\gamma}_0} e_i(t) \equiv 0$; see Part I, §29.1), and orthonormal at each point $\gamma_0(t)$ of γ_0. Writing $v(t) = v^i e_i(t)$, equation (8), rewritten in terms of the $v^i = v^i(t)$, becomes the system

$$\frac{d^2 v^i}{dt^2} + \sum_{j=1}^{n} R_j^i(t) v^j(t) = 0, \qquad \text{where} \quad R_j^i(t) = \langle R(\dot{\gamma}_0, e_j)\dot{\gamma}_0, e_i \rangle. \tag{9}$$

(This can be seen as follows: at each time t, using the orthonormality of the basis $e_i(t)$, we have

$$\begin{aligned} \langle R(\dot{\gamma}_0, e_j)\dot{\gamma}_0, e_i \rangle v^j &= \langle (R^r_{pkq} \dot{\gamma}_0^p \dot{\gamma}_0^k \delta_j^q), \delta_i^s \rangle v^j \\ &= g_{rs} R^r_{pkq} \dot{\gamma}_0^p \dot{\gamma}_0^k \delta_j^q \delta_i^s v^j = \delta_{rs} R^r_{pkq} \dot{\gamma}_0^p \dot{\gamma}_0^k \delta_j^q \delta_i^s v^j \\ &= R^i_{pkj} \dot{\gamma}_0^p \dot{\gamma}_0^k v^j = [R(\dot{\gamma}_0, v)\dot{\gamma}_0]^i.) \end{aligned}$$

It now becomes clear that a Jacobi field $v(t)$, as a solution of that system, is determined uniquely by the initial data $v(0)$ and $\nabla_{\dot{\gamma}_0} v(0) \in T_{\gamma_0(0)}(M^n)$.

We recall, finally, the definition of a *pair of conjugate points of a geodesic* γ_0: these are points A, B of γ_0 such that there exists some non-zero Jacobi field v along the arc of γ_0 from A to B with $v|_A = v|_B = 0$ (see Part I, §36.2). By the *multiplicity* of a pair of conjugate points A, B of γ_0, we shall mean the dimension of the linear space consisting of all such Jacobi fields along the arc of γ_0 joining A and B.

Considering the Hessian $d^2E(v_1, v_2)$ (see (5)) as a symmetric bilinear form on pairs (v_1, v_2) of piecewise-smooth fields along γ_0 vanishing at the end-points of γ_0, i.e. on the tangent space $T_{\gamma_0}\Omega$, we denote by $W_{\gamma_0} \subset T_{\gamma_0}\Omega$ the "kernel" (or "nullspace") of d^2E, i.e. the set of $v_1 \in T_{\gamma_0}\Omega$ such that $d^2E(v_1, v_2) \equiv 0$ for all $v_2 \in T_{\gamma_0}\Omega$. We then define the *degeneracy degree of the Hessian* d^2E at γ_0 to be $\dim W_{\gamma_0}$.

21.5. Theorem. *Let γ_0 be a (smooth) geodesic on M joining the point p to the point q. A piecewise-smooth vector field v along γ_0 belongs to the kernel W_{γ_0} of the Hessian d^2E of the functional $E(\gamma)$ if and only if v is a Jacobi field of E along γ_0 satisfying $v|_p = v|_q = 0$. (It follows that the kernel W_{γ_0} of the Hessian d^2E is non-zero if and only if the end-points p and q of the geodesic γ_0 form a pair of conjugate points of γ_0 (cf. Part I, Lemma 36.2.3), and furthermore that then the dimension of the vector subspace $W_{\gamma_0} \subset T_{\gamma_0}\Omega$, i.e. the degeneracy degree of the Hessian d^2E, coincides with the multiplicity of the pair of conjugate points p, q on γ_0.)*

PROOF. If v is a Jacobi field along γ_0 satisfying $v|_p = v|_q = 0$, then by definition v is smooth (see (9)), whence certainly $v \in T_{\gamma_0}\Omega$. Since the path γ_0 is smooth (by Theorem 21.3), it follows that $\Delta(\nabla_{\dot\gamma_0} v(t)) = 0$ for all $t \in [0, 1]$. In view of this together with the assumption that v is Jacobi and so satisfies (8), the formula for the second variation (6) yields

$$d^2E(v, \tilde{v}) = -\sum_i \langle \tilde{v}(t), 0\rangle - \int_0^1 \langle \tilde{v}(t), 0\rangle\, dt = 0,$$

for every field $\tilde{v} \in T_{\gamma_0}\Omega$. Hence $v \in W_{\gamma_0}$.

For the converse, suppose $v \in W_{\gamma_0}$. We wish to show that v is a Jacobi field along γ_0. Since v is piecewise-smooth there is a finite subdivision of the interval $[0, 1]$ by means of points

$$0 = t_0 < t_1 < \cdots < t_k = 1,$$

such that $v(t)$ is smooth on each subinterval $[t_{i-1}, t_i]$. Imitating the proof of Theorem 21.3, we now take any smooth function $f(t)$ on $[0, 1]$ vanishing at the points t_i, $i = 0, \ldots, k$, and positive at all other points, and consider the field

$$q(t) = f(t)(\nabla^2_{\dot\gamma_0} v + R(\dot\gamma_0, v)\dot\gamma_0).$$

Taking α to be a 2-parameter variation of γ_0 with $v_1(t) = v(t)$, $v_2(t) = q(t)$ (see (4)), from the formula for the second variation and the fact that $v \in W_{\gamma_0}$, we obtain

$$\begin{aligned} \frac{1}{2}\frac{\partial^2 E(\tilde{\alpha})}{\partial u_1 \partial u_2}(0, 0) &= \tfrac{1}{2} d^2E(v, q) \\ &= -\sum_i \langle q(t_i), \Delta(\nabla_{\dot\gamma_0} v(t_i))\rangle \\ &\quad - \int_0^1 f \cdot |\nabla^2_{\dot\gamma_0} v + R(\dot\gamma_0, v)\dot\gamma_0|^2\, dt = 0. \end{aligned} \tag{10}$$

Since $q(t_i) = 0$ for $i = 0, \ldots, k$, the first of the two terms on the right-hand side of (10) is zero. Since $f(t) > 0$ for all $t \neq t_i$, it then follows that

$$(\nabla_{\dot\gamma_0})^2 v + R(\dot\gamma_0, v)\dot\gamma_0 \equiv 0$$

on each open interval (t_{i-1}, t_i), so that v is a Jacobi field on each of the corresponding open segments of γ_0.

To show that v is in fact a Jacobi field along the whole of γ_0, it suffices to prove that $\nabla_{\dot\gamma_0} v$ is continuous on $[0, 1]$ (i.e. has no "jumps" at the t_i). (That this suffices again follows from (10), however with f now taken to be a non-zero constant function.) Writing, as usual, $\Delta(\nabla_{\dot\gamma_0} v(t_i))$ for the "jump" in the value of $\nabla_{\dot\gamma_0} v(t)$ at t_i, and taking $g(t)$ to be any smooth vector field on γ_0 satisfying $g(t_i) = \Delta(\nabla_{\dot\gamma_0} v(t_i))$ (cf. proof of Theorem 21.3), we obtain from the formula for the second variation (6), invoking once again the assumption that v is in the kernel of the Hessian d^2E, that

$$\tfrac{1}{2}d^2E(v, g) = -\sum_{i=1}^{k-1} |\Delta(\nabla_{\dot{\gamma}_0}v(t_i))|^2 - \int_0^1 \langle g, (\nabla_{\dot{\gamma}_0})^2 v + R(\dot{\gamma}_0, v)\dot{\gamma}_0\rangle\, dt = 0.$$

Here the integral term is zero (in view of the already-established fact that v is Jacobi on the open intervals between the t_i), so that the summation term must also vanish. Hence $\Delta(\nabla_{\dot{\gamma}_0}v(t_i)) = 0$ for all i, and the proof is complete. □

Remark. Since the vector space of all Jacobi fields along γ_0 (vanishing at p and q) is finite-dimensional (each such Jacobi field v being uniquely determined, as noted earlier, by the initial data $v(0)$ ($= 0$), $\nabla_{\dot{\gamma}_0}v(0)$), it follows from this theorem that the kernel of the Hessian d^2E has finite dimension.

Amongst the different permissible variations (with or without fixed end-points) of a geodesic γ_0 one may distinguish the subclass of so-called *geodesic variations*: these are smooth maps α: $[-\varepsilon, \varepsilon] \times [0, 1] \to M$ such that $\alpha(0, t) = \gamma_0(t)$ and each path $\tilde{\alpha}(u)$ (defined as usual by $\tilde{\alpha}(u)(t) = \alpha(u, t)$) is again a geodesic (cf. Definition 21.1), i.e. variations where the perturbations of the geodesic γ_0 remain geodesics. The "velocity vectors" of such trajectories $\tilde{\alpha}$ in Ω turn out to be precisely the Jacobi fields on γ_0.

21.6. Proposition. *Given a geodesic variation $\tilde{\alpha} = \tilde{\alpha}(u)$ of a geodesic $\gamma_0 \in \Omega$, the velocity vector $(\partial\alpha/\partial u)(0, t)$, $0 \le t \le 1$, of $\tilde{\alpha}$ is a Jacobi field along γ_0. Conversely, given a Jacobi field $v = v(t)$ on γ_0, there exists a geodesic variation of γ_0 having v as its initial velocity vector.*

PROOF. In some sufficiently small neighbourhood of almost every point of M in the range of α, we may choose co-ordinates $x^1, \ldots, x^n$ with $x^1 = u$, $x^2 = t$. In terms of such co-ordinates we shall then have

$$\frac{\partial\alpha^1}{\partial x^1} = 1 = \frac{\partial\alpha^2}{\partial x^2}, \qquad \frac{\partial\alpha^2}{\partial x^1} = 0 = \frac{\partial\alpha^1}{\partial x^2},$$

whence it follows easily that the commutator $[\partial\alpha/\partial t, \partial\alpha/\partial u]$ of the vector fields $\partial\alpha/\partial u$, $\partial\alpha/\partial t$ (see Part I, §23.2) vanishes identically. From this and the formula of Lemma 30.1.3 of Part I (recalling also that the connexion defined by the Riemannian metric is symmetric) we obtain immediately

$$\nabla_{\partial\alpha/\partial u}\left(\nabla_{\partial\alpha/\partial t}\left(\frac{\partial\alpha}{\partial t}\right)\right) = \nabla_{\partial\alpha/\partial t}\left(\nabla_{\partial\alpha/\partial u}\left(\frac{\partial\alpha}{\partial t}\right)\right) + R\left(\frac{\partial\alpha}{\partial t}, \frac{\partial\alpha}{\partial u}\right)\frac{\partial\alpha}{\partial t},$$

$$\nabla_{\partial\alpha/\partial u}\left(\frac{\partial\alpha}{\partial t}\right) = \nabla_{\partial\alpha/\partial t}\left(\frac{\partial\alpha}{\partial u}\right),$$

whence

$$\nabla_{\partial\alpha/\partial u}\left(\nabla_{\partial\alpha/\partial t}\left(\frac{\partial\alpha}{\partial t}\right)\right) = (\nabla_{\partial\alpha/\partial t})^2\left(\frac{\partial\alpha}{\partial u}\right) + R\left(\frac{\partial\alpha}{\partial t}, \frac{\partial\alpha}{\partial u}\right)\frac{\partial\alpha}{\partial t}. \tag{11}$$

Since for each u the path $\tilde{\alpha}(u)$ is a geodesic, we have, by definition of a geodesic, $\nabla_{\partial\alpha/\partial t}(\partial\alpha/\partial t) \equiv 0$, whence by (11)

$$(\nabla_{\partial\alpha/\partial t})^2\left(\frac{\partial\alpha}{\partial u}\right) + R\left(\frac{\partial\alpha}{\partial t}, \frac{\partial\alpha}{\partial u}\right)\frac{\partial\alpha}{\partial t} = 0,$$

Hence in particular $(\partial\alpha/\partial u)(0, t)$ is a Jacobi field.

For the converse, consider to begin with points p', q' on γ_0, lying in a disc $D^n \subset M^n$ sufficiently small for there to exist, for each pair of points x, y of D^n, exactly one geodesic from x to y. (See [8] for verification that such discs exist around every point of M.) We shall show that given any tangent vectors a_0, b_0 to M at p', q' respectively, there exists a Jacobi field along the segment of γ_0 from p' to q' taking the value a_0 at p' and b_0 at q' (see Figure 99). To this end, let $a(u)$ be a smooth arc in $D^n \subset M$ beginning at p' with initial velocity vector $da/du|_{u=0} = a_0$, and similarly let $b(u)$ be a smooth arc in D^n beginning at q' with initial velocity vector b_0. The desired family of geodesics (constituting a one-parameter geodesic variation $\alpha = \alpha(u, t)$ with parameter u) is then obtained by joining, for each value of u, the point $a(u)$ of the first arc to the corresponding point $b(u)$ of the second arc, by means of a geodesic (unique by choice of D^n) (see Figure 100). Having constructed in this way a geodesic variation $\alpha = \alpha(u, t)$ of the arc of γ_0 from p' to q' satisfying $(\partial\alpha/\partial u)(0, 0) = a_0$, $(\partial\alpha/\partial u)(0, 1) = b_0$, we obtain the desired Jacobi field v on that arc by simply differentiating α partially with respect to u:

$$v(t) = \frac{\partial\alpha}{\partial u}(0, t).$$

That this field v is indeed Jacobi follows from the first part of the theorem, already established.

Every Jacobi field on an arc of γ_0 joining a pair of non-conjugate points of γ_0, is uniquely determined by its values at those end-points (since if there were two distinct Jacobi fields with the same boundary values, their difference

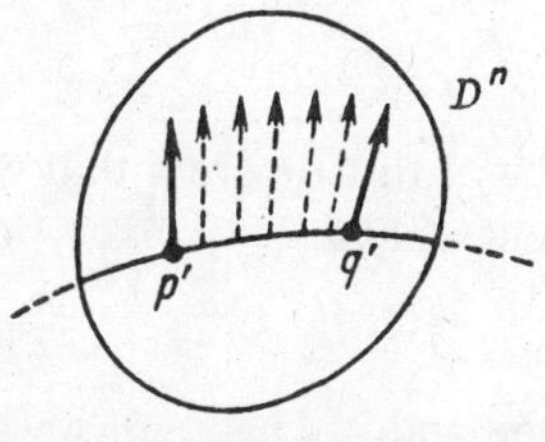

Figure 99

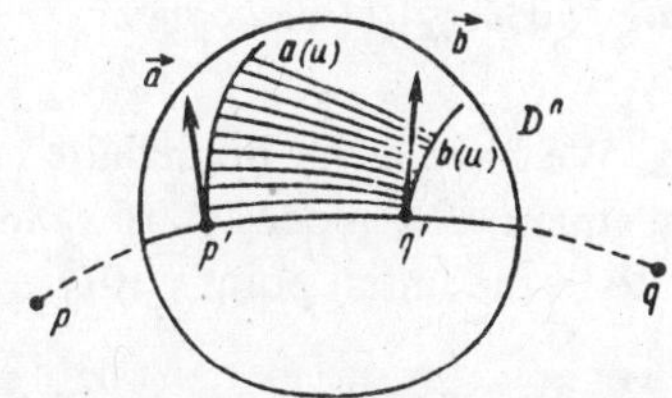

Figure 100

would furnish a non-zero Jacobi field vanishing at the end-points, contrary to the assumption that those points do not form a conjugate pair). Since p' and q' are not conjugate (essentially by the results of Part I, §36.2), we therefore have that every Jacobi field on the arc from p' to q', being uniquely determined by its values at p' and q', is obtainable via the above-described construction. (It also follows that the linear space of all Jacobi fields along the arc of γ_0 from p' to q', is isomorphic to the $2n$-dimensional vector space $T_{p'}(M^n) \times T_{q'}(M^n)$.)

Denoting by v any Jacobi field on γ_0, we now construct a geodesic variation of γ_0 with initial velocity vector v, as follows. Let p', q' be any points on γ_0 lying in a disc $D^n \subset M$ with the same property as before, and construct, as above, a geodesic variation β of the arc of γ_0 from p' to q', using as boundary values at p' and q' the vectors $a_0 = v|_{p'}$, $b_0 = v|_{q'}$. From the above-noted uniqueness of the Jacobi field on this arc with these boundary values, we infer that the initial velocity vector of this geodesic variation must coincide with the field v (restricted to that arc). The desired geodesic variation on the whole of γ_0 is then obtained by simply extending (smoothly) the geodesic arcs of the family β, beyond the disc D^n as far as necessary. □

Analogously to the index of a (non-degenerate) critical point of a function on a finite-dimensional manifold (as defined in §15 above, or Part II, §10.4), we define the *index of the Hessian* d^2E at a critical point γ_0, denoted by $\lambda = \lambda(\gamma_0)$, to be the largest dimension of a subspace of $T_{\gamma_0}\Omega$ on which d^2E, as a quadratic form on $T_{\gamma_0}\Omega$, is negative definite. (If the end-points p, q of the geodesic γ_0 do not form a conjugate pair, then by Theorem 21.5 the kernel of d^2E is null, so that, again by analogy with the finite-dimensional case, it is appropriate in this situation to call the point γ_0 a *non-degenerate* critical point of the functional E.)

In the following important result a connexion is established between the index of d^2E at γ_0 and the number of conjugate pairs of points (of a certain type) occurring on the geodesic γ_0.

21.7. Theorem (Index Theorem). *The index $\lambda = \lambda(\gamma_0)$ of the Hessian d^2E at a critical point γ_0 is finite, and coincides with the number of points of the form $\gamma_0(t)$, $0 < t < 1$ (i.e. in the interior of γ_0) conjugate to the initial point $p = \gamma_0(0)$, where each such point $\gamma_0(t)$ is counted according to the multiplicity of the conjugate pair p, $\gamma_0(t)$ (see above).*

Remark. We have as an immediate consequence of this theorem that every initial segment of a geodesic arc γ_0 in M contains only finitely many points conjugate to the initial point $\gamma_0(0)$.

Before embarking on the proof proper of the Index Theorem, we shall sketch an intuitive preliminary argument, showing that with each conjugate pair of the form $\{p, \gamma_0(t)\}$ on γ_0 there is associated a variation $\tilde{\alpha}(u)$ of γ_0 along which the "quadratic part" of the functional $E = E(\gamma)$ initially decreases. (This will be exploited in the proof of the index theorem decreases. (This will be exploited in the proof of the index theorem which follows.) (We remind the reader that, as before, the manifold M comes endowed with a Riemannian metric, and that the operation of covariant differentiation is that determined by the connexion compatible with this metric.)

Thus let x_0 be an interior point of γ_0 forming with $p = \gamma_0(0)$ a conjugate pair of points of γ_0, and denote by $\lambda(x_0)$ the multiplicity of this conjugate pair, i.e. the dimension of the vector space of Jacobi fields along the segment $[p, x_0]$ of γ_0 vanishing at the points p, x_0. (Note that such a field may vanish also at some interior point or points of the arc $[p, x_0]$.) By Proposition 21.6, given any such Jacobi field v on the arc $[p, x_0]$, there is a geodesic variation $\tilde{\alpha}(u)$ of that arc, leaving the end-points fixed, with the field v as its initial velocity vector: $\partial\tilde{\alpha}/\partial u|_{u=0} = v$. (In other, more intuitive, words, each non-zero Jacobi field on $[p, x_0]$ vanishing at p and x_0, represents a non-zero "infinitesimal" perturbation of the geodesic arc $[p, x_0]$ leaving the end-points fixed; see Figure 101.) We use the geodesic arcs $\tilde{\alpha}(u)(t)$, $0 \leq t \leq t_0$, comprising this variation (where t_0 is the co-ordinate of x_0 on γ_0) to define a new smooth variation $\tilde{\varphi}(u)$ of γ_0 (or trajectory in Ω beginning at γ_0), each path $\tilde{\varphi}(u)(t)$ of which coincides with $\tilde{\alpha}(u)(t)$ for $0 \leq t \leq t_0$, and with $\gamma_0(t)$ for $t_0 \leq t \leq 1$ (see Figure 102):

$$\tilde{\varphi}(u)(t) = \begin{cases} \tilde{\alpha}(u)(t), & 0 \leq t \leq t_0, \\ \gamma_0(t), & t_0 \leq t \leq 1. \end{cases}$$

Clearly, by taking u sufficiently small we may ensure that the value of the functional E at γ_0 is, to whatever accuracy desired, approximately equal to its value on the curve from p to q comprised of the geodesic arc $\tilde{\alpha}(u)(t)$ from p to x_0 followed by the segment of γ_0 from x_0 to q (and likewise for the length functional), i.e. that the functional E does not change significantly under sufficiently small perturbations of γ_0 along the trajectory $\tilde{\varphi}(u)$, $0 \leq u \leq \varepsilon$.

Since a geodesic emanating from a point is completely determined (to within a constant multiple of the (natural) parameter) by the direction of its tangent vector at the point, the angle between the tangent vectors at x_0 to the geodesic arcs γ_0 and $\tilde{\alpha}(u)(t)$ (for each $u > 0$) must be non-zero (see Figure 103). This allows us to construct a new trajectory $\tilde{\psi}(u)$ in the space Ω, starting from

Figure 101

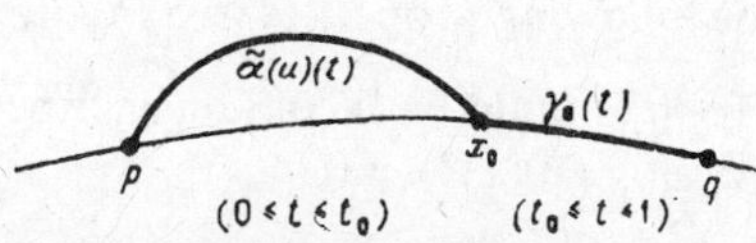

Figure 102

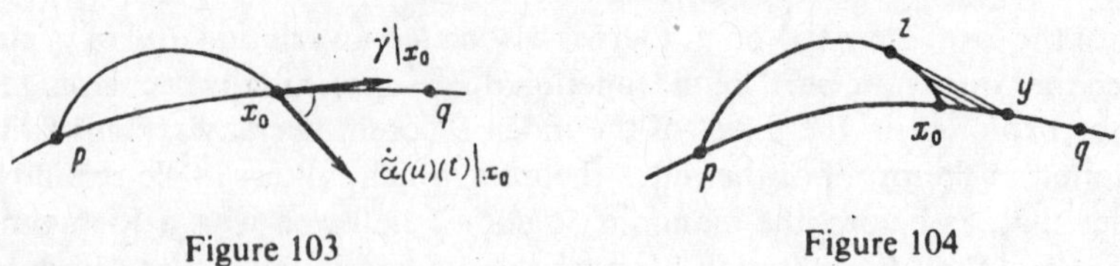

Figure 103 Figure 104

the point $\gamma_0 \in \Omega$, along which the "quadratic part" of the functional E strictly decreases, i.e. such that the velocity vector $\dot{\tilde{\psi}}(u)|_{u=0}$ belongs to a subspace of $T_{\gamma_0}\Omega$ on which the Hessian d^2E is negative definite (cf. the Taylor series expansion of an ordinary function about a critical point). The way in which the variation $\tilde{\varphi}(u)$ is to be modified to obtain $\tilde{\psi}(u)$ is indicated in Figure 104: For a sufficiently small geodesic triangle x_0yz (as in the figure) we shall have, in view of the positive-definiteness of the Riemannian metric on M, the strict triangle inequality

$$\text{length}\,[x_0, y] + \text{length}\,[x_0, z] > \text{length}\,[z, y],$$

and likewise for the functional E (since $L^2 = E$ if the parameter t is natural), so that on the path $\tilde{\psi}(u)(t)$ formed of the arcs $[p, z]$, $[z, y]$, $[y, q]$ traced out successively, the functional E will take on a smaller value than on the path $\tilde{\varphi}(u)(t)$, and therefore also, for sufficiently small u, smaller than its value on the path $\tilde{\alpha}(u)$. Hence the functional $E = E(\gamma)$ initially decreases along the trajectory $\tilde{\psi}(u)$ (with increasing u). In fact, the "quadratic part" of the functional E is initially decreasing on $\tilde{\psi}(u)$ (essentially by the positive definiteness of d^2E along $[z, y]$; cf. Part I, §36.2), i.e. the quadratic form d^2E is negative at the velocity vector $\dot{\tilde{\psi}}(u)|_{u=0} \in T_{\gamma_0}\Omega$. Thus we conclude that each member of a basis for the linear space of Jacobi fields on the segment $[p, x_0]$, vanishing at p and x_0, gives rise in this way to a separate contribution of 1 to the index of the Hessian d^2E at γ_0.

Proof of the Theorem. Consider a subdivision of the interval $[0, 1]$ by points $0 = t_0 < t_1 < \cdots < t_k = 1$, sufficiently fine for each segment $[\gamma_0(t_{i-1}), \gamma_0(t_i)]$ of the geodesic γ_0 to be contained in an open ball B_i in M with the property that each pair of points in the ball is joined by a unique geodesic contained in B_i, on which the functional E takes its least value among all curves joining those points (see [8] and cf. Part I, Theorem 36.2.5). Denote by $T_{\gamma_0}\{t_i\}$ the subspace of $T_{\gamma_0} = T_{\gamma_0}\Omega$ consisting of all piecewise-smooth vector fields $v(t)$ along γ_0 with the following properties (see Figure 105):

(i) $v(t)$ should be Jacobi on each segment $[\gamma_0(t_{i-1}), \gamma_0(t_i)]$, $i = 1, \ldots, k$;
(ii) $v(0) = 0$, $v(1) = 0$.

Thus $T_{\gamma_0}\{t_i\}$ is the subspace of all (continuous) "broken" Jacobi fields $v(t)$ along the path $\gamma_0(t)$, with the points $\gamma_0(t_i)$ of the given subdivision of γ_0 as the

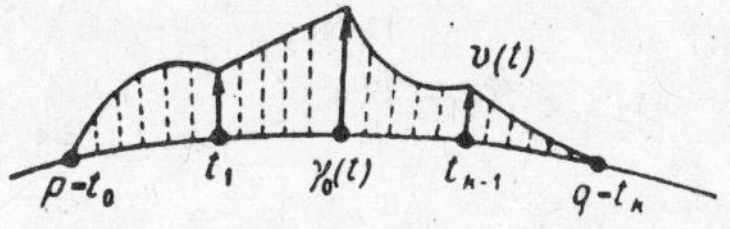

Figure 105

only points where (jump) discontinuities of $\dot{v}(t)$ may occur. We shall also be concerned with the subspace $Q_{\gamma_0} \subset T_{\gamma_0}\Omega$, consisting of all piecewise-smooth vector fields $v(t)$ satisfying $v(t_i) = 0$ for $i = 0, 1, \ldots, k$.

21.8. Lemma. *The tangent space $T_{\gamma_0}\Omega$ decomposes as the direct sum of its subspaces $T_{\gamma_0}\{t_i\}$ and Q_{γ_0}:*

$$T_{\gamma_0}\Omega = T_{\gamma_0}\{t_i\} \oplus Q_{\gamma_0}. \tag{12}$$

These subspaces are mutually orthogonal with respect to the scalar product afforded by the Hessian:

$$d^2E(v_1, v_2) = 0 \qquad \text{whenever} \quad v_1 \in T_{\gamma_0}\{t_i\}, \quad v_2 \in Q_{\gamma_0}.$$

The restriction of the Hessian d^2E to the subspace Q_{γ_0} is, as a quadratic form, positive definite, so that the index of d^2E on $T_{\gamma_0}\Omega$ is equal to the index of its restriction to the subspace $T_{\gamma_0}\{t_i\}$. Since $T_{\gamma_0}\{t_i\}$ is finite-dimensional (in view of the finite-dimensionality of the space of Jacobi fields along a geodesic arc—cf. the remark following the proof of Theorem 21.5), it follows that the index of the Hessian d^2E at γ_0 is finite.

Proof. Let $v = v(t)$ be any element of T_{γ_0}. It follows from the choice of the points of subdivision $\gamma_0(t_i)$ that no two consecutive such points $\gamma_0(t_{i-1}), \gamma_0(t_i)$ form a conjugate pair (see Part I, Theorem 36.2.4). Hence, for reasons noted earlier, there is a unique Jacobi field on each segment $[\gamma_0(t_{i-1}), \gamma_0(t_i)]$ with boundary values $v(t_{i-1}), v(t_i)$, or, in other words, there is a unique broken Jacobi field v_1 on γ_0 satisfying $v_1(t_i) = v(t_i)$, $i = 0, 1, \ldots, k$. The difference $v_2 = v - v_1$ then clearly belongs to the subspace Q_{γ_0} since $v_2(t_i) = v(t_i) - v_1(t_i) = 0$, $i = 0, \ldots, k$. Hence $v = v_1 + v_2$ where v_1 is a uniquely determined (by v) element of $T_{\gamma_0}\{t_i\}$, establishing (12). The orthogonality is immediate from the formula for the second variation of γ_0 (see (6)):

$$\tfrac{1}{2}d^2E(v_1, v_2) = -\sum_{i=1}^{k-1} \langle v_2(t_i), \Delta(\nabla_{\dot{\gamma}_0} v_1(t_i))\rangle - \int_0^1 \langle v_2, 0\rangle\, dt = 0,$$

invoking the facts that $v_2(t_i) = 0$, $i = 1, \ldots, k$, and that v_1 is a (broken) Jacobi field.

We next show that the restriction of the quadratic form d^2E to the subspace Q_{γ_0} is positive definite, i.e. $d^2E(v, v) > 0$ for all $v \in Q_{\gamma_0}$, $v \neq 0$. Let v be any element of Q_{γ_0}, and let $\tilde{\alpha}(u)$ be a variation of γ_0 with initial velocity vector v,

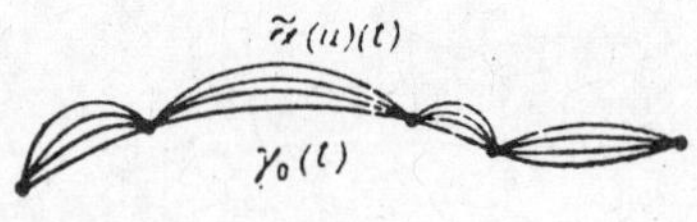

Figure 106

i.e. $\partial\tilde{\alpha}/\partial u|_{u=0} = v$; it is easy to see that since $v(t)$ vanishes at the points t_i, $i = 0, 1, \ldots, k$, we may find such a variation $\tilde{\alpha}(u)$ satisfying $\tilde{\alpha}(u)(t_i) \equiv 0$ for $-\varepsilon < u < \varepsilon$ (and all i) (see Figure 106). In view of the minimality property of each segment $[\gamma_0(t_{i-1}), \gamma_0(t_i)]$ of the geodesic γ_0 we have

$$E_{t_{i-1}}^{t_i}(\tilde{\alpha}(u)(t)) \geq E_{t_{i-1}}^{t_i}(\gamma_0(t)), \qquad i = 1, \ldots, k,$$

where $E_{t_{i-1}}^{t_i}(\tilde{\alpha}(u)(t))$ denotes the value of the functional E on the segment of $\tilde{\alpha}(u)(t)$ from t_{i-1} to t_i (and similarly for $E_{t_{i-1}}^{t_i}(\gamma_0(t))$). Adding these inequalities for $i = 1, \ldots, k$, we obtain

$$E(\tilde{\alpha}(u)(t)) \geq E(\gamma_0(t)) = E(\tilde{\alpha}(0)(t)). \tag{13}$$

Since $d^2E(v, v)$ may be regarded as the "second derivative" of $E(\tilde{\alpha}(u))$ at $u = 0$ (where E has a critical point) we deduce from the presence of a local minimum of $E(\tilde{\alpha}(u)(t))$ at $u = 0$, shown by (13), that $d^2E(v, v) \geq 0$.

It remains to show that in fact $d^2E(v, v) \neq 0$ for non-zero $v \in Q_{\gamma_0}$. Suppose that $v \in Q_{\gamma_0}$ is such that $d^2E(v, v) = 0$; we shall deduce that then $d^2E(\varphi, v) = 0$ for all $\varphi \in T_{\gamma_0}$. With this aim in view, first write $\varphi = \varphi_1 + \varphi_2$, where $\varphi_1 \in T_{\gamma_0}\{t_i\}$ and $\varphi_2 \in Q_{\gamma_0}$; then since $d^2E(\varphi_1, v) = 0$ by the mutual orthogonality of $T_{\gamma_0}\{t_i\}$ and Q_{γ_0} with respect to the form d^2E, we have

$$d^2E(\varphi_1 + \varphi_2, v) = d^2E(\varphi_1, v) + d^2E(\varphi_2, v) = d^2E(\varphi_2, v).$$

For any real α, we of course have $\alpha\varphi_2 + v \in Q_{\gamma_0}$, so that

$$d^2E(\alpha\varphi_2 + v, \alpha\varphi_2 + v) \geq 0,$$

that is

$$\begin{aligned} \alpha^2 d^2E(\varphi_2, \varphi_2) + d^2E(v, v) &+ 2\alpha d^2E(\varphi_2, v) \\ &= \alpha^2 d^2E(\varphi_2, \varphi_2) + 2\alpha d^2E(\varphi_2, v) \geq 0. \end{aligned}$$

Since α is arbitrary (and so may be taken large negative) and $d^2E(\varphi_2, \varphi_2) \geq 0$ (since $\varphi_2 \in Q_{\gamma_0}$), it follows that $d^2E(\varphi_2, v) = 0$, whence $d^2E(\varphi, v) = 0$ for all $\varphi \in T_{\gamma_0}$, as claimed.

Thus v belongs to the kernel of the Hessian d^2E. Since by Theorem 21.5 that kernel consists of the Jacobi fields vanishing at the end-points of γ_0, and since the only Jacobi field in Q_{γ_0} is the zero vector field on γ_0 (essentially by choice of the subdivision $\{t_i\}$; see the beginning of the proof), it follows that $v = 0$. This establishes the positive definiteness of d^2E on Q_{γ_0}, and completes the proof of the lemma. □

Figure 107

We now return to the proof of the theorem. In view of the above lemma we may, in computing the index of d^2E at γ_0, restrict attention to broken Jacobi fields on γ_0 where the breaks in smoothness occur sufficiently near to one another on γ_0. For each $t, 0 \leq t \leq 1$, denote by $\lambda(t)$ the index of the Hessian d^2E on the initial segment $[\gamma_0(0), \gamma_0(t)]$ $(= [p, \gamma_0(t)])$ of γ_0, i.e. restricted to (appropriate broken Jacobi) fields which are arbitrary on this segment and zero on the remainder of γ_0. It is clear that $\lambda(t)$ is an increasing function of t, i.e. $\lambda(t) \leq \lambda(t')$ if $t < t'$, since any (appropriate) field on the segment $[p, \gamma_0(t)]$ can be extended to one on the larger segment $[p, \gamma_0(t')]$ by defining it to be zero on $[\gamma_0(t), \gamma_0(t')]$ (in fact on $[\gamma_0(t), q]$) (see Figure 107).

Now for t sufficiently small, $[p, \gamma_0(t)]$ will be a minimal geodesic arc for E (by Theorem 36.2.5 of Part I), and consequently (essentially by Theorem 36.2.4 of Part I) there will be no conjugate pairs of points on $[p, \gamma_0(t)]$, and so in particular no points conjugate to the initial point p. It follows (by Lemma 36.2.3 of Part I) that the bilinear form d^2E is positive definite, and therefore has index 0, on such a segment. A similar argument shows that as one proceeds along γ_0, i.e. considers the form d^2E on lengthening initial segments $[p, \gamma_0(t)]$, along any interval of γ_0 on which there are no points conjugate to p the index of d^2E remains constant. However, at a value of $t < 1$ such that $\{p, \gamma_0(t)\}$ forms a conjugate pair, by the argument given prior to the present proof the index of d^2E increases by an amount equal to the multiplicity of that conjugate pair (since corresponding to each Jacobi field on $[p, \gamma_0(t)]$ vanishing at p and $\gamma_0(t)$, there is a variation $\tilde{\alpha}(u)$ of γ_0 (as indicated in Figure 108) along which the "quadratic part" of E is initially decreasing; see the preliminary argument, in particular Figure 104).

Thus it is only as t $(0 < t < 1)$ passes through those values for which $\gamma_0(t)$ is conjugate to p, that $\lambda(t)$ increases, and then by an amount equal to the multiplicity of the conjugate pair $\{p, \gamma_0(t)\}$. Hence on reaching the end-point $q = \gamma_0(1)$ (in the case where q is not conjugate to p) we finally obtain that $\lambda(1)$, the index of d^2E at γ_0, is equal to the sum of the multiplicities of all the points

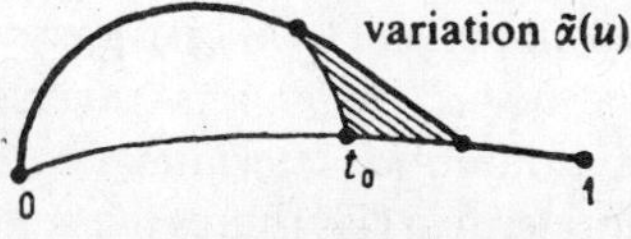

Figure 108

of γ_0 conjugate to the initial point p. (In the case where the end-point q is conjugate to p, one can show that there is no contribution to the index of d^2E from the point q.) This completes the proof of the index theorem. □

§22. Applications of the Index Theorem

We shall now use the Index Theorem (proved in the preceding section) to investigate the topological structure of the path space $\Omega(M) = \Omega(M, p, q)$ where $M = M^n$ is a compact smooth manifold, endowed with a Riemannian metric. We shall actually apply the construction for the finite-dimensional case (whereby we obtained a finite cell complex homotopically equivalent to a given compact finite-dimensional manifold by exploiting a Morse function on the manifold—see §15) in order to obtain an analogous result for the "infinite-dimensional manifold" $\Omega(M, p, q)$ whose points are the piecewise-smooth paths from p to q.

The role of the Morse function is in the present infinite-dimensional context taken over by the action functional $E(\gamma)$, $\gamma \in \Omega(M, p, q)$, with the points p, q so chosen that the critical points of E, i.e. the geodesic arcs joining p to q, are "non-degenerate" for E (i.e. the form d^2E is non-degenerate on such arcs). (As noted in §21 this will occur precisely when the points p, q do not form a conjugate pair on any geodesic joining them.) Each critical point $\gamma_0 \in \Omega(M)$ of the functional E has associated with it its "index" (i.e. the index of the Hessian d^2E at γ_0; see §21), analogous to the index of a non-degenerate critical point of a Morse function. Consequently, by analogy with the finite-dimensional situation, one expects that each critical point (i.e. geodesic γ_0 from p to q) will give rise to a cell of dimension equal to the index of that critical point (i.e. to the index of the geodesic γ_0), yielding ultimately a cell space homotopically equivalent to $\Omega(M)$, whose cells correspond one-to-one to the geodesics γ_0 from p to q, with dimensions equal to the indices of the corresponding geodesics.

Recalling that the underlying manifold M is Riemannian, we define a distance function on $\Omega(M)$ (turning $\Omega(M)$ into a metric space), appropriate for our purposes, by

$$d(\gamma_1, \gamma_2) = \max_{0 \le t \le 1} \rho(\gamma_1(t), \gamma_2(t)) + \left(\int_0^1 \left(\frac{ds_1(t)}{dt} - \frac{ds_2(t)}{dt} \right)^2 dt \right)^{1/2}, \tag{1}$$

where $\gamma_1(t)$, $\gamma_2(t)$ are any two points of $\Omega(M)$, i.e. piecewise-smooth paths from p to q in M, $s_1(t)$, $s_2(t)$ their respective lengths traversed from time 0 to t, and $\rho(x, y)$ is the distance function on M determined by the Riemannian metric. (The presence of the integral term in (1) ensures the continuity of $E: \Omega(M) \to \mathbb{R}$.) For each $a > 0$ consider the region $\Omega^a \subset \Omega(M)$ consisting of all $\gamma \in \Omega(M)$ for which $E(\gamma) \leq a$. It turns out (as we shall see below) that the subspace Ω^a can be approximated (in a certain precise sense) by a smooth finite-dimensional manifold.

For each subdivision $0 = t_0 < t_1 < \cdots < t_k = 1$ of the interval $[0, 1]$, denote by $\Omega(t_0, \ldots, t_k)$ the subspace of $\Omega(M)$ consisting of all piecewise-smooth geodesics joining p and q with their "breaks", i.e. jump-discontinuities in their tangent vectors, occurring among the points $t_0, \ldots, t_k$ only, and with the arcs $[\gamma(t_{i-1}), \gamma(t_i)]$ joining successive points of (possible) discontinuity each of length less than $2L(\gamma)/k$, where $L(\gamma)$ is the length of γ. Write $\Omega^a(t_0, \ldots, t_k)$ for the intersection $\Omega^a \cap \Omega(t_0, \ldots, t_k)$; thus $\Omega^a(t_0, \ldots, t_k)$ is comprised of the above piecewise-geodesic paths γ from p to q for which $E(\gamma) \leq a$.

22.1. Lemma. *Let $M = M^n$ (compact) and $\Omega(M)$ be as above. Then for each subdivision $0 = t_0 < t_1 < \cdots < t_k = 1$ of the interval $[0, 1]$ with k sufficiently large, the subspace*

$$\Omega(t_0, \ldots, t_k) \cap \{\gamma \mid E(\gamma) < a\} \tag{2}$$

may be endowed with the additional structure of a smooth finite-dimensional manifold, i.e. there is a homeomorphism between this subspace and a finite-dimensional (smooth) manifold.

PROOF. Let $\varepsilon > 0$ be sufficiently small that for each pair of points of M whose distance apart is less than ε there is just one geodesic arc joining them in a disc of radius ε containing them (cf. the preceding section, or see [8]). Let γ be any element of the intersection (2), assuming $k > 2\sqrt{a}/\varepsilon$; then each segment $[\gamma(t_{i-1}), \gamma(t_i)]$ of γ has length $< 2L(\gamma)/k < 2\sqrt{a}/k < \varepsilon$. It follows by choice of ε that each such path γ is completely determined by the ordered $(k-1)$-tuple $(\gamma(t_1), \ldots, \gamma(t_{k-1}))$ of points of M. The map from the subspace (2) to the direct product of $k-1$ copies of M, defined by

$$\gamma \mapsto (\gamma(t_1), \ldots, \gamma(t_{k-1}))$$

is therefore one-to-one, and it is not difficult to verify that it is also continuous. Hence $\Omega(t_0, \ldots, t_k) \cap (E < a)$ is homeomorphic to a subspace of a finite-dimensional manifold. Since, as is easily seen, that subspace is open in $M \times \cdots \times M$ $(k-1)$ factors), the lemma follows. □

22.2. Lemma. *The restricton E' of the functional E to the intersection $\Omega(t_0, \ldots, t_k) \cap (E < a)$ is, for sufficiently large k, a Morse function on this space (endowed with the structure of a smooth finite-dimensional manifold in accordance with the preceding lemma). The critical points of this Morse function are precisely those critical points of the functional E (i.e. smooth (unbroken) geodesics from p to q) of length less than $\sqrt{a}$, and the index of each such geodesic γ as a critical point of the Morse function E' coincides with the index of the Hessian d^2E at γ. Finally, for each $b < a$ the manifold $\Omega(t_0, \ldots, t_k) \cap (E \leq b)$ is a "deformation retract" of the subspace Ω^b.*

(A subspace X of a topological space Y is a *deformation retract* of Y if Y can be deformed to X by means of a homotopy throughout which X is fixed identically.)

PROOF. That E' is a Morse function with critical points as stated follows from Lemma 22.1 and results of the preceding section. (The nature of the critical points is established using the formula for the first variation of a path as in the proof of Theorem 21.3, and their non-degeneracy follows from the (implicit) assumption that p and q do not form a conjugate pair of points on any geodesic joining them—see the definition of "non-degeneracy" in §21.) The coincidence of the indices of a geodesic γ from p to q, regarded as a critical point on the one hand of E', and on the other of E, is in essence a consequence, via the Index Theorem (21.7), of the local character of the definition of a Jacobi field along a geodesic.

The final assertion of the lemma is established as follows. Let $\hat{\gamma} \in \Omega^b$, i.e. $E(\hat{\gamma}) \leq b < a$, and let $0 = t_0 < t_1 < \cdots < t_k = 1$ be a fixed subdivision of $[0, 1]$ with k sufficiently large that $2\sqrt{a/k} < \varepsilon$ (where ε is as in the proof of Lemma 22.1). First deform $\hat{\gamma}$: $[0, 1] \to M$ within Ω^b by means of a homotopy which merely re-parametrizes $\hat{\gamma}$ (i.e. throughout which the image $\hat{\gamma}[0, 1] \subset M$ remains fixed) in such a way that the end result is a path $\gamma \in \Omega^b$ having each segment $[\gamma(t_{i-1}), \gamma(t_i)]$ of length $< \varepsilon$. Write $r(\gamma)$ for the unique piecewise-geodesic arc in $\Omega(t_0, \ldots, t_k) \cap (E \leq b)$ determined by the ordered $(k - 1)$-tuple of points $(\gamma(t_1), \ldots, \gamma(t_{k-1}))$. The desired deformation retraction is then completed as indicated in Figure 109, where the final image $r(\gamma)$ of the original path $\hat{\gamma}$ is represented by the path made up of straight-line segments. □

The following corollary is immediate from these lemmas and Theorem 15.4 (given the compactness of $\Omega(t_0, \ldots, t_k) \cap (E \leq a)$).

22.3. Corollary. *Let M^n be a compact, smooth manifold (in fact, completeness can be shown to suffice), and let p, q be two points of M^n not conjugate on any geodesic arc joining them, of length at most $\sqrt{a}$. Then the space Ω^a is homotopically equivalent to a finite cell complex each of whose cells arises from a unique geodesic from p to q of length $\leq \sqrt{a}$, and of index equal to the dimension of the cell.*

The full analogue for $\Omega(M)$ of Theorem 15.4 is obtained from this result essentially by letting $a \to \infty$: *The space $\Omega(M)$ is homotopically equivalent to a countable cell complex each of whose cells of dimension λ corresponds (in a*

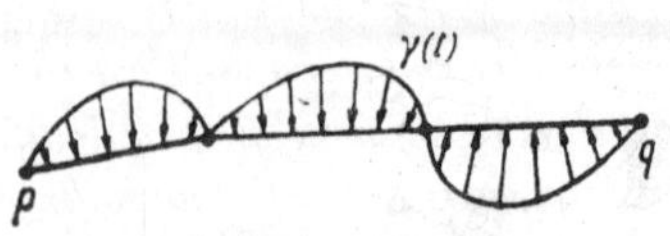

Figure 109

one-to-one manner) to a geodesic from p to q of index λ. (The formal argument, which we shall not give here, for deducing this from the above corollary, involves taking the appropriate ascending "homotopy direct limit" of topological spaces; see [44].)

We now consider the space $\Omega^*(M, p, q)$ of all continuous paths $\gamma: [0, 1] \to M$ from p to q. We shall show that $\Omega^*(M)$ and $\Omega(M)$ are homotopically equivalent, so that the above conclusion (concerning the homotopic equivalence of $\Omega(M)$ to a countable cell complex) applies of $\Omega^*(M)$ also. As the topology on the space $\Omega^*(M)$ one usually takes that determined by the metric

$$d^*(\gamma_1, \gamma_2) = \max_{0 \leq t \leq 1} \rho(\gamma_1(t), \gamma_2(t)), \qquad \gamma_1, \gamma_2 \in \Omega^*(M), \tag{3}$$

where ρ is as before the distance function defined by the Riemannian metric on M. (This topology is in fact the "compact-open" topology; cf. §7, Example (γ).) The natural embedding $i: \Omega(M) \to \Omega^*(M)$ is readily seen to be continuous relative to the metrics (1) and (3) on $\Omega(M)$ and $\Omega^*(M)$ respectively.

22.4. Lemma. *The spaces $\Omega(M)$ and $\Omega^*(M)$ are homotopically equivalent.*

Proof. It can be shown (see above, or [8]) that every point of M has an open neighbourhood U with the property that any two points of U are joined by a unique minimal geodesic lying entirely in U and depending differentiably on those points. Choose a covering of M by such open sets U_α. For each $k \geq 1$ define Ω_k^* to be the subspace of $\Omega^*(M)$ consisting of all continuous paths γ from p to q such that the segments $[\gamma((i-1)/2^k), \gamma(i/2^k)]$, $i = 1, \ldots, 2^k$, are each contained in some U_α. Clearly each Ω_k^* is an open subset of $\Omega^*(M)$, which is the union of the ascending sequence

$$\Omega_1^* \subset \Omega_2^* \subset \Omega_3^* \subset \cdots,$$

and the sets $\Omega_k = i^{-1}(\Omega_k^*)$ are open sets of $\Omega(M)$ with union $\Omega(M)$.

We now define a homotopy inverse $h: \Omega_k^* \to \Omega_k$ of the inclusion $i|_{\Omega_k}$, as follows: for each $\gamma \in \Omega_k^*$ take $h(\gamma) \in \Omega_k$ to be the piecewise-geodesic path which coincides with γ at the points $\gamma(i/2^k)$, $i = 0, 1, \ldots, 2^k$, and on each interval $[(i-1)/2^k, i/2^k]$ is the unique minimal geodesic joining $\gamma((i-1)/2^k)$ and $\gamma(i/2^k)$ (see Figure 110). We leave to the reader the verification that h is continuous, and that the maps $i|_{\Omega_k} \circ h$ and $h \circ i|_{\Omega_k}$ are homotopic to the identity maps of

Figure 110

Ω_k^* and Ω_k respectively. The remainder of the proof then involves showing that $\Omega^*(M)$ and $\Omega(M)$ are (ascending) "homotopy direct limits" of their subspaces Ω_k^* and Ω_k respectively, whence it follows that i is a homotopy equivalence (we refer the reader once again to [44] for the details). □

From this lemma and our earlier conclusion concerning $\Omega(M)$, we have the following culminating result.

22.5. Theorem (Fundamental Theorem of Morse Theory). *Let M^n be a compact (or merely complete) Riemannian manifold, and let p, q be two points of M^n which are not conjugate on any geodesic joining them. Then the space $\Omega^*(M^n, p, q)$ of continuous (parametrizations of) paths from p to q, has (in view of its homotopic equivalence to $\Omega(M, p, q)$) the homotopy type of a countable cell complex whose cells of dimension λ correspond one-to-one with the geodesics of index λ from p to q.*

Remark. It can be shown that given any particular such geodesic γ_0 of index λ, the corresponding λ-dimensional cell is represented by the set of paths obtained by perturbing γ_0 in the directions of the various Jacobi fields on the appropriate initial segments of γ_0 vanishing at the end-points of those initial segments (see Figure 111).

We now give some applications of this theorem. To begin with we shall use it to compute the integral homology and cohomology groups of the path space $\Omega(S^n)$ of the n-sphere S^n. (Note incidentally that $\Omega(S^n)$ and the loop space of S^n are easily seen to be homotopically equivalent.) Take the usual Riemannian metric on S^n (defined via the standard embedding of S^n in Euclidean $\mathbb{R}^n$), and let p, q be any two points of S^n sufficiently close for them not to be conjugate along any geodesic joining them. (Since each point of S^n is conjugate only to the antipodal point, it suffices to choose p, q to be any non-antipodal pair of points.) Clearly the points p and q are joined by the countably many geodesic arcs $\gamma_0, \gamma_1, \ldots$ (and only these), where γ_0 is the shorter arc of the great circle through p and q (see Figure 112), γ_1 is the longer, and for each $i = 1, 2, \ldots,$

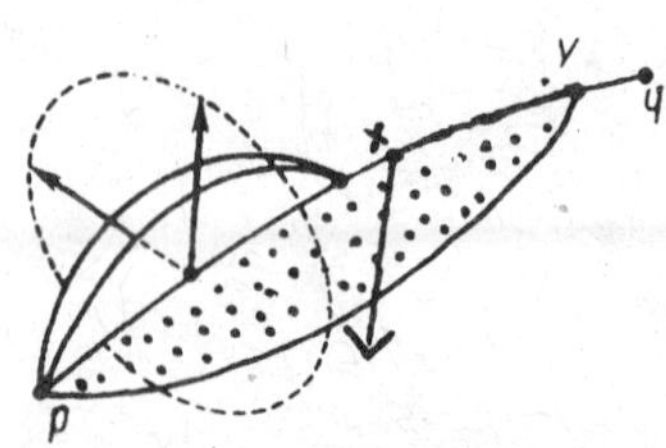

Figure 111. Here the index $\lambda = 3$, so that the geodesic corresponds to a 3-dimensional cell σ^3.

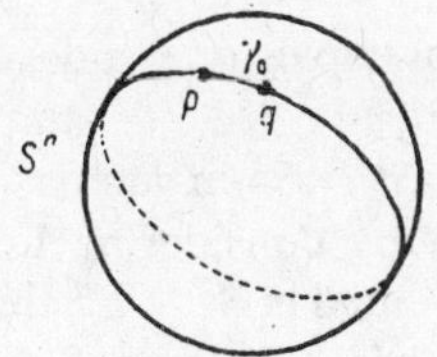

Figure 112

γ_{2i}, γ_{2i+1} are obtained by traversing that great circle i times (in the direction of γ_0 for γ_{2i}, and in the other direction for γ_{2i+1}) and following with γ_0 (in the case of γ_{2i}) or γ_1 (for γ_{2i+1}). Noting that p is conjugate to itself on the great-circle path starting and ending at p, and also to the antipodal point p', it is easily verified that for each k the geodesic γ_k has on it exactly k points (namely p and p' with their repetitions counted) conjugate to p. Since there are $(n - 1)$ independent geodesic variations of the great-circle path, and also of the great-semi-circle from p to p', obtained essentially by revolving the great circle about the axis pp' in each of $(n - 1)$ possible independent directions, it follows (via Proposition 21.6) that each pair (p, p'), (p, p) of conjugate points on γ_k has multiplicity $(n - 1)$. Thus the index of the geodesic γ_k is $k(n - 1)$. From Theorem 22.5 it follows that $\Omega(S^n)$ is homotopically equivalent to a cell complex with exactly one cell of each of the dimensions

$$0, n - 1, 2(n - 1), 3(n - 1), \ldots,$$

and no cells of the other dimensions. If $n > 2$ it is immediate that each of the cells $\delta^{k(n-1)}$ is a (co)cycle (there being no cells of neighbouring dimensions), whence

$$H^i(\Omega(S^n); \mathbb{Z}) \simeq H_i(\Omega(S^n); \mathbb{Z}) \simeq \begin{cases} \mathbb{Z} & \text{if } i = k(n-1), \quad k = 0, 1, 2, \ldots, \\ 0 & \text{for all other } i. \end{cases} \tag{4}$$

The case $n = 2$ requires further analysis, since here the cell complex contains just one cell of each dimension 0, 1, 2, ..., so that the triviality of the boundary operator is not immediate. Consider the 3-skeleton of $\Omega(S^2)$:

$$\Omega(S^2)^{(3)} = \sigma^0 \cup \sigma^1 \cup \sigma^2 \cup \sigma^3.$$

Recall (from Part II, §22.2) that the homotopy exact sequence of the fibre space $E(x_0) \to S^2$ (where $E(x_0)$ is the space of all paths on S^2 beginning at the particular point x_0) with fibre $\Omega = \Omega(S^2)$, yields $\pi_i(S^2) \simeq \pi_{i-1}(\Omega)$, $i \geq 1$. Since $\pi_2(S^2) \simeq \mathbb{Z} \simeq \pi_3(S^2)$ (see Part II, §§21.1, 23.3, 24.3), we infer that $\pi_1(\Omega) \simeq \mathbb{Z} \simeq \pi_2(\Omega)$, and then since $H_1(\Omega; \mathbb{Z})$ is isomorphic to the commutator quotient group of $\pi_1(\Omega)$ (see §4), it follows in turn that also $H_1(\Omega; \mathbb{Z}) \simeq \mathbb{Z}$. From this we deduce that the boundary of σ^2 must be attached to $\sigma^0 \cup \sigma^1$ ($\cong S^1$) in such a way as to be contractible in $\sigma^0 \cup \sigma^1$ (since otherwise some non-zero integer multiple of σ^1 would be a boundary). Hence the 2-skeleton $\Omega^{(2)}$ is homotopically equivalent to the bouquet $S^1 \vee S^2$. Since $\pi_1(S^1) \simeq \mathbb{Z}$ acts on

$\pi_2(S^2 \vee S^1)$, the direct sum of copies of $\mathbb{Z}$ indexed by the integers, by translating these copies of $\mathbb{Z}$ (see Part II, §22.3, Example (b)), while $\pi_2(\Omega) \simeq \mathbb{Z}$, the 3-cell σ^3 must be attached to $S^1 \vee S^2$ in such a way as to nullify this action; it is not too difficult to see, by considering how the immersed 2-spheres generating $\pi_2(S^2 \vee S^1)$ are situated in $S^2 \vee S^1$, that in fact $\Omega^{(3)}$ must be homotopically equivalent to the product $S^1 \times S^2$. Since the 2-dimensional homology and cohomology groups of $\Omega(S^2)$ are completely determined by the structure of the 3-skeleton $\Omega(S^2)^{(3)}$, we infer that

$$H_2(\Omega(S^2); \mathbb{Z}) \simeq \mathbb{Z} \simeq H^2(\Omega(S^2); \mathbb{Z}).$$

Let a and b denote generators of the cohomology groups $H^1(\Omega(S^2); \mathbb{Z})$ and $H^2(\Omega(S^2); \mathbb{Z})$ respectively ($\deg(a) = 1$, $\deg(b) = 2$). By working in $\Omega^{(2)} \sim S^1 \vee S^2$ it is easy to see that we must have $a^2 = 0$ in the cohomology ring $H^*(\Omega(S^2); \mathbb{Z})$ (see §7). Now, as observed in Part I, §22.4 (see also §7 above, in particular Example (γ)), the loop space $\Omega(p, M)$ ($\sim \Omega(p, q, M)$) is a "generalized H-space", i.e. a space X in which there is defined a continuous multiplication μ with respect to which there is a "homotopic identity element" x_0 in the sense that the maps

$$\mu(x_0, \ ): X \to X, \qquad \mu(\ , x_0): X \to X$$

are both homotopic to the identity map. In a loop space the multiplication μ is given by the product of paths, and the homotopic identity element is the constant path $[0, 1] \to \{p\}$. By Hopf's theorem (7.5) and the structure theorem for Hopf algebras (Theorem 7.6), the rational cohomology algebra of a generalized H-space is isomorphic to a tensor product of the form

$$\bigwedge(x_1, \ldots, x_r) \otimes \mathbb{Q}[y_1, \ldots, y_s] \tag{5}$$

where $\bigwedge(x_1, \ldots, x_r)$ is the free exterior algebra on generators $x_1, \ldots, x_r$ of odd degree, and $\mathbb{Q}[y_1, \ldots, y_s]$ is the polynomial algebra over $\mathbb{Q}$ in generators $y_1, \ldots, y_s$ of even degree. Hence, in particular, the algebra $H^*(\Omega(S^2); \mathbb{Q})$ has the form (5). The elements a, b of this algebra (of degrees 1 and 2 respectively) which we have already obtained must therefore generate a subalgebra of the form $\bigwedge(a) \otimes \mathbb{Q}[b]$. In fact, since for each $k = 0, 1, 2, \ldots$, this algebra contains exactly one additive generator of degree k (namely b^l if $k = 2l$, and ab^l if $k = 2l + 1$), and since on the other hand the cell complex homotopically equivalent to $\Omega(S^2)$, inferred above from the Fundamental Theorem (22.5), has exactly one cell of each dimension, it follows that

$$H^*(\Omega(S^2); \mathbb{Q}) \simeq \bigwedge(a) \otimes \mathbb{Q}[b].$$

From this one can infer, in particular, that the integral homology and cohomology groups of $\Omega(S^2)$ also satisfy (4) (verify this!), i.e.

$$H^i(\Omega(S^2); \mathbb{Z}) \simeq H_i(\Omega(S^2); \mathbb{Z}) \simeq \mathbb{Z}, \qquad i = 0, 1, 2, \ldots.$$

Applying the Hopf theorem more generally to $\Omega(S^n)$ yields similarly

$$H^*(\Omega(S^{2n+1}); \mathbb{Q}) \simeq \mathbb{Q}[b_{2n}], \qquad H^*(\Omega(S^{2n}); \mathbb{Q}) \simeq \bigwedge(a_{2n-1}) \otimes \mathbb{Q}[b_{4n-2}].$$

From our knowledge of the homology groups of $\Omega(S^n)$ (just obtained) we can infer the existence of infinitely many geodesics joining each pair of non-conjugate points of certain Riemannian manifolds.

22.6. Proposition. *If M is a Riemannian manifold with the same homotopy type as the sphere S^k, $k \geq 2$, then there are infinitely many geodesic arcs joining each pair of points p, q of M not conjugate along any geodesic through them.*

PROOF. To see this observe first that a homotopy equivalence between M and S^k induces one between $\Omega^*(M) = \Omega^*(M, p, q)$ and $\Omega^*(S^k)$, whence, in view of Lemma 22.4, the spaces $\Omega(M)$ and $\Omega(S^k)$ are also homotopically equivalent. Hence the cell complex K homotopically equivalent to $\Omega(M)$ given by the Fundamental Theorem (22.5), also has the same homotopy type as $\Omega(S^k)$. Since the homology groups $H_i(\Omega(S^k); \mathbb{Z})$ are non-zero for infinitely many i (see above), it follows that the complex K must contain infinitely many cells, whence in view of the provenance (the Fundamental Theorem) of this complex, there must be infinitely many geodesics in $\Omega(M, p, q)$. □

Remark. Although the geodesics whose existence this proposition establishes are distinct points of the path space $\Omega(M, p, q)$, i.e. distinct maps $[0, 1] \to M$, $0 \mapsto p$, $1 \mapsto q$, some of them may coincide "geometrically" as smooth curves on M, i.e. may not correspond to essentially distinct curves on M (as, for instance, in the case of the sphere S^k). The problem of ascertaining the number of "geometrically distinct" geodesic arcs from p to q requires, generally speaking, further investigation.

We conclude the section by indicating a connexion between the homotopy groups of a compact Riemannian manifold M and the behaviour (and existence: cf. the above proposition) of the geodesics joining a pair p, q of non-conjugate points of M. *If $i > 0$ is the least positive integer such that $\pi_i(M) \neq 0$, then each pair p, q of non-conjugate points of M are joined by a geodesic arc of index $i - 1$.* To see this observe first that since $\pi_j(M) \simeq \pi_{j-1}(\Omega(M))$ (see Part II, Corollary 22.2.3), it follows via Hurewicz' theorem (Corollary 4.9) that $H_{i-1}(\Omega M; \mathbb{Z})$ is also non-zero. Hence the cell complex $K \sim \Omega(M)$ given by the Fundamental Theorem must contain an $(i - 1)$-dimensional cell, so that, by that theorem, there is a geodesic arc from p to q of index $(i - 1)$. □

We shall show in the next section (see Lemma 23.4) that if the manifold M has negative (or non-positive) sectional curvature in every two-dimensional direction, then every critical point of the functional E on $\Omega(M, p, q)$, is non-degenerate and has index zero (and so corresponds to a local minimum of E).

EXERCISE

Deduce that in this case the geodesic arcs from p to q are in natural one-to-one correspondence with the elements of $\pi_1(M)$.

§23. The Periodic Problem of the Calculus of Variations

In Parts I and II (and above) we examined in some detail the one-dimensional variational problem (on a Riemannian manifold M^n) of extremizing, in particular, the length and action functionals $L(\gamma)$ and $E(\gamma)$, $\gamma \in \Omega(M^n, p, q)$, p, q fixed points of M^n. This type of problem might be termed "a variational problem with fixed end-points" to distinguish it from the important related "periodic" problem, where it is rather the closed extremal curves that are sought, with initial and terminal point left unspecified. The investigation of the latter curves, which will be our main concern in the present section, requires techniques differing somewhat from those employed in connexion with the "problem with fixed end-points".

The periodic problem for a compact smooth Riemannian manifold M^n naturally involves the space $\Pi(M^n)$ whose points are the closed smooth curves in M^n with initial (and terminal) point unspecified, i.e. smooth maps $\gamma: S^1 \to M^n$, where S^1 is, for definiteness, taken to be the unit circle parametrized by the standard angular co-ordinate t, $0 \le t \le 2\pi$, with the point of S^1 corresponding to $t = 0, 2\pi$ arbitrary, i.e. the corresponding map $[0, 2\pi] \to M^n$, giving the parametization of γ, is determined only up to an additive constant (modulo 2π). (Note, however, that parametizations of smooth closed paths in M^n not differing merely by an additive constant, represent distinct elements of $\Pi(M^n)$.) The topology on $\Pi(M^n)$ is defined via a natural metric analogous to that imposed on $\Omega(M)$ (see §22(1)).

Remark. The topological space $\Pi(M^n)$ must be distinguished from the space $\bigcup_p \Omega(M^n, p, p)$: Under the naturally-defined map

$$\bigcup_p \Omega(M^n, p, p) \to \Pi(M^n),$$

(which incidentally is not a fibre-space projection!) the complete inverse image of each point of $\Pi(M^n)$ is topologically a circle.

In the present section we shall use as needed, without detailed comment, the apparatus developed in the preceding two sections for investigating the extremals of the functonals L and E on the path space $\Omega(M^n, p, q)$, modifying it appropriately for application to the periodic problem.

Thus the space $\Pi(M^n)$ is endowed with the differentiable structure of a smooth "infinite-dimensional manifold" by means analogous to those used for $\Omega(M^n, p, q)$, essentially by taking the "tangent vectors" to $\Pi(M^n)$ at a point γ to be the smooth vector fields along γ (and so periodic) tangential to M^n, together comprising the "tangent space" $T_\gamma \Pi(M^n)$ to the "manifold" $\Pi(M^n)$ at the "point" γ. "Trajectories" in $\Pi(M^2)$ are also defined much as for $\Omega(M^n, p, q)$, namely as one-parameter variations. (See Definition 21.1.)

Remark. The path-connected components of the space $\Pi(M^n)$ are clearly the "free" homotopy classes of maps $S^1 \to M^n$, which, by Theorem 17.3.1 of Part II, are in natural one-to-one correspondence with the conjugacy classes of the group $\pi_1(M^n)$. We conclude that: *For a functional of closed paths in M^n which attains a local minimum value on each path-connected component of the space $\Pi(M^n)$, there will be at least as many minimal curves in $\Pi(M^n)$ as the group $\pi_1(M^n)$ has conjugacy classes.*

The length and action functionals $L(\gamma)$ and $E(\gamma)$ are defined on the space $\Pi(M^n)$ just as they were on $\Omega(M^n, p, q)$ (see §21). The following result, characterizing the critical points (which include the extremals) of the functional E on $\Pi(M^n)$, is proved in essentially the same way as the corresponding Theorem 21.3.

23.1. Lemma. *A closed smooth path $\gamma_0 \in \Pi(M^n)$ is a critical point of the functional E if and only if γ_0 is a (closed) geodesic parametrized by the natural parameter (i.e. proportional to arc-length) determined by its length. Hence, in particular, extremal smooth closed curves for E are geodesics.*

Note that if $\gamma(t)$ is a periodic extremal (or, more generally, critical point) for the length functional L on $\Pi(M^n)$, then, arc-length being of course independent of (smooth) parametrization, every trajectory $\gamma(t')$ obtained from $\gamma(t)$ by means of a smooth change of parameter $t \to t'$, will also be an extremal for L, so that the critical points of the functional L are never isolated in $\Pi(M^n)$. It is essentially for this reason that the functional E (rather than L) is used in extending Morse theory to the space $\Pi(M^n)$. (This remark applies of course also to the Morse theory developed in the preceding two sections for the path space $\Omega(M^n, p, q)$, where E was chosen for the role of the "Morse function" rather than L.)

In connexion with the above lemma, observe also that a closed geodesic $\gamma_0(t) \in \Pi(M^n)$ may be "multiple" in the sense that as t varies from 0 to 2π, the image set $\{\gamma_0(t)\} \subset M^n$ is traced out two or more times (see Figure 113). If that set is traced out only once as t varies over the full interval $[0, 2\pi]$, we call the closed geodesic *simple*. A given simple closed geodesic γ_0 determines an infinite sequence of closed geodesics, constituting an infinite discrete sequence of distinct points of the space $\Pi(M^n)$, obtained by means of multiple successive circuits of γ_0 (at speeds which are appropriate integer multiples of the

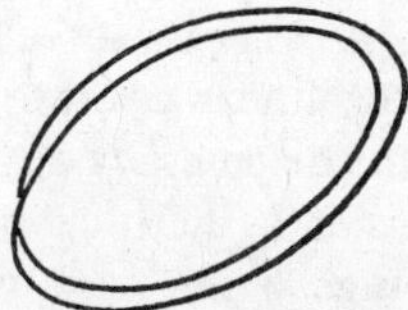

Figure 113. A double geodesic.

(constant) speed with which γ_0 itself is traced out). It is not difficult to see that if the simple closed geodesic γ_0 represents a non-trivial element of the fundamental group $\pi_1(M^n)$ (or, more precisely, corresponds to a non-trivial conjugacy class of $\pi_1(M^n)$), then the above-described multiples of γ_0 will correspond to distinct conjugacy classes of $\pi_1(M^n)$.

The "degree of degeneracy" of a critical point of the functional E on $\pi(M^n)$, i.e. of a closed geodesic γ_0, is defined essentially as for geodesic arcs in $\Omega(M^n, p, q)$ (see §21), namely as the degree of degeneracy of the Hessian d^2E on the tangent space $T_{\gamma_0}\Pi(M^n)$. The Hessian d^2E is the symmetric bilinear form defined (much as in §21) in terms of the "formula for the second variation" of γ_0, by (cf. §21(5), (6))

$$\tfrac{1}{2}d^2E(v_1, v_2) = \frac{1}{2}\frac{\partial^2 E(\tilde{\alpha})}{\partial u_1 \partial u_2}(0, 0) = -\int_0^1 \langle v_2, \nabla_{\dot{\gamma}_0}\nabla_{\dot{\gamma}_0}v_1 + R(\dot{\gamma}_0, v_1)\dot{\gamma}_0\rangle\, dt,$$

where v_1, v_2 are now smooth periodic vector fields on γ_0 (i.e. tangent vectors to $\Pi(M^n)$ at γ_0), $\tilde{\alpha}$ is the 2-parameter variation of γ_0 (with parameters u_1, u_2) determined by v_1 and v_2, R is the Riemann curvature tensor, and $\dot{\gamma}_0$ is the velocity vector of γ_0. Just as a symmetric bilinear form $B(x, y)$ $(= g_{ij}x^iy^j)$ on the tangent space at each point of a finite-dimensional manifold is determined by (and determines) a linear operator $D(y)$ $(= g_{ij}y^j)$ satisfying $B(x, y) = \langle x, D(y)\rangle$ (where here $\langle\ ,\ \rangle$ denotes the Euclidean scalar product), so, analogously, is the symmetric bilinear form d^2E defined above determined by the linear differential operator

$$D = -(\nabla_{\dot{\gamma}_0})^2 - R(\dot{\gamma}_0,\ \ )\dot{\gamma}_0,$$

acting on vectors v in the infinite-dimensional tangent space $T_{\gamma_0}\Pi(M^n)$ according to the formula

$$D(v) = -(\nabla_{\dot{\gamma}_0})^2 v - R(\dot{\gamma}_0, v)\dot{\gamma}_0.$$

Similarly to the situation considered in §21 of geodesics with fixed end-points, a tangent vector $v \in T_{\gamma_0}\Pi(M^n)$ (i.e. a periodic smooth vector field along the closed geodesic γ_0) is defined to be *Jacobi* if it is annihilated by the operator D:

$$D(v) = -(\nabla_{\dot{\gamma}_0})^2 v - R(\dot{\gamma}_0, v)\dot{\gamma}_0 = 0,$$

i.e. if it belongs to the nullspace of the linear operator D on $T_{\gamma_0}\Pi(M^n)$. Once again as in §21, it follows from the fact that a Jacobi field is completely determined by two initial data, that the nullspace of D is finite-dimensional.

23.2. Definition. The *degree of degeneracy* of a closed geodesic γ_0 is the dimension of the nullspace of the linear operator D on the tangent space $T_{\gamma_0}\Pi(M^n)$. (Thus, as just noted, the degeneracy degree of γ_0 is finite.) A closed geodesic is said to be *non-degenerate* (as a critical point of the functional E on $\Pi(M^n)$) if its degeneracy degree is zero.

Partly for the sake of simplicity we shall, by and large, restrict attention to the non-degenerate closed geodesics. We now seek an appropriate definition of the "index" of such a geodesic. By way of motivating our eventual definition, observe that, essentially as in the situation of non-degenerate geodesics with fixed end-points, we might define the index of a non-degenerate closed geodesic γ_0 to be the number of negative squared terms appearing in the quadratic form d^2E on $T_{\gamma_0}\Pi(M^n)$, after it has been brought into canonical diagonal form. (That it is diagonalizable requires some form of the "spectral theorem" for (suitable) linear operators on infinite-dimensional spaces, in view of the infinite-dimensionality of the vector space $T_{\gamma_0}\Pi(M^n)$.) If $v \in T_{\gamma_0}\Pi(M^n)$ is a member of the corresponding basis (with respect to which the quadratic form d^2E is diagonal) for which $d^2E(v, v) < 0$, then v will be an eigenvector of the associated linear operator D, corresponding to a negative eigenvalue λ:

$$D(v) = \lambda v, \qquad \lambda < 0.$$

We are thus led to the following.

23.3. Definition. The *index* of a non-degenerate, closed geodesic γ_0 is the largest number of linearly independent, smooth, periodic vector fields v along γ_0 satisfying a system of differential equations of the form

$$D(v) = -(\nabla_{\dot{\gamma}_0})^2 v - R(\dot{\gamma}_0, v)\dot{\gamma}_0 = \lambda v$$

for some negative value of the parameter λ (which is allowed to vary with v).

(Note that a version of this definition for the index of a non-degenerate geodesic with fixed end-points, might also have been used in §21 as an (equivalent) alternative to the one given there.)

Important Remark. As might be expected, there is a result analogous to the index theorem (21.7) valid in the present context, relating the index of a non-degenerate closed geodesic to the number of points of the geodesic conjugate to a chosen initial point on it. However, the precise relationship is more complicated than in the case of geodesics with fixed end-points, and we shall not enter into the details here beyond the following

EXERCISE

1. Prove that the index of a non-degenerate, closed geodesic γ_0 on M^n is not less than the number of points of γ_0 conjugate to any particular point of γ_0.

In certain respects the investigation of the present "periodic problem" of the calculus of variations is beset with more difficulties than in the case of the "problem with fixed end-points". While the presence of multiple geodesics is illustrative of one kind of difficulty, the problem of finding the simple closed geodesics presents other difficulties. We shall for this reason confine ourselves to a particular case, namely where the given Riemannian manifold M^n has negative sectional curvature, i.e. in every 2-dimensional direction at each point (see Part I, §30.1). We met with examples of such Riemannian manifolds in Parts I and II, namely: the Lobachevskian plane L^2 with the standard metric of constant negative curvature (see Part I, §10); the smooth, closed orientable surfaces of genus > 1, each obtainable (as Riemannian manifolds) from the Lobachevskian plane L^2 as the orbit space of the action of a discrete group of isometries of L^2 (isomorphic to the fundamental group of the surface) (see Part II, §20, where these groups were called "non-Euclidean crystallographic groups").

23.4. Theorem. *Let M be a compact, smooth Riemannian manifold with negative sectional curvature. Then there is exactly one closed geodesic in each free homotopy class of maps $S^1 \to M$.*

PROOF. We first indicate briefly the line of argument for establishing the existence of such geodesis. Consider an arbitrary free homotopy class of closed loops in M. By Theorem 12.1.3 of Part II, there will be smooth loops in the class; let c be the greatest lower bound of the values E takes on all these smooth loops. There is then an infinite sequence of smooth loops γ_i, $i = 1, 2, \ldots$ (not necessarily all distinct) such that $E(\gamma_i) \to c$ as $i \to \infty$. Using the compactness of M it can be shown that there exists a subsequence of the sequence $\{\gamma_i\}$, converging pointwise to a smooth loop γ_0. It then follows relatively easily that in fact E takes the value c on γ_0, and thence that γ_0 is a closed geodesic.

To establish the uniqueness we need the following important result, whose significance geos beyond its use in proving the present theorem.

23.5. Lemma. *Every closed geodesic γ_0 on a (not necessarily compact) Riemannian manifold M of negative sectional curvature (in every 2-dimensional direction), is non-degenerate and has index zero. i.e. for every value of $\lambda \leq 0$, the differential equation $D(v) = \lambda v$ has only the zero solution.*

PROOF. Let $v(t)$ be any vector field along γ_0 satisfying $D(v) = \lambda v$ for some $\lambda \leq 0$, i.e.

$$-(\nabla_{\dot\gamma_0})^2 v - R(\dot\gamma_0, v)\dot\gamma_0 = \lambda v, \qquad \lambda \leq 0.$$

Taking the scalar product (determined by the metric on M) of this equation with v, we obtain at each t where $v(t) \neq 0$.

$$\langle (\nabla_{\dot\gamma_0})^2 v, v\rangle = -\langle R(\dot\gamma_0, v)\dot\gamma_0, v\rangle - \lambda\langle v, v\rangle > 0,$$

since $\lambda \leq 0$, and for each t the quantity $\langle R(\dot\gamma_0, v)\dot\gamma_0, v\rangle$ is what is meant by "the sectional curvature in the 2-dimensional direction" determined by the vectors $\dot\gamma_0(t)$ and $v(t)$ at each point $\gamma_0(t)$ of the path γ_0 (cf. Part I, §30.1(7)). It follows that for those t where $v(t) \neq 0$, we have

$$\frac{d}{dt}\langle \nabla_{\dot\gamma_0} v, v\rangle = \nabla_{\dot\gamma_0}\langle \nabla_{\dot\gamma_0} v, v\rangle = \langle (\nabla_{\dot\gamma_0})^2 v, v\rangle + \langle \nabla_{\dot\gamma_0} v, \nabla_{\dot\gamma_0} v\rangle$$

$$= \langle (\nabla_{\dot\gamma_0}{}^2 v, v\rangle + |\nabla_{\dot\gamma_0} v|^2 \rangle > 0,$$

where the first equality follows from the fact that $\langle \nabla_{\dot\gamma_0} v, v\rangle$ is a scalar, and the second from the compatibility of the connexion (determining the covariant directional derivative) with the metric on M. Hence the scalar function

$\langle \nabla_{\dot{\gamma}_0} v, v \rangle$ of t is increasing along γ_0 with increasing t, and strictly increasing at those t where $v(t) \neq 0$. Since $\langle \nabla_{\dot{\gamma}_0} v, v \rangle$ is a periodic continuous function of t (v being a smooth periodic vector field along γ_0), and therefore takes on the same value at $t = 0$ and $t = 2\pi$, it follows that $v(t) \equiv 0$, as claimed. □

We now return to the proof of the uniqueness assertion of the theorem. By the lemma just proved every critical point (i.e. closed geodesic) in $\Pi(M)$ is non-degenerate, and is therefore an isolated critical point (verify this—cf. Part II, §10.4), and furthermore has index zero and therefore corresponds to an (isolated) local minimum value of E (essentially by the preliminary definition of the index as the number of negative squares in the diagonal form of d^2E; cf. §15). Hence if γ_0 and γ_0' are two distinct closed geodesics on M, we shall have the situation depicted schematically in Figure 114, each of γ_0, γ_0' yielding an isolated local minimum value of E. If γ_0 and γ_0' are in the same free homotopy class of maps $S^1 \to M$, then there will exist a smooth homotopy deforming γ_0 into γ_0' (see Part II, Theorem 12.1.4), yielding a smooth trajectory in $\Pi(M)$ from the point γ_0 to γ_0'. However, as indicated schematically in Figure 115, such a trajectory will then inevitably have on it a critical point (i.e. closed geodesic) which does not correspond to a local minimum value of E. Since we have shown above that every closed geodesic is a local minimum point for E, we have reached a contradiction. Hence each free homotopy class of maps $S^1 \to M$ contains at most one, and therefore, by the first part of the proof, exactly one, closed geodesic. □

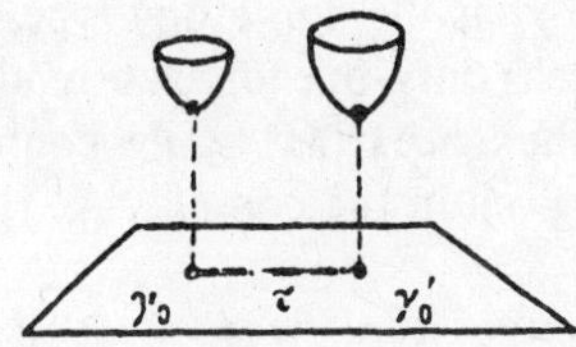

Figure 114

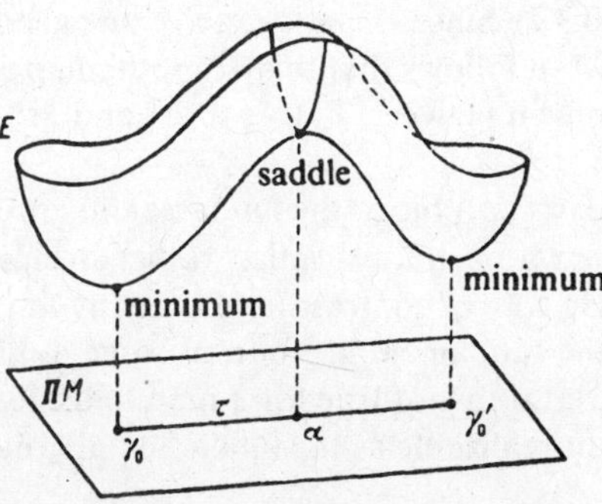

Figure 115

From Lemma 23.5 (or rather its proof) we deduce the following two interesting results concerning (not necessarily compact) manifolds of negative curvature.

23.6. Theorem. *A smooth Riemannian manifold M with negative curvature in every two-dimensional direction, contains no conjugate pairs of points.*

PROOF. Recall from §21 that points p, q on a geodesic form a conjugate pair if there is a non-zero Jacobi field v along the arc of the geodesic from p to q (i.e. a field satisfying $D(v) = 0$) vanishing at the points p and q. Now just as in the proof of Lemma 23.5 it follows from the hypothesis that M has negative sectional curvature that the only solution of the equation $D(v) = 0$ is the zero solution. Hence there can be no conjugate pairs of points on M. □

EXERCISE

2. Show that this result remains valid if "negative" is replaced by "non-positive".

23.7. Theorem. *Let M^n be a connected, simply-connected, smooth, complete Riemannian manifold with negative (or non-positive) sectional curvature (in every two-dimensional direction). Then each pair of points of M are joined by a unique geodesic. It follows that the manifold M^n is diffeomorphic to Euclidean n-space.*

PROOF. Since there are no conjugate pairs of points on M^n (by the preceding theorem), every geodesic on M^n has index 0 (by the Index Theorem (21.7)). From the Fundamental Theorem of Morse theory (22.5) it then follows that the path space $\Omega(M^n, p, q)$ has the homotopy type of a cell complex K comprised of 0-dimensional cells only, one for each geodesic arc from p to q. Since M^n is simply-connected, the space $\Omega(M^n, p, q)$ is connected, and therefore there can in fact be only one 0-cell in the complex K. Hence there is exactly one geodesic arc from p to q.

For the final assertion of the theorem we need the *exponential map* $\exp_p: T_pM^n \to M^n$, defined as follows: For each $v \in T_pM^n$, the tangent space to M^n at the point p, let $\gamma: [0, 1] \to M^n$ be the (unique) geodesic arc starting at p with initial velocity vector v (see Part I, §29.2), and set $\exp_p(v) = \gamma(1)$ (cf. Part I, §14.2, Theorem 30.3.7). Since there is exactly one geodesic arc joining p to any other point of M^n, it follows that this (smooth) map is bijective, and hence defines a diffeomorphism between T_pM^n ($\cong \mathbb{R}^n$) and M^n. □

For a group to be isomorphic to the fundamental group of a manifold with negative sectional curvature, it must satisfy rather stringent conditions, as the next result shows. (By way of contrast note that any finitely presented group can be realized as the fundamental group of some 4-dimensional manifold. Incidentally, this is certainly not true for 3-manifolds: for instance, the group $\mathbb{Z} \oplus \mathbb{Z} \oplus \mathbb{Z} \oplus \mathbb{Z}$ is not realizable as the fundamental group of any 3-manifold (Stallings).)

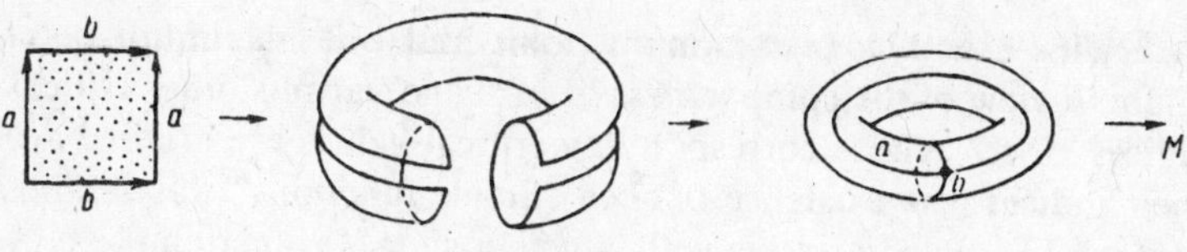

Figure 116

23.8. Theorem. *Let M be a smooth Riemannian manifold with negative sectional curvature. If two elements a and b of the fundamental group $\pi_1(M)$ commute, then they lie in a single cyclic subgroup of $\pi_1(M)$.*

PROOF. Consider the smooth map of the torus T^2 to M determined by the commutativity relation $aba^{-1}b^{-1} = 1$, where the elements a, b are represented by the images of the standard meridian and latitude of T^2 (see Figure 116). It can be shown that if there is no cyclic subgroup of $\pi_1(M)$ containing both a and b, then the assumption of negative sectional curvature of M allows this map to be smoothly deformed into an embedding, where the embedded torus so obtained is a "totally geodesic" submanifold of M, i.e. has the property that each geodesic on it (with respect to the metric induced from M) is at the same time a geodesic in the ambient Riemannian manifold M. (The proof of this involves first deforming the image of T^2 so as to minimize its surface area; the result that such an area-minimizing poisition exists is itself non-trivial, representing as it does a particular case of "Plateau's problem" (see Part I, §37.5, and cf. Appendix 2 below). In terms of conformal co-ordinates on this torus the embedding of it into M is harmonic (this property is specific to the 2-dimensional case, for which a uniformization theorem is valid; see Part I, Proposition 37.5.3). From this point on it becomes relatively easy to see that the embedded minimal torus is indeed completely geodesic.)

Since the ambient manifold M has negative sectional curvature, the metric induced on this completely geodesic, embedded torus must be such that the Gaussian curvature it determines is negative (since the Gaussian curvature coincides, to within a constant factor, with the sectional curvature of M in the tangent plane to the torus; cf. Part I, Theorem 30.3.3). However, no such metric can exist on the torus, since otherwise the integral of the corresponding Gaussian curvature would be negative, in violation of the Gauss–Bonnet theorem (Part II, Theorem 14.3.4), according to which this integral must be zero. This contradiction completes the proof. □

§24. Morse Functions on 3-Dimensional Manifolds, and Heegaard Splittings

Let M^3 be an orienable, connected, closed (i.e. compact and without boundary), 3-dimensional, smooth manifold, and let f be a Morse function on M^3 with exactly one minimum point and one maximum point (both absolute in view of the compactness of M^3) and various other critical points of indices 1 or 2, whose corresponding critical values are ordered according to their indices (the existence of such "Smale functions" was established in Theorem 17.1). By composing with a further suitable function $\mathbb{R} \to \mathbb{R}$ we may clearly assume that, in fact, $f(M^3) = [0, 1]$ with $f(p) = 0$, $f(p') = 1$ where p and p' are the minimum and maximum points, and that all the critical points of index 1 lie on the level surface $f = \frac{1}{3}$, and all those of index 2 on the level surface $f = \frac{2}{3}$. It follows almost immediately from §16(2) and the fact (of Poincaré duality—see Theorem 18.1) that the Betti numbers b_1 and b_2 are equal, that the number of critical points of index 1 is equal to the number of index 2.

Consider the level surface $\{x \mid f(x) = \frac{1}{2}\}$. Since there are no critical points of f on this surface, it follows (via the Implicit Function Theorem) that is is a 2-dimensional, closed, smooth submanifold $M^2 \subset M^3$, which is moreover connected (in view, essentially, of the fact that f has exactly one local-maximum point and one local-minimum point) and orientable (since it forms the boundary of the orientable manifold-with-boundary $\{x \mid 0 \le f(x) \le \frac{1}{2}\}$). Hence, by the classification theorem for closed 2-manifolds (see §3), M^2 is homeomorphic to a sphere-with-handles. Let r denote the genus of this surface, i.e. the number of handles. Thus we have discovered embedded in our manifold M^3 an orientable surface M_r^2 of genus r, which is simultaneously the boundary of the two 3-dimensional manifolds-with-boundary

$$\Pi_1 = \{x \mid 0 \le f(x) \le \tfrac{1}{2}\} \qquad \text{and} \qquad \Pi_2 = \{x \mid \tfrac{1}{2} \le f(x) \le 1\}.$$

By considering the restriction of f to the interior of Π_1 and applying Theorem 15.4 (or rather examining the details of its proof), it can be shown that Π_1 has the homotopy type of a bouquet of as many circles (q say) as f has critical points of index 1 (and similarly for Π_2), and thence that $q = r$. Hence Π_1 and Π_2 each have the homotopy type of the closed region of $\mathbb{R}^3$ bounded by the surface of genus $q = r$ embedded in a standard way in $\mathbb{R}^3$. (Since such a manifold-with-boundary is clearly homeomorphic to a 3-ball with r solid handles attached, it is often called a "solid handlebody".) Thus we have decomposed our manifold M^3 as a union of two manifolds-with-boundary each homotopically equivalent to a solid handlebody of genus r, with their boundaries ($\cong M_r^2$ the surface of genus r) identified. (A decomposition of this type can alternatively be obtained as follows: Since M^3 can be triangulated (Moise), take some triangulation of it, and consider a suitably small, closed, tubular neighbourhood of the 1-skeleton; this tubular neighbourhood and the closure of its complement are then handlebodies with union M^3.) We have thus adumbrated the proof of the following

24.1. Theorem. *Every orientable, connected, closed, 3-dimensional smooth manifold can be represented (non-uniquely) as the union of two 3-dimensional solid handlebodies of some genus r, with their boundaries identified by means of some*

diffeomorphism (each of the boundaries being diffeomorphic to a smooth surface of genus r).

Such a decomposition of M^3 as a union of two handlebodies of genus r "glued" together along their boundaries via a diffeomorphism, is often called a *Heegaard splitting* (of genus r) of M^3. It is not difficult to see that a manifold M^3 can be so decomposed in many distinct ways, i.e. that its representation

$$M^3 = \Pi_1 \cup_\alpha \Pi_2,$$

where $\alpha: M_r^2 \to M_r^2$ is a diffeomorphism determining the identification of the boundaries of the handlebodies Π_1 and Π_2 of genus r, is not unique; in fact, the genus r may vary with the choice of the Morse function f. It is clear that the diffeomorphism α essentially determines $M^3 = M^3(\alpha)$, and that if α_1, α_2 are homotopic within the class of diffeomorphisms $M_r^2 \to M_r^2$, then the corresponding 3-manifolds $M^3(\alpha_1)$, $M^3(\alpha_2)$ will be diffeomorphic.

In the above construction it was indicated how, starting with a Smale function on M^3, one might produce a Heegaard splitting of M^3. Conversely, given some Heegaard splitting, i.e. a self-diffeomorphism $\alpha: M_r^2 \to M_r^2$ of a surface of genus r, determining the 3-manifold $M^3(\alpha) = \Pi_1 \cup_\alpha \Pi_2$, one can construct from it a Smale function determining, in the above-described manner, exactly this Heegaard splitting of $M^3(\alpha)$. This is done by first constructing functions $f_1: \Pi_1 \to [0, \frac{1}{2}]$, $f_2: \Pi_2 \to [\frac{1}{2}, 1]$, such that f_1 has exactly one non-degenerate critical point of index 0, and r other non-degenerate critical points, all of index 1, and similarly for f_2 (with 0, 1 replaced by 3, 2 respectively), and both f_1 and f_2 take the value $\frac{1}{2}$ on the boundaries (verify the existence of such functions!) On identifying the boundaries of Π_1 and Π_2 via the self-diffeomorphism α, we obtain a Smale function $f = f_1 \cup f_2$ on $M^3(\alpha)$, of the required form.

In the case $r = 0$, where $\alpha: S^2 \to S^2$ is a self-diffeomorphism of the 2-sphere, the manifold $M^3(\alpha)$ is obtained from two 3-balls by identifying their boundaries via α. Since there is only one homotopy class of self-diffeomorphisms of S^2 (see Part II, Theorem 13.3.1), it follows that there is only one closed 3-manifold with a Heegaard splitting of genus 0, namely the 3-sphere S^3.

EXERCISE

Using the Morse Lemma (15.1) show that if there is a Morse function on M^3 with just two critical points then M^3 is diffeomorphic to S^3.

In the case $r = 1$, an exhaustive list of possibilities can also be compiled (though with greater difficulty), i.e. a list of all 3-manifolds obtainable by glueing together two solid tori Π_1, Π_2 ($\cong S^1 \times D^2$) along their boundaries ($\partial\Pi_1 = \partial\Pi_2 \cong S^1 \times S^1 = T^2$) in a manner prescribed by some diffeomorphism $\alpha: T^2 \to T^2$.

24.2. Theorem. *An orientable, closed, smooth 3-manifold which admits a Heegaard splitting of genus 1 is homeomorphic (and consequently diffeomorphic) to* *one of the following 3-manifolds*:

(i) *the 3-sphere* S^3;
(ii) $S^1 \times S^2$;
(iii) *a lens space* $L_m^3(1, k)$, $(m, k) = 1$.

(It is known that if two smooth 3-manifolds are homeomorphic, then they are diffeomorphic. The reader may also like to be reminded that the lens space $L_m^3(1, k)$ can be defined as the orbit space of the action of $\mathbb{Z}_m$ on

$$S^3 = \{(z, w) | z, w \in \mathbb{C}, |z|^2 + |w|^2 = 1\},$$

where the generator 1 of $\mathbb{Z}_m$ acts according to the formula

$$(z, w) \to (e^{2\pi i/m}z, e^{2\pi ik/m}w); \tag{1}$$

in particular, $L_2^3(1, 1) = S^3/\mathbb{Z}_2$ is diffeomorphic to $\mathbb{R}P^3$; see §4.)

PROOF (in outline only). By the foregoing discussion it is sufficient to consider just one diffeomorphism from each isotopy class of self-diffeomorphisms of the torus T^2. Since the torus has the homotopy type of an Eilenberg–MacLane space $K(\pi_1, 1)$ (see §§9, 10.2), the homotopy classes of continuous self-maps of T^2 are each determined by an endomorphism $\pi_1(T^2) \to \pi_1(T^2)$ (see §9, Exercise 5), whence it follows that the isotopy classes of self-diffeomorphisms of T^2 are each determined by an automorphism of $\pi_1(T^2)$. (Verify this; i.e. show that if two self-diffeomorphisms of T^2 are homotopic then they are isotopic.) Since $\pi_1(T^2) \simeq \mathbb{Z} \oplus \mathbb{Z}$, it follows in turn that to within an isotopy each self-diffeomorphism α of T^2 is determined by an integer 2×2 matrix $\alpha_* = \begin{bmatrix} a & b \\ c & d \end{bmatrix}$, where $ad - bc = 1$ if the diffeomorphism is orientation-preserving, and $ad - bc = -1$ in the contrary case. (Here we may suppose for definiteness that the generators of the direct factors of $\pi_1(T^2) \simeq \mathbb{Z} \oplus \mathbb{Z}$, relative to which the self-diffeomorphisms are associated with such integer matrices, are represented by the standard meridian and latitude on the torus T^2.)

We shall now compute the fundamental group of the manifold $M^3(\alpha)$ where $\alpha_* = \begin{bmatrix} a & b \\ c & d \end{bmatrix}$, $ad - bc = \pm 1$. The two solid tori Π_1, Π_2, whose union is $M^3(\alpha)$, are each homotopically equivalent to a circle; let γ_1, γ_2 denote generators of the fundamental groups ($\simeq \mathbb{Z}$) of each of those solid tori. A presentation of the fundamental group of $M^3(\alpha)$ is then obtained by taking γ_1, γ_2 as generators, with relations imposed between them by virtue of the fact that the boundaries of the two solid tori Π_1, Π_2 are actually identified in $M^3(\alpha)$. (The precise general result that we wish to apply, in our present particular situation, is the "Seifert–Van Kampen Theorem", which asserts that a presentation of the fundamental group of a union of two path-connected topological spaces X, Y, with path-connected intersection $X \cap Y$, is afforded by the disjoint union of presentations of $\pi_1(X)$ and $\pi_1(Y)$ with additional relations imposed by appropriately identifying generators of the subgroups of $\pi_1(X)$ and $\pi_1(Y)$ forming the images under the inclusion homomorphisms $i_*^{(1)}: \pi_1(X \cap Y) \to \pi_1(X)$ and $i_*^{(2)}: \pi_1(X \cap Y) \to \pi_1(Y)$, where $i^{(1)}$ and $i^{(2)}$ are the respective inclusion maps of $X \cap Y$.) It is easy to see that in the present situation the relations on the generators γ_1, γ_2 arising from the identification determined by α are (in multiplicative notation)

$$\gamma_1^a = \gamma_2, \qquad \gamma_2^d = \gamma_1, \qquad \gamma_1^c = \gamma_2^c = 1, \tag{2}$$

so that if $c \neq 0$ then $\pi_1(M^3(\alpha))$ is cyclic of order $|c|$, while if $c = 0$ then $\pi_1(M^3(\alpha)) \simeq \mathbb{Z}$. For instance, if $\alpha_* = \begin{bmatrix} \pm 1 & 0 \\ 0 & \pm 1 \end{bmatrix}$, then $\pi_1(M^3(\alpha)) \simeq \mathbb{Z}$ and $M^3(\alpha) \cong S^1 \times S^2$, since the boundaries of Π_1 and Π_2 ($\cong S^1 \times D^2$) are in this case identified essentially by means either of the identity map on $\partial(S^1 \times D^2) = S^1 \times S^1$, or a canonical orienation-reversing self-map of $S^1 \times S^1$. If (to take a further instance) $\alpha_* = \begin{bmatrix} 0 & \pm 1 \\ \pm 1 & 0 \end{bmatrix}$, then $\pi_1(M^3(\alpha)) = \{1\}$ and $M^3(\alpha) \cong S^3$, since in this case the boundaries of Π_1 and Π_2 are identified by means of a diffeomorphism interchanging the meridian and latitude; such decompositions of the 3-sphere $S^3 = \{(z, w) | |z|^2 + |w|^2 = 1\} \subset \mathbb{C}^2$ can be realized by taking

$$\Pi_1 = S^3 \cap \{|z| \geq |w|\}, \qquad \Pi_2 = S^3 \cap \{|z| \leq |w|\}, \tag{3}$$

with the identification of the boundaries determinded for instance by $\alpha_* = \begin{bmatrix} 0 & -1 \\ 1 & 0 \end{bmatrix}$, given by the map $(z, w) \to (w, z)$ (which, as a map of the whole 3-sphere, interchanges the two solid tori Π_1 and Π_2). These two special cases are in fact typical for $c = 0$ and $c = \pm 1$ respectively: If, on the one hand, $c = 0$, then $ad = \pm 1$, whence $a = \pm 1$, $d = \pm 1$, and it is not difficult to verify (much as above) that the manifold $M^3(\alpha)$ with α_* of the form $\begin{bmatrix} \pm 1 & b \\ 0 & \pm 1 \end{bmatrix}$ is homeomorphic to $S^1 \times S^2$. If, on the other hand, $c = \pm 1$, then since $\pi_1(M^3(\alpha))$ ($\simeq H_1(M^3(\alpha))$) is trivial, so is $H_2(M^3(\alpha))$ (by Poincaré duality), and therefore also $\pi_2(M^3(\alpha)) = \{1\}$ (by Hurewicz' theorem), and thence, similarly, $\pi_3(M^3(\alpha)) \simeq \mathbb{Z}$. It can be inferred from this (e.g. by constructing a Morse function with just two critical points using in addition the assumption that $M^3(\alpha)$ is a union of solid tori) that $M^3(\alpha)$ must in fact be homeomorphic to the 3-sphere.

For the final case, namely $c \neq 0, \pm 1$, where we have $\pi_1(M^3(\alpha))$ cyclic of order $|c| > 1$, we consider the universal cover $\tilde{M}^3(\alpha)$ (see Part II, 18.2). Since the universal cover of a solid torus is itself (being simply-connected), it follows that each of the solid tori Π_1, Π_2 is covered in $\tilde{M}^3(\alpha)$ by a disjoint union of solid tori. Since the complement in $\tilde{M}^3(\alpha)$ of this union is connected and is likewise a disjoint union of solid tori it follows that Π_1 and Π_2 are in fact each covered by just one solid torus. Hence the given Heegaard splitting of $M^3(\alpha)$ induces one of $\tilde{M}^3(\alpha)$ (as a union again of two solid tori). Since $\tilde{M}^3(\alpha)$ has trivial fundamental group, it follows (much as in the case $c = \pm 1$ above) that $\tilde{M}^3(\alpha)$ is homeomorphic to the 3-sphere. From the construction of the universal cover as a regular covering (see Part II, Definition 18.4.2, and [36]) it follows that $M^3(\alpha)$ can be realized as the orbit space of an action of the cyclic group of order $|c|$ on the 3-sphere. It can be shown that such an action must essentially be as in (1) (with $m = |c|$). With this we conclude our sketch of the proof. □

In going from the genus-one case examined above to genus > 1, there is a dramatic increase in the difficulty of the classification problem, and no comparably explicit classification of 3-manifolds with Heegaard splittings of genus > 1 is known.

Note that all of the manifolds listed in Theorem 24.2 do in fact admit Heegaard splittings of genus 1. For the manifolds S^3 and $S^1 \times S^2$ this was shown in the course of proving that theorem. For the lens space $L^3_m(1, k)$, $(m, k) = 1$, an explicit Heegaard splitting of genus 1 can be obtained as follows. Note first that since the action (1) of $\mathbb{Z}_m$ on S^3 is free (in view of the coprimality of m and k) and discrete, the corresponding projection $p: S^3 \to L^3_m(1, k)$ to the orbit space, is a covering map (see Part II, §18.4), and $L^3_m(1, k)$ is a manifold. Now the toroidal surface T^2 *in* S^3 determined by the equation $|z| = |w|$ separates S^3 into two solid tori: $S^3 = \Pi_1 \cup \Pi_2$ (see (3) above). Under the action (1) of $\mathbb{Z}_m$ on S^3 the torus T^2 is clearly sent to itself, and therefore projects under p to a torus in $L^3_m(1, k)$ determining a Heegaard splitting of genus 1 for this manifold.

The problem of classifying all 3-dimensional manifolds up to homeomorphism (or, equivalently, diffeomorphism) is not only unsolved, it is even unknown whether or not it is "algorithmically soluble" (in a certain precise sense of the theory of recursive decidability). (Note by way of contrast that the corresponding problem for 2-manifolds *is* "algorithmically soluble", in view, essentially, of the clssification derived in §3 (see in particular Theorem 3.20).) On the other hand, the simpler problem of compiling a list containing all (orientable) 3-manifolds, with possible repetitions, can be solved: in view of the above discussion, it suffices to list, for each $g \geq 0$, the isotopy classes of self-diffeomorphisms of the surface of genus g, and it turns out that such an enumeration can be given in the following way.

Corresponding to each embedded circle s in the surface M^2_r of genus r, define two self-diffeomorphisms $T^\lambda_s: M^2_r \to M^2_r$, $\lambda = \pm 1$, as follows. Let U_s be a small closed ε-neighbourhood of the circle s in M^2_r (thus $U_s \cong S^1 \times [0, 1]$), and define T^λ_s to be the identity map on the complement $M^2_r \setminus U_s$, and to act on U_s by rotating each circle $S^1 \times t$ in $S^1 \times [0, 1] \cong U_s$ through the angle $2\pi t$ if $0 \leq t \leq \frac{1}{2}$, and $2\pi(1 - t)$ if $\frac{1}{2} \leq t \leq 1$, in a direction determined by the sign of $\lambda = \pm 1$. There is then an important and highly non-trivial result (whose proof we omit) to the effect that every isotopy class of self-diffeomorphisms of the surface M^2_r contains a product (under compositon) of diffeomorphisms of the form T^λ_s for various of the particular circles s shown in Figure 117 (labelled by c_i, e_i, f_i). From this we conclude that: *A list containing all orientable, closed,*

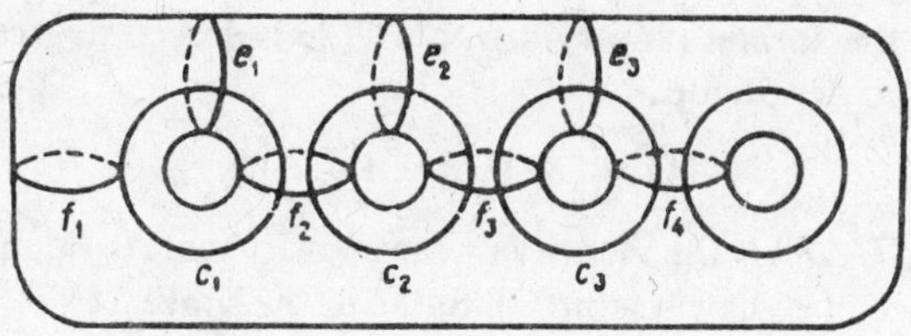

Figure 117

connected 3-manifolds (with repetitions) can in effect be obtained by enumerating, for each $r \geq 0$, all isotopy classes of self-diffeomorphisms of the surface of genus r, and therefore by enumerating (in any order) all possible finite products of the form

$$\prod_j T_{s_j}^{\lambda_j}, \qquad s_j \in \{c_i, e_i, f_i\}.$$

§25. Unitary Bott Periodicity and Higher-Dimensional Variational Problems

In this section we shall prove the important topological result usually referred to as "Bott periodicity". For the sake of simplicity we shall, for the most part, concentrate our attention on the unitary groups: although "orthogonal Bott periodicity" is established along the same general lines as "unitary Bott periodicity", there are certain rather substantial technical difficulties to overcome in the former case. (We shall nonetheless outline a proof of "orthogonal periodicity" in the final subsection (§25.3).)

25.1. The Theorem on Unitary Periodicity

We shall prove this theorem in its "classical" form, namely as a result on the periodicity of the homotopy groups of the unitary groups, without considering its role as a "periodic" theorem for vector bundles.

25.1. Theorem (On Unitary Periodicity). *The following isomorphisms hold between the homotopy groups of the special unitary groups:*

$$\pi_{i-1}(SU(2m)) \simeq \pi_{i+1}(SU(2m)) \qquad \textit{for} \quad 1 \leq i \leq 2m.$$

It follows that for the "stable" unitary group $U = \lim_{m\to\infty} U(m)$, the direct limit of the unitary groups $U(m)$ with respect to the standard embeddings $U(m) \subset U(m+1)$, we have

$$\pi_{i-1}(U) \simeq \pi_{i-1}(U) \qquad \textit{for} \quad i \geq 1,$$

whence

$$\pi_{2n}(U) = 0, \qquad \pi_{2n+1}(U) \simeq \mathbb{Z} \qquad \textit{for} \quad n \geq 0.$$

Consider the special unitary group $SU(2m)$ of even degree, regarded as a Lie group, and denote up

$$\Omega = \Omega(SU(2m); I_{2m}, -I_{2m}),$$

where $I_{2m} \in SU(2m)$ is the identity linear transformation (identity matrix), the function space of piecewise-smooth paths in the space $SU(2m)$ from the point I_{2m} to $-I_{2m}$, and by

$$\Omega^* = \Omega^*(SU(2m); I_{2m}, -I_{2m})$$

the full space of all continuous paths in $SU(2m)$ from I_{2m} to $-I_{2m}$. (Recall from Lemma 22.4 that the inclusion $\Omega \to \Omega^*$ is a homotopy equivalence.) We shall be particularly concerned with the subspace

$$\tilde{\Omega} = \tilde{\Omega}(SU(2m); I_{2m}, -I_{2m})$$

of Ω consisting of all minimal geodesics γ (i.e. geodesics of least length among all piecewise-smooth paths) joining the points I_{2m} and $-I_{2m}$, with respect to the invariant metric on $SU(2m)$ determined by the Killing form $\langle X, Y\rangle = \operatorname{Re}\operatorname{tr}(X\bar{Y}^T)$, $X, Y \in su(2m)$, on the Lie algebra $su(2m)$ (see Part I, §24.4). (See [44] for general conditions under which such minimal geodesics exist.)

25.2. Lemma. *The space $\tilde{\Omega}$ is homeomorphic to the complex Grassmannian manifold $G^{\mathbb{C}}_{2m,m}$, i.e. the manifold whose points are the m-dimensional complex planes through the origin in complex 2m-dimensional space $\mathbb{C}^{2m}$ (see Part II,* §5.2).

PROOF. By Theorem 30.3.7 of Part I, in the Lie group $SU(2m)$ the geodesics with respect to the above-mentioned Killing metric are precisely the one-parameter subgroups (and their translates (i.e. multiples) by elements of $SU(2m)$). Hence in order to characterize the geodesics in $SU(2m)$ joining the points I_{2m} and $-I_{2m}$, it suffices to describe all one-parameter subgroups emanating from the identity I_{2m} of $SU(2m)$ and reaching the point $-I_{2m}$. By Part I, Theorem 24.3.1 (and the definition of an invariant metric on a Lie group given in Part I, §24.4), the one-parameter subgroups $\gamma(t)$ passing through the identity I_{2m} have the form $\gamma(t) = \exp(tX)$ for some skew-Hermitian matrix X with zero trace (i.e. element of the Lie algebra $su(2m)$; see Part I, §14.1). Since the parameter can always be chosen to vary from 0 to 1 along the geodesic arc from I_{2m} to $-I_{2m}$, we infer the conditions $\gamma(0) = I_{2m}$ (as required), and $\gamma(1) = \exp X = -I_{2m}$, from which we can ascertain X. To this end, recall the well-known result (a consequence of the classical process for bringing a matrix into its Jordan canonical form, or, if you like, of an orthogonalization process exploiting the operator Ad applied to the unitary case) to the effect that X is conjugate, by means of a matrix in $SU(2m)$, to a diagonal matrix, i.e. there is an element $g_0 \in SU(2m)$ such that $g_0 X g_0^{-1} = X_0$ with X_0 of the form

$$X_0 = \begin{pmatrix} i\varphi_1 & & 0 \\ & \ddots & \\ 0 & & i\varphi_{2m} \end{pmatrix}, \qquad \text{where} \quad \varphi_1 + \quad + \varphi_{2m} = 0.$$

(Thus X_0 belongs to a "Cartan subalgebra", i.e. maximal commutative subalgebra, of $su(2m)$.) Under the transformation $\mathrm{Ad}(g_0)$ (see Part II, §3.1), applied to (the terminal point of) the geodesic $\gamma(t)$, the above condition $\exp(X) = -I_{2m}$ becomes

$$g_0(\exp X)g_0^{-1} = \exp(g_0 X g_0^{-1}) = \begin{pmatrix} e^{i\varphi_1} & & 0 \\ & \ddots & \\ 0 & & e^{i\varphi_{2m}} \end{pmatrix} = g_0(-I_{2m})g_0^{-1} = -I_{2m},$$

whence we have, for $i = 1, \ldots, 2m$, that $\varphi_i = \pi k_i$ where the k_i are odd integers satisfying $k_1 + \cdots + k_{2m} = 0$.

Having thus completely described the geodesics in $SU(2m)$ from I_{2m} to $-I_{2m}$, it remains to pick out those of least length. Since, in view of the definition of the metric on $SU(2m)$ in terms of the Killing form on $su(2m)$, the exponential map sends the line segment $\{tX \mid 0 \le t \le 1\}$ in the tangent space $su(2m)$ isometrically onto the geodesic arc $\gamma(t) = \exp(tX)$, the length of this latter arc is equal to that of the line segment in $su(2m)$. Now, as noted above, the Killing form on $su(2m)$ is given by

$$\langle A, B \rangle = \mathrm{Re}\ \mathrm{tr}(A\bar{B}^{\mathrm{T}}),$$

so that the length of the latter line segment is

$$\sqrt{\langle X, X \rangle} = \sqrt{\mathrm{tr}\ X\bar{X}^{\mathrm{T}}} = \pi \sqrt{\sum_{i=1}^{2m} (k_i)^2}.$$

Hence the least length of a geodesic from I_{2m} to $-I_{2m}$ is $\pi\sqrt{2m}$, attained when $k_i = \pm 1$ for all $i = 1, \ldots, 2m$. Since $k_1 + \cdots + k_{2m} = 0$, it follows that the corresponding matrices X_0 must have the same number of i's as $(-i)$'s on the diagonal. The upshot of our argument is, therefore, that the minimal geodesics is $SU(2m)$ joining the points I_{2m} and $-I_{2m}$ are precisely those of the form $\gamma(t) = \exp(tX)$ where X is conjugate in $SU(2m)$ to the matrix

$$X_0 = \pi \begin{pmatrix} i & & & & & 0 \\ & i & & & & \\ & & \ddots & & & \\ & & & i & & \\ & & & & -i & \\ & & & & & \ddots \\ 0 & & & & & & -i \end{pmatrix} = \pi \begin{pmatrix} iI_m & 0 \\ 0 & -iI_m \end{pmatrix}, \tag{1}$$

i.e. $X = gX_0g^{-1}$ for some $g \in SU(2m)$.

We have thus established a one-to-one correspondence between the space $\tilde{\Omega}$ of all minimal geodesics from I_{2m} to $-I_{2m}$, and the conjugacy class of the matrix X_0 (see (1)) in $SU(2m)$, and it is intuitively clear that in fact this correspondence is a homeomorphism. Since the conjugacy class of X_0 in $SU(2m)$ is in turn homeomorphic to the coset space $SU(2m)/C(X_0)$, where $C(X_0)$ is the centralizer of X_0 in $SU(2m)$, and since clearly $C(X_0) = S(U(m) \times U(m))$ (with the obvious interpretation), we finally conclude that $\tilde{\Omega}$

is homeomorphic to

$$SU(2m)/S((U(m) \times U(m)),$$

which (as indicated in Part II, §5.2) is identifiable with the Grassmannian manifold $G^{\mathbb{C}}_{2m,m}$. This completes the proof of the lemma. □

25.3. Lemma. *Each minimal geodesic arc $\gamma(t)$, $0 \le t \le 1$, in $SU(2m)$, from the identity I_{2m} to $-I_{2m}$, is uniquely determined by its midpoint, i.e. by the point $\gamma(\frac{1}{2})$, so that the space $\tilde{\Omega}$ of such minimal geodesics (shown in the preceding lemma to be homeomorphic to the Grassmannian manifold $G^{\mathbb{C}}_{2m,m}$) may be identified with the subspace $\{\gamma(\frac{1}{2}) | \gamma \in \tilde{\Omega}\}$ of $SU(2m)$. The latter subspace coincides with the intersection of $SU(2m)$ with its Lie algebra $su(2m)$ (both considered as subspaces of the space ($\simeq \mathbb{R}^{8m^2}$) of all $2m \times 2m$ complex matrices).*

PROOF. The first assertion follows easily from the form of the minimal geodesics $\gamma(t)$, established in the proof of Lemma 25.2 (see (1)):

$$\gamma(t) = \exp(tX) = (\cos \pi t) I_{2m} + (\sin \pi t) X;$$

putting $t = 0, 1$ we obtain $\gamma(0) = I_{2m}$, $\gamma(1) = -I_{2m}$, while at $t = \frac{1}{2}$, we have $\gamma(\frac{1}{2}) = X$, i.e. the midpoint of the geodesic arc is just X.

The remainder of the lemma then follows from the observation that those unitary $2m \times 2m$ complex matrices X which are at the same time skew-Hermitian, i.e. satisfy both $X\bar{X}^{\mathrm{T}} = 1$, and $X + \bar{X}^{\mathrm{T}} = 0$, are precisely the solutions of the matrix equation $X^2 = -I_{2m}$, and in $SU(2m)$ these are just X_0 (as in (1)) and its conjugates. (Note incidentally the consequence of this, that the elements of Grassmannian manifold $G^{\mathbb{C}}_{2m,m}$ correspond one-to-one to the possible "complex structures" in $SU(2m)$.) □

25.4. Lemma. *Every non-minimal geodesic γ in $SU(2m)$ from I_{2m} to $-I_{2m}$ has index at least $2m + 2$.*

PROOF. By the Index Theorem (21.7) the index of a geodesic arc γ in Ω is equal to the number of points in the interior of γ (counted according to their multiplicities) conjugate to the initial point I_{2m}. Since by definition two points on a geodesic arc are conjugate along that arc if there is a non-zero Jacobi field along the arc vanishing at those points, we are led to consider Jacobi's differential equation (see §21 (8)).

By the proof of Lemma 25.2, we may assume that $\gamma(t) = \exp(tX), 0 \le t \le 1$, where

$$X = \begin{pmatrix} i\pi k_1 & & 0 \\ & \ddots & \\ 0 & & i\pi k_{2m} \end{pmatrix}, k_1 + \cdots + k_{2m} = 0,$$

$$k_1 \ge k_2 \ge \cdots \ge k_{2m}, \quad k_i \text{ odd}. \tag{2}$$

In particular $\dot{\gamma}(0) = X$. Consider the (real) linear transformation

$$K_X: su(2m) \to su(2m),$$

defined in terms of the curvature by

$$K_X(Y) = R(X, Y)X \quad (= \tfrac{1}{4}[[X, Y], X] \text{ by Part I, Corollary 30.3.6}).$$

It follows from the symmetry relation

$$\langle R(X, Y)Z, W\rangle = \langle R(Z, W)X, Y\rangle$$

(see Part I, Corollary 30.3.6 or Theorem 30.2.1 (iv)) that K_X is a self-adjoint linear transformation:

$$\langle K_X(Y), W\rangle = \langle Y, K_X(W)\rangle.$$

We may therefore choose an orthonormal basis $e_1, \ldots, e_k$ for $su(2m)$ such that

$$K_X(e_i) = \mu_i e_i, \tag{3}$$

where the μ_i are the eigenvalues of K_X. If we extend the e_i to vector fields $e_i(t)$ along γ by parallel transport, then an arbitrary vector field $v(t)$ along γ can be expressed uniquely in the form

$$v(t) = \sum_{i=1}^{k} v_i(t)e_i(t),$$

and the Jacobi differential equation takes the form of the system (see §21 (9))

$$\frac{d^2v_i}{dt^2} + \sum_{j=1}^{k} \langle R(\dot{\gamma}, e_j)\dot{\gamma}, e_i\rangle v_j = 0, \qquad i = 1, \ldots, k. \tag{4}$$

Since there is an isometry ("symmetry") of the Lie group $SU(2m)$ which interchanges $\gamma(0)$ with any particular point $\gamma(t)$ and preserves the geodesic γ as a whole (see §1 (19) above, or Part II, §6.4), it follows that (3) holds at every point of γ:

$$K_{\dot{\gamma}(t)}(e_i(t)) = \mu_i e_i(t),$$

so that the system (4) becomes

$$\frac{d^2v_i}{dt^2} + \lambda_i v_i = 0, \qquad i = 1. \ldots, k.$$

We are interested in solutions of this system vanishing at $t = 0$. If $\mu_i > 0$, then $v_i(t) = c_i \sin \sqrt{\mu_i}\, t$ for some constant c_i, whence the zeros of $v_i(t)$ are at the integer multiples of $t = \pi/\sqrt{\mu_i}$. If $\mu_i = 0$, then $v_i(t) = c_i t$, and if $\mu_i < 0$, then $v_i(t)$ has the form $c_i \sinh \sqrt{|\mu_i|}\, t$; hence if $\mu_i \leq 0$ then $v_i(t)$ is either identically zero or vanishes only at $t = 0$. We conclude that the points on the geodesic $\exp(tX)$ conjugate to the initial point $\gamma(0)$, are determined by the positive eigenvalues μ_i of the linear transformation K_X: $su(2m) \to su(2m)$, given by

$$K_X(Y) = \tfrac{1}{4}[[X, Y], X], \tag{5}$$

occurring at the integer multiples of $t = \pi/\sqrt{\mu_i}$ in the open interval $(0, 1)$.

With X as in (2) it is easy to show that

$$[X, Y] = (i\pi(k_i - k_j)y_{ij}), \qquad Y = (y_{ij}),$$

whence

$$[X, [X, Y]] = (-\pi^2(k_i - k_j)^2 y_{ij}),$$

or (see (5))

$$K_X(Y) = \left(\frac{\pi^2}{4}(k_i - k_j)^2 y_{ij}\right).$$

Now it is easy to check that for each pair i, j with $i < j$ the $2m \times 2m$ matrices E_{ij} with entry $+1$ in the (i, j)th place and -1 in the (j, i)th place and zeros elsewhere, and also the matrices E'_{ij} with (i, j)th and (j, i)th entries both i (and zeros elsewhere), are eigenvectors of the linear transformation K_X, corresponding to the eigenvalue $(\pi^2/4)(k_i - k_j)^2$. Furthermore, each diagonal matrix in $su(2m)$ is clearly also an eigenvector of K_X (corresponding to the eigenvalue 0). Since a basis for $su(2m)$ can be chosen from among these eigenvectors, it follows that the non-zero eigenvalues of K_X are just those numbers $(\pi^2/4)(k_i - k_j)^2$ for which $k_i > k_j$. By the first part of the proof each of these positive eigenvalues μ gives rise to a sequence of conjugate points of γ corresponding to the parameter values $t = n\pi/\sqrt{\mu}$, $n = 1, 2, 3, \ldots$, i.e.

$$t = \frac{2}{k_i - k_j}, \frac{4}{k_i - k_j}, \frac{6}{k_i - k_j}, \ldots.$$

Since the number of such values of t in the open interval (0, 1) is $(k_i - k_j)/2 - 1$, and since each eigenvalue $(\pi^2/4)(k_i - k_j)^2 > 0$ has multiplicity 2, the Index Theorem tells us that the index λ of γ is given by the formula

$$\lambda = \sum_{k_i > k_j} (k_i - k_j - 2). \tag{6}$$

(Note that if γ were minimal, then by Lemma 25.2 (see (1)), we should have $k_i = \pm 1$ for all i, so that the index would be zero (as might be expected!).) Now if, on the one hand, at least $(m + 1)$ of the k_i have the same sign (negative say), then at least one of the positive k_i must be ≥ 3, whence

$$\lambda \geq \sum_{1}^{m+1} (3 - (-1) - 2) = 2m + 2.$$

On the other hand, if m of the k_i are negative and m positive, then, since they cannot all be ± 1 (γ being non-minimal), there must be one ≤ -3 and one ≥ 3, whence

$$\begin{aligned}\lambda &\geq \sum_{1}^{m+1} [(3 - (-1) - 2) + (1 - (-3) - 2)] + (3 - (-3) - 2) \\ &= 4m \geq 2m + 2.\end{aligned}$$

This completes the proof of the lemma. □

The theorem on unitary periodicity can now be deduced relatively easily, in two steps (the first of which involves the Fundamental Theorem of Morse Theory (22.5)), as follows.

25.5. Lemma. *The inclusion of the space $\tilde{\Omega}$ of minimal geodesics (homeomorphic to the Grassmannian manifold $G^{\mathbb{C}}_{2m,m}$) in the path space $\Omega = \Omega(SU(2m); I_{2m}, -I_{2m})$, induces an isomorphism between the corresponding homotopy groups in all dimensions up to and including 2m. In view of the isomorphism $\pi_i(\Omega M) \simeq \pi_{i+1}(M)$ (see Part II, Corollary 22.2.3) it then follows that*

$$\pi_i(G^{\mathbb{C}}_{2m,m}) \simeq \pi_{i+1}(SU(2m)), \qquad 0 \le i \le 2m.$$

PROOF. By Theorem 21.3 the action functional E on the path space Ω has as its critical points the geodesic arcs joining I_{2m} and $-I_{2m}$. By the preceding lemmas the critical points of index 0 (the minimal geodesics) form a subspace homeomorphic to $G^{\mathbb{C}}_{2m,m}$, while the index of all other critical points is at least $2m + 2$. It follows via Theorem 22.5 that $\Omega(SU(2m); I_{2m}, -I_{2m})$ has the homotopy type of a space obtained by attaching cells of dimension at least $2m + 2$ to the Grassmannian manifold $G^{\mathbb{C}}_{2m,m}$. Since attachment of cells of dimensions $\ge 2m + 2$ has no effect on the homotopy groups of dimensions $\le 2m$, we conclude that

$$\pi_i(\Omega) \simeq \pi_i(G^{\mathbb{C}}_{2m,m}) \qquad \text{for} \quad 0 \le i \le 2m,$$

as claimed. □

25.6. Lemma. *There is an isomorphism*

$$\pi_{i-1}(U(m)) \simeq \pi_i(G^{\mathbb{C}}_{2m,m}) \qquad \text{for} \quad 1 \le i \le 2m.$$

PROOF. As noted in Part II, §24.3(c), from the homotopy exact sequence of the standard fibration $U(m + 1) \to S^{2m+1}$ with fibre $U(m)$:

$$\cdots \to \pi_j(S^{2m+1}) \xrightarrow{\partial} \pi_{j-1}\, U(m) \xrightarrow{i_*} \pi_{j-1}\, U(m + 1) \to \pi_{j-1}(S^{2m+1}) \to \cdots,$$

it follows readily that the inclusion $i\colon U(m) \to U(m + 1)$ induces an isomorphism

$$i_*\colon \pi_{j-1}\, U(m) \simeq \pi_{j-1}\, U(m + 1) \qquad \text{for} \quad j \le 2m$$

(and also that the homomorphism $i_*\colon \pi_{2m}\, U(m) \to \pi_{2m}\, U(m + 1)$ is onto). Since the maps

$$\pi_{j-1}\, U(m) \to \pi_{j-1}\, U(m + 1) \to \pi_{j-1}\, U(m + 2) \to \cdots \tag{7}$$

induced by the inclusions are all isomorphisms, these groups are all isomorphic to $\pi_{j-1}(U)$, the $(j - 1)$st "stable homotopy group" of the unitary group.

From (7) and the homotopy exact sequence of the fibre bundle with total space $U(2m)$, fibre $U(m) \subset U(2m)$, and base the coset space $U(2m)/U(m)$ (the "complex Stiefel manifold"; cf. Part II, §5.2(d)):

$$\cdots \to \pi_j U(m) \xrightarrow{i_*} \pi_j U(2m) \to \pi_j(U(2m)/U(m)) \xrightarrow{\partial} \pi_{j-1}\, U(m) \to \cdots,$$

it follows that

$$\pi_j(U(2m)/U(m)) = 0 \qquad \text{for} \quad j \le 2m. \tag{8}$$

Considering, finally, the homotopy exact sequene of the fibration

$$U(2m)/U(m) \to G^{\mathbb{C}}_{2m,m} = U(2m)/(U(m) \times U(m)),$$

with fibre $U(m)$, namely

$$\cdots \to \pi_j(U(2m)/U(m)) \to \pi_j(G^{\mathbb{C}}_{2m,m}) \xrightarrow{\partial} \pi_{j-1}(U(m)) \xrightarrow{i_*} \pi_{j-1}(U(2m)/U(m)) \to \cdots,$$

and invoking (8), we obtain

$$\partial\colon \pi_j(G^{\mathbb{C}}_{2m,m}) \simeq \pi_{j-1}(U(m)) \qquad \text{for} \quad j \le 2m, \tag{9}$$

as required. □

The theorem is now immediate from this lemma and the preceding one, together with (7) and the isomorphism

$$\pi_j(SU(m)) \simeq \pi_j(U(m)) \qquad \text{for} \quad j \ne 1,$$

which follows easily by considering the homotopy exact sequence of the bundle $U(m) \to S^1$ with fibre $SU(m)$ (see Part II, §24.4). □

We wish now to exhibit explicitly the isomorphism $\pi_{j-1}(U(m)) \simeq \pi_{j+1}(U(2m))$, $j \le 2m$, obtained above as the composite of the chain of isomorphisms

$$\pi_{j-1}(U(m)) \underset{\simeq}{\overset{\partial}{\leftarrow}} \pi_j(G^{\mathbb{C}}_{2m,m}) \underset{\simeq}{\to} \pi_j(\Omega(SU(2m); I_{2m}, -I_{2m})) \simeq \pi_{j+1}(SU(2m))$$
$$\simeq \pi_{j+1}(U(2m)). \tag{10}$$

To this end, let $f_{j-1}\colon S^{j-1} \to U(m)$ be a continuous map (representing the homotopy class $[f_{j-1}] \in \pi_{j-1}(U(m))$); starting with f_{j-1}, we wish to construct a corresponding map $f_{j+1}\colon S^{j+1} \to SU(2m)$. We first distinguish in $SU(2)$, considered in its matrix guise as the group of complex 2×2 matrices of the form

$$x(\alpha, \beta) = \begin{bmatrix} \alpha & \beta \\ -\bar\beta & \bar\alpha \end{bmatrix}, \qquad \text{where} \quad |\alpha|^2 + |\beta|^2 = 1, \tag{11}$$

the subspace D^2, homeomorphic to a 2-dimensional closed disc, consisting of those matrices (11) with β non-negative real. We embed $SU(2)$ (and therefore D^2) into $SU(2m)$ by means of the map

$$\varphi\colon x \to x \otimes I_m = \begin{bmatrix} \alpha I_m & \beta I_m \\ -\bar\beta I_m & \bar\alpha I_m \end{bmatrix}. \tag{12}$$

We next consider the smooth curve $\hat\gamma$ in D^2, defined by

$$\hat\gamma(\beta, \tau) = \{x(\alpha, \beta) \mid \alpha = i\tau, \tau \text{ real}\},$$

and write $\gamma = \varphi(\hat\gamma)$. By Lemma 25.3 the complex Grassmannian manifold $G^{\mathbb{C}}_{2m,m}$ may be identified with the set of solutions in $SU(2m)$ of the matrix equation $X^2 = -I_{2m}$; since the points of $\hat\gamma$ all satisfy $(\hat\gamma(\beta, \tau))^2 = -I_{2m}$, it follows that, assuming this identification made, we have

$$\hat\gamma(\beta, \tau) \in G^{\mathbb{C}}_{2m,m} \subset SU(2m) \qquad \text{for} \quad 0 \le \beta \le 1.$$

We now distinguish in $SU(2m)$ the subset consisting of all conjugates of the matrices in (12), of the form

$$g(y, \alpha, \beta) = \begin{bmatrix} I_m & 0 \\ 0 & f_{j-1}(y)^{-1} \end{bmatrix} \begin{bmatrix} \alpha I_m & \beta I_m \\ -\beta I_m & \bar{\alpha} I_m \end{bmatrix} \begin{bmatrix} I_m & 0 \\ 0 & f_{j-1}(y) \end{bmatrix}$$

$$= \begin{bmatrix} \alpha I_m & \beta f_{j-1}(y) \\ -\beta f_{j-1}(y)^{-1} & \bar{\alpha} I_m \end{bmatrix}, \tag{13}$$

where $\beta \geq 1$, $|\alpha|^2 + \beta^2 = 1$ and $y \in S^{j-1}$. This defines a map $g\colon S^{j-1} \times D^2 \to SU(2m)$, and since

$$g(y, \alpha, 0) = \begin{bmatrix} \alpha I_m & 0 \\ 0 & \bar{\alpha} I_m \end{bmatrix},$$

for all y, it is not difficult to see that, in fact, the image in $SU(2m)$ is a $(j+1)$-dimensional sphre $S^{j+1} \subset SU(2m)$; thus g determines a map

$$f_{j+1}\colon S^{j+1} \to SU(2m). \tag{14}$$

Restricting β to be real, $0 \leq \beta \leq 1$, and α to have the form $i\tau$, τ real, we obtain from (13) a map

$$S^{j-1} \times I \to G^{\mathbb{C}}_{2m,m} \subset SU(2m),$$

whose image is a j-dimensional sphere

$$S^j \subset G^{\mathbb{C}}_{2m,m} \subset SU(2m), \tag{15}$$

obtained essentially as the quotient space of $S^{j-1} \times \hat{\gamma}$ with, on the one hand, all points of the form $(y, \hat{\gamma}(1, 0))$ (i.e. with $\alpha = 1$, $\beta = 0$, forming the lid of the cylinder) identified, and, on the other hand, all points $(y, \hat{\gamma}(-1, 0))$ (constituting the base of the cylinder) identified. Putting $\beta = 1$ (whence $\alpha = 0$) in (13) we obtain the map

$$h\colon y \mapsto \begin{bmatrix} 0 & f_{j-1}(y) \\ -f_{j-1}(y)^{-1} & 0 \end{bmatrix}, \qquad S^{j-1} \to G^{\mathbb{C}}_{2m,m} \subset SU(2m); \tag{16}$$

it can be shown that in fact $\partial[S^j] = [h]$, where $S^j \subset G^{\mathbb{C}}_{2m,m}$ is as in (15), and ∂ is (essentially) the boundary map determining the isomorphism (9). The upshot is that with the element $[f_{j-1}] \in \pi_{j-1}(U(m))$ (which may be identified with $[h]$), we have associated the element $[f_{j+1}] \in \pi_{j+1}(SU(2m))$ given by (14), i.e. by the formula

$$\{g(y, \alpha, \beta)\} = f_{j+1}(S^{j+1}) = \left\{\begin{bmatrix} \alpha I_m & \beta f_{j-1}(y) \\ -\beta f_{j-1}(y)^{-1} & \bar{\alpha} I_m \end{bmatrix}\right\}.$$

In fact, the map $[f_{j-1}] \mapsto [f_{j+1}]$ so defined coincides with the isomorphism of unitary periodicity given by (10): The first step, whereby $[f_{j-1}]$ is associated (via ∂^{-1}) with the S^j of (15), has already been noted. The next step is achieved by means of the embedding of $G^{\mathbb{C}}_{2m,m}$ in $SU(2m)$ as the set of mid-points of the minimal geodesics from I_{2m} to $-I_{2m}$ (see Lemma 25.3); this embedding associates the sphere S^j (or, more properly, "spheroid", i.e. map of the sphere) in $G^{\mathbb{C}}_{2m,m}$ with the spheroid in $\Omega(SU(2m))$ consisting of all the minimal geodesics whose mid-points are in S^j, which in turn, via the standard isomorphism of Part II, Corollary 22.2.3, is associated with $S^{j+1} \subset SU(2m)$. We have thus outlined the proof of the following

25.7. Theorem (Fomenko). *Let $f_{j-1}: S^{j-1} \to U(m)$ represent an arbitrary element of the homotopy group $\pi_{j-1}(U(m))$. The isomorphism of unitary periodicity (10) is given by the explicit formula $[f_{j-1}] \to [f_{j+1}]$, where $f_{j+1}: S^{j+1} \to SU(2m)$ is defined by*

$$f_{j+1}: S^{j+1} \to \{g(y, \alpha, \beta) | y \in S^{j-1}, \beta \geq 1, |\alpha|^2 + \beta^2 = 1\} \subset SU(2m),$$

with $g(y, \alpha, \beta)$ as in (13).

Thus, to repeat somewhat, the isomorphism of unitary periodicity (10) is, from a visual, geometric point of view, constructed in rather a simple way in the following two steps:

Step 1. Starting with an arbitrary spheroid f_{j-1} in $U(m)$ one transforms this spheroid by means of the boundary homomorphism $\partial: \pi_j(G^{\mathbb{C}}_{2m,m}) \to \pi_{j-1}(U(m))$ (or rather its inverse; here it is in fact an isomorphism—see (9)), into a spheroid S^j of one higher dimension in the Grassmannian manifold (see (16) for the explicit formula).

Step 2. By identifying the Grassmannian manifold $G^{\mathbb{C}}_{2m,m}$ with the intersection of the group $SU(2m)$ with its Lie algebra $su(2m)$ (both regarded as subspaces of the space of all complex $2m \times 2m$ matrices), which intersection consists precisely of the mid-points of the minimal geodesics in $SU(2m)$ from I_{2m} to $-I_{2m}$ (see Lemma 25.3), one can associate the spheroid S^j in $G^{\mathbb{C}}_{2m,m}$ obtained in Step 1, with the spheroid $S^{j+1} \subset SU(2m)$ obtained as the union of those minimal geodesics whose mid-points are points of the spheroid $S^j \subset G^{\mathbb{C}}_{2m,m}$. This spheroid, now in $SU(2m)$ and of dimension $j + 1$, is then the final image of the initial spheroid f_{j-1} under the isomorphism of unitary periodicity (the explicit formula for which is given in the above theorem).

In the case $m = 2, j = 4$, if the initial spheroid $f_{j-1} = f_3: S^3 \to SU(2) \subset U(2)$ is taken to be the map identifying the sphere S^3 with $SU(2)$:

$$f_3(y) = \begin{bmatrix} \alpha & \beta \\ -\bar{\alpha} & \bar{\beta} \end{bmatrix}, \qquad |\alpha|^2 + |\beta|^2 = 1,$$

then the homotopy class $[f_3]$ is a generator of $\pi_3(SU(2)) \simeq \mathbb{Z}$. Applying repeatedly the above explicit formula for the isomorphism of unitary periodicity, starting with this $f_{j-1} = f_3$, one obtains in succession generators $[f_5]$, $[f_7], \ldots$, in the cases $m = 2^2, 2^3, \ldots$, i.e. for each $k = 1, 2, \ldots$, a map

$$f_{2k+1}: S^{2k+1} \to SU(2^k), \tag{17}$$

representing a generator of $\pi_{2k+1}(SU(2^k)) \simeq \mathbb{Z}$. For each k the map f_{2k+1} so obtained is in fact the complex analogue $\alpha_{2k+1}^{\mathbb{C}}$ of the (real) "duality" map α_{2k+1} occurring in the theory of "Clifford algebras" and spinor representations of the orthogonal group (cf. [5], where the "isomorphism of orthogonal periodicity" is related to the structure of certain Clifford algebras).

We conclude this subsection by providing some of the details of this analogy, since for the above pairs of values $m = 2^k$, $j = 2k + 2$ $(k = 1, 2, \ldots)$ it furnishes another explicit formula for the isomorphism of unitary periodicity, simplifying still further the geometric picture. Thus the map $\alpha_{2k+1}^{\mathbb{C}}$ may be defined as follows: Let

$$f: S^{n-1} \to GL(N, \mathbb{C}), \qquad g: S^{m-1} \to GL(M, \mathbb{C}),$$

be any two continuous maps. Regarding S^{n-1} and S^{m-1} as embedded in the standard way in $\mathbb{R}^n$ and $\mathbb{R}^m$ respectively, we may clearly extend these maps to homogeneous maps of $\mathbb{R}^n$ and $\mathbb{R}^m$ (to the respective general linear groups). One then defines a map

$$f * g = \omega: \mathbb{R}^{n+m} \backslash \{0\} \to GL(2MN, \mathbb{C}),$$

by setting

$$(f * g)(x, y) = \begin{bmatrix} f(x) \otimes I_M & I_N \otimes g(y) \\ -I_N \otimes g^*(y) & f^*(x) \otimes I_M \end{bmatrix},$$

where $f^*(x)$ and $g^*(y)$ are the conjugate transposes of $f(x)$ and $g(y)$, and $(x, y) \in \mathbb{R}^n \times \mathbb{R}^m$, $(x, y) \neq (0, 0)$. The restriction of this map to the standard unit sphere $S^{n+m-1} \subset \mathbb{R}^{n+m}$, then yields a map $S^{n+m-1} \to GL(2MN, \mathbb{C})$. Taking, in particular, the map $\alpha: S^1 \to GL(1, \mathbb{C}) = \mathbb{C} \backslash \{0\}$ to the unit circle: $\alpha(z) = z$, $|z| = 1$, one then defines $\alpha_{2k+1}^{\mathbb{C}}$ to be the restriction to $S^{2k+1} \subset \mathbb{R}^{2k+2}$ of the map

$$\alpha * \cdots * \alpha: \mathbb{R}^{2k+2} \backslash \{0\} \to GL(2^k, \mathbb{C}).$$

It is now not difficult to verify directly that $\alpha_{2k+1}^{\mathbb{C}} \equiv f_{2k+1}$, where f_{2k+1} is as defined above (see (17) *et seqq.*).

25.2. Unitary Periodicity via the Two-Dimensional Calculus of Variations

The proof of unitary periodicity given in the preceding subsection is based on the one-dimensional calculus of variations applied to the action functional defined on (piecewise-smooth) paths in the unitary group. It turns out that the isomorphism of unitary periodicity can be established more naturally by considering instead an appropriate 2-dimensional variational problem.

In the above, "classical", approach to unitary periodicity, the proof is carried out in two quite distinct steps, in each of which the dimension of the homotopy groups under consideration increases by 1. The fact that the proof breaks up in this way into two parts (the required increase of 2 in the dimension being achieved by means of two successive increases of 1) is a natural consequence of the method used, inasmuch as that method involves the one-dimensional calculus of variations of the action (and length) functionals on paths, i.e. maps of the one-dimensional disc D^1 (a line segment). We now recapitulate the details of that method, in order to divine, as it were, the appropriate 2-dimensional analogue: Thus we take as the disc D^1 the interval $[0, 1]$, with boundary $\partial D^1 = S^0$ (the zero-dimensional sphere, consisting of the points 0, 1), and now denote by Π_1^* the space $\Omega^*(SU(2m); I_{2m}, -I_{2m})$ of all continuous maps f of D^1 to $SU(2m)$ satisfying $f|_{S^0} = i_0|_{S^0}$, where $i_0(S^0) = \{I_{2m}, -I_{2m}\}$, i.e. every f sends the end-points 0, 1 of D^1 to the same points $I_{2m}, -I_{2m}$ (respectively) of $SU(2m)$. The action functional E on Π_1, the subspace of Π_1^* consisting of all piecewise-smooth paths in Π_1^* (and homotopically equivalent to Π_1^*) is given by

$$E(\omega) = E_0^1(\omega) = \int_0^1 \left|\frac{d\omega}{dt}\right|^2 dt, \qquad \omega \in \Pi_1.$$

(The associated length functional

$$L_0^1(\omega) = \int_0^1 \left|\frac{d\omega}{dt}\right| dt,$$

was shown (essentially) in Part I, §31.2, to have the same extremal paths as the action functional E_0^1 in the class of all (piecewise-) smooth paths.) The set of points (i.e. paths) on which the action E_0^1 (and therefore also L_0^1) attains an absolute minimum, forms a subspace $W = \tilde{\Pi}_1$, shown in Lemma 25.2 to be homeomorphic to the Grassmannian manifold $G_{2m,m}^{\mathbb{C}}$, whence it was deduced (using Morse theory; see in particular Lemma 25.5) that the spaces Π_1 (and hence also Π_1^*) and $G_{2m,m}^{\mathbb{C}}$ have the homotopy type of cell complexes with identical $(2m+1)$-dimensional skeletons. Thus one might say that the "analytic part" of the isomorphism of unitary periodicity is given by the consequent isomorphism

$$\pi_i(G_{2m,m}^{\mathbb{C}}) \simeq \pi_i(\tilde{\Pi}_1) \simeq \pi_i(\Pi_1) \simeq \pi_{i+1}SU(2m), \qquad i \leq 2m,$$

since the subsequent step, $\pi_i(G_{2m,m}^{\mathbb{C}}) \simeq \pi_{i-1}(U(m))$, does not involve the action functional E_1^0, being of a purely homotopy-theoretic character (see Lemma 25.6).

The above-described geometrical procedure for obtaining the isomorphism of unitary periodicity prompts the idea of trying to construct that isomorphism in a single step, rather than two distinct steps, by utilizing some 2-dimensional functional rather than the one-dimensional action functional on paths. It turns out that there is a feasible such method of obtaining the isomorphism of unitary periodicity, which moreover leads to a further simplification of the "geometrical picture" of that isomorphism. We shall in the present subsection expound this "one-step" procedure. (Note, however, that the construction we give ultimately invokes the explicit formula (13), so that it does not constitute an independent proof of unitary periodicity.)

We first need to describe the appropriate 2-dimensional problem; thus we seek to define a suitable functional on a suitable specially chosen class of maps of a 2-dimensional disc. To begin with consider in $SU(2m)$ the embedded circle (a one-parameter subgroup)

$$S_0^1 = \left\{ \begin{bmatrix} \alpha I_m & 0 \\ 0 & \bar{\alpha} I_m \end{bmatrix} \middle| |\alpha| = 1 \right\}. \tag{18}$$

Let $D^2 \subset \mathbb{R}^2$ be the disc with centre $(0, 0)$ and radius 1, and let $j_0: S^1 \to SU(2m)$ be the (fixed) map identifying the boundary $S^1 = \partial D^2$ of the disc D^2 with the circle S_0^1 in the obvious way. Denote by Π_2^* the topological space (topologized by means of an appropriate distance function, i.e. metric) consisting of all continuous maps $f: D^2 \to SU(2m)$ satisfying $f|_{S^1} = j_0$. (Thus $S_0^1 \subset SU(2m)$ plays, in the present 2-dimensional context, the role analogous to that played previously by the zero-dimensional sphere consisting of the two points I_{2m}, $-I_{2m}$.) It can be shown that Π_2^* has the homotopy type of a cell complex.

We next consider the subspace $\Pi_2 \subset \Pi_2^*$ consisting of all maps f in Π_2^*, belonging to a certain function space $H_1^2(D^2)$ which we shall now define: Given a region X of Euclidean space $\mathbb{R}^n(x^1, \ldots, x^n)$, we shall say that a function $u: X \to \mathbb{R}$ *belongs to the class of functions* $H_m^p(X)$, if it satisfies the following two conditions:

(i) $u \in L_p(X)$, i.e. $|u|^p$ is integrable;
(ii) corresponding to each n-tuple $\alpha = (\alpha_1, \ldots, \alpha_n)$ of non-negative integers with $0 \le |\alpha| \le m$ if $m > 1$ and $|\alpha| = 1$ if $m = 1$, where $|\alpha| = \alpha_1 + \cdots + \alpha_n$, there should exist a "generalized derivative" $D^\alpha u$ of u, i.e. a function $r_\alpha \in L_p(X)$ with the property that for every function $g: X \to \mathbb{R}$ of class C^∞

$$\int_X g(x) r_\alpha(x)\, dx = (-1)^{|\alpha|} \int_X |\bar{D}^\alpha g(x)| u(x)\, dx,$$

where $\bar{D}^\alpha g = \partial^{|\alpha|} g / (\partial x^1)^{\alpha_1} \ldots (\partial x^n)^{\alpha_n}$.

(Thus, in particular, $H_1^2(D^2)$ consists of those square integrable functions $u: D^2 \to \mathbb{R}$ for which for each $i = 1, 2$ there exists a "generalized derivative" $D_i(u)$, i.e. a square integrable function $\bar{r}_i$ such that for every C^∞-function $g: D^2 \to \mathbb{R}$

$$\int_{D^2} g(x) \bar{r}_i(x)\, dx = \int_{D^2} \left| \frac{\partial g}{\partial x^i} \right| u(x)\, dx,$$

where x_1, x_2 are the standard co-ordinates on $\mathbb{R}^2$.) To say that a map $f: D^2 \to SU(2m)$ is in $H_1^2(D^2)$ is then understood to mean that each of the $8m^2$ real co-ordinate functions is in $H_1^2(D^2)$. Thus the condition that $f: D^2 \to SU(2m)$ belong to $H_1^2(D^2)$ (as well as to Π_2^*) to qualify as a member of Π_2, replaces, in the 2-dimensional context, the previous condition that paths $D^1 \to SU(2m)$ be piecewise-smooth to qualify for membership in Π_1 $(= \Omega)$, the appropriate condition for developing a "one-dimensional" Morse theory applicable to path spaces.

Having specified Π_2, we now choose a suitable functional on Π_2; the Dirichlet functional $D[f]$, which associates with each $f \in \Pi_2$ the "Dirichlet integral" of f (see the definition below), turns out to be appropriate for our purpose. As was noted in Part I, §37.5, the 2-dimensional Dirichlet functional is the analogue of the one-dimensional action functional in much the same sense that the area functional A is the 2-dimensional analogue of the length functional. In particular, the value of the functional D at f, like the action E, depends also on the parameters, i.e. on the co-ordinates chosen on the domain of f, whereas the values taken on by A and L are parameter-independent. We now give the definition of the general n-dimensional Dirichlet functional (since the 8-dimensional version will be needed in the following subsection). Thus let M and V be Riemannian manifolds with metric tensors $g_{ij}(y)$, $y \in M$, and $\hat{g}_{kl}(v)$, $v \in V$. With each map $f: V \to M$, $f \in H_1^p(V, M)$ (see above) we associate the "mixed" tensor $y_k^i = D^{\alpha_k}(y^i)$, where the y^i are the local co-ordinates on M of the image under f of each $v \in V: (y^i) = y = f(v)$, and where $\alpha_k = (0, \ldots, 1, \ldots, 0)$ is the n-tuple ($n = \dim V$) with kth component 1 and all other components 0, and D^{α_k} is the corresponding generalized differential operator defined above. (Thus the upper index i of the tensor y_k^i ranges from 1 to $\dim M$, and the lower, k, from 1 to $\dim V = n$.) Given two such mixed tensors y_k^i, $\bar{y}_l^j$, arising from maps $f, \bar{f}: V \to M$, we define their (mixed) *scalar product* by

$$(y_k^i, \bar{y}_l^j) = \hat{g}^{kl} g_{ij} y_k^i \bar{y}_l^j,$$

where $(\hat{g}^{kl}) = (\hat{g}_{kl})^{-1}$. Then at each $f \in H_1^p(V, M)$ we define the *Dirichlet functional* D on $H_1^p(V, M)$ by

$$D[f] = \int_V \left[\frac{1}{n}(y_k^i, y_l^j)\right]^{n/2} dV, \tag{19}$$

where dV denotes the element of volume of the Riemannian manifold V. A map $f \in H_1^p(V, M)$ is then said to be *harmonic* if $\delta D[f; \eta] = 0$ for every vector field η on $f(V)$ belonging to H_1^p, i.e. if the first-order variation of $D[f]$ resulting from a variation of f by means of any such field η is zero (see, e.g. Part I, §37, and cf. in particular Proposition 37.5.3 there). Thus the harmonic maps are the n-dimensional analogues of the geodesics. It can be shown by direct calculation that the resulting Euler–Lagrange equations for the harmonic maps are of the form $\nabla^k \nabla_k y^i = 0$ (see Part I, Theorem 37.1.2), where ∇_k denotes the operation of covariant differentiation with respect to the connexion on V compatible with its Riemannian metric.

In the particular case of present interest, the manifold V is the 2-dimensional unit disc in its standard position in Euclidean $\mathbb{R}^2$, so that $\hat{g}_{kl} = \delta_{kl}$, and the Dirichlet functional (19) takes the form

$$D[f] = \frac{1}{2}\int_{D^2} [(y_1^i, y_1^j) + (y_2^i, y_2^j)]\, dV = \frac{1}{2}\int_{D^2} g_{ij}(y_1^i y_1^j + y_2^i y_2^j)\, dV, \tag{20}$$

where $f \in \Pi_2$, and g_{ij} is the metric on $M = SU(2m)$ induced from the Euclidean metric on some $\mathbb{R}^N$ with $SU(2m)$ embedded in a standard way in $S^{N-1} \subset \mathbb{R}^N$ (see below). If the Euclidan co-ordinates on the disc $D^2 \subset \mathbb{R}^2$ are u, v, then (20) beocmes (cf. Part I, end of §37.5)

$$D[f] = \frac{1}{2}\int_{D^2} [(y_u, y_u) + (y_v, y_v)]\, du\, dv, \qquad y = (y^i) \in SU(2m), \quad (21)$$

and the first variation δD of the functional D corresponding to the addition to f of a vector field $\eta \in H_1^2(D^2)$ on $f(D^2)$, is readily seen to be given by

$$\delta D[f; \eta] = \int_{D^2} \left[\left(\frac{\partial \eta}{\partial u}, y_u\right) + \left(\frac{\partial \eta}{\partial v}, y_v\right)\right] du\, dv. \quad (22)$$

The related functional A on the space Π_2 assigns to each map $f \in \Pi_2$, $f(u, v) = (y^i(u, v))$, the integral (cf. Part I, §37.5)

$$\int_{D^2} \left[\det\begin{pmatrix} (y_u, y_u) & (y_u, y_v) \\ (y_u, y_v) & (y_v, y_v) \end{pmatrix}\right]^{1/2} du\, dv;$$

thus $A[f]$ is the 2-dimensional surface-area functional. As was noted in Part I, §37.5, for all f one has $A[f] \leq D[f]$, with equality occurring precisely if the map f is (generalized) conformal, i.e. if u, v furnish conformal co-ordinates on the surface $f(D^2) \subset SU(2m)$ (so that in terms of these co-ordinates the induced Riemannian metric on $f(D^2)$ has diagonal form). It was also shown (see loc. cit.) that under this condition on u, v as co-ordinates of $f(D^2)$, the Euler–Lagrange equations for a critical (in particular area-minimizing) surface $f(D^2)$ of the functional A, are equivalent to harmonicity (in the usual sense) of the radius vector $y = (y^i(u, v))$. This is analogous to the one-dimensional situation (of paths γ in $SU(2m)$ from I_{2m} to $-I_{2m}$) where the action and length functionals E and L satisfy $L^2(\gamma) \leq E(\gamma)$, with equality holding precisely when γ is parametrized by means of a natural parameter, in which case the critical paths for both functionals are the geodesic arcs from I_{2m} to $-I_{2m}$.

Thus the use of the Dirichlet functonal D (rather than the area functional A) serves the purpose, analogous to that served by E, of enabling us to ignore those maps f which, though not harmonic, can be made harmonic by means of a continuous change of co-ordinates on the disc D^2: although such a change of parameters has of course no effect on the area functional, it will in general cause the value of the Dirichlet functional to change.

Before stating the theorem we are leading up to, we mention the following auxiliary facts:

(i) there is a (natural) isomorphism

$$\beta_2\colon \pi_j(\Pi_2^*) \simeq \Pi_{j+2}(SU(2m)); \quad (23)$$

(ii) the space Π_2^* is homotopically equivalent to the space $\hat{\Pi}_2$ consisting of all continuous maps $S^2 \to SU(2m)$ with the north pole (say) mapped always to the same point of $SU(2m)$.

(The isomorphism (23) is established by means of two applications of Corollary 22.2.3 of Part II: if $\hat{\Omega}$ denotes the loop space of $SU(2m)$, i.e. the space of all loops beginning and ending at I_{2m}, then that corollary gives immediately $\pi_{j+1}(\hat{\Omega}) \simeq \pi_{j+2}(SU(2m))$; on the other hand, by applying it to the path space in $\hat{\Omega}$ consisting of all trajectories in $\hat{\Omega}$ with initial "point" the circle S_0^1 of (18) (such trajectories being given by maps $D^2 \to SU(2m)$; cf. Definition 21.1 *et seqq.*), on noting that the fibre above S_0^1 itself is precisely Π_2^*, one obtains $\pi_j(\Pi_2^*) \simeq \pi_{j+1}(\hat{\Omega})$.)

25.8. Theorem (Fomenko). *Let Π_2^*, Π_2 be the spaces of maps $D^2 \to SU(2m)$ defined above, and let W be the subspace of Π_2 consisting of all points (i.e. maps) $f \in \Pi_2$ at which the Dirichlet functional $D[f]$ attains its absolute minimum value. The following assertions hold:*

(i) *the subspace $W \subset \Pi_2$ is homeomorphic to the group $U(m)$;*
(ii) *the inclusion $i\colon W \to \Pi_2 \to \Pi_2^*$ induces an isomorphism*

$$i'_*\colon \pi_j(U(m)) \simeq \pi_j(\Pi_2^*) \qquad \textit{for} \quad j \le 2m,$$

(whence it follows that the $2m$-dimensional skeletons of $U(m)$ and Π_2^*, or rather of some realizations of these spaces as cell complexes, are homotopically equivalent).

The composite of the isomorphism i'_* with the isomorphism β_2 of (23) coincides with the isomorphism of unitary periodicity (see (10)):

$$\beta_2 \circ i'_*\colon \pi_j(U(m)) \simeq \pi_{j+2}(SU(2m)), \qquad j \le 2m. \tag{24}$$

(Thus by considering the set W of absolute minimum points of the 2-dimensional Dirichlet functional, the isomorphism of unitary periodicity can be achieved in a single step (involving an increase of 2 in the dimension of the homotopy groups), instead of the two steps required in the proof using the one-dimensional functionals of action and length.)

We give the proof in the form of a sequence of lemmas. First consider the 2-dimensional sphere

$$S_0^2 = \left\{ \begin{bmatrix} \alpha I_m & \beta I_m \\ -\beta I_m & \bar{\alpha} I_m \end{bmatrix} \middle| \; \beta \in \mathbb{R},\ |\alpha|^2 + |\beta|^2 = 1 \right\} \subset SU(2m), \tag{25}$$

whose equator, defined by $\beta = 0$, coincides with the circle S_0^1 of (18). The upper hemisphere, $\beta \ge 0$, which we shall now denote by D_0^2, played a significant role in the earlier explicit construction of the isomorphism of unitary periodicity (see (11) *et seqq.*). Using the fact that the inclusion $S_0^2 \to SU(2m)$ extends to a natural embedding of $SU(2)$ into $SU(2m)$ (see (12)), it can be shown that the sphere S_0^2 is a "totally geodesic" submanifold of $SU(2m)$, and therefore a minimal submanifold (i.e. area-minimizing) (see the discussion following Lemma 25.10 below). (A submanifold of a manifold M is said to be *totally geodesic* if every geodesic in M tangent to the submanifold at any point of it lies wholly in the submanifold or, equivalently, if every geodesic of the submanifold is a geodesic in M. That such submanifolds are locally volume-minimizing follows from the explicit form taken by the Riemannian curvature tensor on M when restricted to the submanifold; in particular, in the (present) case where M is a Lie group, this restriction is a direct summand of the full Riemannian curvature tensor on the ambient group M.)

Thus we may take it that the disc D_0^2, the upper hemisphere of S_0^2, is a totally geodesic submanifold of $SU(2m)$. Let $\hat{W}$ denote the set of (totally geodesic) discs $D^2(x) \subset SU(2m)$ of the form $D^2(x) = xD_0^2x^{-1}$, where $x \in SU(2m)$ and $xsx^{-1} = s$ for all $s \in S_0^1$.

25.9. Lemma. *The subspace $\hat{W} \subset \Pi_2$ is homeomorphic to $U(m)$.*

PROOF. Let x be any element of $SU(2m)$ satisfying $xs = sx$ for all $s \in S_0^1$. Since $S_0^1 = \{\alpha I_m \oplus \bar{\alpha} I_m \mid |\alpha| = 1\}$ (see (18)), it follows easily that x must also have block diagonal form:

$$x = A \oplus D, \qquad A, D \in U(m),$$

whence

$$x = (I_m \oplus DA^{-1})(A \oplus A) = x_1(A \oplus A) \qquad \text{where} \quad x_1 = I_m \oplus DA^{-1}.$$

From the fact that $(A \oplus A)d = d(A \oplus A)$ for all $d \in D_0^2$, $A \in U(m)$, we infer that

$$D^2(x) = D^2(x_1) = \left\{ \begin{bmatrix} \alpha I_m & \beta C^{-1} \\ -\beta C & \bar{\alpha} I_m \end{bmatrix} \middle| \beta \geq 1, |\alpha|^2 + \beta^2 = 1 \right\} \quad \text{where } C = DA^{-1}.$$

Since $\beta \geq 0$, the matrix $C \in U(m)$ is uniquely determined by the disc $D^2(x)$; in fact, as is easily verified, the map $D^2(x) \mapsto C$ is a bijection from $\hat{W}$ to $U(m)$, and therefore, bicontinuity being obvious, a homeomorphism. This completes the proof. □

We use this homeomorphism to construct an embedding $i'\colon U(m) \to \Pi_2$. That homeomorphism associates with each $g \in U(m)$ the 2-dimensional disc

$$D^2(I_m \oplus g) = \left\{ \begin{bmatrix} \alpha I_m & \beta g^{-1} \\ -\beta g & \bar{\alpha} I_m \end{bmatrix} \middle| \beta \geq 1, |\alpha|^2 + \beta^2 = 1 \right\}.$$

(Note incidentally that if $g_1 \neq g_2$, then

$$D^2(I_m \oplus g_1) \cap D^2(I_m \oplus g_2) = S_0^1.)$$

Denoting by $f_0\colon D^2 \to D_0^2$ the standard identification with $D_0^2 \subset SU(2m)$ of the standard disc $D^2 \subset \mathbb{R}^2$, we define, for each $g \in U(m)$, a map $f_g\colon D^2 \to SU(2m)$ in Π_2 by setting

$$f_g(y) = (I_m \oplus g) f_0(y) (I_m \oplus g^{-1}), \qquad y \in D^2.$$

The embedding $i'\colon U(m) \to \Pi_2$ we are seeking to define, is then given by $i'\colon g \mapsto f_g$.

It follows from the above lemma that the set of maps $i'(U(m)) \subset \Pi_2$, coincides with the set of maps of the form $\mathrm{Ad}_x \circ f_0$, where x ranges over the group $G = \{I \oplus C\} \subset U(2m)$, $G \simeq U(m)$, and Ad_x denotes conjugation by x; thus $i'(U(m))$ is the orbit containing the point f_0 of Π_2 under the conjugating action of the group G.

25.10. Lemma. *The homomorphism*

$$\beta_2 \circ i'_*: \pi_j(U(m)) \to \pi_{j+2}(SU(2m))$$

coincides, for $j \leq 2m$, with the isomorphism of unitary periodicity. It follows that the homomorphism

$$i'_*: \pi_j(U(m)) \to \pi_j(\Pi_2^*)$$

is an isomorphism for $j \leq 2m$.

PROOF. Let $\varphi: S^j \to U(m)$ represent a typical element $[\varphi]$ of $\pi_j(U(m))$. Writing $\varphi(y) = g$, for $y \in S^j$, we have

$$(\beta_2 \circ i'_*)(\varphi)(y) = \beta_2(f_g), \qquad f_g: D^2 \to SU(2m),$$

by definition of i'. By analysing the map β_2 in detail (see (23) *et seqq.*) it can be shown that

$$\beta_2(f_g) = D^2(I_m \oplus \varphi(y)) = D^2(I_m \oplus g)$$
$$= \left\{ \begin{bmatrix} \alpha I_m & \beta g^{-1} \\ -\beta g & \bar{\alpha} I_m \end{bmatrix} \middle| \beta \geq 1, |\alpha|^2 + \beta^2 = 1 \right\},$$

which coincides, in its essentials, with the explicit formula (13) (of Theorem 25.7) for the isomorphism of unitary periodicity (valid for $j \leq 2m$).

Since β_2 is an isomorphism (for all j), it follows that the homomorphism $i'_*: \pi_j(U(m)) \to \pi_j(\Pi_2^*)$ is an isomorphism for $j \leq 2m$. This completes the proof of the lemma. □

To complete the proof of the theorem it only remains to show that $i'(U(m)) = W$. To this end, consider the Euclidean space $\mathbb{R}^{8m^2}$ identified with the complex space $\mathbb{C}^{4m^2}$ of all $2m \times 2m$ complex matrices endowed with the bilinear form $\psi(A, B) = \mathrm{Re}(\mathrm{tr}\, A\bar{B}^{\mathrm{T}})$. The group $SU(2m)$ then embeds (isometrically) in the sphere $S^{8m^2-1} \subset \mathbb{R}^{8m^2}$ of radius $\sqrt{2m}$, as a smooth submanifold equipped with the special Riemannian metric (in fact, just the Euclidean metric on $\mathbb{R}^{8m^2}$) induced from the above bilinear form, invariant under both right and left translations by elements of $SU(2m)$ (the Killing form; cf. the beginning of §25.1).

Many of the metrical properties of the group $SU(2m)$ are more conveniently examined in the larger context of the ambient sphere S^{8m^2-1}, for instance, the property that there are no variations ("infinitesimal perturbations") of the embedded 2-dimensional disc $D_0^2 \subset SU(2m)$, leaving the boundary $S_0^1 = \partial D_0^2$ pointwise fixed, such that the perturbed disc $\tilde{D}_0^2$, while remaining minimal in

$SU(2m)$, is no longer totally geodesic (see above). To see this suppose that such a variation exists. Note that the circle $S_0^1 \subset SU(2m) \subset S^{8m^2-1}$ is a "great circle" in the sphere S^{8m^2-1}, in fact the equator of the "great 2-sphere" S_0^2, obtained as the cross-section of the sphere S^{8m^2-1} by a 3-dimensional subspace of $\mathbb{R}^{8m^2}$, i.e. by a 3-dimensional "plane" through the origin. Since by hypothesis the disc $\tilde{D}_0^2$, obtained from D_0^2 by means of a small perturbation, is not totally geodesic in $SU(2m)$, it will not be totally geodesic as a submanifold of the sphere S^{8m^2-1}. Hence the disc $\tilde{D}_0^2$ cannot form part of a "great 2-sphere" in S^{8m^2-1} (with equator S_0^1), in consequence of which its area must to a linear approximation exceed that of D_0^2, i.e. $\delta A > 0$. Thus every variation of a disc $D^2(x) \in \hat{W}$ either preserves the property of total geodesicity (in which case the perturbed disc is again a hemisphere, with boundary S_0^1, of a great 2-sphere in S^{8m^2-1}, and the transformation effecting the perturbation is a "rotation about S_0^1" and can be achieved by mans of an inner automorphism of the group $SU(2m)$), or destroys the property of local minimality enjoyed by $D^2(x)$ (at least in a neighbourhood of some point).

25.11. Lemma. *The inclusion $i'(U(m)) \subset W$ holds.*

PROOF. Since each map $f \in i'(U(m))$ has the form $\mathrm{Ad}_x \circ f_0$, where Ad_x is the map effecting conjugation by $x \in G$ (see above), it suffices to verify that f_0, the standard map $D^2 \to D_0^2$, is an absolute minimum point for the Dirichlet functional D. Now, as already made clear in the preceding discussion, the disc D_0^2, being a hemisphere of the spherical cross-section of the sphere S^{8m^2-1} by a 3-dimensional plane through the origin of $\mathbb{R}^{8m^2}$, has least area among all (appropriate) discs in S^{8m^2-1} with boundary S_0^1, so that the area functional of such discs takes on its least value at f_0. Since $A[f_0] = D[f_0]$ provided $D[f_0]$ is evaluated with respect to conformal co-ordinates, and since we always have $A[f] \leq D[f]$, the fact that $A[f_0]$ is least for A implies that $D[f_0]$ is least for D. □

25.12. Lemma. *We have $i'(U(m)) = W$.*

PROOF. Let $f: D^2 \to SU(2m)$, $f|_{S^1} = j_0$, be an element of W, i.e. a map in Π_2 at which the Dirichlet functional D attains its least value (among all values taken on Π_2). It follows from the inequality $A[f] \leq D[f]$ together with the fact that equality holds precisely when the surface defined by f is parametrized by conformal co-ordinates (on D^2), that in fact the present absolute-minimum function f is so parametrized, whence

$$D[f] = D[f_0] = A[f] = A[f_0]$$

(cf. the proof of the preceding lemma). (Note incidentally that f must then be harmonic as a function of those co-ordinates.) Since the metric on $SU(2m)$ is induced from the standard metric on $S^{8m^2-1} \subset \mathbb{R}^{8m^2}$ (see above), it follows that, like D_0^2, the image $f(D^2)$ must be a "great hemisphere" in S^{8m^2-1}, i.e. a hemisphere of a great 2-sphere $\tilde{S}^2 \subset S^{8m^2-1}$. Thus we have in S^{8m^2-1} the two great 2-spheres $\tilde{S}^2$ and S_0^2 (therefore totally geodesic in S^{8m^2-1}, and hence certainly in $SU(2m)$) whose intersection contains S_0^1 (which contains the point I_{2m}). Since each of these 2-spheres is the cross-section of S^{8m^2-1} by a 3-dimensional plane through the origin of $\mathbb{R}^{8m^2-1}$, they can be transformed into one another by means of a linear transformation of $\mathbb{R}^{8m^2}$ interchanging the two planes. Since the smallest subgroup of $SU(2m)$ containing S_0^2 is the isomorphic copy G_1 of $SU(2)$ obtained from (25) by letting β take on all complex values consistent with $|\alpha|^2 + |\beta|^2 = 1$, it can be deduced that the smallest subgroup G_2 containing $\tilde{S}^2$ is likewise an isomorphic copy of $SU(2)$. In both of the groups G_1, G_2, the circle S_0^1 is the image of a maximal torus $T^1 \cong S^1$ in $SU(2)$ (i.e. a maximal connected commutative Lie subgroup), and $S_0^1 \subset T^{2m-1}$ for some maximal torus $T^{2m-1} \subset SU(2m)$. Letting $\alpha_0, \alpha\colon SU(2) \to SU(2m)$ denote the maps with images G_2, G_1, extending f_0, f respectively, it follows from Lie theory, in particular as it pertains to maximal tori (see e.g. [51]), that the representations α_0, α are equivalent, i.e. there is an element x of $SU(2m)$ such that $\alpha = \mathrm{Ad}_x \circ \alpha_0$. Restricting to $D^2 \subset SU(2)$ we infer that $f = \mathrm{Ad}_x \circ f_0$, whence it follows (via the proof of Lemma 25.9 and the ensuing definition of i') that $f \in i'U(m)$. This completes the proof of the lemma, and with it the proof of Theorem 25.8. □

In connexion with the above proof, observe that the points of the set W (the absolute-minimum points for both functionals A and D) turned out to be "totally geodesic" (i.e. the surfaces they defined in $SU(2m)$ turned out to be totally geodesic in $SU(2m)$). This is somewhat analogous to the one-dimensional situation since, as we saw in §21, a (piecewise-smooth) minimal path $[0, 1] \to SU(2m)$ is automatically a geodesic in $SU(2m)$. However, in the present 2-dimensional context, minimality (as opposed to absolute minimality) of a map $D^2 \to SU(2m)$ does not in general entail that the resulting surface will be totally geodesic in $SU(2m)$. In fact, it can be shown that the only totally geodesic discs in $SU(2m)$ with boundary S_0^1 are those in $\hat{W}$, or, more precisely, if $f \in \Pi_2$ is a critical point of the functional D and if $f(D^2)$ is totally geodesic in $SU(2m)$, then $f \in W$.

25.3. Orthogonal Periodicity via the Higher-Dimensional Calculus of Variations

There is a periodic isomorphism also for the orthogonal group. This result, the "theorem on orthogonal periodicity", is also due to Bott.

25.13. Theorem (On Orthogonal Periodicity). *For the stable orthogonal group $O = \lim_{n\to\infty} O(n)$, where $O(n)$ is embedded in the standard way in $O(n + 1)$ for $n = 1, 2, \ldots$, there are isomorphisms*

$$\pi_i(O) \simeq \pi_{i+8}(O), \qquad i = 0, 1, 2, \ldots. \tag{26}$$

It follows that each stable homotopy group $\pi_i = \pi_i(0)$ is isomorphic to the appropriate one of the following groups:

$$\pi_0 \simeq \mathbb{Z}_2,\ \pi_1 \simeq \mathbb{Z}_2,\ \pi_2 = 0,\ \pi_3 \simeq \mathbb{Z},\ \pi_4 = \pi_5 = \pi_6 = 0,\ \pi_7 \simeq \mathbb{Z},\ \pi_i \simeq \pi_{i+8}. \tag{27}$$

We shall in the present subsection sketch the proof of (26) only. (Some of the isomorphisms in (27) may however be gleaned from Part II, §24.4.) The standard proof of the isomorphism of orthogonal periodicity utilizes one-dimensional Morse theory and breaks up into eight steps (analogously to the way in which the standard proof of the isomorphism of unitary periodicity expounded above splits up naturally into two steps). However, as we shall indicate below, by using the 8-dimensional calculus of variations as applied to the Dirichlet functional on a certain space of 8-dimensional balls (rather than the 2-dimensional balls, i.e. discs, that it was natural to consider in the context of unitary periodicity) the isomorphism of orthogonal periodicity $\pi_i(0) \simeq \pi_{i+8}(0)$ may be achieved (with some sacrifice of rigour) in a single step.

Proceeding much as in the unitary case, we consider the Euclidean space $\mathbb{R}^{p^2}$ of real $p \times p$ matrices with the usual Euclidan scalar product, which in terms of matrices is given by the bilinear form $\varphi(A, B) = \operatorname{tr}(AB^{\mathrm{T}})$. The group $SO(p)$ is then a smooth Riemannian submanifold of the standard sphere $S^{p^2-1} \subset \mathbb{R}^{p^2}$ of radius $\sqrt{p}$, centre the origin, with the (two-sided invariant) Killing metric induced on it from the Euclidean metric on $\mathbb{R}^{p^2}$ (cf. Part II, Corollary 6.4.4). The Lie algebra $so(p)$ of the group $SO(p)$ is also embedded in $\mathbb{R}^{p^2}$, as the subspace of matrices X satisfying $X^{\mathrm{T}} = -X$. We denote the intersection $so(p) \cap SO(p)$ by $\Omega_1(p)$. (Note incidentally that if p is even this intersection can be identified with the compact symmetric space $O(p)/U(p/2)$ (see Part II, §6, for the definition of symmetric space).) Clearly $\Omega_1(p)$ consists precisely of those elements g of $SO(p)$ satisfying the equation $g^2 = -I$, i.e. is in one-to-one correspondence with the set of "complex structures" on $\mathbb{R}^p$.

We now take p in the form $p = 16r$; it can be shown (see [44]) that in $\Omega_1(16r)$ there exist eight anti-commuting complex structures, i.e. there are eight matrices $J_1, \ldots, J_8$ in $\Omega_1(16r)$ satisfying

$$J_iJ_k = -J_kJ_i, \quad i \neq k, \qquad J_j^2 = -I.$$

It follows that the J_i, as position vectors of points in $SO(16r)$, are pairwise orthogonal (with respect to the Euclidean metric on $\mathbb{R}^{(16r)^2}$), and moreover together with the identity matrix I form an orthonormal set. Hence the sphere

$$S_0^8 = \{x \in SO(16r) | x = a^0I + a^1J_1 + \cdots + a^8J_8, (a^0)^2 + \cdots + (a^8)^2 = 1\}$$

is a cross-section of the sphere $S^q \subset \mathbb{R}^{(16r)^2}$ of dimension $q = (16r)^2 - 1$, by a 9-dimensional plane through the origin of $\mathbb{R}^{(16r)^2}$. Consequently, S_0^8 is a totally geodesic submanifold of S^q and therefore certainly of $SO(16r) \subset S^q$. Clearly

$$S_0^8 \cap so(16r) = S_0^8 \cap \Omega_1(16r) = \bar{S}_0^7,$$

where $\bar{S}_0^7$ is the 7-dimensional "great sphere" or "equator" of S_0^8, defined by

$a^0 = 0$, again totally geodesic (in S^q and therefore $SO(16r)$). Consider, on the other hand, the "equator"

$$S_0^7 = \{x \in S_0^8 | x = a^0 I + a^1 J_1 + \cdots + a^7 J_7\}$$

defined by $a^8 = 0$, the boundary of the totally geodesic 8-dimensional ball $D_0^8 \subset S_0^8$, the "upper hemisphere" of S_0^8, given by

$$D_0^8 = \{x \in S_0^8 | a^8 \geq 0\}.$$

Let D^8 denote the standard 8-dimensional ball (in standard position in the Euclidean space $\mathbb{R}^8$), let $S^7 = \partial D^8$, let $\hat{i}\colon D^8 \to D_1^8$ be the standard bijection from the ball D^8 to the upper hemisphere of the sphere S^8 (in standard position in $\mathbb{R}^9$), and let i'' be the standard isometric bijection identifying $\hat{i}(D^8) \subset S^8$ with $D_0^8 \subset S_0^8 \subset SO(16r)$ and coinciding on $\hat{i}(S^7)$ with a standard fixed isometry $j_0\colon \hat{i}(S^7) \to S_0^7$. Write

$$f_0 = i'' \circ \hat{i}\colon D^8 \to SO(16r).$$

Denote by Π_8^* (in notation reminiscent of that used in the unitary case) the space of all continuous maps $f\colon D^8 \to SO(16r)$ satisfying $f|_{S^7} = j_0 \circ \hat{i}$, and by $\Pi_8 \subset \Pi_8^*$ the subspace of all such maps $f\colon D^8 \to SO(16r)$ in the class $H_1^8(D^8)$ (see the preceding subsection). On the function space Π_8 we consider the (parameter-independent) 8-dimensional-volume functional $A[f]$ (the analogue of the area functional utilized in the unitary case) defined by

$$A[f] = \int_{D^8} \sqrt{\det(g_{ij})}\, dV,$$

(see Part I, §18.2) where (g_{ij}) is the metric on $SO(16r) \subset \mathbb{R}^{(16r)^2}$, and the Dirichlet functional (see (19))

$$D[f] = \int_{D^8} \left[\frac{1}{8}(y_k^i, y_l^j)\right]^4 dV = \int_{D^8} \left[\frac{1}{8}\sum_{k=1}^{8} g_{ij} y_k^i y_k^j\right]^4 dV,$$

which, as before, depends on the co-ordinatization of D^8. It is not difficult to show that for all $f \in \Pi_8$ we have $A[f] \leq D[f]$. Finally, let

$$\beta_8\colon \pi_j(\Pi_8^*) \simeq \pi_{j+8}(SO(16r)) \tag{28}$$

be the standard isomorphism constructed analogously to β_2 (see (23) *et seqq.*).

25.14. Theorem (Fomenko). *Let $SO(16r)$ be embedded in the Euclidean space $\mathbb{R}^{(16r)^2}$ as above, and let the spaces Π_8^* and Π_8 of maps of the 8-dimensional ball D^8 into $SO(16r)$ be defined also as above. Denote by W the subspace of Π_8 consisting of all those points (i.e. maps) $f \in \Pi_8$ at which the Dirichlet functional $D[f]$ attains its absolute minimum value on Π_8. The following statements hold:*

(i) *the subspace W is homeomorphic to the orthogonal group $O(r)$;*
(ii) *the inclusion $i\colon W \to \Pi_8 \to \Pi_8^*$ induces isomorphisms*

$$i'_*\colon \pi_j(O(r)) \simeq \pi_j(\Pi_8^*) \qquad \text{for} \quad j \leq r-2,$$

(whence it follows that the $(r-2)$-dimensional skeletons of the spaces $O(r)$

and Π_8^*, or rather of some realizations of these spaces as cell complexes, are homotopically equivalent).

The composite of the isomorphism i'_* with the isomorphism β_8 of (28) *coincides with the* (*standard*) *"isomorphism of orthogonal periodicity":*

$$\beta_8 \circ i'_*\colon \pi_j(O(r)) \simeq \pi_{j+8}(SO(16r)), \qquad j \le r-2.$$

We shall now indicate the proof of this theorem. Observe first that, just as the triviality of $\pi_2(U(2m))$ entails the connectedness of the space Π_2^* of continuous maps $D^2 \to U(2m)$ with prescribed restriction to ∂D^2, so the fact that $\pi_8(SO(16r)) \simeq \mathbb{Z}_2$ (a consequence of the theorem) implies that Π_8^* must have precisely two connected components. It will appear below that the space W is also made up of two connected components, one contained in each of the connected components of Π_8^*; the connected components of Π_8^* are in fact contractible (in the limit as $r \to \infty$) onto the respective components of W which they contain.

The present proof (like the more standard ones—see, e.g. [44]) involves the subset $\Omega_8 \subset SO(16r)$ of all complex structures (i.e. solutions of $J^2 = -I$) which anti-commute with the fixed complex structures $J_1, \dots, J_7$ whose existence was noted above, and hence with every point of the 6-dimensional standard sphere $S_0^6 \subset S_0^7$ defined in S_0^7 by the equation $a^0 = 0$ (see above). It is clear that $J_8 \in \Omega_8$, that Ω_8 is contained in the vector subspace of $\mathbb{R}^{(16r)^2}$ orthogonal to the subspace spanned by the vectors $E, J_1, \dots, J_8$, that $S_0^8 \cap \Omega_8 = \{\pm J_8\}$, and, consequently, that $D_0^8 \cap \Omega_8$ is the one-point space $\{J_8\}$. By means of a largely computational algebraic argument it can be shown that the space Ω_8 has two connected components, and in fact is diffeomorphic to $O(r)$ (see [44]).

With each point $x \in \Omega_8$ we associate the totally geodesic 8-sphere $S^8(x)$, having the standard 7-sphere S_0^7 as equator, given by

$$S^8(x) = \{a^0 I + a^1 J_1 + \cdots + a^7 J_7 + a^8 x \mid (a^0)^2 + \cdots + (a^8)^2 = 1\}.$$

Since the vectors $I, J_1, \dots, J_7, x$ form an orthonormal subsystem in $\mathbb{R}^{(16r)^2}$, it follows that $S^8(x)$ is the cross-section of $S^q \subset \mathbb{R}^{(16r)^2}$ by a 9-dimensional plane through the origin, and is therefore certainly a totally geodesic submanifold of $SO(16r) \subset \mathbb{R}^{(16r)^2}$. For each $x \in \Omega_8$ we then denote by $D^8(x)$ the upper hemisphere of $S^8(x)$:

$$D^8(x) = \{a \in S^8(x) \mid a = a^0 I + \cdots + a^7 J_7 + a^8 x,\ a^8 \ge 0\}. \tag{29}$$

Having in this way associated with each point $x \in \Omega_8$ a totally geodesic 8-dimensional disc (i.e. ball) $D^8(x)$ such that $\partial D^8(x) = S_0^7$ and

$$D^8(x_1) \cap D^8(x_2) = S_0^7 \qquad \text{if} \quad x_1 \ne x_2,$$

we are now in a position to construct a canonical embedding $i'\colon O(r) \to \Pi_8^*$ (the analogue of the embedding $i'\colon U(m) \to \Pi_2$ of the preceding subsection). For each $x \in \Omega_8$ let $\omega(x)\colon D^8 \to D^8(x)$ be the unique isometry whose restriction to the boundary is the standard map $j_0 \circ \hat{i}\colon S^7 \to S_0^7$ defined above ($= f_0|_{\partial D^8}$);

this then determines an embedding

$$\omega\colon \Omega_8 \to \Pi_8 \subset \Pi_8^*. \tag{30}$$

Denoting by $\varphi\colon O(r) \to \Omega_8$ the above-mentioned diffeomorphism, we now define i' by

$$i' = \omega \circ \varphi\colon O(r) \to \Pi_8 \subset \Pi_8^*. \tag{31}$$

25.15. Lemma. *The embedding $i'\colon O(r) \to \Pi_8^*$ induces isomorphisms*

$$\beta_8 \circ i'_*\colon \pi_j(O(r)) \simeq \pi_{j+8}SO(16r) \qquad \textit{for} \quad j \le r-2,$$

coinciding with the usual "isomorphisms of orthogonal periodicity".

PROOF (sketch only). Let $f\colon S^j \to O(r)$ be a map representing an arbitrary element of the jth homotopy group of $O(r)$: $[f] \in \pi_j(O(r))$, and consider the family of maps $\omega(x)\colon D^8 \to D^8(x)$ (see (30)) indexed by the elements x of $\varphi(f(S^j)) \subset \Omega_8$ where $\varphi\colon O(r) \cong \Omega_8$ is the above-mentioned diffeomorphism). Since each map $\omega(x)$ restricts to the standard map $S^7 \to S_0^7$ on the boundary $\partial D^8 = S^7$, and the $D^8(x)$ intersect pairwise precisely in S_0^7, the union of these maps $\omega(x)$ yields a map

$$F\colon S^{j+8} \to \bar{S} = \bigcup_{x \in \varphi(f(S^j))} D^8(x) \subset SO(16r), \tag{32}$$

with the property that $F|_{S^j} = \varphi \circ f$, where S^j is canonically situated (equatorially) in S^{j+8}. We claim that the correspondence $f \mapsto F$ induces an isomorphim

$$\pi_j(O(r)) \simeq \pi_{j+8}(SO(16r)), \qquad j \le r-2,$$

coinciding with the isomorphism of orthogonal periodicity as usually constructed. We now summarize very briefly the usual construction (see, e.g. [44]).

For each $k = 0, 1, \ldots, 8$, denote by Ω_k the set of all complex structures $J \in SO(16r)$ which anti-commute with the fixed complex structures $J_1, \ldots, J_{k-1}$. (The particular set Ω_8 has already figured in our discussion.) Starting at $k = 7$ (with a view to working our way down to $k = 0$), we define a map

$$\gamma_7\colon \Omega_8 \to \Omega(SO(16r); J_7, -J_7)$$

by setting $\gamma_7(x) = D^1(x)$, the minimal geodesic in $SO(16r)$ with midpoint x, lying in Ω_7. (It can be shown that $D^1(x)$ is uniquely defined by these conditions.) By taking the union of those $D^1(x)$ for which $x \in \varphi(f(S^j))$, we obtain a map $F_7\colon S^{j+1} \to \Omega_7$ extending $\varphi \circ f$:

$$F_7\colon S^{j+1} \to \bigcup_{x \in \varphi(f(S^j))} D^1(x) \subset \Omega_7, \qquad F_7|_{S^j} = \varphi \circ f.$$

It follows via one-dimensional Morse theory (this and the similar steps succeeding it contain the burden of the usual proof), that the correspondence $f \mapsto F_7$ induces an isomorphism

$$\pi_j(O(r)) \simeq \pi_{j+1}(\Omega_7).$$

Continuing in the same vein, one next defines a map $\gamma_6: y \mapsto D^1(y)$, from Ω_7 to the space of minimal geodesics in $SO(16r)$ joining J_6 to $-J_6$ and lying in Ω_6, whence one obtains as before a map

$$F_6: S^{j+2} \to \bigcup_{x \in F_7(S^{j+1})} D^1(y) \subset \Omega_6, \qquad F_6|_{S^{j+1}} = F_7,$$

with the property (established using Morse theory) that the correspondence $F_7 \mapsto F_6$ induces an isomorphism

$$\pi_{j+1}(\Omega_7) \simeq \pi_{j+2}(\Omega_6).$$

Iterating this procedure, one defines further maps $\gamma_5, \ldots, \gamma_0$ (taking $J_0 = I$), and $F_5, \ldots, F_0$, where in particular F_0 maps S^{j+8} to $\Omega_0 = SO(16r)$. It can now be verified (we omit the details) that

$$\bar{S} = \bigcup_{x \in \varphi(f(S^j))} [\gamma_0 \circ \gamma_1 \circ \cdots \circ \gamma_7(x)]$$

(where $\bar{S}$ is as in (32)), and that F_0 essentially coincides with the map F of (32). Since the composite correspondence

$$f \mapsto F_7 \mapsto \cdots \mapsto F_0$$

induces an isomorphism

$$\pi_j O(r) \simeq \pi_{j+8} SO(16r), \qquad j \leq r-2,$$

(the isomorphism of orthogonal periodicity as usually constructed), it follows that the correspondence $f \mapsto F$ induces the same isomorphism. Since $i' = \omega \circ \varphi$ the lemma now follows from the definition of the map F in terms of ω. □

To complete the proof of Theorem 25.14 it remains to show that $i'(O(r)) = W$. As in the unitary case we verify this in two steps.

25.16. Lemma. *The map $i': O(r) \to \Pi_8^*$ given by* (31) *above, satisfies $i'(O(r)) \subset W$, where W consists of all those functionals in Π_8 at which the Dirichlet functional D has its least value on Π_8.*

This is established by an argument analogous to that used in proving Lemma 25.11, exploiting the inequality $A[f] \leq D[f]$ together with the fact that for each $g \in O(r)$ the disc $i'(g) = D^8(\varphi(g))$ is a hemisphere of the cross-section of the sphere $S^q \subset \mathbb{R}^{(16r)^2}$ by a plane through the origin.

25.17. Lemma. *We have $i'(O(r)) = W$.*

PROOF. Initially the argument proceeds as in the proof of the analogous Lemma 25.12. Let $f \in W$, i.e. suppose $D[f]$ is least among all values taken by D on Π_8, and let $f_0: D^8 \to D_0^8 \subset SO(16r)$ be as before. In terms of standard co-ordinates on D^8 it is easy to see that $A[f_0] = D[f_0]$. Furthermore, in view of the boundary conditions on admissible maps $D^8 \to SO(16r)$ (namely that $\partial D^8 = S^7$ be mapped canonically onto $S_0^7 \subset SO(16r)$), the value $A[f_0]$ is the least taken by A on Π_2. Hence we can infer, via the general inequality $A[h] \leq D[h]$, that

$$D[f] = A[f] = A[f_0].$$

It follows that $f(D^8)$, like $f_0(D^8)$, must be a hemisphere of a central-plane cross-section of $S^q \subset \mathbb{R}^{(16r)^2}$ (of course having S_0^7 as its boundary: $\partial(f_0(D^8)) = S_0^7$).

Now let x be a vector in $f(D^8)$ orthogonal to the eight vectors $I, J_1, \ldots, J_7$. It follows from the above description of $f(D^8)$ that

$$f(D^8) = \{a^0 I + a^1 J_1 + \cdots + a^7 J_7 + a^8 x \mid (a^0)^2 + \cdots + (a^8)^2 = 1, a^8 \geq 0\}. \tag{33}$$

Since $f(D^8)$ is a central-plane cross-section of S^q with the points $\pm I$ as antipodal points of its boundary S_0^7, there is a unique minimal geodesic arc $\gamma(t)$, $0 \leq t \leq 1$, of $SO(16r)$ (actually of S^q), such that $\gamma(0) = I$, $\gamma(\frac{1}{2}) = x$, $\gamma(1) = -I$ (namely half the circle obtained by sectioning S^q by the 2-dimensional plane spanned by the vectors I and x). Now any geodesic γ in $SO(16r)$ with $\gamma(0) = I$, has the form $\gamma(t) = \exp(\pi t A)$ for some $A \in so(16r)$. By means of a fairly direct matrix computation (see [44]) it can be shown that if $\gamma(1) = -I$ and γ is minimal, then A is a complex structure: $A^2 = -I$. It follows that

$$\gamma(t) = (\cos \pi t)I + (\sin \pi t)A,$$

whence $\gamma(\frac{1}{2})$ $(= A)$ is a complex structure. Since in our situation $\gamma(\frac{1}{2}) = x$, we infer that x is a complex structure, i.e. $x \in \Omega_1(16r)$. By (33)

$$\frac{1}{\sqrt{2}}(x + J_k) \in f(D^8) \subset SO(16r) \qquad \text{for} \quad k = 1, \ldots, 7,$$

whence

$$\frac{1}{\sqrt{2}}(x + J_k) \in SO(16r) \cap so(16r) = \Omega_1(16r), \qquad k = 1, \ldots, 7,$$

i.e. $\frac{1}{2}(x + J_k)^2 = -I$. Hence $xJ_k + J_k x = 0$, i.e. $x \in \Omega_8$. We conclude that $f(D^8) = D^8(x)$ (see (29)), so that $f \in i'(O(r))$, as claimed. This completes the proof of the lemma and with it our sketch of the proof of Theorem 25.14. □

From the proof of orthogonal periodicity given for instance in [44], it emerges that the stable orthogonal group O has the homotopy type of the 4-fold loop space $\Omega\Omega\Omega\Omega\, Sp$, and that the stable symplectic group Sp has the homotopy type of the 4-fold loop space $\Omega\Omega\Omega\Omega\, O$, whence one obtains an analogous "symplectic periodicity" theorem. We leave to the reader the detailed formulation of this theorem incorporating an approach to its validation via the higher-dimensional calculus of variations, analogous to that described above in the unitary and orthogonal cases.

EXERCISE

Derive the following homotopy equivalences:
(i) $BSp \sim \Omega\Omega\Omega\, SO$;
(ii) $BO \sim \Omega\Omega\Omega\, Sp$ (see §10.4).
Find using these the first eight homotopy groups $\pi_i O$, $i = 0, \ldots, 7$, namely, in order, $\mathbb{Z}_2, \mathbb{Z}_2, O, \mathbb{Z}, O, O, O, \mathbb{Z}$.

In the preceding subsection, in the course of establishing unitary periodicity we showed that the space $i'(U(m)) \subset \Pi_2$ is actually the orbit of the particular point $f_0 \in \Pi_2$ under the conjugating action of the group $G = \{I \oplus C \mid C \in U(m)\} \subset U(2m)$. In the present situation of orthogonal periodicity a version of this result holds also (not however needed in the above proof of orthogonal periodicity).

25.18. Proposition. *The two connected components of the space $W = i'(O(r)) \subset \Pi_8$ are orbits of the conjugating action on Π_8 of the group*

$$G = J_8\Omega_8 \subset SO(16r) \qquad (G \simeq O(r)). \tag{34}$$

PROOF. We need to show that given any one of the totally geodesic balls $D^8(x)$ defined in (29), where $x \in \Omega_8$ is in the same connected component of Ω_8 as J_8, there is an element $g \in G$ such that $gxg^{-1} = J_8$, since then

$$gD^8(x)g^{-1} = D^8(J_8) = D_0^8,$$

so that the corresponding map $\omega(x)\colon D^8 \to D^8(x)$ (see (30)) is obtained from the map $f_0\colon D^8 \to D_0^8 \subset SO(16r)$, by conjugating by g. Since each $g \in G$ has the form $g = J_8 y$, $y \in \Omega_8$, the equation $gxg^{-1} = J_8$ is equivalent to $yxy^{-1} = J_8$, or $yxy = -J_8$ (since $y^{-1} = -y$).

Now in [44] the aforementioned diffeomorphism $\Omega_8 \cong O(r)$ is constructed by finding two r-dimensional subspaces X_1, X_2 of $\mathbb{R}^{16r}$ with the property that the matrices of the form $J_7 y$, $y \in \Omega_8$, act as (distinct) orthogonal transformations $X_1 \to X_2$, and in fact account for all such transformations. It follows readily that G, as defined in (34), is indeed a group, since it is identifiable with the group of all orthogonal self-transformations of X_1 (or X_2). The equation $yxy = -J_8$ for which we seek a solution $y \in \Omega_8$, is equivalent to

$$(J_7 y)(J_7 x^{-1})^{-1} J_7 y = -J_7 J_8,$$

whence we see that the existence of such a solution is equivalent to the existence of a solution $z \in O(r)$ of the equation

$$zAz = B$$

for each (fixed) pair A, B of elements in the same connected component of $O(r)$. Since every such equation does indeed have a solution (we lay the onus of verification of this on the reader), the proposition follows. □

In summary, the idea underlying the above approach to unitary and orthogonal periodicity (and borne out by the above discussion), is that both these results should arise via the same mechanism (namely the higher-dimensional calculus of variations applied to the Dirichlet functional), the particular outcome depending only on the space on which that functional is considered—the space of maps of 2-dimensional discs in the unitary case, and of 8-dimensional discs in the orthogonal case.

The reader will have observed that, this uniform approach notwithstanding, our proofs of both unitary and orthogonal periodicity ultimately relied on the more standard arguments of one-dimensional Morse theory. It would be of considerable interest if a direct uniform proof (using the Dirichlet functional) could be found which avoided any appeal to the Morse theory of the one-dimensional functionals of action and length. Such a proof would in fact be feasible if one could establish directly the contractibility of the $2m$-dimensional skeleton of Π_2 (or the $(r-2)$-skeleton of Π_8 in the orthogonal case) onto the space $i'(U(m))$ (resp. $i'(O(r))$) of points where the Dirichlet functional has its least value. It is precisely a contractibility result of this kind in the classical (i.e. one-dimensional) Morse theory, of the path space $\Pi_1 = \Omega(SU(2m); I_{2m}, -I_{2m})$ which allows one to obtain the crucial isomorphism $\pi_{j-1}(G^{\mathbb{C}}_{2m,m}) \simeq \pi_{j-1}(\Pi_1)$. However, the analogous result for the higher-dimensional calculus of variations applied to the Dirichlet functional is not available, a lack essentially due to the typical sort of difficulty which arises in higher-dimensional problems of the "Plateau" type, whereby the higher-dimensional functional in question may be degenerate on certain sets of positive measure contained in the extremal submanifolds.

§26. Morse Theory and Certain Motions in the Planar n-Body Problem

In this section we shall show how Morse theory may be applied to the analysis of certain motions in the many-body problem of celestial mechanics, i.e. the problem of describing the motions of n objects acting on each other by means of mutual (gravitational) forces. It is well known that to a first approximation the planets of our solar system move in a plane, the so-called "plane of the ecliptic". Furthermore, to a large degree of accuracy the centre of mass of the whole system may be identified with the sun's position, and the motion of the system may be assumed to be governed by the Newtonian gravitational potential of classical mechanics. The motion is then determined in the usual way by initial conditions, namely the positions and velocities of the gravitating masses (considered as point-particles) at some chosen initial instant of time. As is also well known, the general solution of the resulting system of differential equations is exceedingly complex: for instance, by the classical results of Bruns and Poincaré, there are no single-valued analytic integrals of the system (i.e. expressions in the position variables and their derivatives which are constant at all times) other than the "classical" ones (the integrals of energy, angular momentum, and of motion of the centre of mass). (However, in the case $n = 2$ a complete solution can be given; see Part I, §32.2.)

It has nevertheless proved possible to delineate among all solutions certain natural subclasses of solutions admitting a relatively simple description. One such subclass is that of all so-called "rigid-body" solutions, i.e. those particular solutions obtained under the additional assumption that all the masses of the system revolve together about the sun, at the same angular speed, in the plane of the ecliptic. Thus, in this special case, the relative positions of the masses remain unchanged and the whole system revolves like a (planar) rigid body about its centre of mass. In the literature such periodic solutions are sometimes referred to as "circular orbits" (with all particles having the same angular velolcity). It is a remarkable fact that the description of such planar "rigid-body" solutions of the many-body problem, reduces to the description of the critical points of a certain function, possibly Morse, on a smooth manifold, and that the resulting topological information about the manifold which we have seen (in for instance §15) to be obtainable from knowledge of the critical points, can then be used to draw important qualitative inferences about the geometric structure of the "circular" solutions under investigation.

For instance, the following question is of considerable interest: Given a system of point-masses, what planar configurations of the system are compatible with some "rigid-body" solution? (It is intuitively clear that far from every arrangement of the particles will quality for some planar "rigid-body" solution.) Clearly the admissible configurations will be determined by the masses of the n particles of the system. (In the particular case where all but one of the particles have the same mass, it turns out that these configurations are linked by the action of a certain discrete group of symmetries.) Such configurations are sometimes called "relative equilibria" of the system.

We now turn to the precise formulation of the problem. The planar n-body problem of celestial mechanics is determined essentially by a sequence of n real, positive numbers $m_1, \ldots, m_n$, representing the masses of the n bodies, or rather point-particles, situated at n points of the (2-dimensional) Euclidean plane. We may clearly suppose that the origin O coincides at all times with the centre of mass of the system of particles. The position of the jth particle on the plane will then be denoted, in terms of the standard Euclidean co-ordinates, by (x_j, y_j), or by the complex co-ordinate $z_j = x_j + iy_j$. From the assumption that the origin is the centre of mass we infer immediately the relation $\sum_{j=1}^{n} m_j z_j = 0$. The "configuration space" of the system is a certain subspace (see below) contained in the linear subspace $M = M^{2n-2}$ (or complex hyperplane) of the Euclidean space $\mathbb{R}^{2n} = \mathbb{C}^n$, given by

$$M^{2n-2} = \left\{(z_1, \ldots, z_n) \in \mathbb{R}^{2n} \,\middle|\, \sum_{j=1}^{n} m_j z_j = 0\right\}, \tag{1}$$

and the "phase space" of the system is (a certain submanifold of) the cotangent bundle $T^*M \cong M \times M$ (the direct square of M). (Note that since the co-ordinates are Euclidean, T and T^* may be identified; see Part I, §19.1.)

The *kinetic energy* K of the system is defined by the formula

$$K(v) = \frac{1}{2}\sum_{j=1}^{n} m_j|v_j|^2, \tag{2}$$

where $v = (\dot{z}_1, \ldots, \dot{z}_n) = (v_1, \ldots, v_n)$ is the velocity vector, and $|v_j|$ is the Euclidean length of the vector $v_j = \dot{z}_j = (\dot{x}_j, \dot{y}_j)$ in $\mathbb{R}^2$. (Note that $\Sigma\, m_j v_j = 0$.)

Denoting by Δ the subset of M^{2n-2} comprised of the points of the "bisector hyperplanes"

$$\Delta_{ij} = \{(z_1, \ldots, z_n) \in \mathbb{C}^n | z_i = z_j\}; \qquad \Delta = \bigcup_{i,j} \Delta_{ij}, \tag{3}$$

we define the *potential energy* to be the function on the subspace $M\backslash\Delta$ of M defined by

$$V(z_1, \ldots, z_n) = -\sum_{i \neq j} \frac{m_i m_j}{|z_i - z_j|}. \tag{4}$$

We now define the *configuration space* of the system to be $M\backslash\Delta$, and as usual (see Part II, §28.1) identify the *phase space* with the cotangent bundle $T^*(M\backslash\Delta)$ (which in the present context, as noted above, may be identified with the tangent bundle $T = T(M\backslash\Delta)$). Newton's second law yields the system of equations

$$\frac{\partial V}{\partial z_j} = m_j \ddot{z}_j \,(= m_j \dot{v}_j), \qquad j = 1, \ldots, n, \tag{5}$$

where $\partial V/\partial z_j = (\partial V/\partial x_j, \partial V/\partial y_j)$. This system determines a vector field X on the phase space with the vector attached at each point $(z, v) \in T$ taken to be $((\partial V/\partial z_j), \dot{v})$.

The *total energy* $E\colon T \to \mathbb{R}$ is given as usual by the formula $E(z, v) = V(z) + K(v)$, and represents a first integral of the flow X on T, i.e. $E(z, v)$ is constant on each integral trajectory $(z(t), v(t))$ of the dynamical system X determined by (5). That system admits a further integral, functionally independent of the integral E (at the points "in general position" on T), namely the *angular momentum*

$$J(z, v) = \sum_{i=1}^{n} m_i(z_i \wedge v_i), \tag{6}$$

where $z_i \wedge v_i$ denotes, as usual, the wedge product of the 1-forms z_i, v_i, or, equivalently, their cross product $z_i \times v_i$ (see Part I, §18.3):

$$z_i \wedge v_i = x_i v_i^2 - y_i v_i^1 \qquad \text{where} \quad z_i = (x_i, y_i), \quad v_i = (v_i^1, v_i^2) = (\dot{x}_i, \dot{y}_i).$$

Consider now the standard action of the group $G = S^1$ of rotations of the plane $\mathbb{R}^2$ about the origin (i.e. the centre of mass of the system), and the action induced co-ordinate-wise on $M \subset \mathbb{R}^{2n} = \mathbb{R}^2 \times \cdots \times \mathbb{R}^2$ (n times) and on the tangent space $T(M)$. Under this action G clearly preserves each "bisector" hyperplane Δ_{ij} (see (3)), and can therefore be regarded as acting on the spaces $M\backslash\Delta$, $T(M\backslash\Delta)$, leaving invariant K, V, E, J and X. It follows that the flow X determines a dynamical system on the orbit space $T(M\backslash\Delta)/G$ (identifiable with

$T^*((M\backslash\Delta)/G))$. Since, clearly, we may also factor by the action of the group of dilations $z \mapsto \lambda z$ of $\mathbb{C}^n$, the dynamical system may be further reduced to one on $T(\mathbb{C}P^{n-1}\backslash\tilde{\Delta})$, where $\tilde{\Delta}$ is the orbit space of Δ under the actions of the groups of rotations and dilations. We shall exploit this reduction a little later on; for the present we return to the original system on $T(M\backslash\Delta)$.

Our knowledge of the two integrals E and J permits us to define a map $I: T \to \mathbb{R}^2 = \mathbb{R} \times \mathbb{R}$, by setting

$$I(\xi) = (E(\xi), J(\xi)) \in \mathbb{R}^2, \qquad \xi = (z, v) \in T(M\backslash\Delta) = T. \tag{7}$$

The map I is obviously smooth, and affords a fibration of T with fibres the preimages

$$I_{c,p} = I^{-1}(c, p), \qquad (c, p) \in \mathbb{R}^2, \qquad E(\xi) = c, \qquad J(\xi) = p.$$

The fibres $I_{c,p}$ are, for almost all points $(c, p) \in \mathbb{R}^2$, smooth submanifolds of codimension 2 in $T = T(M\backslash\Delta)$ (by Sard's theorem; see Part II, Theorem 10.2.1), i.e. of dimension $4n - 4 - 2 = 4n - 6$, T having dimension $4n - 4$. It is clear from the definition of I that each fibre $I_{c,p}$ is a joint level set of the two integrals E and J. The following lemma is immediate from the definition of I and the invariance of E and J under the action of $G = S^1$.

26.1. Lemma. *The manifold $I_{c,p}$ (i.e. fibre in T above a point "in general position" in $\mathbb{R}^2$) is invariant under the action of the group $G = S^1$ on T, and relative to the flow X (i.e. any integral curve of the dynamical system X is contained in some $I_{c,p}$).*

From this invariance of $I_{c,p}$, the surface of constant energy $E = c$ and angular momentum $J = p$, it now follows that the orbit space $\tilde{I}_{c,p} = I_{c,p}/S^1$ is correctly defined. One of the problems soluble within the framework of classical celestial mechanics is that of determining the structure of the surfaces $I_{c,p}$ and $\tilde{I}_{c,p}$; we shall state the solution of this problem in a special case below (Proposition 26.2).

However, as noted earlier, we are mainly interested in the circular trajectories in the n-body problem (with all the "bodies" orbiting with the same angular speed about 0). A configuration $z = (z_1, \ldots, z_n)$ of the point-masses $m_1, \ldots, m_n$ (satisfying $\sum m_i z_i = 0$) is called a *relative equilibrium* if the standard action of S^1 on $\mathbb{R}^2$, and thence, co-ordinate-wise, on $\mathbb{C}^n$, determines a motion $z(t) = (z_1(t), \ldots, z_n(t))$ which satisfies Newton's law (5); or, in other words, if the circles $z_i(t)$ described by each of the points z_i when the plane is rotated about 0 (so that the relative positions of the points $z_i(t)$ remains fixed), together form an integral trajectory of the dynamical system X. We denote the set of all relative equilibria by R_e.

It is easy to see that the set $R_e \subset M\backslash\Delta$ is invariant not only under the action of S^1 but also under the group of dilations $z \mapsto \lambda z$, $\lambda \neq 0$. It follows that it makes sense to speak of the set Φ_n of equivalence classes, or orbits, of R_e under the joint action of S^1 and the group of dilations; thus we regard two configuration R_e as equivalent if one can be obtained from the other by means of a rotation followed by a dilation. It turns out that for small n the set Φ_n can be "effectively described" (in a certain precise sense givn in Theorem 26.3 below).

In Theorem 26.3, the main result of this section, the relative equilibria are characterized in terms of the critical points of the potential function V. Consider in $M \subset \mathbb{C}^n = \mathbb{R}^{2n}$ the scalar product $\langle\ ,\ \rangle$ defined by the symmetric bilinear form (afforded by the formula (2) for the kinetic energy of the system)

$$K(\xi, \eta) = \sum_{i=1}^{n} m_i \xi^i \eta^i, \tag{8}$$

and let $S_K = S_K^{2n-3}$ be the unit sphere in M relative to this scalar product:

$$S_K = \{z \in M \mid K(z, z) = 1\}.$$

(Here we are using the fact that M (as also $M \backslash \Delta$) is a $(2n-2)$-dimensional subspace of the linear space $\mathbb{R}^{2n}$, so that M is identifiable isometrically with any of its tangent spaces.) It turns out that the level surfaces $I_{c,p}$ may be described topologically in terms of the manifold $S_K \backslash \Delta$, consisting of those points of S_K not lying in any bisector hyperplane. We illustrate this by examining, by way of an example, the special case where the system moves along an integral trajectory in a level surface of the form $I_{c,o}$, i.e. where the angular momentum of the system is zero:

$$J(z, v) = \sum_{i=1}^{n} m_i(z_i \wedge v_i) = \sum_{i=1}^{n} m_i(x_i v_i^2 - y_i v_i^1) \equiv 0.$$

One can in this case prove the following geometrical

26.2. Proposition (Smale). *In the planar n-body problem, involving point-masses $m_1, \ldots, m_n$, under the assumption of zero angular momentum, the level surface $I_{c,0}$ (of the two prime integrals $E = c =$ const., $J = p = 0$) in which the motion of the system takes place, has the following topological structure:*

(i) *If the energy $E = c$ is non-negative, then $I_{c,o}$ is diffeomorphic to the direct product*

$$S^{2n-4} \times (S_K \backslash \Delta) \times \mathbb{R}^1$$

(which, it may be noted, has the appropriate dimension $(2n-4) + (2n-3) + 1 = 4n - 6$).

(ii) *If the energy $E = c$ is negative, then $I_{c,o}$ is diffeomorphic to the product*

$$\mathbb{R}^{2n-3} \times (S_K \backslash \Delta)$$

(again of the right dimension $(2n-3) + (2n-3) = 4n - 6$).

It can, in fact, be shown more generally that the manifolds $I_{c,p}$ corresponding to constant values of the energy, and (no longer necessarily zero) angular momentum, lend themselves to a quite simple description as certain Riemannian fibre bundles over $S_K \backslash \Delta$. However, since knowledge of the topological structure of the $I_{c,p}$ will not be required in what follows, we shall omit the details of this general description.

We now formulate the main result of this section (concerning "circular orbits").

26.3. Theorem (Smale). *Let $m_1, \ldots, m_n$ be the (arbitrary) masses of n point-particles in the planar n-body problem, and let V_S denote the restriction to the submanifold $S_K \backslash \Delta \subset M \backslash \Delta$ of the potential V (see (4)). A point (i.e. configuration) $z \in S_K \backslash \Delta$ (i.e. $z \in M \backslash \Delta$ and $K(z, z) = 1$) is a point of relative equilibrium precisely if it is a critical point of the function V_S: $S_K \backslash \Delta \to \mathbb{R}$. Since each point $z' \in M \backslash \Delta$ goes under an appropriate dilation to some point z satisfying $K(z, z) = 1$, it follows that the critical points of the function V_S on $S_K \backslash \Delta$ exactly determine Φ_n, the set of equivalence classes of relative equilibria.*

Prior to beginning the proof of this theorem, we state (without proof) a few consequences of a classificatory nature concerning the equivalence classes of relative equilibria. In the (simplest) case of two bodies ($n = 2$) it is not difficult to see directly that there is exactly one equivalence class of relative equilibria. In the 3-body problem ($n = 3$) it turns out that there are in general just five equivalence classes of relative equilibria. Two of these classes correspond to an arrangement of the three masses at the vertices of an equilateral triangle, and are distinguished only by the orientation of the masses around the vertices (the classical situation of Lagrange). The three remaining classes correspond to the so-called "collinear" relative equilibria discovered by Euler, where the three points z_1, z_2, z_3 lie in a straight line; there are in general three such collinear arrangements of the masses (up to equivalence) which are relative equilibria (i.e. are such that the masses revolve about the centre of gravity of the system, maintaining their relative positions).

It remains an unsolved problem as to whether, for every n point-masses $m_1, \ldots, m_n$, the corresponding set Φ_n of distinct equivalence classes of relative equilbria is finite. In all cases exhaustively investigated to date Φ_n has in fact turned out to be finite.

We now turn to the proof of Theorem 26.3. We shall show that it follows from a rather general result of the theory of Hamiltonian systems. With a view to formulating the latter result, let M now denote any smooth manifold (the configuration space of some mechanical system) and $T = TM$ the phase space (taken to be the tangent bundle over M) of the system. As before we interpret the kinetic energy K as a Riemannian metric on the manifold M, i.e. at each point z of M the form $K = K_z = g_{\alpha\beta}(z)\dot{z}^\alpha\dot{z}^\beta$ (cf. (8)) is to be regarded as affording a scalar product on the tangent space T_zM to M at z. Taking the total energy E to be given (as the Hamiltonian of the system), we may then if we wish define the potential energy via the equation $E = K + V$. (Alternatively, the potential may be give almost everywhere (e.g. as in (4)) and then E defined by $E = K + V$.) We can then write down in Hamiltonian or Lagrangian form the usual system of partial differential equations on the tangent (or cotangent) bundle over M. This system may be re-expressed, if one so desires, as a system of second-order partial differential equations on the manifold M itself. (For the details see Part I, Chapter 5, and Part II, §28.)

Suppose now that this Lagrangian (or Hamiltonian) system has some group of "configuration symmetries", i.e. there is a Lie transformation group acting smoothly on M in such a way as to preserve the Riemannian metric K and the potential energy V (perhaps defined only almost everywhere on M), so that V is constant on each orbit of G. Thus G is assumed to be a (Lie) subgroup of the group of isometries of the Riemannian manifold M (with metric K) preserving the Hamiltonian system determined by $E = K + V$.

26.4. Proposition. *Consider as above a mechanical system with associated smooth configuration manifold M, kinetic energy K (affording a Riemannian metric on M), and potential V on $M \backslash \Delta$, where Δ is a subset of M of measure zero, and let G be a Lie group of isometries of M leaving V (as well as K) invariant. It is intuitively clear that each element X of the Lie algebra of G gives rise in a natural way to a vector field $X(z)$ (of "infinitesimal displacements" in the direction X) on M ($z \in M$). (In fact, this field is given by $X(z) = (d/dt)(T_{g(t)}(z))|_{t=0}$, where $g(t)$ is any curve in G satisfying $g(0) = 1$, $g'(0) = X$, and $T_{g(t)}$ here denotes the transformation determined by $g(t) \in G$.) Let $\psi_t(z_0)$ be an integral trajectory of the field $X(z)$, passing through z_0, i.e. a solution of the dynamical system $\dot{z} = X(z)$, and let $\varphi_t(z_0)$ denote a typical integral trajectory on M, passing through z_0, of the given mechanical system (i.e. a solution of the appropriate system of second-order* p.d.e.'s *equivalent to the Hamiltonian system with Hamiltonian $K + V$). Then the trajectory $\psi_t(z_0)$ coincides with some $\varphi_t(z_0)$ precisely when the point z_0 is a critical point of the function f on M given by*

$$f(z) = V(z) - K(X(z), X(z)).$$

(If $V = 0$, then the Hamiltonian system becomes a geodesic flow (see Part II, §28.3), and we obtain a characterization of those geodesics relative to the given Riemannian metric K, which coincide with orbits of the action of some one-parameter subgroup of the isometry group of M.)

Thus the condition for tangency of the field $X(z)$ at a point z_0, to a trajectory through z_0, in M, of the Hamiltonian flow determined by $K + V$, is precisely that $\operatorname{grad}[V(z_0) - K(X(z_0), X(z_0))] = 0$. Although the proof is relatively straightforward, we shall omit it. (Note, however, that the proof can be reduced to the case of a geodesic flow using the "Maupertuis principle" (see, e.g. Part II, §28.3).)

We shall now show how Theorem 26.3 follows from this proposition. For each value p of the angular momentum of the system of n bodies with which that theorem is concerned, let V_p denote the function on $M \backslash \Delta$ defined by

$$V_p(z) = V(z) + \frac{p^2}{4K(z)}, \tag{9}$$

(where we have written $K(z)$ for $K(z, z)$). It follows from the definition of M that $M\backslash\{0\}$ is diffeomorphic to the space $\mathbb{R}^+ \times S_K$, where $\mathbb{R}^+$ denotes the space of positive reals and, as in the theorem, $S_K = \{z | K(z) = 1\}$. (In fact, a diffeomorphism $g: M\backslash\{0\} \to \mathbb{R}^+ \times S_K$ is given by the formula

$$g(z) = (\sqrt{K(z)}, z/\sqrt{K(z)}), \qquad \sqrt{K(z)} \in \mathbb{R}^+, \quad z/\sqrt{K(z)} \in S_K.)$$

Hence the restriction of g to $M\backslash\Delta$ yields a diffeomorphism

$$M\backslash\Delta \cong \mathbb{R}^+ \times (S_K\backslash\Delta). \tag{10}$$

In what follows we shall consider $M\backslash\Delta$ and $\mathbb{R}^+ \times (S_K\backslash\Delta)$ as identified under this diffeomorphism. Recall that in the theorem V_S denotes the restriction of the potential V to the submanifold $S_K\backslash\Delta \subset M\backslash\Delta$. We first establish the following two set equalities (where $\sigma(h)$ denotes the set of critical points of a map h):

(i) $\sigma(V_p) = \{z = (\tau, v) \in \mathbb{R}^+ \times S_K\,(= M\backslash\Delta) | v \in \sigma(V_S), \tau = -p^2/2V(v)\}$;

(ii) $\sigma(V - K(X)) = \{z = (\tau, v) \in \mathbb{R}^+ \times S_K\,(= M\backslash\Delta) | v \in \sigma(V_S),$
$\tau = \sqrt[3]{-V(v)/2K(X(v))}\}$,

where X is any element of the Lie algebra of the Lie group G consisting of those isometries of M (endowed with the metric K) which preserve V, and $K(X(v)) = K(X(1, v))$.

Observe to begin with that under the above-mentioned identification of $M\backslash\Delta$ with $\mathbb{R}^+ \times (S_K\backslash\Delta)$, we have $K(z) \doteq \tau^2$, and also, in view of the explicit formula (4) for the potential

$$V(z) = V(\tau, v) = V(v)/\tau.$$

We now give the verificaton of (i). A point $z = (\tau, v)$ is of course a critical point of the function V_p precisely if both $\partial_\tau(V_p) = 0$ and $\partial_v(V_p) = 0$ at that point (where $\partial_\tau = \partial/\partial\tau$, $\partial_v = (\partial/\partial v^1, \ldots)$). Now from $\partial_\tau V_p(\tau, v) = 0$ and the definition (9) we obtain

$$\partial_\tau V_p(\tau, v) = \partial_\tau\left[V(z) + \frac{p^2}{4K(z)}\right] = \partial_\tau\left[\frac{V(v)}{\tau} + \frac{p^2}{4\tau^2}\right] = -\frac{V(v)}{\tau^2} - \frac{p^3}{2\tau^3} = 0,$$

whence

$$\tau = -p^2/2V(v). \tag{11}$$

The second condition, $\partial_v V_p(\tau, v) = 0$, yields

$$\partial_v V_p(z) = \frac{1}{\tau}\left[\partial_v V(v) + \partial_v\left[\frac{p^2}{4\tau^2}\right]\right] = \frac{1}{\tau}\partial_v V(v) = 0. \tag{12}$$

From (11) and (12) we conclude that grad $V_p(\tau, v) = 0$ precisely if $\partial_v V(v) = 0$ and $\tau = -p^2/2V(v)$, which establishes (i).

Turning to the verification of (ii), observe first that as a function on $M\backslash\Delta$, $V - K(X)$ is to be interpreted as follows:

$$(V - K(X))(z) = V(\tau, v) - K(X(\tau, v)),$$

where $X(z) = X(\tau, \nu)$ denotes the vector field on M determined by X (see Proposition 26.4). Now

$$\partial_\tau[V(\tau, \nu) - K(X(\tau, \nu))] = \partial_\tau\left[\frac{V(\nu)}{\tau} - \tau^2 K(X(1, \nu))\right]$$
$$= -\frac{V(\nu)}{\tau^2} - 2\tau K(X(\nu)), \tag{13}$$

where we have written $X(\nu)$ for $X(1, \nu)$. (The first equality here follows from the fact that the group G preserves each sphere $\{\tau\} \times S_K$, so that $X(\tau, \nu) = \tau X(1, \nu)$, whence $K(X(\tau, \nu)) = \tau^2 K(X(1, \nu))$.) From (13) we see that the condition $\partial_\tau(V - K(X)) = 0$ is equivalent to

$$\tau^3 = -\frac{V(\nu)}{2K(X(\nu))}.$$

On the other hand, the condition $\partial_\nu(V - K(X)) = 0$ yields

$$\partial_\nu\left[\frac{V(\nu)}{\tau} - \tau^2 K(X(\nu))\right] = \frac{1}{\tau}\partial_\nu V(\nu) - \tau^2 \partial_\nu[K(X(\nu))] = 0. \tag{14}$$

Now since X is an element of the Lie algebra of the group of isometries of M (i.e. transformations preserving K), we have $\partial_\nu K(X(\nu)) = 0$. Taking this into account in (14) then yields $\partial_\nu V(\nu) = 0$. Hence $\operatorname{grad}(V - K(X))(\tau, \nu) = 0$ precisely when the two conditions

$$\tau^3 = -\frac{V(\nu)}{2K(X(\nu))}, \qquad \partial_\nu V(\nu) = 0,$$

hold simultaneously. The equality (ii) now follows from the observation that the condition $\partial_\nu V(\nu) = 0$ is equivalent to the condition $\nu \in \sigma(V_S)$, since $V(\nu)$ is in fact just the restriction of V to $S_K \backslash \Delta$, i.e. $V(\nu) = V_S(\nu)$, $\nu \in S_K \backslash \Delta$.

PROOF OF THEOREM 26.3. Let $z_0 = (\tau_0, \nu_0)$ be such that $K(z_0) = 1$. By Proposition 26.4 an orbit containing z_0 of a one-parameter subgroup of the Lie group G (of isometries preserving V) coincides with some integral trajectory of the mechanical system under consideration precisely if z_0 is critical for the function $V(z) - K(X(z))$. Now by the equality (ii), the set of critical points of $V(z) - K(X(z))$ (satisfying $K(z) = 1$) coincides with the set of critical points of the function V_S on $S_K \backslash \Delta$. Since the sphere S_K is invariant under the group G of isometries, the orbit of a one-parameter subgroup of G containing z_0 will be a great circle through z_0 on S_K. Hence the integral trajectories of the mechanical system corresponding to points of relative equilibrium (with $K(z) = 1$) are precisely the great circles on S_K that are orbits of one-parameter subgroups of G and pass through critical points of V_S. We thus conclude that a point $z_0 \in M \backslash \Delta$, satisfying $K(z_0) = 1$, is a point of relative equilibrium if and only if it is a critical point of the restriction V_S of the potential V to $S_K \backslash \Delta$. This completes the proof of the theorem. □

We next investigate the special subclass of *collinear* relative equilibria, i.e. configurations in relative equilibrium where the n bodies lie along a line in the plane. To be specific, we shall find, for arbitrary n, the precise number of such collinear relative equilibrium points, using the above characterization of the relative equilibria generally in terms of the critical points of the potential function.

26.5. Theorem (Moulton). *For any system of masses $m_1, \ldots, m_n$ in the planar n-body problem, there exist precisely $n!/2$ equivalence classes of collinear relative equilibria of the system, i.e. classes of relative equilibrium configurations with the points z_i (giving the positions of the n particles) distributed on a straight line through the centre of mass (the origin O), so that the motion of the system amounts to the rotation of this line about O (the particles individually describing circles with centre O).*

The proof of this result will occupy most of the remainder of this section. To begin, observe that any straight line l through the origin in the plane $\mathbb{R}^2$ determines the subset M_l of M (see (1)) consisting of all l-collinear configurations, i.e. of those points $z = (z_1, \ldots, z_n)$ such that $z_1, \ldots, z_n$ all lie on l and $\sum m_i z_i = 0$. The sets $S_l = S_K \cap M_l$ and $S_l \backslash \Delta = S_l \backslash (S_l \cap \Delta)$ (where S_K and Δ are as before) will be important in the proof. It is clear that under the action of S^1 on S_K induced by rotating the plane $\mathbb{R}^2$ about the origin, the set S_l is preserved (as a whole) only by the rotation of the plane through the angle π (apart from the trivial rotation), and this determines a natural action of $\mathbb{Z}_2$ on S_l. Given a fixed line l through the origin of $\mathbb{R}^2$, a collinear configuration of points $z_1, \ldots, z_n$ on $l \backslash \{0\}$ satisfying $K(z) = 1$, is fully determined (to within the rotation through π) by the ordered $(n-1)$-tuple

$$\frac{z}{z_n} = \left[\frac{z_1}{z_n}, \ldots, \frac{z_{n-1}}{z_n}\right] \in \mathbb{R}^{n-1}.$$

(To see this note first that since $\sum m_i |z_i|^2 = 1$, we have

$$\sum_{j=1}^{n-1} m_j \left|\frac{z_j}{z_n}\right|^2 + m_n = \frac{1}{|z_n|^2},$$

whence $|z_n|$ may be obtained; the argument of the complex number z_n is given (to within $\pm\pi$) by the direction of l. Once we have z_n, the other z_j may be retrieved from z/z_n.) The condition $\sum m_i z_i = 0$ ensures that there are no points of $\mathbb{R}^{n-1}$ on the line L_z in $\mathbb{R}^{n-1}$ passing through the origin and z/z_n, other than $\pm z/z_n$, yielding a point of S_l. It follows that the correspondence $z \mapsto L_z$ induces a bijection (in fact, a diffeomorphism) between the orbit space $S_l/\mathbb{Z}_2$ and the real projective space $\mathbb{R}P^{n-2}$:

$$S_l/\mathbb{Z}_2 \cong \mathbb{R}P^{n-2}. \tag{15}$$

The surjection $\varphi\colon S_K \to \mathbb{C}P^{n-2}$ sending each $z \in S_K$ to the complex line ($\cong \mathbb{R}^2$) in $\mathbb{C}^{n-1}$ through the origin and the point $(z_1, \ldots, z_{n-1}) \in \mathbb{C}^{n-1}$, determines a (fixed) embedding of $S_l/\mathbb{Z}_2 \cong \mathbb{R}P^{n-2}$ into complex $(n-2)$-dimensional projective space $\mathbb{C}P^{n-2}$, and thence (writing $\tilde{\Delta} = \Delta/S^1$ and identifying $S_l/\mathbb{Z}_2$ with $\mathbb{R}P^{n-2}$ via the diffeomorphism (15)) an embedding

$$\mathbb{R}P^{n-2}\backslash(\tilde{\Delta} \cap \mathbb{R}P^{n-2}) \xrightarrow{\subset} \mathbb{C}P^{n-2}\backslash\tilde{\Delta}. \tag{16}$$

The surjection $\varphi: S_K \to \mathbb{C}P^{n-2}$ also induces from the restricted potential $V_S: S_K\backslash\Delta \to \mathbb{R}$, a smooth function

$$\tilde{V}: \mathbb{C}P^{n-2}\backslash\tilde{\Delta} \to \mathbb{R}. \tag{17}$$

26.6. Lemma. *The number of equivalence classes of relative equilibria is equal to the number of critical points of the smooth function* $\tilde{V}$.

PROOF. It follows from the definition of an equivalence class of points of relative equilibrium (see above) that each such class C contains a normalized relative equilibrium point, i.e. a relative equilibrium point in $S_K\backslash\Delta$, and that all points in $C \cap (S_K\backslash\Delta)$ are obtained from a single one via the action of S^1. In other words, modulo the action of S^1 there is just one relative equilibrium point in each $C \cap (S_K\backslash\Delta)$. Since the map $\varphi: S_K \to \mathbb{C}P^{n-2}$ identifies each orbit of the action of S^1 on S_K with a point, it follows from Theorem 26.3 that the classes of relative equilibrium points correspond one-to-one to the critical points of $\tilde{V}$, as claimed. □

26.7. Proposition (Smale). *The equivalence classes of collinear relative equilibrium points are in one-to-one correspondence with those critical points of the function* $\tilde{V}: \mathbb{C}P^{n-2}\backslash\tilde{\Delta} \to \mathbb{R}$, *which lie in the submanifold* $\mathbb{R}P^{n-2}\backslash(\tilde{\Delta} \cap \mathbb{R}P^{n-2})$ *(considered as a submanifold of* $\mathbb{C}P^{n-2}\backslash\tilde{\Delta}$ *under the fixed embedding* (16)).

PROOF. Since a collinear relative equilibrium point $z = (z_1, \ldots, z_n)$ is by definition a relative equilibrium with the points $z_1, \ldots, z_n$ on a single straight line through the origin of $\mathbb{R}^2$, we may by means of a suitable rotation of $\mathbb{R}^2$ (which will of course preserve the equivalence class of z) bring these points onto the initially chosen straight line l. Hence under the map φ, collinear relative equilibrium points (and, among the relative equilibrium points, only these) will be sent to the subspace $\mathbb{R}P^{n-2}\backslash(\tilde{\Delta} \cap \mathbb{R}P^{n-2})$ (see (15) and (16)). (They are of course critical points of $\tilde{V}$ by the preceding lemma.) This completes the proof. □

We see therefore that the problem of determining the number of inequivalent collinear relative equilibria is equivalent to the enumeration of those critical points of the "potential" function $\tilde{V}$ lying in the real submanifold $\mathbb{R}P^{n-2}$ of $\mathbb{C}P^{n-2}$. Now, of course, in general, a critical point of the restriction of a function to a submanifold of a manifold need not be a critical point of the unrestricted function. However, in the present special context, it turns out that every critical point of the restriction of $\tilde{V}$ to the submanifold $\mathbb{R}P^{n-2}\backslash(\tilde{\Delta} \cap \mathbb{R}P^{n-2})$ is in fact a critical point of $\tilde{V}$ (unrestricted), so that the problem of investigating the critical points of $\tilde{V}$ lying in that submanifold reduces to the simpler one of describing the critical points of the restriction of $\tilde{V}$ to the submanifold.

26.8. Proposition. *If $z \in \mathbb{R}P^{n-2}\backslash(\tilde{\Delta} \cap \mathbb{R}P^{n-2})$ is a critical point of the restriction of $\tilde{V}$ to the submanifold*

$$\mathbb{R}P^{n-2}\backslash(\tilde{\Delta} \cap \mathbb{R}P^{n-2}) \subset \mathbb{C}P^{n-2}\backslash\tilde{\Delta},$$

then z is in fact a critical point of the unrestricted function $\tilde{V}: \mathbb{C}P^{n-2}\backslash\tilde{\Delta} \to \mathbb{R}$.

PROOF. The functon $\tilde{V}$ defined above arises from the potential (4):

$$V(z) = -\sum_{i \neq j} \frac{m_i m_j}{|z_i - z_j|},$$

where, as usual, $m_1, \ldots, m_n$ are the masses of the n particles under consideration. By differentiating with respect to t one easily obtains the following formula for the (first) differential of V evaluated at $v \in T_z(M\backslash\Delta)$:

$$dV(z)(v) = \sum_{i \neq j} \frac{m_i m_j}{|z_i - z_j|^3} \langle z_i - z_j, v_i - v_j \rangle. \tag{18}$$

A further differentiation with respect to t yields (after neglecting terms in higher-order derivatives) the following formula for the second differential of V evaluated at $(v, w) \in T_z(M\backslash\Delta) \times T_z(M\backslash\Delta)$:

$$d^2V(z)(v, w) = \sum_{i \neq j} \frac{m_i m_j}{|z_i - z_j|^3} \left[\frac{3}{|z_i - z_j|^2} \langle z_i - z_j, v_i - v_j \rangle \langle z_i - z_j, w_i - w_j \rangle - \langle v_i - v_j, w_i - w_j \rangle \right]. \tag{19}$$

(In both (18) and (19), $\langle \cdot, \ \rangle$ denotes the ordinary Euclidean scalar product in $\mathbb{R}^2$.) The second differential of the restriction $V|_{(S_l\backslash\Delta)}$ is then of course also given by (19) with z restricted to lie in $S_l\backslash\Delta$ and v and w in $T_z(S_l\backslash\Delta)$.

Proceeding with the proof of the proposition, we write $v_i = (v_i', v_i'')$ for each $v_i \in T\mathbb{R}^2$ $(= \mathbb{R}^2)$, where v_i' is parallel to the chosen straight line l, and v_i'' is perpendicular to l (with respect to the Euclidean metric on $\mathbb{R}^2$). This determines a decomposition $v = (v', v'')$ of each $v \in TM$, where

$$v' = (v_1', \ldots, v_n'), \qquad v'' = (v_1'', \ldots, v_n'').$$

Clearly, at each $z \in S_l \subset S_K$, we then have

$$T_zS_K = \{v \in TM \mid v \perp z\}, \qquad T_zS_l = \{v' \in TM_l \mid v' \perp z\},$$

where now the perpendicularity is that determined by the scalar product afforded by the kinetic energy K (which is determined in turn, of course, by the given system of masses $m_1, \ldots, m_n$). Hence if $v \in T_zS_k$ and $v = (v', v'')$, then $v' \in T_zS_l$. Now if $z \in S_l\backslash\Delta$ and $v \in T_z(S_K\backslash\Delta)$, $v = (v', v'')$, then since in (Euclidean)

$\mathbb{R}^2$ we have $\langle z_i - z_j, v_i'' - v_j'' \rangle = 0$ for all i, j, it follows from (18) that

$$dV(z)(v) = dV(z)(v').$$

Hence if z is a critical point of $V|_{(S_l \backslash \Delta)}$, i.e. if $dV(z)(v') = 0$ for all $v' \in T_z(S_l \backslash \Delta)$, then $dV(z)(v) = 0$ for all $v \in T_z(S_K \backslash \Delta)$, i.e. z is critical for V, whence the proposition. □

26.9. Lemma. *The manifold* $\mathbb{R}P^{n-2} \backslash (\tilde{\Delta} \cap \mathbb{R}P^{n-2})$ *has precisely* $n!/2$ *(path-) connected components.*

PROOF. This (geometrical) assertion derives essentially from the disposition of the bisector hyperplanes Δ_{ij} (see (3)) in M. Taking the real axis of $\mathbb{R}^2$ to be l, let $z = (z_1, \ldots, z_n)$ be a point of $S_l \backslash \Delta$ such that $z_1 < \cdots < z_n$ $(z_i \in \mathbb{R})$. (Note that since $z \notin \Delta$, the z_i must be distinct.) Now the result of applying the $n!$ permutations $\begin{bmatrix} 1 \ldots n \\ i_1 \ldots i_n \end{bmatrix}$ in turn to the indices of the z_i, i.e. replacement of $(z_1, \ldots, z_n)$ by $(z_{i_1}, \ldots, z_{i_n})$, yields precisely one point in each connected component of $S_l \backslash \Delta$; for this it suffices to observe that the points z in any one component of $S_l \backslash \Delta$ yield the same permutation of indices upon having their components z_i ordered according to size, since for the order of two components z_i and z_j to change along a path in S_l, that path must clearly pass through Δ_{ij}. Hence the space $S_l \backslash \Delta$ has exactly $n!$ connected components, whence, in view of (15), its orbit space $\mathbb{R}P^{n-2} \backslash (\tilde{\Delta} \cap \mathbb{R}P^{n-2})$ has $n!/2$ components. □

26.10. Lemma. *Every critical point* $z \in \mathbb{R}P^{n-2} \backslash (\tilde{\Delta} \cap \mathbb{R}P^{n-2})$ *of the restriction of the function* $\tilde{V}$ *to* $\mathbb{R}P^{n-2} \backslash (\tilde{\Delta} \cap \mathbb{R}P^{n-2})$ *is a non-degenerate maximum point of that restriction.*

PROOF. Choosing the real axis in $\mathbb{R}^2$ $(= \mathbb{C})$ to coincide with our fixed line l, so that the components of z_i of z (considered as a point of $S_l \backslash \Delta$) are real (lying as they do on l), as are also the components v_i of every tangent vector v to M_l (being parallel to l), we have from (19) that

$$d^2V(z)(v, v) = -\sum_{i \neq j} \frac{m_i m_j}{|z_i - z_j|^3} \left[\frac{3}{(z_i - z_j)^2} (z_i - z_j)^2 (v_i - v_j)^2 - (v_i - v_j)^2 \right]$$

$$= -\sum_{i \neq j} \frac{2 m_i m_j (v_i - v_j)^2}{|z_i - z_j|^3}.$$

Thus the quadratic form $d^2V(z)$ on $T_z(S_l \backslash \Delta)$ is negative definite at each $z \in S_l \backslash \Delta$, as claimed. □

PROOF OF THEOREM 26.5. Since clearly the potential $V(z)$ (see (4)) approaches $-\infty$ as z approaches Δ, it follows that the function $\tilde{V}$ (see (17)) restricted to $\mathbb{R}P^{n-2} \backslash (\tilde{\Delta} \cap \mathbb{R}P^{n-2})$ also approaches $-\infty$ out towards the boundary of each connected component of that space, and hence that this restriction achieves an (absolute) maximum value on each such component. However this (restricted) function cannot have more than one maximum point since otherwise, as is intuitively plausible (and can in fact be proved along the lines of Morse theory, e.g. by using the appropriate extensions of the Morse inequalities and Poincaré duality to the present context; see §§16, 18), the non-degeneracy of these maximum points would then imply the existence of at least one critical (non-maximum) saddle point, contradicting Lemma 26.10. Hence each connected component of $\mathbb{R}P^{n-2}\backslash(\tilde{\Delta} \cap \mathbb{R}P^{n-2})$ contains exactly one critical point, and that a non-degenerate maximum point, of the restriction to this space of $\tilde{V}$. By Lemma 26.9, therefore, that restriction has precisely $n!/2$ critical points altogether, and the theorem now follows from Propositions 26.7 and 26.8. □

Observe that in the above proof of Moulton's theorem the assumption of collinearity of the component points z_i of the relative equilibria was exploited in an essential way at several crucial junctures, so that our ability to actually calculate the number of such relative equilibria would seem to depend heavily on just this assumption.

It is also clear from the above that to answer the question of the number of relative equilibrium classes generally (i.e. without the collinearity assumption) one needs to determine (in some sense) the critical points, together with their indices, of the "potential" function $\tilde{V}$ on the whole of the complex projective space $\mathbb{C}P^{n-2}\backslash\tilde{\Delta}$, rather than just its real subspace $\mathbb{R}P^{n-2}\backslash(\tilde{\Delta} \cap \mathbb{R}P^{n-2})$. This, however, turns out to be a decidedly more difficult problem.

As noted above, the group of rotations of $\mathbb{R}^2$ about the origin induces an action of S^1 on S_K leaving the potential V (and the singular subset Δ) invariant, with orbit space S_K/S^1 identifiable in a natural way with the complex projective space $\mathbb{C}P^{n-2}\backslash\tilde{\Delta}$. Recall also that we earlier denoted by $\tilde{V}\colon \mathbb{C}P^{n-2}\backslash\tilde{\Delta} \to \mathbb{R}$ (see (17)) where $\tilde{\Delta} = \Delta/S^1$ (which may be considered as a union of complex projective subspaces), the function induced by the potential V.

Conjecture. For almost all n-tuples $(m_1, \ldots, m_n)$ of masses in the planar n-body problem, the smooth function $\tilde{V}$ induced by the original potential function $V(z)$ (see (4)), is a Morse function, i.e. all of its critical points are non-degenerate.

This conjecture has as yet been neither proved nor disproved. Its significance derives chiefly from its relevance to the above-noted question of whether, for almost every n-tuple of masses $m_1, \ldots, m_n$, the number of classes of relative equilibria is finite. It can be shown (we shall not give the proof here) that there is an open neighbourhood of the singular subset $\tilde{\Delta} \subset \mathbb{C}P^{n-2}$ where the function $\tilde{V}$ has no critical points. Assuming the above conjecture true, it follows immediately that the number of critical points of the function $\tilde{V}$ (which by Theorem 26.3 coincides with number of relative equilibrium classes) is finite (for almost all n-tuples of masses $m_1, \ldots, m_n$).

We conclude by describing a further consequence of the conjecture. If the conjecture is true, then the number of relative equilibrium classes can be estimated from below as follows. With each relative equilibrium class one can associate the index of the critical point (assumed non-degenerate) of $\tilde{V}$ to which it corresponds in accordance with Theorem 26.3. From the Morse inequalities (see §16) extended to the present context, we would then have that for each index λ the number of relative equilibrium classes of that index is bounded below by the corresponding Betti number (i.e. rank of the λth real homology group) of the space $\mathbb{C}P^{n-2}\backslash\tilde{\Delta}$. The homology of this space turns out to be sufficiently "rich" to then allow the existence, for $n > 3$, of non-trivial relative equilibrium classes (other than the collinear ones) to be established (of course conditionally on the truth of the above conjecture). In fact, the cohomology ring of the space $\mathbb{C}P^{n-2}\backslash\tilde{\Delta}$ has been computed explicitly (by Arnol'd): it is isomorphic to the cohomology ring of a rather easily described topological space, namely the direct product of $(n - 2)$ bouquets of circles consisting respectively of $2, 3, \ldots, n - 1$ circles, i.e.

$$H^*(\mathbb{C}P^{n-2}\backslash\tilde{\Delta}) \simeq H^*((S^1 \cup S^1) \times (S^1 \cup S^1 \cup S^1) \times \cdots) \quad (n - 2 \text{ factors}).$$

It follows that the Poincaré polynomial (see §16) of the space $\mathbb{C}P^{n-2}\backslash\tilde{\Delta}$ is $\prod_{\alpha=2}^{n-1}(1 + \alpha t)$.

CHAPTER 3

Cobordisms and Smooth Structures

§27. Characteristic Numbers. Cobordisms. Cycles and Submanifolds. The Signature of a Manifold

27.1. Statement of the Problem. The Simplest Facts About Cobordisms. The Signature

Using the machinery developed in the preceding chapters we shall now investigate the following questions (already considered briefly at the end of §6) concerning smooth manifolds.

(1) *The cobordism problem*: Under what conditions is a smooth, closed manifold M^n realizable as the boundary of a smooth, compact manifold-with-boundary W^{n+1}, i.e. $\partial W^{n+1} \cong M^n$? What, in particular, is the situation if M^n and W^{n+1} are required to be orientable?

(2) *Realizability of cycles as submanifolds*: Given an element $x \in H_i(M^n; \mathbb{Z})$ (or $x \in H_i(M^n; \mathbb{Z}_2)$, if M^n is non-orientable), under what conditions is x represented by some closed submanifold $M^i \subset M^n$?

(3) *Cycles that are continuous images of manifolds*: Given an arbitrary cell complex X and an element $x \in H_i(X; \mathbb{Z})$ (or $x \in H_i(X; \mathbb{Z}_2)$ in the non-orientable case), when does there exist a "singular bordism" (M^i, f) (i.e. a manifold M^i and a map $f: M^i \to X$) such that $f_*[M^i] = x$?

The generalization of problem (3) to the relative situation is also of interest: Given a subcomplex Y of the cell complex X, and an element $x \in H_i(X, Y; \mathbb{Z})$ (or $x \in H_i(X, Y; \mathbb{Z}_2)$ in the non-orientable case), one seeks conditions under which there exists a manifold W^i with boundary M^{i-1} and a map

$f:(W^i, M^{i-1}) \to (X, Y)$ of pairs such that, with the obvious interpretation, $f_*[W^i, M^{i-1}] = x$.

The "i-dimensional bordism group" of a cell complex X (or more general topological space) is defined in the natural way (cf. end of §6): A *singular i-dimensional bordism* of X is a pair (M^i, f) as above (i.e. f is a continuous map from the i-dimensional closed manifold M^i to X), and an *i-dimensional cycle* is then in this context a finite formal $\mathbb{Z}$-linear combination ($\mathbb{Z}_2$-linear if the manifolds M^i involved are not assumed orientable) of i-dimensional singular bordisms

$$\sum_k (M_k^i, f_k),$$

(to be regarded as a single i-dimensional bordism $(\bigcup_k M_k^i, \bigcup_k f_k)$, where the unions are disjoint). A *singular film* in X is a pair (W^{i+1}, g) where W^{i+1} is a compact manifold-with-boundary and g is a map $W^{i+1} \to X$, and its *boundary* is the i-dimensional singular bordism (or cycle)

$$\partial(W^{i+1}, g) = (\partial W^{i+1}, g|_{\partial W^{i+1}}).$$

(As before, linear combinations of i-boundaries are to be considered—via disjoint unions—as i-boundaries.) The quotient group of the group of all i-dimensional cycles (i.e. singular bordisms) by that of all boundaries of $(i + 1)$-dimensional films in X is then the *i-dimensional bordism group* of X. It is denoted by $\Omega_i^{SO}(X)$ when the manifolds M^i and W^{i+1} are (for all i) restricted to being orientable (in which case one speaks of the ith *orientable-bordism group*), and by $\Omega_i^O(X)$ in the general case where the M^i and W^{i+1} are not necessarily orientable.

The i-dimensional relative bordism group $\Omega_i^O(X, Y)$ (or $\Omega_i^{SO}(X, Y)$ in the orientable case) of a pair (X, Y) is defined analogously: here the cycles are maps of manifolds possibly with boundary, where the boundary is mapped into $Y \subset X$, and films are defined similarly appropriately.

Note that in the groups Ω_i^O we have of course the identity

$$2x = 0,$$

arising from the fact that in this, the general, case the cycles were defined over $\mathbb{Z}_2$, in essence since

$$\partial(M^i \times I) = M^i \cup M^i = 2M^i$$

for a closed manifold M^i. On the other hand, in the orientable case since

$$\partial(M^i \times I) = M_+^i \cup M_-^i, \tag{1}$$

where the signs indicate opposite orientations of the (closed) orientable manifold M^i, it follows that the most we can say is that for each map $f: M^i \to X$, the pairs (M_+^i, f), (M_-^i, f) define mutually inverse elements in Ω_i^{SO}.

It follows almost directly from the definition that the bordism groups are homotopy invariants, i.e. that homotopically equivalent spaces have (canonically) isomorphic groups Ω_i^O (and Ω_i^{SO}).

For a contractible space X (or, equivalently, a one-point space), the groups

Ω_i^O and Ω_i^{SO} are non-trivial for certain (in fact almost all) i; these are the *classical cobordism groups*. The operation of forming the direct product of manifolds induces in this case on each of the direct sums

$$\Omega^O = \sum_{i \geq 0} \Omega_i^O, \qquad \Omega^{SO} = \sum_{i \geq 0} \Omega_i^{SO},$$

a multiplicative operation turning them into graded rings, commutative in the general case and skew-commutative in the orientable case:

$$\Omega_i^O \Omega_j^O \subset \Omega_{i+j}^O, \qquad xy = yx;$$

$$\Omega_i^{SO} \Omega_j^{SO} \subset \Omega_{i+j}^{SO}, \qquad xy = (-1)^{ij} yx.$$

The identity $2x = 0$ of course holds in Ω^O, while for Ω^{SO} equation (1) signifies that the two orientations of an orientable closed manifold M^i determine mutual (additive) inverses.

The first few $\Omega_i^O(*)$, $\Omega_i^{SO}(*)$ are easy to calculate:

(i) $\Omega_0^O \simeq \mathbb{Z}_2, \qquad \Omega_0^{SO} \simeq \mathbb{Z};$
(ii) $\Omega_1^O \simeq \Omega_1^{SO} = 0;$ (2)
(iii) $\Omega_2^{SO} = 0.$

(The fact that $\Omega_2^{SO} = 0$ is a consequence of the classification of closed surfaces (Theorem 3.20): it is immediately clear from that classification that every orientable closed surface M^2 is embeddable in $\mathbb{R}^3$ in such a way as to bound a region W^3.)

To compute Ω_2^O we use the following

27.1. Lemma. *If a closed manifold M^i is a boundary, i.e. $M^i = \partial W^{i+1}$ for some compact manifold W^{i+1}, then its Euler characteristic is even: $\chi(M^i) = 2m$.*

PROOF. By the concluding remark of §18, if i is odd and M^i closed, we have $\chi(M^i) = 0$ by Poincaré duality. We may therefore suppose that i is even, say $i = 2k$. Suppose then that $M^i = M^{2k}$ is the boundary of a compact, smooth manifold-with-boundary W^{2k+1}, and consider the "double" of W^{2k+1}:

$$V^{2k+1} = W^{2k+1} \bigcup_{M^{2k}} W^{2k+1}, \tag{3}$$

obtained by taking two copies of W^{2k+1} and identifying their boundaries M^{2k}. Now it follows from the definition of the Euler characteristic for simplicial complexes (see §3(7)) that for simplicial complexes X, Y intersecting in a subcomplex L, we have

$$\chi(X \cup_L Y) = \chi(X) + \chi(Y) - \chi(L). \tag{4}$$

Applying this to an appropriate triangulation of the manifold (3), and noting that $\chi(V^{2k+1}) = 0$ (again since V^{2k+1} is a closed, odd-dimensional manifold) we obtain

$$0 = \chi(V^{2k+1}) = 2\chi(W^{2k+1}) - \chi(M^{2k}),$$

whence the lemma. □

Since $\chi(\mathbb{R}P^2) = 1$ (this may be verified directly by triangulating $\mathbb{R}P^2$, or gleaned from various places in §§3, 4, using the definition of Euler characteristic given in the final remark of §18), it is immediate from this lemma that $\mathbb{R}P^2$ is not realizable as a boundary:

$$\mathbb{R}P^2 \neq \partial W^3, \qquad \text{whence} \quad \Omega_2^O \neq 0.$$

On the other hand, it is easy to construct a film W^3 having the Klein bottle K^2 as its boundary: $\partial W^3 = K^2$. (Find such a film!) Since by the classification theorem for surfaces (Theorem 3.20) every closed non-orientable manifold has the form $\mathbb{R}P^2$ + (handles) or K^2 + (handles), it follows that in fact

$$\Omega_2^O \simeq \mathbb{Z}_2, \qquad \text{with } [\mathbb{R}P^2] \text{ as generator.} \tag{5}$$

By means of an elaboration of geometrical techniques it can be shown further (this is due to Rohlin) that

$$\Omega_3^O \simeq \Omega_3^{SO} = 0 \qquad \text{and} \qquad \Omega_4^{SO} \simeq \mathbb{Z}.$$

However, as will be shown below (in §27.2), these results, among many others, can be obtained instead by means of a theory due to Thom, based on homological techniques.

The following result extends (or rather supplements) the above lemma.

27.2. Lemma (Pontrjagin)

(i) *If a closed manifold M^k is a boundary, i.e. represents zero in the 0-cobordism group Ω_i^O, then all of its* mod 2 *"stable" characteristic numbers or "Stiefel–Whitney numbers" (i.e. the values taken on $[M^k]$ by its* mod 2 *"stable" characteristic classes of dimension k—see below for the definition) are zero.*

(ii) *If a closed orientable manifold M^k is a boundary in SO-cobordism theory (i.e. is the boundary of a compact, orientable manifold W^{k+1}), then in addition all of its rational "stable" characteristic numbers (i.e. the values taken at $[M^k]$ by certain classes in the kth cohomology group of M^k over the rationals $\mathbb{Q}$—see below) vanish.*

PROOF. Embed $M = M^k$ in $\mathbb{R}^N$ for suitably large N. Recall that the Grassmannian manifold $G_{N,k}$ has as its points the k-dimensional vector subspaces of $\mathbb{R}^N$, and that in the limit as $N \to \infty$ we obtain (as a direct limit) $G_{\infty,k} = BO(k)$, the universal classifying space for the orthogonal group $O(k)$ (see §10.4 above and Part II, §24.4). Consider the "generalized Gauss map"

$$\tau_M \colon M^k \to G_{\infty,k} = BO(k), \tag{6}$$

sending each point x of M^k to the tangent space T_x to $M^k \subset \mathbb{R}^\infty$ (translated back to the origin in $\mathbb{R}^\infty$). Note that in accordance with the classification theorem for fibre bundles of Part II, §24.4, the map τ_M induces from the universal bundle $V_{\infty,k} \to G_{\infty,k}$, the tangent bundle on M (actually the bundle of orthonormal tangent k-frames on M with respect to the induced metric). Each element $w \in H^l(G_{\infty,k}; \mathbb{Z}_2)$ then determines a corresponding *mod* 2 *characteristic class* $w(M^k)$ given by (cf. Part II, §25.4)

$$w(M^k) = \tau_M^*(w),$$

and the *stable* mod 2 characteristic classes of M^k (with respect to the group $O(k)$) are then those determined by elements $w \in H^l(BO(k); \mathbb{Z}_2)$ which are pullbacks of elements $\bar{w} \in H^l(BO(k+1); \mathbb{Z}_2)$ via the natural embedding $\lambda: BO(k) \to BO(k+1)$, induced by the standard embedding $O(k) \to O(k+1)$ (see §10.4):

$$w = \lambda^* \bar{w}.$$

(The stable mod 2 characteristic classes of M^k with respect to the groups $SO(k)$, $U(k)$ and $Sp(k)$ are defined analogously in terms of $BSO(k)$, $BU(k)$ and $BSp(k)$.)

Thus if $w(M^k)$ is a stable mod 2 characteristic class of M^k, then

$$w(M^k) = \tau_M^* \lambda^*(\bar{w})$$

for some $\bar{w} \in H^l(BO(k+1); \mathbb{Z}_2)$. Bringing in the hypothesis $M^k = \partial W^{k+1}$, we then have for $\bar{w}$:

$$\bar{w}(W^{k+1}) = \tau_W^*(\bar{w}),$$

and, writing $i: M^k \to W^{k+1}$ for the inclusion map, the restriction to M^k of the map $\tau_W: W^{k+1} \to BO(k+1)$ defined by (6), satisfies

$$\tau_W|_{M^k} = \tau_W \circ i = \lambda \circ \tau_M \quad (= \tau_M \oplus 1)$$

(where we are using $\tau_M \oplus 1$ to denote the map $M^k \to BO(k+1)$ inducing the Whitney sum (see §10.4 or Part II, §24.5) of the tangent bundle of M^k with the trivial line bundle over M^k.) Hence

$$w(M^k) = \tau_M^* \lambda^*(\bar{w}) = i^* \tau_W^*(\bar{w}).$$

Now since $M^k = \partial W^{k+1}$, it follows that for the fundamental homology class $[M^k]$ (see, e.g. §3) we have $i_*[M^k] = 0$, whence assuming that the cohomology class $w(M^k) \in H^l(M^k; \mathbb{Z}_2)$ has dimension k, i.e. $l = k$, we obtain on evaluating it at $[M^k]$:

$$(w(M^k), [M^k]) = (i^* \tau_W^*(\bar{w}), [M^k]) = (\tau^*(\bar{w}), i_*[M^k]) = 0.$$

Since $[M^k]$ generates $H_k(M^k; \mathbb{Z}_2)$ (see Theorem 3.9), the assertion (i) now follows (via Theorem 2.9).

The proof of (ii) is identical to the above, with $\mathbb{Q}$-homology replacing $\mathbb{Z}_2$-homology, using the fact that in the orientable case also, we have $i_*[M^k] = 0$ in $H_k(W^{k+1}; \mathbb{Q})$. This completes the proof of the lemma. □

Remark. In fact the converse of this lemma is also true: the vanishing of all characteristic numbers of a closed manifold suffices for it to be realizable as a boundary (see the proof of Theorem 27.14 below).

It can be shown that the Euler characteristic $\chi(M^k)$ is not a stable characteristic number. An exhaustive list of mod 2 stable characteristic classes of a manifold M^k is afforded by the polynomials of dimension k in the Stiefel–Whitney classes $w_q \in H^q(M^k; \mathbb{Z}_2)$ (see Definition 9.3), and of rational stable characteristic classes by the polynomials of dimension $k = 4s$ in the Pontrjagin characteristic classes $p_q \in H^{4q}(M^k; \mathbb{Q})$ (see Definition 9.5, and cf. §9, Exercise 25, and §10.4(12)).

Examples

(a) $M^2 = \mathbb{R}P^2$. The Stiefel–Whitney polynomial (over $\mathbb{Z}_2$) of the tangent bundle of $\mathbb{R}P^2$ is (see §9, Exercise 13)

$$w(z) = (1 + w_1 z)^3 = 1 + w_1 z + w_2 z^2,$$

where w_1 is the generator of $H^1(\mathbb{R}P_2; \mathbb{Z}_2) \simeq \mathbb{Z}_2$, and $w_2 = w_1^2$ is the generator of $H^2(\mathbb{R}P_2; \mathbb{Z}_2)$. Hence we obtain a non-zero mod 2 stable characteristic number, confirming the nontriviality of Ω_2^O (see (5)).

(b) $M^4 = \mathbb{C}P^2$ with the natural orientation. The Pontrjagin polynomial of $\mathbb{C}P^2$ is (see §10.4(23))

$$p(z) = (1 + z^2 t^2)^3 = 1 + p_1 z^2,$$

where t is a generator of $H^2(\mathbb{C}P^2; \mathbb{Z}) \simeq \mathbb{Z}$, regarded as an element of $H^2(\mathbb{C}P^2; \mathbb{Q})$. Hence we obtain the rational stable characteristic number $p_1(\mathbb{C}P^2) = 3$, so that by part (ii) of the above theorem $\mathbb{C}P^2$ is not a boundary, and $\Omega_4^{SO} \neq 0$.

(c) (i) $M_1^8 = \mathbb{C}P^4$, with the natural orientation. Again by §10.4(23) the Pontrjagin polynomial is

$$p(z) = 1 + p_1 z^2 + p_2 z^4 = (1 + z^2 t^2)^5 = 1 + 5t^2 z^2 + 10t^4 z^4,$$

where t is a generator of $H^2(\mathbb{C}P^4; \mathbb{Z}) \simeq \mathbb{Z}$, considered as an element of $H^2(\mathbb{C}P^4; \mathbb{Q})$. Hence we obtain the rational stable characteristic numbers

$$p_1^2 = 25, \qquad p_2 = 10.$$

(ii) $M_2^8 = \mathbb{C}P_1^2 \times \mathbb{C}P_2^2$. Here by §10.4(23) and using the formula $p(z) = \bar{p}(z)\bar{\bar{p}}(z)$ for a product of manifolds (cf. §9, Exercises 12, 15), we have

$$p(z) = 1 + p_1 z^2 + p_2 z^4 = (1 + t_1^2 z^2)^3 (1 + t_2^2 z^2)^3$$
$$= 1 + 3(t_1^2 + t_2^2) z^2 + 9 t_1^2 t_2^2 z^4,$$

where t_i generates $H^2(\mathbb{C}P_i^2; \mathbb{Z})$ (but is considered as an element of $H^2(\mathbb{C}P_i^2; \mathbb{Q})$. Since $(t_1^2)^2 = 0$, $(t_2^2)^2 = 0$ (by dimensional considerations), it follows that $(t_1^2 + t_2^2)^2 = 2t_1^2 t_2^2$. Hence we obtain the stable rational characteristic numbers

$$p_1^2 = 18, \qquad p_2 = 9.$$

Aside from the stable characteristic numbers we have been considering, there is another interesting invariant of SO-cobordism for orientable manifolds of dimension $4k$, called the "signature" of the manifold. (Two smooth, closed (oriented) manifolds belong to the same (*SO*-)*cobordism class* if their disjoint union is the boundary of a smooth compact (oriented) manifold.) In §18, in the course of establishing Poincaré duality (Theorem 18.1), a non-degenerate, integer-valued, unimodular, bilinear form was obtained on the homology group of "middle" dimension $n/2$ of an even-dimensional orientable manifold M^n, symmetric if $n = 4k$ and alternating if $n = 4k + 2$ (in particular of an orientable surface ($k = 0$)—see §18, Example (b)). This form arose from the intersection index of representative cycles of elements of the homology group $H_{2k}(M^{4k}; \mathbb{Q})$ (taking $n = 4k$), and according to §18, Exercise 2, may equivalently be defined via the cohomological product of the corresponding elements of $H^{2k}(M^{4k}; \mathbb{Q}) \simeq H_{2k}(M^{4k}; \mathbb{Q})$, i.e.

$$\langle \tilde{x}, \tilde{y} \rangle = \tilde{x} \circ \tilde{y} \quad \text{(the intersection index)}, \qquad \tilde{x}, \tilde{y} \in H_{2k}(M^{4k}; \mathbb{Q}), \tag{7}$$

or, equivalently,

$$\langle x, y \rangle = (xy, [M^{4k}]), \qquad x, y \in H^{2k}(M^{4k}; \mathbb{Q}), \qquad x = D\tilde{x}, \quad y = D\tilde{y}. \tag{7'}$$

27.3. Definition. The *signature* $\tau[M^{4k}]$ of an orientable manifold M^{4k} of dimension $4k$ is the difference in the number of positive and negative squares in the (diagonalized) quadratic form $\langle\ ,\ \rangle$ on $H_{2k}(M^{4k}; \mathbb{Q})$. (Note that under a change of orientation $M^{4k} \to -M^{4k}$, the form, and hence the signature, changes sign.)

27.4. Lemma (Rohlin). *The signature of a bounding manifold (orientable, of dimension* $4k$*) is zero. Since the signature of a disjoint union of manifolds is the sum of the signatures of the component manifolds, it follows that the signature defines a group homomorphism*

$$\tau\colon \Omega_{4k}^{SO} \to \mathbb{Z}.$$

Since the signature of a (direct) product of manifolds is the product of the signatures of the factors, the map τ *in fact extends to a ring homomorphism*

$$\tau\colon \Omega_*^{SO} = \sum_{j \geq 0} \Omega_j^{SO} \to \mathbb{Z},$$

where one defines $\tau(1) = 1$, *and* $\tau(\Omega_j^{SO}) = 0$ *if* j *is not divisible by* 4.

(Note that the multiplicative identity $1 \in \Omega_*^{SO}$ is the *SO*-cobordism class of the one-point manifold $\{*\}$ of dimension 0—cf. (2)(i).)

Exercise

1. Prove the assertion concerning the signature of a product of manifolds. (*Hint*. Use the Künneth isomorphism (4) of §7.)

Proof of Lemma. The statement concerning the signature of a disjoint union of manifolds being trivial, it remains to show only that the signature of a (closed, smooth, orientable) manifold M^{4k} is zero if it is realizable as a boundary: $M^{4k} = \partial W^{4k+1}$, W compact and orientable.

Denoting by i the inclusion map $M^{4k} \to W^{4k+1}$, we have (as already observed in the proof of Lemma 27.2) $i_*[M^{4k}] = 0$ in the group $H_{4k}(W^{4k+1}; \mathbb{Q})$. Hence if two $2k$-dimensional cocycles x, y in M^{4k} are obtained by restricting cocycles $\bar{x}$, $\bar{y}$ in W^{4k+1}, i.e. $x = i^*(\bar{x})$, $y = i^*(\bar{y})$, then $\langle x, y\rangle = 0$ since

$$\langle x, y\rangle = (xy, [M^{4k}]) = (i^*(\overline{xy}), [M^{4k}]) = (\overline{xy}, i_*[M^{4k}]) = 0.$$

(In terms of the intersection index of corresponding cycles $\tilde{x}$, $\tilde{y}$ (see (7)) this is geometrically clear: if both cycles are null-homologous in W^{4k+1}, then their intersection index is zero.)

The next step is to show that the dimension (as a vector space) of the subgroup $i^*H^{2k}(W^{4k+1}; \mathbb{Q}) \subset H^{2k}(M^{4k}; \mathbb{Q})$ is exactly half that of the containing vector space $H^{2k}(M^{4k}; \mathbb{Q})$. To this end, consider the exact rational homology and cohomology sequences of the pair (W^{4k+1}, M^{4k}) with the Poincaré-duality isomorphisms D (actually "Lefschetz-duality" isomorphisms—see §18, Exercise 3) between appropriate terms of the two sequences:

$$\begin{array}{ccccccc}
\to H_{2k+1}(W^{4k+1}, M^{4k}) & \xrightarrow{\partial_*} & H_{2k}(M^{4k}) & \xrightarrow{i_*} & H_{2k}(W^{4k+1}) \to \\
\wr\!\wr D & & \wr\!\wr D & & \wr\!\wr D \\
\to H^{2k}(W^{4k+1}) & \xrightarrow{i^*} & H^{2k}(M^{4k}) & \xrightarrow{\delta^*} & H^{2k+1}(W^{4k+1}, M^{4k}) \to
\end{array} \tag{8}$$

It is not difficult to verify (e.g. using the version given in §18(7) of the definition of the Poincaré-duality map) that this diagram commutes, i.e., more specifically, $D \circ i^* = \partial_* \circ D$ (and $D \circ \delta^* = i_* \circ D$), or in other words, that ∂_* and i^* (and i_* and δ^*) coincide once the identifications are made under the isomorphisms D indicated in (8). Since i_* and i^* are mutually dual vector-space maps, it follows that the dimensions of $\operatorname{Im} i^*$ and $\operatorname{Im} \delta^*$ are equal, and thence by exactness that $\operatorname{Im} i^*$ has dimension half of that of $H^{2k}(M^{4k}; \mathbb{Q})$, as we wished to show.

Since by the first part of the proof the form $\langle\ ,\ \rangle$ is zero on $\operatorname{Im} i^*$, it now follows from the non-degeneracy of that form that the signature is zero, completing the proof. □

We shall take for granted for the moment the result $\Omega_4^{SO} \simeq \mathbb{Z}$ mentioned earlier. (We shall show below at least that $\Omega_4^{SO} \otimes \mathbb{Q} \simeq \mathbb{Q}$.) From our knowledge of the structure of the cohomology ring $H^*(\mathbb{C}P^2; \mathbb{Q})$ (see the first example in §7), namely

$$H^*(\mathbb{C}P^2; \mathbb{Q}) \simeq \mathbb{Q}[x]/(x^3), \qquad \deg x = 2,$$

we readily infer that the form $\langle\ ,\ \rangle$ on $H^2(\mathbb{C}P^2; \mathbb{Q}) \simeq \mathbb{Q}$ satisfies $\langle x, x\rangle = 1$, whence $\tau(\mathbb{C}P^2) = 1$. Hence the SO-bordism class $[\mathbb{C}P^2]$ is not a proper multiple of any other such class, and so generates the group $\Omega_4^{SO} \simeq \mathbb{Z}$. Thus every element of Ω_4^{SO} has the form $\lambda[\mathbb{C}P^2]$ for some integer λ. Now since from Example (b) above we know that the characteristic number $p_1(\mathbb{C}P^2) = 3$, it follows that the "character" $p_1 - 3\tau$ vanishes at $[\mathbb{C}P^2]$, and hence at every element of Ω_4^{SO} (since every such class is a multiple of $[\mathbb{C}P^2]$). Hence we have the "Thom–Rohlin formula" for any orientable manifold M^4:

$$\tau(M^4) = \tfrac{1}{3}p_1(M^4). \tag{9}$$

(In the following subsection (§27.2), as well as showing in particular that $\mathbb{Q} \otimes \Omega_4^{SO} \simeq \mathbb{Q}$, we shall indicate a generalization of the formula (9)—"Hirzebruch's formula".)

Like the Euler characteristic, the signature may be defined also for manifolds-with-boundary: if $M = M^{4k}$ is a compact, smooth, orientable manifold-with-boundary, where

$$\partial M = V = V^{4k-1} = V_1 \cup \cdots \cup V_m$$

(the V_i being the connected components of the boundary V), then the intersection index of $2k$-cycles in M defines as before a bilinear form on $H_{2k}(M^{4k}; \mathbb{Q})$, which now, however, may be degenerate; the *signature of the manifold-with-boundary* M^{4k} is defined, as in the closed case, as the signature of this form, and denoted by $\tau(M^{4k})$.

The signature (as just extended to manifolds-with-boundary) has the following "additive property" (Novikov–Rohlin):

27.5. Theorem (On the Additivity of the Signature). *Let M_1^{4k} and M_2^{4k} be compact, smooth manifolds-with-boundary where*

$$\partial M_1^{4k} = \bigcup_j V_j, \qquad \partial M_2^{4k} = \bigcup_q W_q,$$

which meet in a single connected boundary component (the first say) of each:

$$V_1^{4k-1} = W_1^{4k-1}.$$

We then have

$$\tau\left(M_1^{4k} \bigcup_{V_1 = W_1} M_2^{4k}\right) = \tau(M_1^{4k}) + \tau(M_2^{4k}). \tag{10}$$

Thus one has additivity of signatures in the situation of a union of two manifolds glued together along a full connected component of their respective boundaries. We remark that this property is shared by the Euler characteristic of even-dimensional manifolds since by (4)

$$\chi\left(M_1^{2q} \bigcup_V M_2^{2q}\right) = \chi(M_1^{2q}) + \chi(M_2^{2q}) - \chi(V),$$

and $\chi(V) = 0$ by virtue of the odd-dimensionality of the closed manifold V (see the concluding remark of §18).

Proof of Theorem 27.5. We first decompose the rational homology groups $H_{2k}(M_1^{4k})$ and $H_{2k}(M_2^{4k})$ as follows:

$$H_{2k}(M_s^{4k}) = A_s \oplus B_s,$$

where $B_s = \operatorname{Im}(i_s)_*$, $i_s\colon V_1 = W_1 \to M_s^{4k}$ denoting the inclusion map ($s = 1, 2$), and A_s is any direct complement. Since the "intersection form" on $H_{2k}(M_s^{4k})$, i.e. the symmetric bilinear form defined by the intersection index of $2k$-cycles in M_s^{4k}, is zero on B_s (see, for instance, the beginning of the proof of Lemma 27.4), it follows that $\tau(M_s^{4k}) = \tau(A_s)$ (with the obvious interpretation of the latter notation) for $s = 1, 2$.

The group (actually rational vector space) $H_{2k}(M_1^{4k} \bigcup_{V_1=W_1} M_2^{4k})$ has the direct decomposition

$$H_{2k}(M_1 \cup M_2) = A_1 \oplus A_2 \oplus C_1 \oplus C_2 \oplus D \oplus F, \tag{11}$$

where

$$D = \operatorname{Im}(H_{2k}(V_1) \to H_{2k}(M_1 \cup M_2))$$

(i.e. the image under the inclusion homomorphism induced by the inclusion $V_1 = W_1 \subset M_1 \cup M_2$),

$$B_s = C_s \oplus D = \operatorname{Im}(i_s)_*, \qquad s = 1, 2 \text{ (see above)},$$

$$C_1 \oplus C_2 \oplus D \oplus E = H_{2k}(V_1) \quad (= H_{2k}(W_1)),$$

(where $E = \operatorname{Ker}(i_1)_* \cap \operatorname{Ker}(i_2)_* \subset H_{2k}(V_1)$), and, finally, F is comprised of those elements of $H_{2k}(M_1 \cup M_2)$, each of which is obtained from some $(2k-1)$-cycle in $V_1 = W_1$ whose images under both maps $i_s\colon V_1 = W_1 \to M_s$ are null-homologous (in the respective M_s), as the union of a pair of films spanning the cycle, one in M_1 and the other in M_2.

We leave as an exercise for the reader the verification of the following simple properties of the (symmetric) "intersection form" $\langle\ ,\ \rangle$ on $H_{2k}(M_1 \cup M_2)$ relative to the decomposition (11):

(i) $C_1 \oplus C_2$ is the kernel of the form;
(ii) the restriction of the form to each of the subspaces D, F (as well as C_1, C_2 of course) is trivial (i.e. zero), and the subspaces D, F are mutually dual with respect to the form, i.e. the map sending each $f \in F$ to the functional in D^* defined by $d \mapsto \langle f, d\rangle$ is a vector-space isomorphism;
(iii) relative to this form the subspaces A_1 and A_2 are each orthogonal to their respective direct complements in the decomposition (11).

The additivity property (10) now follows readily from (i), (ii) and (iii), together with the earlier observation that $\tau(M_s^{4k}) = \tau(A_s)$. □

27.2. Thom Complexes. Calculation of Cobordisms (Modulo Torsion). The Signature Formula. Realization of Cycles as Submanifolds

Consider a smooth vector bundle ξ over a connected, smooth, closed manifold B, with fibre $\mathbb{R}^n$, bundle group $G = O(n)$ (or $SO(n)$, $U(n/2)$, $U(n/2)$, $Sp(n/4)$) and projection map $p: E \to B$:

$$\xi = (E, B, p, F, G), \qquad F \cong \mathbb{R}^n, \quad G = O(n).$$

Restricting consideration to the vectors of length ≤ 1 in each fibre, we obtain a fibre bundle $\tilde{E} \to B$, now with fibre $\tilde{F} \cong D^n \subset \mathbb{R}^n$, and with boundary $\partial\tilde{E}$ a fibre bundle with fibre S^{n-1} (a "sphere bundle"). Note that being a compact, smooth manifold the space $\tilde{E}$ is finitely triangulable (by means of a triangulation naturally extending one of B).

27.6. Definition. The *Thom complex* $M(\xi)$ of a vector bundle ξ (as above) is the factor complex

$$M(\xi) = \tilde{E}/\partial\tilde{E},$$

obtained by identifying the boundary $\partial\tilde{E}$ of $\tilde{E}$ with a point.

27.7. Lemma. *For each $i \geq 0$ there are natural isomorphisms*

$$\begin{aligned} \varphi &: H_i(B) \to H_{n+i}(M(\xi)), \\ \varphi &: H^i(B) \to H^{n+i}(M(\xi)), \end{aligned} \tag{12}$$

where $n =$ dim F, valid in mod 2 *homology in the case $G = O(n)$ and over $\mathbb{Z}$, $\mathbb{Q}$, $\mathbb{Z}_p$ (p prime) if $G = SO(n)$. (The first isomorphism φ arises in a geometrically obvious way, essentially sending each (i-dimensional) cycle z in B to its complete inverse image $p^{-1}(z)$ modulo $\partial\tilde{E}$.)*

The proof of this lemma is in its essentials the same as that of Lemma 20.4 (used in §20 to establish an analogue of the Morse inequalities in the situation of critical manifolds). As in that proof one obtains, for instance, the first of the isomorphisms φ in (12) essentially as the composite $D_{\tilde{E}}D_B$ of the Poincaré-duality isomorphism

$$D_B: H_i(B) \to H^{m-i}(B), \qquad m = \dim B \qquad (\text{cf. §20(5)}),$$

with the Lefschetz-duality isomorphism (see §18, Exercise 3)

$$\begin{array}{ccc} D_{\tilde{E}}: H^{m-i}(\tilde{E}) & \to & H_{n+m-(m-i)}(\tilde{E}, \partial\tilde{E}) \qquad (\text{cf. §20(4)}) \\ \wr\wr & & \wr\wr \\ H^{m-i}(B) & & H_{n+i}(M(\xi)), \end{array}$$

where the isomorphism $H^{m-i}(\tilde{E}) \simeq H^{m-i}(B)$ follows from the homotopic equivalence of $\tilde{E}$ and B. □

Note that the cohomology class $\varphi(1) \in H^n(M(\xi))$ (where 1 is the multiplicative identity of $H^*(B)$) arises from the fundamental class of the fibre $\tilde{F}$ (modulo $\partial\tilde{E}$), and so in this sense is also "fundamental". Observe also that the natural identification of B with the zeroth cross-section of the fibre bundle ξ determines a natural embedding of B in $M(\xi)$. The normal bundle on B in $M(\xi)$ (see Part II, §7.2) is then just ξ itself, and the complement $M(\xi)\backslash B$ is contractible (over itself to a point, namely to the point with which $\partial\tilde{E}$ was identified).

As our first application of Thom complexes we shall establish a relationship between the Stiefel–Whitney characteristic classes $w_j \in H^j(B; \mathbb{Z}_2)$ of the (arbitrary) vector bundle ξ with base B and group $O(n)$ (see Definition 9.3), and the Steenrod squares Sp^i (see Theorem 10.2(i)), namely

$$w_j = \varphi^{-1} Sp^j \varphi(1), \tag{13}$$

where the first map $\varphi\colon H^0(B; \mathbb{Z}_2) \to H^n(M(\xi); \mathbb{Z}_2)$ is as in (12) (so that $\varphi(1)$ is the "fundamental class" noted above), Sp^j is a map (cf. loc. cit.)

$$H^n(M(\xi); \mathbb{Z}_2) \to H^{n+j}(M(\xi); \mathbb{Z}_2),$$

and the final map φ^{-1} is the inverse of the appropriate map in (12):

$$\varphi\colon H^j(B; \mathbb{Z}_2) \to H^{n+j}(M(\xi); \mathbb{Z}_2).$$

The verification of (13) (and its analogue for $SO(n)$) proceeds via calculations (outlined in some of the exercises below) involving the mod 2 cohomology of the universal classifying space $BO(n)$ (and $BSO(n)$), i.e. in $H^*(BO(n); \mathbb{Z}_2)$ (and $H^*(BSO(n); \mathbb{Z}_2)$).

Recall first that the diagonal matrices in $O(n)$ are just those of the form

$$\begin{pmatrix} \pm 1 & & 0 \\ & \pm 1 & \\ & & \ddots & \\ 0 & & & \pm 1 \end{pmatrix},$$

together comprising a subgroup $D(n) \simeq \mathbb{Z}_2 \oplus \cdots \oplus \mathbb{Z}_2$ (n summands). The inclusion $D(n) \subset O(n)$ determines a natural map

$$i\colon BD(n) = \mathbb{R}P_1^\infty \times \cdots \times \mathbb{R}P_n^\infty \to BO(n)$$

(recall from Part II, §24.4(a) that the classifying space of $O(1) \simeq \mathbb{Z}_2$ is $\mathbb{R}P^\infty$), inducing in turn a ring homomorphism

$$i^*\colon H^*(BO(n); \mathbb{Z}_2) \to H^*(BD(n); \mathbb{Z}_2).$$

EXERCISE

2. (Cf. §10, Exercise 6, where the reader was asked to establish the analogous facts for $U(n)$, over the rationals.) Show that i^* has trivial kernel, i.e. is a monomorphism, and that the image Im i^* consists precisely of the symmetric polynomials in the generators x_i ($i = 1, \ldots, n$) of the groups $H^1(\mathbb{R}P_i^\infty; \mathbb{Z}_2)$ ($\simeq \mathbb{Z}_2$; see e.g. §9, Exercise 5, *et seqq.*).

From this exercise one can infer the representation of the qth Stiefel–Whitney class w_q of the canonical vector bundle $\hat{\xi}$ over $BO(n)$ as the elementary symmetric polynomial of degree q in the x_i, i.e.

$$i^* w_q = \sum_{i_1 < \cdots < i_q} x_{i_1} \ldots x_{i_q}. \tag{14}$$

(By the "canonical vector bundle over $BO(n)$" we mean the bundle $\hat{\xi}$ consisting of all pairs

$$(n\text{-plane in } BO(n) = G_{\infty,n}, \text{ vector in that } n\text{-plane}) \qquad (15)$$

with the obvious projection map; this bundle is "universal" in the sense that for every $O(n)$-vector bundle ξ (over a manifold) there is a bundle map $\xi \to \hat{\xi}$ (see Part II, Definition 24.1.2).)

EXERCISES (continued)

3. The (ring) homomorphism
$$j^*: H^*(BO(n); \mathbb{Z}_2) \to H^*(BSO(n); \mathbb{Z}_2)$$
induced by the natural surjection $j: BSO(n) \to BO(n)$ (which "forgets" the orientation on the oriented n-dimensional planes representing the points of $\hat{G}_{\infty,n} = BSO(n)$), is surjective (i.e. an epimorphism), and its kernel is generated as ideal by the first Stiefel–Whitney class $w_1 \in H^1(BO(n); \mathbb{Z}_2)$.

4. Consider the Thom complex $M(\hat{\xi})$ of the canonical vector bundle $\hat{\xi}$ over $BO(n)$ (see (15)). Show that the map
$$f: BO(n) \to M(\hat{\xi})$$
(essentially identifying $BO(n)$ with the zeroth cross-section of $\hat{\xi}$—see above), induces a (ring) monomorphism
$$f^*: H^*(M(\hat{\xi}); \mathbb{Z}_2) \to H^*(BO(n); \mathbb{Z}_2),$$
and that, furthermore, the image $\operatorname{Im} f^*$ consists precisely of those polynomials in the Stiefel–Whitney classes w_q divisible by the nth such class $w_n \in H^n(BO(n); \mathbb{Z}_2)$ (for which we have by (14) $i^*w_n = x_1, \ldots, x_n$). Show also that $f^*\varphi(1) = w_n$ (where $1 = w_0$ is the multiplicative identity of $H^*(BO(n); \mathbb{Z}_2)$), and, more generally, using the properties of the Steenrod squares given in Theorem 10.2(i), that
$$f^*\varphi(w_i) = Sq^i(w_n) = w_i w_n. \qquad (16)$$
(The reader may like to verify that in fact $f^*\varphi(x) = xw_n$ for every $x \in H^i(BO(n); \mathbb{Z}_2)$, i arbitrary.)

Equation (13) now follows for $\hat{\xi}$ since
$$\varphi^{-1}Sq^i\varphi(1) = (f^*\varphi)^{-1}Sq^i(f^*\varphi)(1) = (f^*\varphi)^{-1}Sq^i w_n = w_i \quad \text{(by (16))},$$
and thence for the original vector bundle ξ over B via a bundle map $\xi \to \hat{\xi}$ (see above), by using the "naturality" property
$$w_i(\xi) = g^*(w_i(\hat{\xi})),$$
where g is the map between base spaces induced by that bundle map.

EXERCISES (continued)

5. Establish the results analogous to those of Exercise 4, for $H^*(BSO(n); \mathbb{Z}_2)$. Calculate the operations Sq^i explicitly for the $H^q(M(\hat{\xi}); \mathbb{Z}_2)$ where $\hat{\xi}$ is the canonical "universal" $\mathbb{R}^n$-bundle over $BO(n)$ defined above (see (15)). Investigate the homotopy groups
$$\pi_{n+j}(M(\xi)), \qquad j < n-1,$$
using the results of §10.

6. Starting with the formula $w_i = \varphi^{-1}Sq^i\varphi(1)$, prove that within the class of closed manifolds M^n, the Stiefel–Whitney classes $w_j \in H^j(M^n; \mathbb{Z}_2)$ of the tangent bundle of M^n, are homotopy invariants, by exploiting the relationship between the tangent bundle of M^n and a neighbourhood of the diagonal of $M^n \times M^n$.

7. The Stiefel–Whitney class w_1 of (the tangent bundle of) a manifold M^n (which by §9, Exercise 11, vanishes precisely if M^n is orientable) satisfies

$$Dw_1 = \partial_1[M^n],$$

where D is the appropriate Poincaré-duality isomorphism and $\partial_1: H_n(M^n; \mathbb{Z}_2) \to H_{n-1}(M^n; \mathbb{Z}_2)$ is the Bockstein homomorphism (cf. §10.1, Example (b), and §3(10)). Establish this formula independently of the preceding discussion.

For the base spaces $B = BG$ with $G = O(n)$, $SO(n)$, $U(n/2)$, $SU(n/2)$ or $Sp(n/4)$, the Thom complexes $M(\xi)$ of the respective canonical "universal" $\mathbb{R}^n$-vector bundles ξ introduced above (see (15)) will be denoted by $MO(n)$, $MSO(n)$, $MU(u/2)$, $MSU(n/2)$, $MSp(n/4)$. We shall now identify the Thom complexes $MO(n)$, $MSO(n)$ for small n.

Beginning with the trivial group $G = \{e\}$, in which case the only vector bundle over $BG = \{*\}$ (the one-point space) with fibre $\mathbb{R}^n$ is the trivial one (with total space $\cong \mathbb{R}^n$), one easily verifies that

$$M\{e\} \cong S^n.$$

Hence since $SO(1) = \{e\}$, we deduce in particular $MSO(1) \cong S^1$.

Since $O(1) = \{\pm 1\} \simeq \mathbb{Z}_2$, we have (see Part II, §24.4(a)) $BO(1) = \mathbb{R}P^\infty$ (or $\mathbb{R}P^N$ for "arbitrarily large" N). Here the canonical line bundle ξ with structure group $O(1)$ is (equivalent to) the normal bundle over $\mathbb{R}P^N$ in $\mathbb{R}P^{N+1}$:

$$p: E \to \mathbb{R}P^N, \qquad \text{with fibre} \quad F \cong \mathbb{R}^1.$$

The space $\tilde{E}$ consisting of those vectors in the fibres of lengths ≤ 1 then has fibre $\tilde{F} \cong D^1 = I$, a closed interval (so that $\tilde{E}$ is the "generalized Möbius band"—cf. Part II, §24.5, Exercise 2). Its boundary $\partial\tilde{E}$ is diffeomorphic to the N-sphere S^N and covers the base $\mathbb{R}P^N$ in standard fashion. Hence the Thom complex $MO(1)$ can be identified with $\mathbb{R}P^{N+1}$:

$$MO(1) = \tilde{E}/\partial\tilde{E} \cong \mathbb{R}P^{N+1} \supset \mathbb{R}P^N = BO(1), \qquad N \to \infty.$$

In the case $G = SO(2)$, one obtains analogously

$$\begin{array}{l} MSO(2) \cong \mathbb{C}P^{N+1} \supset \mathbb{C}P^N = BSO(2), \qquad N \to \infty. \\ \quad \wr\!\| \\ MU(1) \end{array}$$

In each of these three cases the "fundamental class" $\varphi(1)$ (see (12)) generates the group to which it belongs:

$$\begin{aligned} &\varphi(1) \in H^1(S^1; \mathbb{Z}) && \text{for} \quad MSO(1) \cong S^1; \\ &\varphi(1) \in H^1(\mathbb{R}P^\infty; \mathbb{Z}_2) && \text{for} \quad MO(1) \cong \mathbb{R}P^\infty; \\ &\varphi(1) \in H^2(\mathbb{C}P^\infty; \mathbb{Z}) && \text{for} \quad MSO(2) \cong \mathbb{C}P^\infty. \end{aligned} \tag{17}$$

In fact, the space $MSO(1)$, $MO(1)$, $MSO(2)$ are Eilenberg–MacLane complexes $K(\pi, n)$ for $(\pi, n) = (\mathbb{Z}, 1), (\mathbb{Z}_2, 1), (\mathbb{Z}, 2)$ (see §§9.2, 10.2), in each of which $\varphi(1)$ is the "fundamental class" of the complex.†)

We next prove the following simple

27.8. Lemma. *The Thom complex $M(\xi)$ of a vector bundle ξ with fibre $F \cong \mathbb{R}^n$ (as above) is $(n-1)$-connected*:

$$\pi_j(M(\xi)) = 0 \qquad \text{for} \quad 1 \le j < n,$$

(and so in particular, for $n > 1$, $M(\xi)$ is always simply-connected). The nth *homotopy group is as follows*:

$$\pi_n(M(\xi)) \simeq \begin{cases} \mathbb{Z} & \text{if } \xi \text{ is orientable,} \\ \mathbb{Z}_2 & \text{is } \xi \text{ is non-orientable.} \end{cases} \tag{18}$$

PROOF. A cell decomposition of $M(\xi)$ can be obtained from a decomposition of the base B (by assumption a connected, closed, smooth manifold) by taking the products of the cells of B with a single n-dimensional cell, namely the fibre $\tilde{F} \cong D^n$ in $\tilde{E}$:

$$B \supset \sigma^j \mapsto \varphi(\sigma^j) = p^{-1}(\sigma^j) = \sigma^{n+j}. \tag{19}$$

Apart from the cells obtained in this way there is a single 0-dimensional cell σ^0 in $M(\xi)$, arising from the contraction of $\partial\tilde{E}$ to a point in $M(\xi) = \tilde{E}/\partial\tilde{E}$. The first assertion of the lemma, namely that $\pi_j(M(\xi)) = 0$ for $0 < j < n$, is then immediate since in this decomposition there are no cells of those dimensions j.

Turning to the second assertion, concerning $\pi_n(M(\xi))$, we observe first that since B is connected we may assume (by Theorem 4.8) that its cell decomposition has in it just one 0-cell σ^0. It then follows that in the corresponding cell decomposition of $M(\xi)$ there will be exactly one n-dimensional cell σ^n (arising from the fibre above the 0-cell σ^0), so that the group $H_n(M(\xi); \mathbb{Z})$ must be cyclic.

If the bundle ξ is non-orientable, there must exist a closed path beginning and ending at the point $\sigma^0 \in B$, along which transport of a tangent frame to the fibre $F \cong \mathbb{R}^n$ reverses orientation. Choosing this path as representing a (singular) 1-cell σ^1 in our decomposition of B, we obtain a corresponding cell $p^{-1}(\sigma^1) = \varphi(\sigma^1) = \sigma^{n+1}$ in the decomposition of $M(\xi)$ given by (19), with the property (geometrically clear if one thinks, for instance, of the example where ξ is a Möbius band, in conjunction with the definition of the operator ∂ in cellular homology—see §4(3))

$$\partial(\sigma^{n+1}) = 2\sigma^n.$$

If, on the other hand, the bundle ξ is orientable, then the boundaries of all $(n+1)$-dimensional cells $p^{-1}(\sigma_j^1)$ of the decomposition (19) of $M(\xi)$ are zero. (Consider, for example, the case of an annulus ξ in conjunction with the

† There is a more difficult result due to Thom to the effect that every complex $MO(n)$ has, up to dimension $2n-1$, the homotopy type of a direct product of Eilenberg–MacLane complexes $K(\mathbb{Z}_2, m_j)$ with the $m_j \ge n$.

definition of ∂ in §4(3).) Hence, in this case, the cycle $[\sigma^n]$ represents an element of infinite order in $H_n(M(\xi); \mathbb{Z})$.

The assertion (18) of the lemma now follows from Hurewicz' theorem (Corollary 4.9). □

We now prove the following important

27.9. Theorem. *The cobordism groups Ω_i^O, Ω_i^{SO} are canonically isomorphic to the corresponding "stable" homotopy groups of $MO(n)$, $MSO(n)$, i.e.*

$$\pi_{n+i}(MO(n)) \simeq \Omega_i^O, \qquad \pi_{n+i}(MSO(n)) \simeq \Omega_i^{SO} \qquad \textit{for} \quad i < n-1.$$

(Cf. Part II, §23.1 (in particular Theorem 23.1.4) where a one-to-one correspondence was established between the group $\pi_{n+i}(M\{e\}) \simeq \pi_{n+i}(S^n)$ and what amounted to cobordism classes of "framed normal bundles".)

PROOF. Let M^i be any closed, smooth manifold of dimension $i < n - 1$, considered as smoothly embedded in $\mathbb{R}^{n+i}$ (such an embedding having been shown to exist in Part II, §11). Since by Theorem 12.1.6 of Part II all such embeddings are isotopic, it follows that up to equivalence there is just one normal bundle on M^i in $\mathbb{R}^{n+i}$ (independently of the embedding $M^i \subset \mathbb{R}^{n+i}$), which we shall denote by ν. Denoting by $\hat{\xi}$, as before, the canonical "universal" vector bundle over $BO(n)$ with fibre $\mathbb{R}^n$ (see (15)), we consider the map (defined to within a homotopy)

$$\psi: M^i \to BO(n)$$

inducing the vector bundle ν from $\hat{\xi}$ (see Part II, §24.4), and determining in particular a map $\hat{\psi}$ (covering ψ) between the respective total spaces of ν and $\hat{\xi}$. The subspace of the total space of ν consisting of all vectors of length ≤ 1 in the fibres, may be identified with a closed neighbourhood N of M^i in $\mathbb{R}^{n+i} \subset S^{n+i}$; this identification made, we may regard $\hat{\psi}$ (or rather its restriction to this subspace of the total space of ν) as a map from the neighbourhood N of M^i to the corresponding subspace $\tilde{E}$ of the total space E of $\hat{\xi}$. By, in essence, sending the whole of the complement of the neighbourhood N in S^{n+i} to the 0-cell σ^0 of the Thom complex $M(\hat{\xi})$ arising from the contraction of $\partial\tilde{E}$ to a point, we obtain from $\hat{\psi}$ a map

$$f: S^{n+i} \to M(\hat{\xi}) = MO(n). \tag{20}$$

It is easy to see that the map f is transversally regular on the submanifold $BO(n) \subset M(\hat{\xi})$, and that $f^{-1}(BO(n)) = M^i$. (Recall from Part II, Definition 10.3.1, that "transversal regularity" of f on the submanifold $BO(n) \subset M(\hat{\xi})$ (where the embedding is as defined earlier) means that for every point $x \in f^{-1}(BO(n)) = M^i$, the image of the tangent space $\mathbb{R}_x^{n+i}$ to S^{n+i} at x under the differential map df of tangent spaces, is transverse to the tangent plane to $BO(n) \subset M(\hat{\xi})$ at $f(x)$, i.e. the linear subspaces $df(\mathbb{R}_x^{n+i})$ and $T_{f(x)}(BO(n))$ together span the whole of the tangent space to $M(\hat{\xi})$ at the point $f(x)$.)

Now suppose M_0^i and M_1^i are two closed, smooth manifolds which are cobordant, i.e. there is a compact manifold-with-boundary W^{i+1} such that $\partial W^{i+1} = M_0^i \cup M_1^i$ (disjoint union). We can embed W^{i+1} in $\mathbb{R}^{n+i} \times I[0, 1]$ in such a way that $M_0^i \subset \mathbb{R}^{n+i} \times 0$, $M_1^i \subset \mathbb{R}^{n+i} \times 1$, and W^{i+1} "approaches" these boundary components normally. By imitating the above construction starting with the normal bundle to $W^{i+1} \subset \mathbb{R}^{n+i} \times I$, one obtains a map

$$S^{n+i} \times I[0, 1] \to M(\hat{\xi})$$

which defines a homotopy between the two maps

$$f_0: S^{n+i} \to M(\hat{\xi}), \qquad f_1: S^{n+i} \to M(\hat{\xi}),$$

obtained as above (see (20)) starting with M_0^i and M_1^i respectively. In this way we have determined a map

$$\Omega_i^O \to \pi_{n+i}(MO(n)), \qquad i < n - 1, \tag{21}$$

which can readily be shown to be a one-to-one homomorphism, i.e. monomorphism.

The analogous homomorphism

$$\Omega_i^{SO} \to \pi_{n+i}(MSO(n)), \qquad i < n - 1, \tag{22}$$

is constructed similarly.

It remains to show that the map (21) (and (22)) is onto. To this end, let

$$f: S^{n+i} \to MO(n) \tag{23}$$

be any map (representing an arbitrary element of $\pi_{n+i}(MO(n))$. We may assume f to be transversally regular on $BO(n) \subset MO(n)$, since if this is not the case then by Theorem 10.3.2 of Part II, t-regularity can be achieved by means of an arbitrarily small perturbation of f. As noted in Part II, §10.3, the complete inverse image $f^{-1}(BO(n))$ will then be a smooth, closed, i-dimensional, non-singular submanifold M^i of $\mathbb{R}^{n+i} \subset S^{n+i}$, and the images of the normal n-planes to M^i in $\mathbb{R}^{n+i}$ under the differential map df will be transverse to $BO(n)$. By deforming the image under f appropriately, i.e. applying a suitable (readily defined) homotopy of f, we may arrange that in fact these images are all normal to $BO(n)$. It then follows that the map $M^i \to BO(n)$ obtained by restricting f to $M^i = f^{-1}(BO(n))$ induces (from the canonical "universal" bundle $\hat{\xi}$ over $BO(n)$ in terms of which $M(\hat{\xi}) = MO(n)$ is defined) the normal bundle ν on M^i in $\mathbb{R}^{n+i}$, so that the map f of (23) can be reconstructed from the manifold M^i according to the recipe given in the first part of the proof. Hence the homomorphism (21) is indeed onto. The proof that the homomorphism (22) is onto is similar. □

27.10. Theorem

(i) *An element $x \in H_i(M^{n+i}; \mathbb{Z}_2)$ is representable by a closed submanifold $M^i \subset M^{n+i}$ precisely if there exists a map $f: M^{n+i} \to MO(n)$ satisfying*

$f^*\varphi(1) = Dx$, where $\varphi(1) \in H^n(MO(n); \mathbb{Z}_2)$ is the fundamental class of the Thom complex $MO(n)$ (see Lemma 27.7 et seqq.), and $D: H_i(M^{n+i}; \mathbb{Z}_2) \to H^n(M^{n+i}; \mathbb{Z}_2)$ is the Poincaré-duality isomorphism.

(ii) *Let M^{n+i} be an oriented manifold. An element $x \in H_i(M^{n+i}; \mathbb{Z})$ is representable by an oriented submanifold $M^i \subset M^{n+i}$ if and only if there exists a map $f: M^{n+i} \to MSO(n)$ such that $f^*\varphi(1) = Dx$.*

(iii) *An element $x \in H_i(M^{n+i}; \mathbb{Z})$ (M^{n+i} oriented) is representable by an oriented, closed submanifold $M^i \subset M^{n+i}$ on which the normal bundle is trivial (i.e. is defined by a non-singular system of equations $\psi_1 = 0, \ldots, \psi_n = 0$ in M^{n+i}—see Part II, §7.2, Example (a)) if and only if there is a map $f: M^{n+i} \to M\{e\} \cong S^n$ satisfying $f^*(\varphi(1)) = Dx$.*

Remark. There are analogous results concerning the realizability of cycles by submanifolds having prescribed normal bundles with structure groups $U(n/2)$, $SU(n/2)$, $Sp(n/4)$. Such realizations arise, as in the present theorem, from maps of the given manifold M^{n+i} into $MU(n/2)$, $MSU(n/2)$, or $MSp(n/4)$, as the case may be, constructed much as in the proof of the preceding theorem (27.9) (and that of the present one). For appropriate i the groups $\pi_{n+i}(MU(n/2))$, $\pi_{n+i}(MSU(n/2))$ and $\pi_{n+i}(MSp(n/4))$ may, analogously to the case of $O(n)$ and $SO(n)$ (Theorem 27.9), be canonically identified with the complex (or unitary) cobordism group Ω_i^U, the special unitary cobordism group Ω_i^{SU}, and the quaternion cobordism group Ω_i^{Sp}, respectively. The unitary cobordisms are of particular importance: every complex (and "quasicomplex") manifold belongs to a cobordism class in the appropriate group Ω_{2i}^U.

Outline of the Proof of Part (i) of the Theorem. Let $M^i \subset M^{n+i}$ be any closed, i-dimensional submanifold of the given manifold M^{n+i}. The normal bundle on M^i in M^{n+i} gives rise (essentially as in the proof of Theorem 27.9, via the corresponding classifying map $M^i \to BO(n)$) to a map $f: M^{n+i} \to MO(n)$ (cf. (20)), under which the whole of the complement of a neighbourhood of $M^i \subset M^{n+i}$ is sent to the point σ^0 in $MO(n)$ arising from the contraction of $\partial \tilde{E}$ in the construction of $M(\xi) = MO(n)$ (see Definition 27.6). By closely examining the construction of the map f it is not difficult to see from the geometry of the situation that

$$f^*\varphi(1) = D[M^i], \tag{24}$$

where $\varphi: H^0(BO(n); \mathbb{Z}_2) \to H^n(MO(n); \mathbb{Z}_2)$ is as in Lemma 27.7, and D is the appropriate Poincaré-duality isomorphism. Thus if $x \in H_i(M^{n+i}; \mathbb{Z}_2)$ is representable by a submanifold $M^i \subset M^{n+i}$, then by (24) we shall have $f^*\varphi(1) = Dx$.

Conversely, suppose $x \in H_i(M^{n+i}; \mathbb{Z}_2)$ satisfies the latter equation for some map $f: M^{n+i} \to MO(n)$. As noted in the proof of Theorem 27.9, the map f may, by Part II, Theorem 10.3.2, be assumed transversally regular on $BO(n) \subset MO(n)$, with the consequence that the complete inverse image $f^{-1}(BO(n))$ will be a smooth, closed i-dimensional submanifold M^i of M^{n+i}. Equation (24) now follows for this f and this M^i, as in the first part of the proof, whence $x = [M^i]$.

(Part (ii) of the theorem, i.e. the analogous result for $SO(n)$ (as also for the groups $U(n/2)$, $SU(n/2)$ and $Sp(n/4)$) is proved similarly.) □

This theorem has several interesting consequences (some requiring also results from §10).

27.11. Corollary

(i) *For all i, every element x of* $H_i(M^{i+1}; \mathbb{Z}_2)$ *is representable by a closed submanifold* $M^i \subset M^{i+1}$.
(ii) *For all i, every element* $x \in H_i(M^{i+1}; \mathbb{Z})$ *and* $x \in H_i(M^{i+2}; \mathbb{Z})$ $(M^{i+1}, M^{i+2}$ *oriented) is representable by a closed, oriented submanifold.*

PROOF. By §9, Exercise 5, each Eilenberg–MacLane complex $K(\pi, n)$ has the following basic property: given any complex X, there is a natural one-to-one correspondence between $[X, K(\pi, n)]$ and $H^n(X; \pi)$, and hence, in our present particular situation, between $[M^{i+1}, K(\mathbb{Z}_2, 1)]$ and $H^1(M^{i+1}; \mathbb{Z}_2)$. In fact, this correspondence associates with each class $Dx \in H^1(M^{i+1}; \mathbb{Z}_2)$ a homotopy class of maps $f: M^{i+1} \to K(\mathbb{Z}_2, 1)$ satisfying

$$f^*u = Dx, \tag{25}$$

where $u \in H^1(K(\mathbb{Z}_2, 1); \mathbb{Z}_2)$ is essentially the generator of $\pi_1(K(\mathbb{Z}_2, 1))$, i.e. the fundamental class of $K(\mathbb{Z}_2, 1)$. Since, as noted in (17) *et seqq.*, $MO(1) \sim K(\mathbb{Z}_2, 1)$, with $\varphi(1)$ the fundamental class, the corollary is now immediate from the theorem. (The proof of (ii) is analogous, using the identifications $MSO(1) = K(\mathbb{Z}, 1)$, $MSO(2) = K(\mathbb{Z}, 2)$.) □

27.12. Corollary. *Given any element* $x \in H_i(M^{n+i}; \mathbb{Z})$ *where* $i < n - 1$, *there is an integer* $\lambda \neq 0$ *such that* λx *is realizable as a submanifold* $M^i \subset M^{n+i}$.

SKETCH OF PROOF. By Lemma 27.8 the complex $MSO(n)$ is $(n-1)$-connected, whence, as noted in the proof of Corollary 10.11, the hypotheses of the Cartan–Serre theorem (10.7) are satisfied: $H^*(MSO(n); \mathbb{Q})$ is identifiable with a free skew-commutative algebra up to and including dimension $2n - 1$. Hence it follows, as in the proof of the latter theorem, that as far as its stable rational homology and homotopy (i.e. in dimensions $n + i$, $i < n - 1$) are concerned, $MSO(n)$ "has the structure of" a product of Eilenberg–MacLane complexes $K(\pi_j, m_j)$ with $m_j \geq n$. From this point on the proof follows along the same essential lines as the preceding one, exploiting basic properties of this product of complexes $K(\pi_j, m_j)$ rather than just $K(\mathbb{Z}, 1)$ or $K(\mathbb{Z}, 2)$. □

27.13. Corollary. *Let X be any finite cell complex. Given any element* $x \in H_i(X; \mathbb{Z})$, *there is an integer* $\lambda \neq 0$ *such that* λx *is realizable as the image of a* *manifold* M^i, *i.e. there is a map* $g: M^i \to X$ *such that*

$$g_*[M^i] = \lambda x.$$

(Note that if X has no odd-order torsion in its homology then this result is valid without tensoring with $\mathbb{Q}$, i.e. with $\lambda = 1$: The elements of $H_i(X; \mathbb{Z})$ themselves, rather than only certain multiples of them, are realizable as images of manifolds (Novikov).)

Sketch of Proof. One first embeds X in $\mathbb{R}^{N+i}$ (for N sufficiently large) and then "thickens" $X \subset \mathbb{R}^{N+i}$ to obtain a manifold-with-boundary $U \supset X$ which contracts to X, so that $U \sim X$. Then since $H_i(U) \simeq H_i(X)$, the element x may be regarded as lying in $H_i(U)$, and by the preceding corollary (or rather the details of its derivation from Theorem 27.10) there is a map $f: (U, \partial U) \to MSO(n)$ sending ∂U to a point, and satisfying

$$f^*(\varphi(1)) = D(\lambda x) = D[M^i]$$

for some integer λ and some submanifold $M^i \subset U$. □

From this result (or rather the relativized version of it) one obtains almost immediately the following

27.14. Corollary. *The natural homomorphism*

$$\Omega_i^{SO}(X, Y) \otimes \mathbb{Q} \to H_i(X, Y; \mathbb{Q})$$

from the ith rational bordism group to the ith rational homology group is an epimorphism.

We turn next to consequences of Theorem 27.9 in conjunction with the Cartan–Serre theorem (10.7). As observed in §10.4(12), the algebra $H^*(BSO(n); \mathbb{Q})$ is generated by the Pontrjagin classes (together with the Euler —or "Euler–Poincaré"—class when n is even):

$$p_i \in H^{4i}(BSO(n); \mathbb{Q}), \qquad \chi \in H^{2n}(BSO(2n); \mathbb{Q}),$$

and is in fact just the algebra of polynomials in these classes. It therefore follows, via the isomorphism (12), namely

$$\varphi: H^i(BSO(n); \mathbb{Q}) \simeq H^{n+i}(MSO(n); \mathbb{Q}), \tag{26}$$

that for $j < n$ not of the form $4k$, we have

$$H^{n+j}(MSO(n); \mathbb{Q}) = 0, \qquad n > j \neq 4k,$$

while for $j = 4k < n$, the stable group $H^{n+4k}(MSO(n); \mathbb{Q})$ ($\simeq H^{4k}(BSO(n); \mathbb{Q})$) has rank equal to the number of partitions $k = m_1 + \cdots + m_q$ of the integer k, since for $4k < n$ the vector space $H^{4k}(BSO(n); \mathbb{Q})$ has a basis consisting of all monomials $z = p_{m_1} p_{m_2} \cdots p_{m_q}$ in the classes p_i (where $m_l = m_j$ is allowed) of degree $\deg z = 4(m_1 + \cdots + m_q)$.

Recall that in the first subsection of the present section we calculated the (stable) rational characteristic numbers of $\mathbb{C}P^2$, $\mathbb{C}P^4$ and $\mathbb{C}P^2 \times \mathbb{C}P^2$, inferring from the fact that those numbers turned out to be non-zero, the non-triviality of the cobordism classes $[\mathbb{C}P^2] \in \Omega_4^{SO}$ and $[\mathbb{C}P^2 \times \mathbb{C}P^2]$, $[\mathbb{C}P^4] \in \Omega_8^{SO}$ (see §27.1, Examples (b), (c)). The following result, a consequence of Theorem 27.9 and the Cartan–Serre theorem (10.7), provides more precise information.

27.15. Theorem

(i) *The rational cobordism groups* $\Omega_j^{SO} \otimes \mathbb{Q}$ *are trivial for* $j \neq 4k$: $\Omega_j^{SO} \otimes \mathbb{Q} = 0$ *for* $j \neq 4k$.

(ii) *For each* $k = 1, 2, \ldots,$ *the group* $\Omega_{4k}^{SO} \otimes \mathbb{Q}$ *has rank (i.e. dimension as a rational vector space) equal to the number of linearly independent vectors of (stable) characteristic numbers of 4k-dimensional closed manifolds (i.e. vectors having as components the values taken by the monomials* $p_{m_1} p_{m_2} \cdots p_{m_q}$ *(in some fixed order) in the Pontrjagin characteristic classes* $p_i \in H^i(M^{4k}; \mathbb{Q})$, *where* $m_1 + m_2 + \cdots + k$ *(so that each manifold* M^{4k} *has associated with it a single vector of characteristic numbers)).*

27.16. Corollary. *The rational characteristic numbers calculated in the cases* $4k = 4, 8$ *in Examples* (b) *and* (c) *of* §27.1 (*for the manifolds* $\mathbb{C}P^2$, $\mathbb{C}P^4$ *and* $\mathbb{C}P^2 \times \mathbb{C}P^2$) *fully determine the corresponding bordism groups* Ω_{4k}^{SO}, $k = 1, 2$, *modulo torsion. It follows, in fact, that* Ω_4^{SO} *and* Ω_8^{SO} *have torsion-free ranks* 1 *and* 2 *respectively, since the* 1*-vector* $(p_1(\mathbb{C}P^2)) = (3)$ *in the case* $k = 1$, *and the* 2*-vectors* $(p_1^2((\mathbb{C}P^4), p_2(\mathbb{C}P^4)) = (25, 10)$ *and*

$$(p_1^2(\mathbb{C}P^2 \times \mathbb{C}P^2), p_2(\mathbb{C}P^2 \times \mathbb{C}P^2)) = (18, 9)$$

in the case $k = 2$, *determine bases for* $\Omega_4^{SO} \otimes \mathbb{Q}$ *and* $\Omega_8^{SO} \otimes \mathbb{Q}$ *respectively.*†

Exercise

8. Compute the vectors of characteristic numbers of all products of the form $\mathbb{C}P_1^{2n_1} \times \cdots \times \mathbb{C}P_k^{2n_k}$, and show that they are linearly independent.

Sketch of Proof of Theorem 27.15. Since the complex $MSO(n)$ is $(n-1)$-connected, we have by Corollary 10.11 (a consequence of the Cartan–Serre theorem) that

$$\pi_j(MSO(n)) \otimes \mathbb{Q} \simeq H_j(MSO(n); \mathbb{Q}) \quad (\simeq H^j(MSO(n); \mathbb{Q})), \qquad j < 2n - 1.$$

Composed appropriately with the isomorphism (26) and that afforded by Theorem 27.9:

$$\pi_{n+i}(MSO(n) \simeq \Omega_i^{SO}, \qquad i < n - 1,$$

(tensored with $\mathbb{Q}$), this yields an isomorphism

$$F: \Omega_i^{SO} \otimes \mathbb{Q} \simeq H^i(BSO(n); \mathbb{Q}), \qquad i < n - 1.$$

† A description of the complete structure of the rings Ω^{SO} and Ω^U may be found in the survey [57].

It is immediate from this, in view of the remarks concerning the algebra $H^*(BSO(n); \mathbb{Q})$ made prior to the theorem, that $\Omega_j^{SO} \otimes \mathbb{Q} = 0$ for $j \neq 4k$, and that the rank of $\Omega_{4k}^{SO} \otimes \mathbb{Q}$ is equal to the number of partitions of k. In order to establish that this is the same as the number of linearly independent vectors of characteristic numbers of manifolds M^{4k}, it remains to show that if the characteristic numbers of an oriented closed manifold M^i all vanish then $[M^i] = 0$ in $\Omega_i^{SO} \otimes \mathbb{Q}$ (in fact, the converses of both parts of Lemma 27.2 are valid). Now it can be shown that the isomorphism F has the property that for each $w \in H^i(BSO(n); \mathbb{Q})$ and each closed, oriented manifold M^i,

$$(w, F[M_i]^*) = (\tau^* w, [M^i]), \tag{27}$$

the corresponding characteristic number of M^i (see the proof of Lemma 27.2). (Note that in (27) the symbol $[M^i]$ has two distinct meanings, on the one hand, as the SO-cobordism class of M^i (actually $[M^i] \otimes 1$), and on the other, as the fundamental homology class of M^i.) Since (27) in fact determines the cohomology class $F[M_i] \in \Omega_i^{SO} \otimes \mathbb{Q}$, it follows that the vanishing of all the characteristic numbers of M^i entails that $[M^i] = 0$ in $\Omega_i^{SO} \otimes \mathbb{Q}$, as required. □

From Lemma 27.4 we deduce immediately the following

27.17. Corollary. *The signature $\tau[M^{4k}]$ may be regarded as a linear form on the vectors of characteristic numbers of oriented manifolds M^{4k}.*

From our earlier computations in the cases $k = 1, 2$, we infer the following relations:

In the case $k = 1$, we have $p_1(\mathbb{C}P^2) = 3$, $\tau(\mathbb{C}P^2) = 1$, whence we conclude (cf. (9))

$$\tau = \tfrac{1}{3} p_1. \tag{28}$$

In the case $k = 2$, one has the following table of characteristic numbers (from Example (c) of §27.1), and values of τ (inferred from the structure of $H^*(\mathbb{C}P^n; \mathbb{Q})$ given, e.g. in Example (g) of §4, or the first example in §7):

	$[\mathbb{C}P^2] \times [\mathbb{C}P^2]$	$[\mathbb{C}P^4]$
p_1^2	18	25
p_2	9	10
τ	1	1

From this table we derive (via Corollary 27.17) the formula

$$\tau = \tfrac{1}{45}(7p_2 - p_1^2). \tag{29}$$

There is in fact a general formula of this type valid for all k due to Hirzebruch, which may be derived as follows. It turns out to be appropriate to take the following general approach: Let B denote an arbitrary complex "character" of the cobordism ring $\Omega_*^G = \sum \Omega_i^G$ where $G = U$ or SO, i.e. a ring homomorphism

$$B: \Omega_*^G \to \mathbb{C},$$

having, therefore, regarded as a functional of closed manifolds, the following properties:

$$B(1) = 1, \qquad B(M_1^n \cup M_2^n) = B(M_1^n) + B(M_2^n),$$
$$B(M_1^n \times M_2^n) = B(M_1^n)B(M_2^n),$$

i.e. preservation of 1, additivity (as usual with respect to disjoint unions), and multiplicativeness (with respect to direct products of manifolds). We shall, in fact, be concerned with the character B only as it operates on the algebra $\Omega_*^G \otimes \mathbb{Q}$. For $G = U$, as in the case $G = SO$ considered above, this algebra is determined by the appropriate "characteristic numbers" (values of polynomials of appropriate dimension in the Chern classes c_i (of the tangent bundle) in the case $G = U$, and in the Pontrjagin classes p_i when $G = SO$—see §10.4(12)). It follows that (in the case $G = U$) in each even dimension $n = 2k$, for each "unitary" manifold M^{2k} (i.e. manifold on whose "stable" normal bundle in $\mathbb{R}^{2N+2k}$ (where M^{2k} is embedded in $\mathbb{R}^{2N+2k}$—cf. proof of Theorem 27.9) there is defined the structure of a U-bundle, so that in particular the U-structure reflects to some extent the complex (or "quasicomplex") structure of the manifold M^{2k}, "remembering", as it were, the characteristic classes) the value $B[M^k]$ of any character B has the form

$$B[M^{2k}] = (B_k(c_1, \ldots, c_k), [M^{2k}]),$$

i.e. the value of some polynomial $B_k(c_1, \ldots, c_k)$ of dimension $2k$ in the Chern classes $c_1, \ldots, c_k$ of the tangent bundle of M^{2k}, at the fundamental homology class $[M^{2k}]$ of M^{2k}. (For $G = SO$ we have analogously

$$B[M^{4k}] = (B_k(p_1, \ldots, p_k), [M^{4k}]),$$

for appropriate polynomials $B_k(p_1, \ldots, p_k)$ of dimension $4k$ in the Pontrjagin classes of the tangent bundle of M^{4k}. In fact, the case $G = SO$ can be assimilated to that of $G = U$ by imposing the additional requirement $B_{2k+1}(c_1, \ldots, c_{2k+1}) = 0$—see below.)

The sequence of polynomials

$$B_0 = 1, B_1, B_2, \ldots, B_k, \ldots \tag{30}$$

is not arbitrary, but has stringent restrictions imposed on it by the multiplicativeness condition, which comes down to the (formal) requirement that

$$B(c)B(c') = B(c \times c'), \tag{31}$$

where $c = 1 + c_1 + c_2 + \cdots$, $c' = 1 + c_1' + c_2' + \cdots$, and $c \times c'$ (whose form is given by Exercise 15 of §9) are the "total Chern classes" of unitary manifolds M, M' and $M \times M'$ respectively, and where

$$B(c) = 1 + B_1(c_1) + B_2(c_1, c_2) + \cdots.$$

Now it can be shown (see [47]) that given any power series over $\mathbb{C}$ of the form

$$f(z) = 1 + \lambda_1 z + \lambda_2 z^2 + \cdots, \tag{32}$$

for each $k = 0, 1, 2, \ldots$, the homogeneous part of degree k of the product

$$f(z_1)f(z_2)\ldots f(z_n), \qquad n \geq k,$$

expressed as a polynomial in the first k elementary symmetric functions $\sigma_1, \ldots, \sigma_k$ in the independent variables $z_1, \ldots, z_n$, is the kth term of a "multiplicative sequence" of polynomials, i.e. a sequence (30) satisfying (31), and, furthermore, that every multiplicative sequence arises in this way from some power series of the form (32).

Suppose the multiplicative sequence (30) determining our original arbitrary character $B: \Omega_*^U \to \mathbb{C}$, arises in this way from the power series

$$B(z) = 1 + a_1 z + a_2 z^2 + \cdots. \tag{33}$$

Since each class c_i can be identified via the inclusion homomorphism

$$i^*: H^*(BU(n); \mathbb{Q}) \to H^*(BT^n; \mathbb{Q})$$

with the ith elementary symmetric polynomial σ_i in the canonical generators $t_1, \ldots, t_n$ of $H^2(BT^n; \mathbb{Q})$ (see §10.4(15)(iii)), we shall have

$$B_k(c_1, \ldots, c_k) \leftrightarrow B_k(\sigma_1, \ldots, \sigma_k) = [B(zt_1)\ldots B(zt_n)]_k, \qquad n \geq k, \tag{34}$$

i.e. $B_k(c_1, \ldots, c_k)$, regarded as a polynomial in the $\sigma_i(t_1, \ldots, t_n)$, is the coefficient of z^k in the product of the (formal) power series $B(zt_i)$, $i = 1, \ldots, n$, where

$$B(zt) = 1 + a_1 zt + a_2 z^2 t^2 + \cdots, \quad t \in H^2(\mathbb{C}P^\infty; \mathbb{Z}) \quad (BT^1 = \mathbb{C}P^\infty).$$

Hence by Cauchy's integral formula

$$\lambda_k = \frac{f^{(k)}(0)}{k!} = \frac{1}{2\pi i} \oint_{|z|=\varepsilon} \frac{f(z)}{z^{k+1}}\, dz$$

for the coefficients of a power series $f(z) = 1 + \lambda_1 z + \lambda_2 z^2 + \cdots$, convergent in some neighbourhood of the origin, we have (formally)

$$B_k(c_1, \ldots, c_k) = \frac{1}{2\pi i} \oint_{|z|=\varepsilon} B(zt_1)\ldots B(zt_n)\frac{dz}{z^{k+1}}, \qquad n \geq k. \tag{35}$$

(The case $G = SO$ is similar, with the power series (33) replaced by one of the form $B(z^2)$. The analogue in this case of (34) is then

$$B_k(p_1, \ldots, p_k) \leftrightarrow B_k(\sigma_1, \ldots, \sigma_k) = [B(z^2 t_1^2)\ldots B(z^2 t_n^2)]_{2k},$$

since under the inclusion homomorphism

$$i^*: H^*(BSO(2n); \mathbb{Q}) \to H^*(BT^n; \mathbb{Q}),$$

the ith Pontrjagin class p_i is identified with the ith elementary symmetric polynomial $\sigma_i(t_1^2, \ldots, t_n^2)$ in the squares of the canonical generators of $H^2(BT^n; \mathbb{Z})$—see §10.4(15)(i).)

The sequence (30) can be used in the obvious way to define a *multiplicative sequence of characteristic classes* $\hat{B}_k$ of an arbitrary $U(n)$-vector bundle ξ over M^{2k} by setting

$$\hat{B}_k(\xi) = B_k(c_1(\xi), \ldots, c_k(\xi)), \qquad k = 0, 1, 2, \ldots.$$

Recalling from §9, Exercise 16, that for the manifold $\mathbb{C}P^n$ we have

$$\tau \oplus 1 = \eta \oplus \cdots \oplus \eta \quad (n + 1 \text{ summands}),$$

where τ is the tangent bundle of $\mathbb{C}P^n$ and η is the generalized Hopf bundle over $\mathbb{C}P^n$ (with fibre $S^1 \cong U(1)$), we infer from the "Whitney product formula" (see §10.4(16) *et seqq.*) that

$$\hat{B}(\tau) = \hat{B}(\eta)^{n+1} \qquad \left(\text{where } \hat{B}(\xi) = \sum_{k \geq 0} \hat{B}_k(\xi) z^k\right).$$

It follows that the number $B[\mathbb{C}P^n]$ is given by (cf. (35))

$$B[\mathbb{C}P^n] = (\hat{B}(\tau)_n, [\mathbb{C}P^n]) = ((\hat{B}(\eta)^{n+1})_n, [\mathbb{C}P^n]) = \frac{1}{2\pi i} \oint_{|z|=\varepsilon} B(z)^{n+1} \frac{dz}{z^{n+1}}, \tag{36}$$

where we have used $(c_1(\eta)^n, [\mathbb{C}P^n]) = 1$, $c_1(\eta) \in H^2(\mathbb{C}P^n; \mathbb{Z})$.

Note that by Theorem 27.14 and Exercise 8, the algebra $\Omega_*^{SO} \otimes \mathbb{Q}$ is generated by the cobordism classes $[\mathbb{C}P^{2n}]$ (and, analogously, $\Omega_*^U \otimes \mathbb{Q}$ is generated by the classes $[\mathbb{C}P^n]$).

Examples

(a) Since $H^{2n}(\mathbb{C}P^n; \mathbb{Q})$ is generated by the single element c_1^n, where $c_1 \in H^2(\mathbb{C}P^{2n}; \mathbb{Q})$ is the first Chern class of $\mathbb{C}P^n$ (cf. for instance §4, Example (g), and the first example of §7), and $(c_1^{2n}, [\mathbb{C}P^n]) = 1$, it follows from the definition of the signature τ (in the form (7′)) that $\tau(\mathbb{C}P^{2n}) = 1$ (while $\tau(\mathbb{C}P^{2n+1}) = 0$ by fiat). Hence the character $L: \Omega_*^{SO} \otimes \mathbb{Q} \to \mathbb{Q}$, defined by $L[\mathbb{C}P^{2n}] = 1$, $L[\mathbb{C}P^{2n+1}] = 0$ coincides with the signature. It can be shown that the corresponding multiplicative sequence $\{L_k(p_1, \ldots, p_k)\}$ arises (in the manner described above) from the power series of the function

$$f(z) = \frac{\sqrt{z}}{\tanh \sqrt{z}},$$

whence one obtains Hirzebruch's formula expressing the signature in terms of the "L-polynomials" in the Pontrjagin classes p_i:

$$\tau[M^{4k}] = (L_k(p_1, \ldots, p_k), [M^{4k}]). \tag{37}$$

(b) The multiplicative sequence $\{T_k(c_1, \ldots, c_k)\}$ of the character $T: \Omega_*^U \otimes \mathbb{Q} \to \mathbb{C}$ defined by $T[\mathbb{C}P^n] = 1$ for all n "belongs to" the power series of the function

$$f(z) = \frac{z}{1 - e^{-z}}$$

(verify this!). For each complex manifold M^{2k} the value

$$T[M^{2k}] = (T_k(c_1, \ldots, c_k), [M^{2k}])$$

is called the *Todd genus* of the manifold. The first few polynomials T_k are as follows:

$$T_0 = 1, \qquad T_1 = \tfrac{1}{2}c_1, \qquad T_2 = \tfrac{1}{12}(c_1^2 + c_2), \qquad T_3 = \tfrac{1}{24}c_1 c_2.$$

By a result of Hirzebruch

$$T[M^{2k}] = \sum_i (-1)^i r_i,$$

where r_i is the dimension of the space of holomorphic differential i-forms definable on M^{2k}.

We end the subsection by deriving a general formula (in terms of the numbers $B[\mathbb{C}P^n]$) for the function $B(z)$ of (33) determining (and determined by) an arbitrary character $B\colon \Omega_*^U \to \mathbb{C}$. Consider the formal power series

$$\sum_{n \geq 0} [\mathbb{C}P^n] z^n \quad \text{(with coefficients in } \Omega_*^U\text{)},$$

and its "integral"

$$g(z) = \sum_{n \geq 0} \frac{[\mathbb{C}P^n]}{n+1} z^{n+1}.$$

We now bring in our arbitrary character B, defining a power series $(g_B)(z)$ by

$$(g_B)(z) = \sum_{n \geq 0} \frac{B[\mathbb{C}P^n]}{n+1} z^{n+1}.$$

Since by (36)

$$B[\mathbb{C}P^n] = \frac{1}{2\pi i} \oint_{|w|=\varepsilon} \frac{B(w)^{n+1}}{w^{n+1}}\, dw,$$

it follows that

$$\begin{aligned} \frac{d}{dz}(g_B)(z) &= \sum_{n \geq 0} B[\mathbb{C}P^n] z^n = \frac{1}{2\pi i z} \sum_{n \geq 0} \oint_{|w|=\varepsilon} \left(\frac{B(w)}{w} z\right)^{n+1} dw \\ &= \frac{1}{2\pi i} \oint_{|w|=\varepsilon} \frac{B(w)/w}{1 - zB(w)/w}\, dw = \frac{1}{2\pi i} \oint_{|w|=\varepsilon} \frac{dw}{(w/B(w)) - z}, \end{aligned}$$

provided $|zB(w)/w| < 1$. From the formula

$$\frac{d}{dz} f^{-1}(z) = \frac{1}{2\pi i} \oint_{|w|=\varepsilon} \frac{dw}{f(w) - z}, \qquad |z| < \varepsilon, \quad f \text{ analytic},$$

(an easy consequence of the basic Cauchy integral formula), we infer that

$$(g_B)(z) = \left(\frac{z}{B(z)}\right)^{-1} \quad \text{(the inverse function)},$$

whence finally we have the desired formula (due to Novikov):

$$B(z) = \frac{z}{(g_B)^{-1}(z)} \tag{38}$$

where

$$(g_B)(z) = \sum_{n \geq 0} \frac{B[\mathbb{C}P^n]}{n+1} z^{n+1}.$$

27.3. Some Applications of the Signature Formula. The Signature and the Problem of the Invariance of Classes

We begin by indicating how one may define the Pontrjagin characteristic classes of rational cohomology in terms of the signature τ; that is, we shall sketch the procedure whereby, for any element $x \in H_{4k}(M^n; \mathbb{Z})$, one may calculate the scalar product $(p_k, x) = (p_k(M^n), x)$ via the signature. We may suppose for this purpose that $4k < n/2 - 1$; if this is not the case then replace the manifold M^n by $M^n \times S^N$ for suitably large N. Then by Corollary 27.13 and Theorem 27.10(iii), for some multiple $\lambda D(x)$ of the image $D(x) \in H^{n-4k}(M^n)$ under the Poincaré-duality isomorphism D, and some map

$$f: M^n \to S^{n-4k} = M\{e\},$$

we shall have

$$f^*\varphi(1) = \lambda D(x),$$

where φ is as in (12), which condition is equivalent to the representability of x by an oriented, closed submanifold $M^{4k} \subset M^n$ with trivial normal bundle. Denoting by

$$i: M^{4k} \times \mathbb{R}^{n-4k} \to M^n \tag{39}$$

an embedding of this normal bundle in M^n which restricts to the identity on $M^{4k} \subset M^n$, we shall have

$$i_*[M^{4k}] = \lambda x \in H_{4k}(M^n),$$

(and this in fact holds with M^{4k} replaced by the complete inverse image $f^{-1}(x_0)$ of any regular value $x_0 \in S^{n-4k}$ of f; cf. the proof of Theorem 27.10).

Considering now the first case $k = 1$, and denoting the tangent bundles of $M^{4k} = M^4$ and M^n by $L(M^4)$ and $L(M^n)$ respectively, we have

$$(p_1, x) = (p_1 L(M^n), x) = \frac{1}{\lambda}(p_1 L(M^n), \lambda x) = \frac{1}{\lambda}(p_1 L(M^n), j_*[M^4]),$$

where $j: M^4 \subset M^n$ is the inclusion map. This inclusion is covered by the bundle map $j: L(M^4) \oplus \nu(M^4) \to L(M^n)$, where $\nu(M^4)$ denotes the normal bundle of $M^4 \subset M^n$, which is trivial (see above):

$$\nu(M^4) \cong M^4 \times \mathbb{R}^{n-4}.$$

From the triviality of the bundle $\nu(M^4)$ it follows that

$$p_1(L(M^4)) = j^* p_1(L(M^n)),$$

whence

$$(p_1(L(M^n)), j_*[M^n]) = (j^* p_1(L(M^n)), [M^4]) = (p_1 L(M^4), [M^4]) = p_1(M^4).$$

Hence by (28)

$$(p_1(M^n), x) = \frac{1}{\lambda} p_1(M^4) = \frac{1}{\lambda} 3\tau(M^4), \tag{40}$$

which may be regarded as affording an alternative definition of p_1 in terms of the signature. Similarly, from (29) one obtains, in the next case $k = 2$, the formula

$$(p_2(M^n), x) = \frac{1}{\lambda}(p_2, \lambda x) = \frac{1}{7\lambda}[45\tau(M^8) + (p_1^2, \lambda x)], \tag{41}$$

which leads, via (40), to a characterization of p_2 in terms of the signature.

From Hirzebruch's general formula (37) it may be shown along these lines that more generally the class $p_k(M^n)$ can be expressed in terms of the signature $\tau(M^{4k})$ for $M^{4k} \subset M^n$, and products of the p_i of lower dimensions. The alternative definition of the rational Pontrjagin classes thus afforded allows one to establish relatively easily (see below) their invariance under piecewise-linear (or piecewise-smooth) homeomorphisms, and is also crucial in establishing their invariance under arbitrary homeomorphisms. (A map $f: K \to L$ between simplicial complexes is *piecewise linear* if there is a "simplicial subdivision" K' of K such that f maps each simplex of K' linearly to a simplex of L. Recall from §3 that a *smooth triangulation* of a manifold M is a homeomorphism $K \to M$ whose restriction to each (closed) simplex of K is smooth and of maximal rank at each point.) The definition of the Pontrjagin classes via the signature is clearly of an essentially rational character, and certain fractions (as, for instance, $\frac{1}{7}$ in the case of p_2—see (41)) arise in a significant, "inevitable", way. This has certain consequences in particular in connexion with the torsion elements among the integral classes $p_k \in H^{4k}(M^n; \mathbb{Z})$; for instance, the integral class p_2 has order 7, and this 7-torsion is a topological (i.e. homeomorphic) invariant.

The idea of the proof that the rational classes p_k are piecewise-linear invariants is briefly as follows. Let M^n be any triangulated manifold (see above) and consider a simplicial map $f: M^n \to S^{n-4k}$ to the sphere (with a suitable triangulation defined on it; e.g. that determined by the boundary of the standard $n - 4k + 1$ simplex). (The term "simplicial" signifies that each vertex of the triangulation of M^n is sent by f to a vertex of S^{n-4k}, and each smooth simplex of M^n into one of S^{n-4k}.) The complete inverse image of an open simplex (i.e. the interior of a smooth simplex) $\sigma^{n-4k} \subset S^{n-4k}$ then has the form (verify!—cf. [47, p. 236])

$$f^{-1}(\sigma^{n-4k}) \cong \sigma^{n-4k} \times f^{-1}(y_0) = \sigma^{n-4k} \times M^{4k},$$

where y_0 is any point in σ^{n-4k}, and M^{4k} is a triangulated manifold (or at least a subcomplex of M^n) with the property that for each of its points x_0,

$$\begin{aligned} H_i(M^{4k}, M^{4k}\backslash\{x_0\}; \mathbb{Q}) &= 0, \qquad i \neq 4k, \\ H_{4k}(M^{4k}, M^{4k}\backslash\{x_0\}; \mathbb{Q}) &\simeq \mathbb{Q}, \end{aligned} \tag{42}$$

i.e. having the same "local $\mathbb{Q}$-homology" as the $4k$-sphere. Manifolds satisfying (42) are called *rational homology manifolds.*

EXERCISE

9. Prove that for rational homology manifolds the Poincaré-duality theorem (18.1) is valid over $\mathbb{Q}$, and deduce that the signature $\tau(M^{4k})$ of such manifolds is defined and has the usual properties. Thus in particular if $M^{4k} = \partial W^{4k+1}$, where W^{4k+1} is also a homology manifold (-with-boundary), then $\tau(M^{4k}) = 0$ (cf. Lemma 27.4).

On the basis of this exercise it is possible to frame a purely simplicial, combinatorially invariant, definition of the classes $p_k \in H^{4k}(M^n; \mathbb{Q})$ using the above-noted formulae for these in terms of the signature (Thom, Rohlin–Švarc). Note, however, the result (due to Milnor and Kervaire) that the *integral* class $p_2 \in H^8(M^n; \mathbb{Z})$ cannot in general be defined combinatorially, and in fact is combinatorially, and hence certainly topologically, non-invariant.

We now turn briefly to the proof that the rational Pontrjagin classes $p_k \in H^{4k}(M^n; \mathbb{Q})$ are in fact topologically invariant. One first observes that in (39) it may be assumed without loss of generality that M^{4k} is simply-connected. Further, supposing $n - 4k \geq 2$, one considers an embedding of the normal bundle of the torus $T^{n-4k-1} \subset \mathbb{R}^{n-4k}$ into $\mathbb{R}^{n-4k}$ which restricts to the identity map on T^{n-4k-1} itself, thereby obtaining an embedding (see (39))

$$M^{4k} \times T^{n-4k-1} \times \mathbb{R} \subset M^{4k} \times \mathbb{R}^{n-4k} \subset M^n$$

of $M^{4k} \times T^{n-4k-1} \times \mathbb{R}$ as an open region of the given manifold M^n. Any smooth structure on M^n will then restrict to a smooth structure on this region, turning it into a smooth manifold. By utilizing the rather complex machinery of the classification theory of smooth simply-connected manifolds, extended to embrace manifolds with free abelian fundamental group $\pi_1 \simeq \mathbb{Z} \oplus \cdots \oplus \mathbb{Z}$, a result can be proved, the simplest version of which is as follows: If $\pi_1(M^{4k}) = 1$ then the smooth universal cover of the (open) smooth manifold $M^{4k} \times T^{n-4k-1} \times \mathbb{R}$ (with smooth structure induced from M^n) is diffeomorphic to a manifold of the form $\tilde{M}^{4k} \times \mathbb{R}^{n-4k}$:

$$p: \tilde{M}^{4k} \times \mathbb{R}^{n-4k} \to M^{4k} \times T^{n-4k-1} \times \mathbb{R},$$

where $\tilde{M}^{4k}$ is a smooth manifold and p is the projection map of the universal cover. From this result one then infers via an induction on k the desired result (of Novikov), namely that the signature $\tau(M^{4k})$ $(= \tau(\tilde{M}^{4k}))$ determines the rational Pontrjagin classes $p_1, \ldots, p_k$ in topologically invariant fashion. Hitherto, no proof of this result has been given which might allow one to avoid the use of auxiliary regions (as above) with free abelian fundamental group; the exploitation of such regions would appear to be artificial since on the face of it one would expect the structure of π_1 to be irrelevant to the problem as posed, which is of "simply-connected" type.

Under homotopy equivalences, however, the class $p_1 \in H^4(M^n; \mathbb{Q})$, unlike the Stiefel–Whitney classes (see Exercise 6) and the homology and cohomology groups themselves, turns out not to be invariant in general (Dold). To see this consider an arbitrary fibre bundle ξ over S^4 with fibre $F = S^{n-1} \subset \mathbb{R}^n$, group $G = SO(n)$ (with the restriction in the case $n = 4$ that the Euler characteristic class $\chi = 0$), and its first Pontrjagin class $p_1(\xi) \in H^4(S^4; \mathbb{Z})$. The total space E of such a fibre bundle (with projection map $p: E \to S^4$) has a natural cell decomposition with cells $\sigma^0, \sigma^4, \sigma^{n-1}, \sigma^{n+3}$ where $\partial\sigma^4 = \partial\sigma^{n-1} = \partial\sigma^{n+3} = 0$ (see §§4, 8); in fact E has the homotopy type

$$E \sim \sigma^{n+3} \cup_\alpha (S^4 \vee S^{n-1}), \qquad \alpha \in \pi_{n+2}(S^4 \vee S^{n-1}), \tag{43}$$

(essentially since above the interior of the cell $\sigma^4 = D^4$ giving rise to the base $B = S^4$, the bundle is trivial).

EXERCISE

10. Prove that the element α in (43) has the form

$$\alpha = [a_4, a_{n-1}] + b, \tag{44}$$

where a_4 and a_{n-1} are generators of $\pi_4(S^4)$ and $\pi_{n-1}(S^{n-1})$ respectively, $b \in \pi_{n+2}(S^{n-1})$, and [,] denotes Whitehead multiplication (see Part II, §22.5). (In the case $n = 5$, where we have $b \in \pi_7(S^4) = \mathbb{Z} \oplus H$ where H is a finite group, it can be shown that b lies in the finite group H.)

By Corollary 10.10 (a consequence of the Cartan–Serre theorem) for $n > 5$ the groups $\pi_{n+2}(S^{n-1})$ are all finite. (In fact, by Theorem 10.14, $\pi_{n+2}(S^{n-1}) \simeq \mathbb{Z}_{24}$ for $n > 5$.) Hence in view of (43) and (44) for each $n \geq 2$ there are only finitely many distinct homotopy types possible for the closed manifolds E (in fact at most 24 for each $n > 5$). On the other hand, for $n \geq 3$ there are infinitely many bundles ξ (as above) with distinct $p_1(\xi) \in H^4(S^4; \mathbb{Z}) \simeq \mathbb{Z}$ (essentially since there are infinitely many non-homotopic classifying maps $S^4 \to BSO(n)$, $\pi_4(BSO(n))$ being infinite for $n \geq 3$; see Part II, §24.4). Since the tangential Pontrjagin class $p_1(E)$ satisfies $p_1(E) = p^*p_1(\xi)$ (verify!), so that $p_1(\xi)$ is a diffeomorphic invariant of E, it follows that for each $n \geq 3$ there are infinitely many manifolds E with distinct $p_1(E)$. Since these manifolds fall into only finitely many homotopy types, we conclude that p_1 is not a homotopy invariant for manifolds of dimension ≥ 6.

For 4-manifolds M^4 the rational class $p_1(M^4)$ is clearly invariant under homotopy equivalences in view of the formula (9) (see also (28)):

$$p_1(M^4) = 3\tau(M^4).$$

It turns out that 5-manifolds M^5 also have homotopically invariant first (rational) Pontrjagin class $p_1(M^5)$. We indicate a proof of this in the orientable case. By Corollary 27.11(ii) and Theorem 27.10(iii) each non-zero element $x \in H_4(M^5; \mathbb{Z})$ can be represented by an oriented submanifold $M^4 \subset M^5$ which locally (but not globally since $x \neq 0$) separates M^5 into two parts. We now construct a "minimal" covering space

$$\hat{M}^5 \xrightarrow{p} M^5,$$

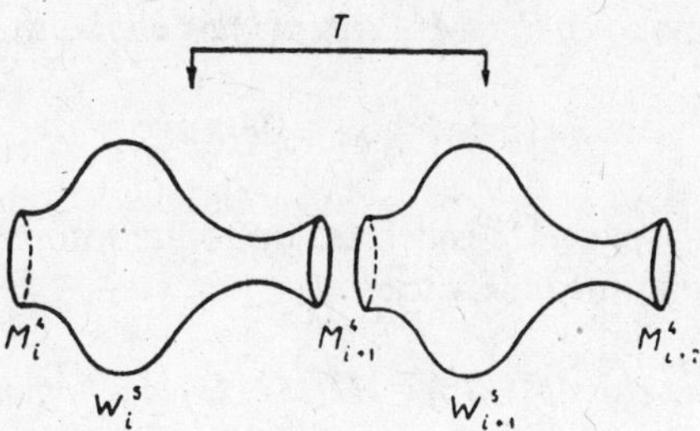

Figure 118

satisfying $(Dx, p_*\pi_1(\hat{M}^5)) = 0$ $(Dx \in H^1(M^5; \mathbb{Z}))$, and determined up to a homotopy equivalence by this condition. One constructs $\hat{M}^5$ geometrically as follows. If the manifold M^5 is cut along its submanifold M^4, it becomes a film W^5 with boundary $\partial W^5 = M^4 \cup M^4$, the disjoint union of two boundary components M^4. The covering space $\hat{M}^5$ is then taken to be the union of an infinite family $\{W_j^5\}$ of distinct copies of W^5 indexed by the integers, with ∂W_j^5 written as $M_j^4 \cup M_{j+1}^4$, where neighbouring copies W_j^5, W_{j+1}^5 are glued together by identifying appropriate components of their respective boundaries (see Figure 118):

$$\hat{M}^5 = \ldots \cup_{M_0^4} W_0^5 \cup_{M_1^4} W_1^5 \cup_{M_2^4} W_2^5 \ldots. \tag{45}$$

The monodromy group $\mathbb{Z}$, with generator T say, acts by translation:

$$T(W_j^5) = W_{j+1}^5, \qquad T(M_j^4) = M_{j+1}^4.$$

Via any embedding $i: M^4 \to \hat{M}^5$ (identifying M^4 with a boundary component of say W_0^5), we obtain from our arbitrary element $x \in H_4(M^5; \mathbb{Z})$, an element

$$\hat{x} = i_*[M^4] \in H_4(\hat{M}^5; \mathbb{Z}),$$

which is T-invariant: $T_*\hat{x} = \hat{x}$. We now define a rational bilinear form $\langle\ ,\ \rangle_{\hat{x}}$ on $H^2(\hat{M}^5; \mathbb{Q})$, associated with $\hat{x}$, by setting, for each pair of elements a, $b \in H^2(\hat{M}^5; \mathbb{Q})$

$$\langle a, b\rangle_{\hat{x}} = (ab, \hat{x}).$$

27.18. Lemma. *The bilinear form $\langle a, b\rangle_{\hat{x}}$ is non-zero only on some finite-dimensional subspace $A \subset H^2(\hat{M}^5; \mathbb{Q})$; i.e. $H^2(\hat{M}^5; \mathbb{Q}) = A \oplus B$ where $\langle B, b\rangle_{\hat{x}} = 0$ for all $b \in H^2(\hat{M}^5; \mathbb{Q})$.*

This is an immediate consequence of the compactness of the manifold M^4 (compact since it is a cycle representing $\hat{x}$), and the equation

$$(ab, \hat{x}) = ((i^*a)(i^*b), [M^4]). \tag{46}$$

27.19. Definition. The *signature* $\tau(\hat{x})$ *of the cycle* $\hat{x}$ is the signature of the form $\langle a, b\rangle_{\hat{x}}$ on the finite-dimensional subspace A.

27.20. Theorem (Novikov). *For M^5, x, $\hat{x}$, $\tau(\hat{x})$ as above, the following formula is valid:*

$$(p_1(M^5), x) = 3\tau(\hat{x}).$$

Since the covering space $\hat{M}^5$ is defined up to homotopy type by the above condition we deduce immediately the

27.21. Corollary. *The class $p_1(M^5) \in H^4(M^5; \mathbb{Q})$ is a homotopy invariant.*

PROOF OF THE THEOREM. The cycle $M^4 \subset \hat{M}^5$ separates $\hat{M}^5$ into two parts: $\hat{M}^5 = M_1 \cup M_2$ (see (45)), into each of which M^4 embeds naturally (as the boundary)

$$i_1: M^4 \to M_1, \qquad i_2: M^4 \to M_2.$$

By (46) the signature $\tau(\hat{x})$ of the element $\hat{x}$ is equal to the signature of the bilinear form $\langle i^*a, i^*b\rangle = ((i^*a)(i^*b), [M^4])$ on $\operatorname{Im} i^* \subset H^2(M^4; \mathbb{Q})$. We shall show that in fact $\tau(\hat{x}) = \tau(M^4)$.

Recall from §27.1 (see (7)) that the signature $\tau(M^4)$ of M^4 is defined in terms of the intersection index $\alpha \circ \beta$ of rational 2-cycles $\alpha, \beta \in H_2(M^4; \mathbb{Q})$. Consider the following subspaces of the vector space $H_2(M^4; \mathbb{Q})$:

$$L_0 = \operatorname{Ker} i_*, \qquad L_1 = \operatorname{Ker} i_{1*} = L_0 \oplus N_1, \qquad L_2 = \operatorname{Ker} i_{2*} = L_0 \oplus N_2,$$
$$L_3 \simeq \operatorname{Im} i_* \subset H_2(\hat{M}_5; \mathbb{Q}), \qquad i_* L_3 = \operatorname{Im} i_*. \tag{47}$$

Since, as is easily verified,

$$\operatorname{Ker} i_* = \operatorname{Ker} i_{1*} \cap \operatorname{Ker} i_{2*},$$

it follows that:

$$H_2(M^4; \mathbb{Q}) = L_0 \oplus N_1 \oplus N_2 \oplus L_3.$$

On each of the spaces L_1 and L_2 the intersection index of pairs of cycles is zero (since a pair of cycles in either M_1 or M_2 which are homologous to zero are certainly homologous to cycles with empty intersection). Hence in terms of a basis for $H_2(M^4; \mathbb{Q})$ made up from bases for L_0, N_1, N_2 and L_3, the matrix of the bilinear form on $H_2(M^4; \mathbb{Q})$, furnished by the intersection index, has the block form

$$\begin{array}{c} \\ L_0 \\ N_1 \\ N_2 \\ L_3 \end{array} \begin{array}{c} \begin{array}{cccc} L_0 & N_1 & M_2 & L_3 \end{array} \\ \begin{bmatrix} O & O & O & X \\ O & O & Q & Y \\ O & Q^{\mathrm{T}} & O & Z \\ X^{\mathrm{T}} & Y^{\mathrm{T}} & Z^{\mathrm{T}} & W \end{bmatrix} \end{array}$$

(where $W = W^{\mathrm{T}}$), whence it is clear that the signature of that form coincides with the signature of its restriction to the subspace L_3 (with matrix W). Since, as noted in (7′)

$$\alpha \circ \beta = (D\alpha D\beta, [M^4]), \qquad \alpha, \beta \in H_2(M^4; \mathbb{Q}),$$

it now follows via (46) and (47) that $\tau(M^4) = \tau(\hat{x})$, as claimed.

Since $3\tau(M^4) = p_1(M^4)$ (the characteristic number), it remains to show that $(p_1(M^4), [M^4]) = (p_1(M^5), x)$ (where here $p_1(M^4)$ denotes the characteristic class). Now by the "naturality" property of p_1 we have.

$$(p_1(M^4), [M^4]) = ((pi)^* p_1(M^5), [M^4]),$$

which in turn is equal to

$$(p_1(M^5), (pi)_*[M^4]) = (p_1(M^5), x),$$

and this completes the proof of the theorem. □

It is apparent from this proof that there is some deep connexion between the fundamental groups of non-simply-connected manifolds M^n and their rational Pontrjagin classes. (This connexion has not, up to the present, been exhaustively investigated.) There is a very general conjecture concerning "higher signatures" which goes as follows: one is concerned with certain elements of the rational cohomology algebra of the manifold M^n with fundamental group $\pi_1(M^n) = \pi$ say, lying in Im j^* where $j: M^n \to K(\pi, 1)$ is the cannonical map; it is conjectured that for each element $x \in H^{n-4k}(\pi; \mathbb{Q})$,† the scalar product of the Hirzebruch polynomial L_k in the Pontrjagin classes (see (37)) with the cycle $Dj^*(x)$, i.e.

$$(L_k(p_1, \ldots, p_k), Dj^*(x)),$$

(where D is the appropriate Poincaré-duality isomorphism), is a homotopy invariant of M^n. (It can be shown that there can exist no other homotopy invariants of closed manifolds, defined in terms of rational (or real) characteristic classes (i.e. of the curvature tensor).) This conjecture has in fact been verified in various cases: when π is free abelian (i.e. when x is a product of one-dimensional classes) (Farrell, Hsiang, Kasparov, Novikov, Rohlin); when π is the fundamental group of a compact Riemannian manifold of negative curvature (Lusztig, Mishchenko); and in several other situations which either reduce to one or the other of those two cases, or are in some sense analogous to them (Cappell, Solov'ev).

§28. Smooth Structures on the 7-Dimensional Sphere. The Classification Problem for Smooth Manifolds (Normal Invariants). Reidemeister Torsion and the Fundamental Hypothesis (Hauptvermutung) of Combinatorial Topology

In this section we shall largely be concerned with manifolds that are differentiable infinitely often, i.e. of class C^∞. By a result of Whitney, any manifold of

† In algebra the cohomology ring of the Eilenberg–MacLane complex $K(\pi, 1)$ is called the "cohomology ring of the group π", and denoted by $H^*(\pi; \mathbb{Q})$.

class C^k with $k \geq 1$, is topologically equivalent to a (unique) C^∞-manifold, in fact to an analytic manifold. One may of course define *continuous manifolds* simply by requiring only that the transition functions between co-ordinate systems on the regions of overlap of charts be continuous (not necessarily differentiable). One also encounters (though less often) purely continuous homeomorphisms between smooth manifolds. Up to the 1950s it was generally regarded as "clear" that any continuous manifold admits a compatible smooth structure, and that any two continuously homeomorphic manifolds would automatically be diffeomorphic. In fact, these assertions *are* clearly true in the one-dimensional case, can be proved without great difficulty in two dimensions, and have been established also in the 3-dimensional case (by Moise), though with considerable difficulty, notwithstanding the elementary nature of the techniques involved.

However in dimensions higher than 3 the situation changes. One of the most surprising applications of the machinery of algebraic topology expounded above, was the discovery (by Milnor), in a class of rather simple manifolds, of some that are (continuously) homeomorphic to the ordinary smooth 7-dimensional sphere S^7, but not diffeomorphic to it. (It follows from more recent work of Freedman and Donaldson that the above statements are false even in dimension 4.)

Recall from §24.3 of Part II, that the "Hopf quaternion bundle"

$$S^7 \xrightarrow{p} S^4, \quad \text{fibre } S^3,$$

is the principal fibre bundle with bundle group $SU(2)$ ($\cong S^3$) identified with the quaternions q of modulus 1 ($|q| = 1$), acting on the 7-sphere

$$S^7 = \{(q_1, q_2) \mid |q_1|^2 + |q_2|^2 = 1\}$$

as follows:

$$(q_1, q_2) \mapsto (qq_1, qq_2), \quad q, q_1, q_2 \text{ quaternions}, \qquad |q| = 1.$$

Since

$$SU(2) \subset SO(4) \quad (\simeq (SU(2) \times SU(2))/\{\pm 1\})$$

(see Part I, §14.3), we may in connexion with this fibre bundle talk of the classes $\chi \in H^4(S^4; \mathbb{Z})$ (see §9, Exercise 9) and $p_1 \in H^4(S^4; \mathbb{Z})$ (see §9, Definition 9.5). Other bundles over S^4 with fibre S^3 can be obtained (by taking boundaries) from bundles over S^4 with fibre D^4 (and group $SO(4)$):

$$E \xrightarrow{p} S^4, \qquad F \cong D^4, \qquad G = SO(4). \tag{1}$$

For fibre bundles of the form (1) the Euler class can be shown to be that element of $H^4(S^4; \mathbb{Z})$ defined by (cf. Part II, §§25.4, 25.5)

$$(\chi, [S^4]) = S^4 \circ S^4,$$

where $S^4 \circ S^4$ denotes the intersection index of $S^4 \subset E$ (embedded as the zeroth cross section $O \times S^4$, where O is the origin in D^4, fixed by $SO(4)$) with itself (or more precisely with the end result of a homotopy bringing it into general position with respect to the original).

28.1. Lemma. *The boundary ∂E of the total space E of a fibre bundle ξ of the form* (1) *(which as total space of the restriction of* (1) *is composed of fibres $S^3 = \partial D^4$), is homeomorphic to the 7-sphere S^7 if and only if $\chi(\xi)$ is a generator of $H^4(S^4; \mathbb{Z})$.*

SKETCH OF THE PROOF. The exact homotopy sequence of the restricted bundle

$$\partial E \xrightarrow{p} S^4, \qquad \text{with fibre} \quad S^3 = \partial D^4, \tag{2}$$

is as follows (see Part II, §24.3):

$$\cdots \to \pi_k(S^3) \xrightarrow{i_*} \pi_k(\partial E) \xrightarrow{j} \pi_k(\partial E, S^3) \xrightarrow{\partial} \pi_{k-1}(S^3) \to \cdots. \tag{3}$$

$$p_*: \pi_k(\partial E) \to \pi_k(S^4) \cong \pi_k(\partial E, S^3)$$

For $k = 4$ the image under the homomorphism ∂ of the generator $[S^4]$ of $\pi_4(S^4)$ $(\simeq \mathbb{Z})$ arises from the appropriate map $\varphi: D^4 \to \partial E$ under which the interior of D^4 is sent to the zeroth cross section of the bundle (2), and the boundary $\partial D^4 = S^3$ is sent to a fibre $(\cong S^3)$, so determining an element of $\pi_3(S^3)$ (see Part II, §§21.2, 22.2). If one takes a suitable non-zero cross-section of the fibre bundle (1) (i.e. homotopic to the zeroth cross-section but in general position with respect to it), then it is not difficult to see that the intersection index $S^4 \circ S^4$ (of the zeroth cross-section with this non-zero one) coincides with the multiplicity with which the boundary of D^4 is sent into the fibre S^3 of ∂E, i.e. with the degree of φ on ∂D^4. Hence

$$(\chi, [S^4]) = \partial[S^4], \tag{4}$$

(in the sense that $\partial[S^4] \in \pi_3(S^3)$ is the multiple of a generator of $\pi_3(S^3)$ by the integer $(\chi, [S^4])$).

Considering now the following segment of the exact sequence (3):

$$\mathbb{Z}_2 \simeq \pi_4(S^3) \to \pi_4(\partial E) \to \pi_4(S^4) \xrightarrow{\partial} \pi_3(S^3) \to \pi_3(\partial E) \to \pi_3(S^4) = 0,$$

we see that if $(\chi, [S^4]) = \pm 1$ (whence by (4) $\partial: \mathbb{Z} \to \mathbb{Z}$ is bijective), then by exactness $\pi_3(\partial E) = \pi_4(\partial E) = 0$. On the other hand, if $(\chi, [S^4]) \neq \pm 1$, then $\pi_3(\partial E) \neq 0$, so that in this case ∂E cannot be homotopically equivalent to S^7. Since, again by exactness of (3), $\pi_1(\partial E) = \pi_2(\partial E) = 0$, we have that $\pi_k(\partial E) = 0$ for $k \leq 4$ exactly if $(\chi, [S^4]) = \pm 1$. The usual cell decomposition of ∂E as fibre bundle consists of just four cells $\sigma^0, \sigma^3, \sigma^4, \sigma^7$. Hence if $\pi_k(\partial E) = 0$ for $k \leq 4$, then (by Hurewicz' theorem in particular) we shall have

$$H_k(\partial E) \simeq \pi_k(\partial E) = 0, \qquad k < 7,$$

$$H_7(\partial E) \simeq \pi_7(\partial E) \simeq \mathbb{Z}.$$

It follows that a map $\alpha: S^7 \to \partial E$ representing a generator of $\pi_7(\partial E) \simeq \mathbb{Z}$, will induce isomorphisms between the homotopy groups of degree ≥ 7, whence it can be shown (this is a standard result) that $\partial E \sim S^7$.

Now there is a general theorem (of Smale, Stallings and Wallace) according to which for $n \geq 5$ a closed n-manifold with the homotopy type of S^n is actually homeomorphic to S^n (see [46]). (Incidentally this has, relatively recently, been established also in the case $n = 4$ by Freedman; the last case, $n = 3$, of this statement remains as yet a conjecture, the "Poincaré conjecture".) From this result the lemma of course now follows. (However, it is to some extent possible to avoid appealing to this theorem by constructing in concrete fashion various of the bundles ξ of (1) (with $(\chi, [S^4]) = \pm 1$) in terms of the quaternions, and then giving explicit homeomorphisms $\partial E \cong S^7$ using Morse functions with exactly one maximum point and one minimum point (see below).) □

The next step is to show that there are infinitely many bundles ξ all having the same $\chi \in H^4(S^4; \mathbb{Z})$, but with distinct p_1.

28.2. Lemma. *For each pair of integers k, l there is a fibre bundle ξ_k of the form (1) with $\chi = lu$, $p_1 = 2ku$ where u is a generator of $H^4(S^4; \mathbb{Z})$.*

Before giving the proof we describe the mechanism for showing that among the smooth 7-manifolds ∂E arising from the bundles ξ of the lemma, there are some that are not diffeomorphic to the usual smooth 7-sphere S^7 (so that S^7 can be endowed with "non-trivial" smooth structures).

Let ξ_k be a bundle as in the lemma. The tangent bundle τ_E to the total space $E = E_k$ (with boundary removed) clearly splits as the Whitney sum of the bundle of vectors tangent to the fibre and the bundle of vectors normal to the fibre (with respect to some Riemannian metric on E). Since these are respectively equivalent to the induced bundles $p^*(\xi_k)$ (or rather its interior) and $p^*(\tau_{S^4})$ (see Part II, §24.4), and since $p_1(S^4) = p_1(\tau_{S^4}) = 0$, it follows (essentially from the Whitney product formula for the Chern classes—see §9, Exercise 15) that $p_1(E) = p^*p_1(\xi_k)$, whence

$$p_1(E) = p^*p_1(\xi_k) = 2kp^*u = 2kv,$$

where $v = p^*u \in H^4(E; \mathbb{Z})$ ($\simeq \mathbb{Z}$), is a generator. Since the homology group $H_4(E; \mathbb{Z}) \simeq \mathbb{Z}$ is generated by $[S^4]$ (i.e. by the cycle represented by the zeroth cross section) and $S^4 \circ S^4 = (\chi, [S^4]) = 1$ by assumption, we infer directly from the definition of the signature τ that

$$\tau(E) = 1.$$

The argument now proceeds by *reductio ad absurdum*: If the smooth closed manifold $\partial E = \partial E_k$ were diffeomorphic to the sphere S^7 (with its usual smooth structure induced from $\mathbb{R}^8 \supset S^7$), then we could form from the manifolds-with-boundary E and D^8 a closed smooth manifold $\bar{E}^8 = \bar{E}^8_k$ by identifying their boundaries by means of a diffeomorphism:

$$\bar{E}^8 = E \cup D^8, \qquad \text{where} \quad \partial E = \partial D^8 = S^7.$$

It is easy to see that for the manifold $\bar{E}^8$ we have $H_i(\bar{E}_8) = H_i(E)$ for $i \leq 7$, and

$$\begin{aligned} p_1(\bar{E}^8) &= p_1(E) = 2kv, \\ \tau(\bar{E}^8) &= \tau(E) = 1. \end{aligned} \tag{5}$$

Now since $\bar{E}^8$ is a smooth, closed, orientable manifold of dimension 8 (having the homological type of the quaternion projective space $\mathbb{H}P^2$), we can apply to it the signature formula of §27.2(29):

$$p_2 = \tfrac{1}{7}(45\tau + p_1^2).$$

Evaluating both sides of this equation at the fundamental class $[\bar{E}^8]$, and bringing in (5), we obtain

$$(p_2, [\bar{E}^8]) = \frac{4k^2 + 45}{7}. \tag{6}$$

For $k = 1$ we obtain $(p_2, [\bar{E}^8]) = 7$ the correct answer for the quaternion projective space $\mathbb{H}P^2$. However, for $k = 0, 2, 3, 5, 7, \ldots$, i.e. such that $k^2 \not\equiv 1 \pmod 7$, we obtain from (6) a non-integral value for $(p_2, [\bar{E}^8])$, which is impossible by definition of p_2 as an element of $H^8(\bar{E}^8; \mathbb{Z})$! From this contradiction we conclude that: *For those k such that $k^2 \not\equiv 1 \pmod 7$ the manifold $\partial E = \partial E_k$, is not diffeomorphic to the ordinary 7-sphere S^7 (although it is homeomorphic to it).*

Since, as we saw in §27.3, the rational classes $p_q \in H^{4q}(M^n; \mathbb{Q})$ are invariant under homeomorphisms, it follows immediately that for these k the manifolds $\bar{E}_k^8$ (where now, of course, we require only that ∂E and ∂D^8 be identified by means of a homeomorphism) cannot be endowed with a smooth structure (compatible with their given continuous structures); for otherwise the invariance of the rational class $p_1(\bar{E}_k^8)$ would be violated (τ being clearly homeomorphically invariant). (There are certain other examples, due to Kervaire, where a detailed analysis allows one to avoid appealing to the topological invariance of the rational classes p_q.)

Sketch of the Proof of Lemma 28.2. We first show that given any even integer there is an $SO(3)$-vector bundle over S^4 whose first Pontrjagin class p_1 has that even integer as its value at the fundamental class $[S^4]$. To this end, observe first that the universal fibre bundle (cf. Part II, §24.4)

$$ESU(2) = V_{\infty,2}^{\mathbb{C}} \to G_{\infty,2}^{\mathbb{C}} = BSU(2),$$

in conjunction with the isomorphism $SO(3) \simeq SU(2)/\mathbb{Z}_2$ (see Part I, §14.3), yields a sequence of maps

$$\begin{aligned} O(1) \simeq \mathbb{Z}_2 \to SU(2) \to SO(3) \simeq SU(2)/\mathbb{Z}_2 \to ESU(2)/\mathbb{Z}_2 &= B\mathbb{Z}_2 \\ \simeq K(\mathbb{Z}_2, 1) &\to BSU(2), \end{aligned}$$

from which one can construct a fibration

$$BSO(3) \xrightarrow{p} K(\mathbb{Z}_2, 2), \qquad \text{with fibre} \quad F \sim BSU(2). \tag{7}$$

(This projection map can also be obtained as a representative of the homotopy class of maps $BSO(3) \to K(\mathbb{Z}_2, 2)$ which corresponds, in accordance with Exercise 5 of §9, to the non-zero element of $H^2(BSO(3); \mathbb{Z}_2) \simeq \mathbb{Z}_2$.)

Considering the homology spectral sequence over $\mathbb{Z}$ (see §8) of the fibration

(7), one sees using the fact that

$$E^{(2)}_{q,j} = H_q(B; H_j(F))$$

(here $B = K(\mathbb{Z}_2, 2)$, $F = BSU(2)$), that the operators

$$d_r\colon E^{(r)}_{q,j} \to E^{(r)}_{q-r,j+r-1},$$

are, for $r \geq 2$, zero on those $E^{(r)}_{q,j}$ with $q + j \leq 5$, whence

$$E^{(\infty)}_{q,j} = E^{(2)}_{q,j} \simeq H_q(B; H_j(F)) \qquad \text{for} \quad q + j \leq 5.$$

(For instance, that $d_5\colon E^{(5)}_{5,0} \to E^{(5)}_{0,4}$ is zero follows from the fact—which we shall not stop to establish—that $H_5(K(\mathbb{Z}_2, 2); \mathbb{Z}) \simeq \mathbb{Z}_2$, while $H_0(K(\mathbb{Z}_2, 2); H_4(F)) \simeq H_4(F) \simeq \mathbb{Z}$.) One obtains the following partial table of generators of the $E^{(\infty)}_{q,j}$:

$j\uparrow$						
4	u_0	0	u_1	0	u_2	u_3
	0	0	0	0	0	0
0	1	0	v	0	w	x
	0	1	2	3	4	5

$q \rightarrow$

It can be shown that $E^{(\infty)}_{4,0} = H_4(K(\mathbb{Z}_2, 2); \mathbb{Z}) \simeq \mathbb{Z}_2$ (with generator denoted by w in the above table). Since by the "naturality" of the Hurewicz homomorphism H (whose definition may be gleaned from the proof of Corollary 4.9) the following diagram commutes (here the upper and lower rows are the homotopy and homology exact sequences of the fibration (7)):

$$\begin{array}{ccccccc} & \mathbb{Z} & & \mathbb{Z} & & & \\ 0 \to & \pi_4(BSU(2)) & \xrightarrow{\simeq} & \pi_4(BSO(3)) & \to 0 & & \\ & \wr\!\wr \downarrow H & & \downarrow H & & & \\ \to & H_4(BSU(2)) & \xrightarrow{i_*} & H_4(BSO(3)) & \to & H_4(K(\mathbb{Z}_2, 2)) & \to 0, \\ & \mathbb{Z} & & \mathbb{Z} & & \mathbb{Z}_2 & \end{array} \tag{8}$$

it follows that the cokernel of $H\colon \pi_4(BSO(3)) \to H_4(BSO(3))$ $(\mathbb{Z} \to \mathbb{Z})$ is $\mathbb{Z}_2$. Since $H^*(BSO(3); \mathbb{Q}) = \mathbb{Q}[p_1]$ (see §10.4(12)), we infer that

$$(p_1, i_* u_0) = 2,$$

where u_0 is a generator of $H\pi_4 \subset H_4(BSO(3); \mathbb{Z})$. Hence corresponding to each even integer $2k$ there is an $SO(3)$-vector bundle ξ over S^4 (classified by the appropriate homotopy class of maps $S^4 \to BSO(3)$, i.e. the appropriate element of $\pi_4(BSO(3))$) such that $(p_1, [S^4]) = 2k$. (Here, of course, $p_1 \in H^4(S^4; \mathbb{Z})$.)

Taking any one of these $SO(3)$-vector bundles ξ one may form $\xi \oplus 1$, the Whitney sum of ξ with the trivial line bundle over S^4. Under the standard embedding $SO(3) \subset SO(4)$ the bundle $\xi \oplus 1$ may be considered as an $SO(4)$-vector bundle. Since for this vector bundle we have

$$p_1(\xi \oplus 1) = p_1(\xi), \qquad \chi(\xi \oplus 1) = 0,$$

(the second equation following from the fact that the Euler class $\chi(\xi \oplus \hat{\xi})$ of a Whitney sum is the product $\chi(\xi)\chi(\hat{\xi})$ in $H^*(B; \mathbb{Z})$), we see that we have established the existence of an $SO(4)$-vector bundle over S^4 with $\chi = 0$ and $(p_1, [S^4])$ equal to any prescribed even integer.

To obtain the lemma as stated we carry out the analogous argument using instead the isomorphism $SO(4) \simeq (SU(2) \times SU(2))/\mathbb{Z}_2$ (see Part I, §14.3). Analogously to (7) one can construct from this a fibration

$$BSO(4) \xrightarrow{p} K(\mathbb{Z}_2, 2), \qquad F \sim BSU(2) \times BSU(2),$$

and once again it turns out that in the homology spectral sequence over $\mathbb{Z}$ (taking account again of $\pi_1(B) = \{1\}$), one has

$$E^{(\infty)}_{q,j} = E^{(2)}_{q,j} \simeq H_q(B; H_j(F)) \qquad \text{for} \quad q + j \leq 5.$$

The following table of generators may be compiled:

$j \uparrow$

4	u_0, x_0	0	u_1, x_1	0	u_2, x_2		
	0	0	0	0	0	0	
0	1	0	v	0	w	x	
	0	1	2	3	4	5	

$q \rightarrow$

From the analogue of diagram (8) (with $BSU(2)$ and $BSO(3)$ replaced by $BSU(2) \times BSU(2)$ and $BSO(4)$ respectively, and $\mathbb{Z}$ by $\mathbb{Z} \oplus \mathbb{Z}$) it follows that the Hurewicz homomorphism

$$H\colon \pi_4(BSO(4)) \rightarrow H_4(BSO(4); \mathbb{Z})$$

has cokernel isomorphic to $\mathbb{Z}_2$. Since p_1 and χ form a basis for $H^4(BSO(4); \mathbb{Z}) \simeq \mathbb{Z} \oplus \mathbb{Z}$ (cf. §10.4(12)), we infer much as before (via classifying maps $S^4 \rightarrow BSO(4)$) that there exist $SO(4)$-vector bundles over S^4 with arbitrarily prescribed $(\chi, [S^4])$ and even $(p_1, [S^4])$. This completes the proof of the lemma. □

We now describe briefly Milnor's direct construction of fibre bundles ξ_k over S^4 with total space M_k^7 homeomorphic, but not diffeomorphic, to S^7. Recall from Part II, §25.5, Example (b)(iv), that inequivalent $SO(4)$-vector bundles over S^4 are determined in a one-to-one manner by the elements of the group $\pi_3(SO(4)) \simeq \mathbb{Z} \oplus \mathbb{Z}$, and so can be labelled in a canonical way by pairs (h, j) of integers. The map $f_{hj}\colon S^3 \to SO(4)$ corresponding to the pair (h, j) is in fact given explicitly in terms of the quaternions by

$$f_{hj}(v)w = v^h w v^j, \tag{9}$$

where $v, w \in \mathbb{H} = \mathbb{R}^4$, and $|v| = 1$ (so that $v \in S^3$). (For each $v \in S^3$, equation (9) defines a proper orthogonal transformation of $\mathbb{H} = \mathbb{R}^4$.) Let ξ_{hj} denote the fibre bundle obtained from the vector bundle over S^4 determined by f_{hj}, by restricting the fibre to $S^3 \subset \mathbb{R}^4$.

EXERCISE

1. Show that

$$\chi(\xi_{hj}) = (h + j)u, \qquad p_1(\xi_{hj}) = \pm 2(h - j)u,$$

where u is a generator of $H^4(S^4)$.

Now let h, j be such that $h + j = 1$, $h - j = k$ (k arbitrary), and denote by M_k^7 the total space of the bundle ξ_{hj} with these conditions imposed. It can be shown that the manifold M_k^7 may be realized as the union of two copies of $\mathbb{R}^4 \times S^3$ with their subspaces $(\mathbb{R}^4 \setminus \{0\}) \times S^3$ identified via the diffeomorphism

$$(u, v) \mapsto (u', v') = \left(\frac{u}{|u|^2}, \frac{u^h v u^j}{|u|}\right), \qquad u, v \in \mathbb{H}, \quad u \neq 0, \quad |v| = 1.$$

(Verify!)

EXERCISE

2. Show that the function defined by

$$f(u, v) = \frac{\operatorname{Re} v}{(1 + |u|^2)^{1/2}} = \frac{\operatorname{Re} u''}{(1 + |u''|^2)^{1/2}}, \qquad u'' = u'(v')^{-1},$$

has just the two critical points $(u, v) = (0, \pm 1)$, and that these are non-degenerate.

It follows from Morse theory that M_k^7 is homeomorphic to the 7-sphere. (For $f(0, 1) = 1$ must be the maximum value of f and $f(0, -1)$ the minimum, whence by the Morse lemma (15.1) for sufficiently small ε the sets $f^{-1}[1 - \varepsilon, 1]$ and $f^{-1}[-1, -1 + \varepsilon]$ will be closed 7-dimensional discs, and it then follows as in the proof of Lemma 15.2 that M_k^7 is the union of two 7-discs with their boundaries identified in a one-to-one manner.) On the other hand, in view of Exercise 1 and the argument following the statement of Lemma 28.2, for each k such that $k^2 \not\equiv 1 \pmod 7$ the manifold M_k^7 is not diffeomorphic to S^7.

Thus there exist "non-trivial" smooth closed manifolds homeomorphic to, and therefore certainly of the same homotopy type as, a sphere. Smooth n-manifolds having the homotopy type of S^n are called *homotopy n-spheres*. Clearly the class of homotopy n-spheres is closed under the operation of forming connected sums (see §4):

$$M_1^n \# M_2^n \sim S^n \qquad \text{if} \quad M_1^n, M_2^n \sim S^n.$$

28.3. Definition. Two closed, smooth n-manifolds M_1^n and M_2^n (of arbitrary homotopy type) are said to be *h-cobordant* (or *J-equivalent*) if there exists a film W^{n+1} (compact manifold-with-boundary) such that $\partial W^{n+1} = M_1^2 \cup M_2^n$ (disjoint sum), and W^{n+1} is contractible onto each of its boundary components M_1^n, M_2^n.

28.4. Lemma. *The oriented h-cobordism classes of (oriented) homotopy n-spheres form a group, denoted by θ^n, under the operation of forming the connected sum.*

Proof. As was noted in §4 the connected sum operation is associative quite generally (not just for homotopy spheres). Consider the connected sum of an oriented homotopy sphere M_+^n with a copy of the same manifold oppositely oriented, M_-^n:

$$M_0^n = M_+^n \# M_-^n.$$

We shall show that this connected sum is h-equivalent to S^n (which represents the identity element of θ^n). To this end, we construct a compact (oriented) smooth manifold-with-boundary W^{n+1} with boundary M_0^n, as follows (see Figure 119): From the cylinder $M_+^n \times I[0, 1]$, we remove the subset $D_\varepsilon^n \times I$, where $D_\varepsilon^n \subset M_+^n$ is a small open n-ball of radius ε; after appropriate "smoothing of corners" we are left with a smooth manifold-with-boundary W^{n+1} for which $\partial W^{n+1} = M_+^n \# M_-^n$, and which is moreover contractible. If we now remove from W^{n+1} a small open ball D_0^n (see Figure 119), then the resulting manifold can be contracted onto the boundary S_0^n of that ball, so that $\partial W^{n+1} = M_0^n$ is h-cobordant to S_0^n. Hence each class of oriented homotopy n-spheres has an inverse (with respect to the identity class represented by S^n itself). □

Recalling from Part II, Theorem 23.1.4, that the equivalence classes of framed normal bundles over n-dimensional submanifolds of $\mathbb{R}^{n+N}$ (i.e. non-degenerate fields of normal N-frames on such manifolds) are in natural one-to-one correspondence with $\pi_{n+N}(S^N)$, we denote by $J_n \subset \pi_{n+N}(S^N)$ $(n < N - 1)$ the subgroup consisting of those elements corresponding to framed normal bundles on the standard sphere $S^n \subset \mathbb{R}^{n+N}$. Now for large enough N every

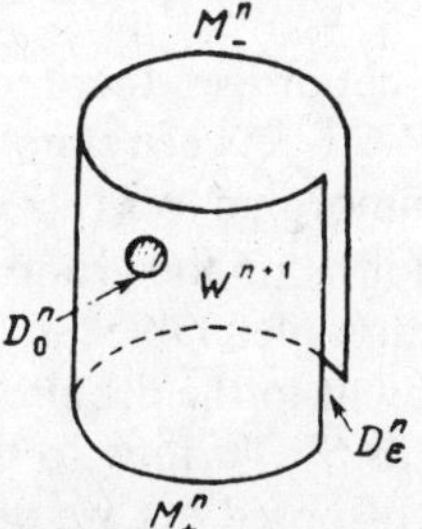

Figure 119

homotopy sphere $M^n \subset \mathbb{R}^{n+N}$ has trivial normal bundle. (For $n \not\equiv 0, 1, 3, 7 \pmod 8$), this follows from the fact that $\pi_n(SO) = 0$ (by Bott periodicity—see Theorem 25.13) and the observation of Part II, §23.2, that the non-degenerate fields of normal N-frames on M^n are classified by homotopy classes of maps $M^n \to SO(N)$; for $n = 3$ or $7 \pmod 8$ it may be deduced from Bott periodicity together with the signature formula (see §27.2) for the appropriate Pontrjagin class, using the fact that $\tau(S^n) = 0$; finally, for $n \equiv 0$ or $1 \pmod 8$ the triviality of the normal bundle to $M^n (\sim S^n)$ is a result of Adams depending on more recent techniques of algebraic topology.) In view of this property (the triviality of the normal bundles of homotopy n-spheres in $\mathbb{R}^{n+N}$), the one-to-one correspondence (noted above) between framed normal bundles (on n-manifolds in $\mathbb{R}^{n+N}$) and $\pi_{N+n}(S^N)$, yields a homomorphism

$$\theta^n \to \pi_{N+n}(S^N)/J_n,$$

whose kernel turns out to coincide with the subgroup ∂P^{n+1} of θ^n consisting of the (classes of) boundaries of compact "parallelizable" manifolds-with-boundary, i.e. having trivial tangent bundles (verify!).

The following is known concerning the groups ∂P^{n+1}:

(i) $\partial P^{n+1} = 0$ for all even n;

(ii) $$\partial P^{n+1} = \begin{cases} 0 & \text{for } n = 1, 5, 13, \\ \mathbb{Z}_2 & \text{for } n = 9, \\ 0 \text{ or } \mathbb{Z}_2 & \text{for } n = 4k+1; \end{cases}$$

(iii) ∂P^{n+1} is cyclic of finite order for $n \neq 3$. (In the case $n = 7$, this order turns out to be 28; earlier in this section we indicated, in essence at least, a homomorphism $\theta^7 \to \mathbb{Z}_7$.)

The groups $\Gamma_n = \pi_{N+n}(S^N)/J_n$ and θ^n are, up to $n = 10$, as in the following table. (These and the above facts about groups of homotopy spheres were discovered by Kervaire and Milnor—see [32].)

$n =$	0	1	2	3	4	5	6	7	8	9	10
$\Gamma_n =$	0	0	Z_2	0	0	0	0	0	Z_2	$(Z_2)^2$	Z_2
$\theta^n =$	0	0	0	?	0	0	0	Z_{28}	$(Z_2)^2$	$(Z_2)^3$	Z_2

An important result (the "h-cobordism theorem") due to Smale (see [46]) states that every h-cobordism W^{n+1} between simply-connected manifolds M_1^n, M_2^n of dimension $n \geq 5$ is diffeomorphic to $M_1^n \times I$ (whence M_1^n is diffeomorphic to M_2^n). This implies that for $n \geq 5$ the group θ^n provides a classification of the distinct smooth structures definable on the n-sphere, so that this classification problem is equivalent to the problem of classifying n-manifolds with the same homotopy type as S^n. Turning to the cases $n = 3, 4$ (the cases $n = 1, 2$ being relatively easily disposed of), we note that while the group θ^3 remains unknown, it *is* known that there are no "non-standard" smooth structures on S^3; on the other hand, while it is known that $\theta^4 = 0$, it is not known whether the 4-sphere S^4 can be endowed with distinct smooth structures.

We shall now expound briefly the classification theory of closed, simply-connected, orientable manifolds of dimension $n \geq 5$, as developed by Browder and Novikov.† Initially the following natural question suggests itself: What invariants, in addition to the obvious ones (namely the homotopy type and the class of equivalent tangent bundles), will suffice to determine a smooth closed n-manifold to within a diffeomorphism? The case of n-manifolds homotopically equivalent to the n-sphere is settled by the theory of Milnor and Kervaire, outlined above. However the closed, smooth n-manifolds of arbitrary (not necessarily spherical) homotopy type, cannot be divided up naturally into classes forming a group. The approach we describe here to the general classification problem of n-manifolds exploits rather the stable normal bundle ν^N on M^n embedded in $\mathbb{R}^{N+n}$, $N > n + 1$ (cf. the beginning of the proof of Theorem 27.9), which is (uniquely) determined by the tangent bundle τ^n in view of the triviality of the Whitney sum $\tau^n \oplus \nu^N$:

$$\tau^n \oplus \nu^N = M^n \times \mathbb{R}^{n+N}. \tag{10}$$

It turns out to be particularly useful to consider the Thom complex $M(\nu^N)$ of the stable normal bundle ν^N: Recall from Definition 27.6 *et seqq.*, that $M(\nu^N)$ is constructed by taking a neighbourhood U of M^n (identifiable with the total space of ν^N) in $\mathbb{R}^{n+N} \subset S^{n+N}$, and identifying the boundary ∂U with a point:

$$U \to M(\nu^N), \qquad \partial U \to \{*\}, \tag{11}$$

and that M^n embeds naturally in $M(\nu^N)$, essentially as the zeroth cross-section, so that we can write $M^n \subset M(\nu^N)$. The map (11) as a map $(U, \partial U) \to (M(\nu^N), *)$ between pairs, extends in the obvious way to a map ψ of the sphere S^{n+N}, where now the whole of the complement of U in S^{n+N} is mapped to a point;

$$\psi = \psi_{M^n}: S^{N+n} \to M(\nu^N).$$

For the map ψ applied to the fundamental class $[S^{N+n}]$ we have

$$\psi_*[S^{N+n}] = \varphi[M^n] \in H_{n+N}(M(\nu^N); \mathbb{Z}), \tag{12}$$

where φ is as in Lemma 27.7. Thus the cycle $\varphi(M^n]$ is spherical. Noting that by Lemma 27.7

$$H_{n+N}(M(\nu^N); \mathbb{Z}) \simeq H_n(M^n; \mathbb{Z}) \simeq \mathbb{Z}$$

(assuming M^n orientable), we have by Corollary 10.11 (and the $(N-1)$-connectedness of $M(\nu^N)$—see Lemma 27.8) that

$$\pi_{N+n}(M(\nu^N)) \simeq \mathbb{Z} \oplus D,$$

for some finite (abelian) group D (provided $N + n < 2N - 1$, i.e. $n < N - 1$). Hence in view of (12) the map ψ_{M^n} represents an element of $\pi_{N+n}(M(\nu^N))$ of

† In the case $n = 4$ this theory yields only the assertion that homotopically equivalent 4-manifolds are h-cobordant.

the form $1 + \alpha$, $\alpha \in D$. Thus we have associated with our initially given orientable, smooth, closed manifold M^n an element

$$[\psi_{M^n}] = 1 + \alpha \in \pi_{N+n}(M(\nu^N)). \tag{13}$$

Further investigation yields the following result (whose proof we omit).

28.5. Proposition (Novikov).

(i) *Let $M^n \subset S^{n+N}$, $M(\nu^N)$, D be as above. To each smooth, closed, oriented (simply-connected) n-manifold M_1^n for which there exists a homotopy equivalence $f: M_1^n \to M^n$ of degree $+1$ (and so orientation-preserving) which preserves normal bundles (i.e. $f^*(\nu_{M^n}^N) = \nu_{M_1^n}^N$, where f^* is the induced bundle map), there corresponds in a canonical fashion an element (not necessarily unique) $\psi_{M_1^n} \in \pi_{n+N}(M(\nu^N))$ of the form $1 + \alpha \in \mathbb{Z} \oplus D$, $\alpha \in D$. For n not of the form $4k + 2$ the converse is also true: each element of $\mathbb{Z} \oplus D$ of the form $1 + \alpha$, $\alpha \in D$, determines a homotopy equivalence with these properties, from some M_1^n to M^n; for $n = 4k + 2$, on the other hand, it may happen that this applies only to those α in a subgroup of index 2 in D, i.e. the elements $1 + \alpha$, $\alpha \in D$, which are "realizable" by such homotopy equivalences, are those for which α lies in some subgroup $\tilde{D}$ where either $\tilde{D} = D$ or $|D : \tilde{D}| = 2$.*

(ii) *If two smooth, closed, oriented, simply-connected n-manifolds M_1^n, M_2^n correspond, in accordance with (i), to the same element $1 + \alpha \in \pi_{n+N}(M(\nu^N))$, $\alpha \in D$, then there is a "Milnor sphere" $\theta \in \partial P^{n+1}$ such that $M_1^n \# \theta \cong M_2^n$.*

In view of (10) and the finiteness of D and ∂P^{n+1}, we immediately infer the following

28.6. Corollary. *For each fixed homotopy type and tangent bundle (or invariants thereof—the classes $p_k \in H^*(M^n; \mathbb{Q})$) of smooth, orientable, closed, simply-connected n-manifolds, $n \geq 5$, there exist only finitely many pairwise non-diffeomorphic such manifolds with those invariants.*

Another result (of Browder and Novikov) gives criteria for a vector bundle ξ over a closed, smooth, simply-connected, orientable manifold M_1^n to be realizable as a stable normal bundle over some other such manifold $M_2^n \subset \mathbb{R}^{n+N}$ of the same homotopy type as M_1^n, namely:

(i) it is necessary (and for $n = 6, 14$ and all odd $n = 2k + 1 \geq 5$ also sufficient) that the cycle $\varphi[M_1^n] \in H_{n+N}(M(\xi))$ (see Lemma 27.7) be spherical (i.e. an image of the sphere S^{N+n}—cf. (12));
(ii) in the case $n = 4k$, for sufficiency it is enough to add the condition that the Hirzebruch polynomial in the classes $p_1(\xi), \ldots, p_k(\xi)$ be equal to the signature $\tau(M_1^n)$. (This condition is obviously necessary in view of (10) and the signature formula (§27.2(37)).)

In fact, as Browder has shown, this result may be given a more general formulation: M_1^n need not be a manifold, but merely a cell complex for which Poincaré duality holds in integral homology (global only, not local), and then the complex M_1^n will have the homotopy type of a smooth, closed, orientable, simply-connected manifold M_2^n if and only if there exists a stable vector bundle ξ over M_1^n for which conditions (i) and (ii) hold.

For n of the form $4k + 2$ there are variants of these results known; however, their formulation is more complicated and we shall not give them here.

EXERCISE

3. In connexion with Proposition 28.5 prove that in the case of the sphere, i.e. $M^n = S^n$, one has

$$M(\nu^N) \cong S^N \vee S^{N+n},$$

$$\pi_{N+n}(M(\nu^N)) \simeq \mathbb{Z} \oplus \pi_{N+n}(S^N), \qquad \text{i.e.} \qquad D \simeq \pi_{N+n}(S^N).$$

Returning to Proposition 28.5, we note that in order to discover to what extent the "normal invariant" $[\psi_{M^n}] = 1 + \alpha \in \pi_{N+n}(M(\nu^N))$ (see (13)) is non-unique, it is clear that one needs to investigate the group of homotopy classes of self-maps of the normal bundle having degree $+1$ on the base:

$$M^n \xrightarrow{f} M^n, \qquad \nu^N \xrightarrow{f^*} \nu^N, \qquad \deg f = 1. \tag{14}$$

This group acts in the obvious way on the Thom complex $M(\nu^N)$, and the orbits under the induced action on the set of admissible elements of the form $1 + \alpha \in \pi_{N+n}(M(\nu^N))$ then correspond (essentially as in the proposition) in a one-to-one manner to the classes of smooth, closed, oriented, simply-connected n-manifolds, two such manifolds M_1^n, M_2^n being defined to be in the same class if one can be obtained from the other by forming the connected sum with a "Milnor sphere" $\theta \in \partial P^{n+1}$: $M_1^n \cong M_2^n \# \theta$.

EXERCISE

4. Prove that in the case $M^n = S^n$ the answer to the question of the "degree of non-uniqueness" depends on how the group $\Gamma_n = \pi_{N+n}(S^N)/J_n$ factorizes as a direct product. Compute the group of homotopy classes of automorphisms (i.e. self-diffeomorphisms) of a smooth, closed, oriented, simply-connected 4-manifold M^4, preserving the stable normal bundle (as in (14)), and show that it acts transitively on the set of elements of the form $1 + \alpha$. Compute the various groups involved, in the cases $M^{2n} = \mathbb{C}P^n$, $M^n = S^k \times S^{n-k}$.

In connexion with the above, we note the following interesting property (which can be established by elementary means) of homotopy equivalences preserving stable normal bundles.

28.7. Theorem (Mazur). *Let M_1^n, M_2^n ($\subset \mathbb{R}^{n+N}$, $N > n + 2$) be smooth, closed manifolds (not necessarily simply-connected). If there exists a homotopy equivalence $f: M_1^n \to M_2^n$ preserving normal bundles, i.e. $f^*(\nu_2^N) = \nu_1^N$, then the total spaces E_1 and E_2 of the normal bundles ν_1^N and ν_2^N are diffeomorphic.*

Remark. In the simply-connected case this theorem follows from the "*h*-cobordism theorem" of Smale (see above); however, the proof below shows that it is valid without the assumption of simple-connectedness.

Proof. Let $\tilde{f}: M_1^n \to E_2$ be a smooth embedding homotopic to and closely approximating $f: M_1^n \to M_2^n \subset E_2$ (see Part II, §§10.1, 12.1, *et seqq.*), where M_2^n is identified with the zeroth cross-section of E_2. Similarly, if $g: M_2^n \to M_1^n \subset E_1$ is a homotopic inverse of the homotopy equivalence f (i.e. $fg \sim 1, gf \sim 1$), we denote by $\tilde{g}: M_2^n \to E_1$ a smooth embedding homotopic to, and closely approximating, g. It follows from Theorem 12.1.6 of Part II, that the normal bundles (with fibre $\mathbb{R}^N$) over $\tilde{f}(M_1^n)$ and $\tilde{g}(M_2^n)$ are equivalent to ν_1^N and ν_2^N respectively (since all embeddings $M_1^n \to \mathbb{R}^{n+N}$, $N > n + 2$, are smoothly isotopic, and similarly for M_2^n). Hence there are diffeomorphisms from the regions $D_1^{(1)} \subset E_1$ and $D_1^{(2)} \subset E_2$, obtained by restricting the vectors in each fibre $\mathbb{R}^N$ to those with length < 1, to ε-neighbourhoods U_1 and U_2 (respectively) of the submanifolds $\tilde{f}(M_1^n) \subset E_2$ and $\tilde{g}(M_2^n) \subset E_1$:

$$\tilde{F}: D_1^{(1)} \to U_1 \subset E_2, \qquad \tilde{G}: D_1^{(2)} \to U_2 \subset E_1;$$

we may moreover assume that (see Part II, Corollary 7.2.2)

$$\tilde{F}|_{M_1^n} = \tilde{f}, \qquad \tilde{G}|_{M_2^n} = \tilde{g}.$$

(Note that the above argument depends crucially on the stability condition $N > n + 2$, which allows Whitney's result, namely that any two embeddings in $\mathbb{R}^{n+N}$ are smoothly isotopic, to be applied.)

Since ε is assumed small and $\tilde{f}$, $\tilde{g}$ closely approximate f, g, we have $U_1 \subset D_1^{(2)}$, $U_2 \subset D_1^{(1)}$, so that the maps

$$\tilde{G}\tilde{F}: D_1^{(1)} \to D_1^{(1)}, \qquad \tilde{F}\tilde{G}: D_1^{(2)} \to D_1^{(2)},$$

are defined.

Since $f: M_1^n \to M_2^n$ is a homotopy equivalence, the image under $\tilde{G}\tilde{F}$ of the zeroth cross-section $M_1^n \subset D_1^{(1)}$ is homotopic in $D_1^{(1)}$ to the zeroth cross-section itself (and similarly for the image under $\tilde{F}\tilde{G}$ of $M_2^n \subset D_1^{(2)}$). Since f preserves normal bundles, it follows that, provided ε is chosen small enough, there is a self-diffeomorphism of the whole of E_1, isotopic to the identity map, which leaves all vectors of length $\geq \frac{1}{2}$ in the fibres fixed and brings $\tilde{G}\tilde{F}(D_1^{(1)}) = \tilde{G}(U_1)$ into coincidence with a neighbourhood of the zeroth cross-section $M_1^n \subset E_1$. Since $\tilde{G}(U_1) \subset U_2$, it follows that we may suppose that U_2 contains, for sufficiently small δ, a δ-neighbourhood $D_\delta^{(1)}$ of the zeroth cross-section $M_1^n \subset E_1$, and likewise that U_1 contains a δ-neighbourhood $D_\delta^{(2)}$ of $M_2^n \subset E_2$. Thus we have the following diagram of containments and diffeomorphisms:

$$\begin{array}{c} D_\delta^{(1)} \subset U_2 \subset D_1^{(1)} \\ \times \\ D_\delta^{(2)} \subset U_1 \subset D_1^{(2)} \end{array}$$

Now the canonical self-diffeomorphism $E_i \to E_i$ $(i = 1, 2)$ which dilates the fibres $\mathbb{R}^N$ by the factor δ^{-1}, sends $D_\delta^{(i)}$ diffeomorphically onto $D_1^{(i)}$, and dilates

U_i also by the factor δ^{-1}, yielding $U_{i,\delta^{-1}}$ say. Iterating this dilation we obtain the following infinite chain of containments and diffeomorphisms:

$$\begin{array}{ccccccccccccc} D_\delta^{(1)} & \subset & U_2 & \subset & D_1^{(1)} & \subset & U_{2,\delta^{-1}} & \subset & D_{\delta^{-1}}^{(1)} & \subset & U_{2,\delta^{-2}} & \subset & D_{\delta^{-2}}^{(1)} \cdots \\ & \nwarrow\!\!\!\!\swarrow & & F^{-1},\ \tilde{G}^{-1} & & \nwarrow\!\!\!\!\swarrow & & F_{\delta^{-1}}^{-1},\ \tilde{G}_{\delta^{-1}}^{-1} & & \nwarrow\!\!\!\!\swarrow & & F_{\delta^{-2}}^{-1},\ \tilde{G}_{\delta^{-2}}^{-1} & \\ D_\delta^{(2)} & \subset & U_1 & \subset & D_1^{(2)} & \subset & U_{1,\delta^{-1}} & \subset & D_{\delta^{-1}}^{(2)} & \subset & U_{2,\delta^{-2}} & \subset & D_{\delta^{-2}}^{(2)} \cdots \end{array}$$

Since

$$\bigcup_j U_{2,\delta^{-j}} = E_1, \qquad \bigcup_j D_{\delta^{-j}}^{(2)} = E_2,$$

the expanding sequence of diffeomorphisms

$$\tilde{G}_{\delta^{-j}}^{-1}\colon U_{2,\delta^{-j}} \to D_{\delta^{-j}}^{(2)},$$

yields in the limit a diffeomorphism $E_1 \to E_2$, as required. □

28.8. Corollary. *The Thom complexes of the stable normal bundles ν_1^N, ν_2^N on the manifolds M_1^n, M_2^n (as in the theorem) are homeomorphic: $M(\nu_2^N) \cong M(\nu_1^N)$.*

EXERCISE

5. Show that all orientable 3-manifolds are parallelizable (i.e. have trivial tangent bundles).

28.9. Corollary. *If $L_p^3(q_1)$, $L_p^3(q_2)$ are homotopically equivalent lens spaces (so that $q_1 \equiv \pm\lambda^2 q_2 \pmod p$ for some λ—see §11), then for each $N \geq 5$ their direct products with $\mathbb{R}^N$ are diffeomorphic:*

$$L_p^3(q_1) \times \mathbb{R}^5 \cong L_p^3(q_2) \times \mathbb{R}^5 \qquad \textit{if} \quad q_1 \equiv \pm\lambda^2 q_2 \pmod p.$$

Hence the Thom complexes $M(\nu_1)$, $M(\nu_2)$ of trivial vector bundles ν_i (with fibre $\mathbb{R}^N$, $N \geq 5$) over $L_p^3(q_1)$, $L_p^3(q_2)$ respectively, are homeomorphic.

However, as Milnor has shown, for odd N and certain primes p, the complexes $M(\nu_1)$ and $M(\nu_2)$ are combinatorially distinct, i.e. do not have essentially identical simplicial decompositions. We conclude with a brief sketch of the argument. The distinguished (singular) point $*$ of the Thom complex $M(\nu)$ of a vector bundle ν (i.e. the point to which $\partial\tilde{E}$ is contracted—see Definition 27.6) has a neighbourhood with the combinatorial structure (i.e. under a simplicial decomposition) of a cone over the boundary of the "star" of the point $*$, identifiable with the total space of the bundle with fibre restricted to S^{N-1}. It turns out that the combinatorial invariants of the boundary of the star are actually invariants (under homeomorphisms) of the complex itself. If $N - 1$ is even and the vector bundles ν_1, ν_2 are as in Corollary 28.9, then the Reidemeister torsion (a combinatorial invariant—see §11) satisfies (verify!)

$$R(L_p^3(q) \times S^{N-1}) = R(L_p^3(q)) \times \chi(S^{N-1}), \tag{15}$$

where $\chi(S^{N-1})(=2)$ is the Euler characteristic. Now, as noted in §11 (Exercise 6, *et seqq.*), for certain p (e.g. $p = 7$) it can happen that although $q_1 \equiv \pm\lambda^2 q_2 \bmod p$, i.e. although $L_p^3(q_1)$ and $L_p^3(q_2)$ have the same homotopy type, nonetheless

$$R(L_p^3(q_1)) \neq R(L_p^3(q_2)).$$

From this, (15) and Corollary 28.9, we draw the following important conclusion (Milnor), giving a negative answer to the "Hauptvermutung" for complexes: *The Thom complexes $M(\nu_1)$, $M(\nu_2)$ (for $p = 7$ and appropriate q_1, q_2, N) are combinatorially inequivalent, though homeomorphic.*

Bibliography

1. Adams, J. F., 1974. *Stable Homotopy and Generalized Homology*. Chicago Lectures in Mathematics. University of Chicago Press: Chicago, Ill.
2. Adams, J. F., 1969. *Stable Homotopy Theory*. Lecture Notes in Mathematics, No. 3 (Third edition). Springer-Verlag: New York.
3. Al'ber, S. I., 1957. 'On the periodic problem of the calculus of variations in the large'. *UMN* **12**, pp. 57–124. (English translation: *Amer. Math. Soc. Translations* **14** (1960), pp. 107–172.)

*4. Aleksandrov, A. D., 1948. *Intrinsic Geometry of a Convex Surface*. Gostehizdat: Moscow–Leningrad.

5. Atiyah, M. F., Bott, R. and Shapiro, A., 1964. 'Clifford modules'. *Topology* **3**, Supplement 1, pp. 3–38.
6. Atiyah, M. F., 1967. *K-theory*. W. A. Benjamin: New York.
7. Atiyah, M. F., 1961. 'Thom complexes'. *Proc. London Math. Soc.* (3), **11**, pp. 291–310.
8. Bishop, R. L. and Crittenden, R. J., 1964. *Geometry of Manifolds*. Pure and Applied Math., Vol. 25. Academic Press: New York–London.

*9. Boltjanskiĭ, V. G. and Yefremovich, V. A., 1982. *Descriptive Topology*. Nauka: Moscow.

10. Borel, A., 1953. 'Sur la cohomologie des espaces fibrés principaux et des espaces homogènes de groupes de Lie compact'. *Ann. Math.* **57**, pp. 115–207.
11. Borel, A. and Serre, J.-P., 1953. 'Groupes de Lie et puissances de Steenrod'. *Amer. J. Math.* **75**, pp. 409–448.
12. Browder, W., 1972. *Surgery on Simply-Connected Manifolds*. Ergebnisse der Math., 65. Springer-Verlag: Berlin–New York.

*13. Bukhshtaber, V. M., Mishchenko, A. S. and Novikov, S. P., 1971. 'Formal groups and their role in the apparatus of algebraic topology'. *UMN* **26**, pp. 131–154.

14. Cartan, H., 1954/55. 'Algèbres d'Eilenberg–MacLane et homotopie'. *Séminaire H. Cartan*, École Norm. Super.
15. Chern, S. S., 1959. *Complex Manifolds*. Textos de Matemática, No. 5. Instituto de Fisica e Matemática, Universidade do Recife.

* In Russian.

16. Dold, A., 1980. *Lectures on Algebraic Topology*. Grundlehren, 200. Springer-Verlag: Berlin–New York.
*17. Finikov, S. P., 1952. *A Course in Differential Geometry*. Gostehizdat: Moscow.
*18. Fomenko, A. T., 1971. 'Bott periodicity from the point of view of the higher-dimensional Dirichlet functional'. *IAN SSSR, Ser. Matem.* **35**, pp. 667–681.
*19. Fomenko, A. T., 1972. 'The higher-dimensional Plateau problem in Riemannian manifolds'. *Matem. Sb.* **89**, pp. 475–520.
*20. Fomenko, A. T. and Mishchenko, A. S., 1971. *A Course in Differential Geometry and Topology*. MGU: Moscow.
*21. Fomenko, A. T., Fuks, D. B. and Gutenmakher, V. L., 1969. *Homotopic Topology*. MGU: Moscow.
*22. Golubyev, V. V., 1953. *Lectures on the Integration of the Equations of Motion of a Heavy Rigid Body about a Fixed Point*. Gostehizdat: Moscow.
23. Griffiths, P. A. and Harris, J., 1978. *Principles of Algebraic Geometry*. Interscience –Wiley: New York.
24. Gromoll, D., Klingenberg, W. and Meyer, W., 1968. *Riemannsche Geometrie im Grossen*. Lecture Notes in Mathematics, No. 55. Springer-Verlag: Berlin–New York.
25. Helgason, S., 1962. *Differential Geometry and Symmetric Spaces*. Academic Press: New York.
26. Hilbert, D. and Cohn-Vossen, S., 1952. *Geometry and the Imagination*. Chelsea: New York. (Translation from the German.)
27. Hilton, P. J. and Wylie, S., 1967. *Homology Theory*. Cambridge University Press: Cambridge, U.K.
28. Hirzebruch, F., 1966. *Topological Methods in Algebraic Geometry*. Grundelehren, 131. Springer-Verlag: Heidelberg.
29. Hu, Szen-Tsen, 1959. *Homotopy Theory*. Academic Press: New York.
30. Husemoller, D. H., 1966. *Fiber Bundles*. McGraw-Hill: New York.
31. Kervaire, M. A., 1960. 'A manifold which does not admit any differentiable structure'. *Comment. Math. Helv.* **34**, pp. 257–270.
32. Kervaire, M. A. and Milnor, J., 1963. 'Groups of homotopy spheres, I'. *Ann. Math.* **77**, pp. 504–537.
33. Klingenberg, W., 1987. *Lectures on Closed Geodesics*. Grundelehren, 230. Springer-Verlag: Berlin–New York.
34. Kobayashi, S. and Nomizu, K., 1963/1969. *Foundations of Differential Geometry*, Vols. I and II. Interscience: New York.
*35. 'Lectures on homotopic topology given in the mathematics seminar'. *UMN* **21**, pp. 117–248 (1972).
36. Lefschetz, S., 1942. *Algebraic Topology*. Amer. Math. Soc. Colloquium Publications, Vol. 27.
37. Lyusternik, L. A. and Shnirelman, L. G., 1930. 'Topological methods in variational problems'. *Proceedings of the Scientific Research Institute of Mathematics and Physics*: Moscow. (French translation: 'Méthodes topologiques dans les problèmes variationnels'. *Actualités Sci. Ind.* Hermann: Paris, 1934.)
*38. Lyusternik, L. A. and Shnirelman, L. G., 1935. 'An application of topology to extremal problems'. *Proceedings of the All-Union Mathematical Congress*, Vol. 1, pp. 224–237.
39. Milnor, J. W., 1956. 'On manifolds homeomorphic to the 7-sphere'. *Ann. of Math.* (2), **64**, pp. 399–405.
40. Milnor, J. W., 1961. 'Two complexes which are homeomorphic but combinatorially distinct'. *Ann. Math.* **74**, pp. 575–590.
41. Milnor, J. W., 1962. 'A survey of cobordism theory'. *Enseign. Math.* **8**, pp. 16–23.
42. Milnor, J. W., 1958. *Differential Topology*. Lecture Notes, Princeton University Press: Princeton, N.J.

43. Milnor, J. W., 1966. 'Whitehead torsion'. *Bull. Amer. Math. Soc.* **72**, pp. 358–426.
44. Milnor, J. W., 1969. *Morse Theory*. Ann. Math. Studies, No. 51. Princeton University Press: Princeton, N.J.
45. Milnor, J. W., 1968. *Singular Points of Complex Hypersurfaces*. Ann. Math. Studies, No. 61. Princeton University Press: Princeton, N.J.
46. Milnor, J. W., 1965. *Lectures on The h-Cobordism Theorem*. Mathematical Notes. Princeton University Press: Princeton, N.J.
47. Milnor, J. W. and Stasheff, J. D., 1974. *Characteristic Classes*. Ann. Math. Studies, No. 76. Princeton University Press: Princeton, N.J.
*48. Mishchenko, A. S., 1976. 'Hermitian *K*-theory, theory of characteristic classes, methods of functional analysis'. *UMN* **31**, pp. 69–134.
49. Morse, M., 1934. *The Calculus of Variations in the Lage*. Amer. Math. Soc. Colloquium Publications, Vol. 18.
50. Mosher, R. E. and Tangora, M. C., 1968. *Cohomology Operations and Applications in Homotopy Theory*. Harper's Series in Modern Mathematics, Harper and Row: New York.
51. Nomizu, K., 1956. *Lie Groups and Differential Geometry*. Math. Soc. Japan.
*52. Norden, A. P., 1956. *The Theory of Surfaces*. Gostehizdat: Moscow.
*53. Novikov, S. P., Mishchenko, A. S., Solov'ev, Ju. P. and Fomenko, A. T., 1978. *Problems in Geometry*. Nauka: Moscow.
*54. Novikov, S. P. (Editor), 1979. *The Theory of Solitons*. Nauka: Moscow.
55. Novikov, S. P., 1964. 'Homotopically equivalent smooth manifolds I'. *IAN SSSR, Ser. Matem.* **28**, pp. 365–474. (English translation: *Amer. Math. Soc. Translations* **48** (1965), pp. 271–396.)
56. Novikov, S. P., 1966. 'On manifolds with free abelian fundamental group and their applications'. *IAN SSSR, Ser. Matem.* **30**, pp. 207–246. (English translation: *Amer. Math. Soc. Translations* **71** (1968), pp. 1–42.)
*57. Novikov, S. P., 1965. 'New ideas in algebraic topology'. *UMN* **20**, pp. 41–66.
58. Novikov, S. P., 1970. 'Pontrjagin classes, the fundamental group and some problems of stable algebra'. In: *Essays on Topology and Related Topics. Mémoires Dediés à Georges de Rham*. Springer-Verlag: Berlin–Heidelberg–New York. (Alternatively, see: *Amer. Math. Soc. Translations* **70** (1968), pp. 172–179.)
59. Pogorelov, A. V., 1967. *Differential Geometry*. P. Noordhoff: Groningen. (Translation from the first Russian edition.)
60. Pogorelov, A. V., 1973. *Extrinsic Geometry of Convex Surfaces*. Translations of Math. Monographs, Vol. 35. Amer. Math. Soc.: Providence, R.I.
61. Pontrjagin, L. S., 1959. *Smooth Manifolds and Their Applications to Homotopy Theory*. Amer. Math. Soc. Translations, Series 2, Vol. 11, pp. 1–114. A.M.S.: Providence, R.I.
62. Pontrjagin, L. S., 1966. *Topological Groups*. Gordon and Breach: New York–London–Paris. (Translation of the second Russian edition.)
*63. Rashevskiĭ, P. K., 1956. *A Course in Differential Geometry*. Nauka: Moscow.
*64. Rashevskiĭ, P. K., 1967. *Riemannian Geometry and Tensor Analysis*. Nauka: Moscow.
65. Rohlin, V. A. and Fuks, D. B., 1984. *Beginner's Course in Topology. Geometric Chapters*. Springer-Verlag: Berlin–New York. (Translation from the Russian.)
66. Rohlin, V. A., 1959. 'Intrinsic homology theory'. *UMN* **14**, pp. 3–20. (English translation: *Amer. Math. Soc. Translations* **30** (1963), pp. 255–271.)
*67. Rohlin, V. A., 1951. 'A 3-dimensional manifold is the boundary of a 4-dimensional one'. *DAN SSSR* **81**, pp. 355–377.
*68. Rozendorn, E. R., 1971. *Problems in Differential Geometry*. Nauka: Moscow.
69. Seifert, H. and Threlfall, W., 1980. *A Textbook of Topology*. Academic Press: New York. (Translation from the German.)

70. Seifert, H. and Threlfall, W., 1932. *Variationsrechnung im Grossen*. Hamburger Math. Einzelschr., No. 24. Teubner: Leipzig. (Reprinted by Chelsea: New York, 1951.)
71. Serre, J.-P., 1964. *Lie Algebras and Lie Groups*. Lectures given at Harvard University. W. A. Benjamin: New York.
72. Serre, J.-P., 1953. 'Cohomologie modulo 2 des complexes d'Eilenberg–MacLane'. *Comment. Math. Helv.* **27**, pp. 198–231.
73. Serre, J.-P., 1951. 'Homologie singulière des espaces fibrés. Applications'. *Ann. Math.* **54**, pp. 425–505.
74. Serre, J.-P., 1953. 'Groupes d'homotopie et classes de groupes abéliens'. *Ann. Math.* **58**, pp. 258–294.
75. Smale, S., 1962. 'On the structure of manifolds'. *Amer. J. Math.* **84**, pp. 387–399.
76. Smale, S., 1970. 'Topology and mechanics I, II'. *Inventiones Mathematicae* **10**, pp. 305–331; **11**, pp. 45–64.
77. Spanier, E. H., 1966. *Algebraic Topology*. McGraw-Hill: New York.
78. Springer, G., 1957. *Introduction to Riemann Surfaces*. Addison-Wesley: Reading, Mass.
79. Steenrod, N. E., 1951. *The Topology of Fibre Bundles*. Princeton Math. Series, Vol. 14. Princeton University Press: Princeton, N.J.
80. Steenrod, N. E. and Epstein, D. B. A., 1962. *Cohomology Operations*. Ann. Math. Studies No. 50, Princeton University Press: Princeton, N.J.
81. Stong, R. E., 1969. *Notes on Cobordism Theory*. Mathematical notes. Princeton University Press: Princeton, N.J.
82. Toda, H., 1962. *Composition Methods in Homotopy Groups of Spheres*. Princeton University Press: Princeton, N.J.
83. Weil, A., 1958. *Introduction à l'étude des variétés kähleriennes*. Actualités Sci. Ind. Hermann: Paris.
*84. Yefimov, N. V., 1971. *Higher Geometry*. Nauka: Moscow.

APPENDIX 1

An Analogue of Morse Theory for Many-Valued Functions. Certain Properties of Poisson Brackets

by S. P. NOVIKOV

Let M be a smooth, closed manifold of finite or infinite dimension (for instance, a path-space consisting of paths satisfying certain conditions, joining points x_0 and x_1 of some smooth manifold W^m, or the space of directed closed curves, i.e. maps of the oriented circle S^1 into W^m; cf. §§21, 23). Given a closed 1-form ω on M, it is easy to see that there exists a (possibly infinite-sheeted) covering space $\hat{M}$ of M (with projection $p: \hat{M} \to M$) such that the pullback $p^*\omega$ of the form ω is exact, i.e. is the differential of a function S on $\hat{M}$ (see Part II, §29.2):

$$p^*\omega = dS. \qquad (1)$$

(The simplest non-trivial example is that where $M = \mathbb{R}^2 \setminus \{0\}$, $\omega = d\varphi$ where φ is the angular co-ordinate of $z \in M$, and $\hat{M}$ is the Riemann surface of the logarithm function $\ln z$.) We call such a function S a *many-valued function* on the manifold M.

It will be assumed throughout that in the infinite-dimensional case the second differential d^2S of S has at the critical (or "stationary") points (i.e. those where $dS = 0$ or $\omega = 0$) finite degeneracy degree (see Theorem 21.5), and finite "Morse index", i.e. finite largest dimension of subspaces of the tangent space on which d^2S is negative definite (cf. Theorem 21.7). (In fact, in the particular situations we shall be considering here, the critical points will either all be non-degenerate, or will together form non-degenerate critical submanifolds; see §20.) We shall further assume that the function S has a well-defined "gradient descent", in the sense that on the covering manifold $\hat{M}$ each compact subset, when translated in $\hat{M}$ in the direction of "steepest descent" for S (i.e. that of the negative of the gradient of S), either encounters a critical point, or else passes successively through all level surfaces S_c ($c <$ some fixed number) in order of decreasing c.

The problem of interest for us is that of constructing an analogue of Morse theory for many-valued functions S (or, equivalently, closed 1-forms ω where $p^*\omega = dS$) yielding an estimate from below of the number of critical points of S of each Morse index i. We denote the number of critical points of index i by $\mu_i(S)$ (or $\mu_i(\omega)$).

It is easy to see that for the group $H_1(M; \mathbb{Z})$ (finitely generated in view of the compactness of M) one may choose a basis $\{\gamma_1, \dots, \gamma_k, \gamma_{k+1}, \dots, \gamma_N\}$ with the property that

$$\int_{\gamma_j} \omega = \begin{cases} 0 & \text{for } j \geq k+1, \\ \varkappa_j \neq 0 & \text{for } j \leq k, \end{cases} \tag{2}$$

where the numbers $\varkappa_j, j = 1, \dots, k$, are linearly independent over the rationals (or, equivalently, over $\mathbb{Z}$). The integer $k - 1$, clearly an invariant of ω, is called the *irrationality degree* of ω. The monodromy group of a minimal covering $\hat{M} \to M$ such that, as above, ω pulls back to the differential of a single-valued function S on $\hat{M}$, is isomorphic to $\mathbb{Z}^k$, the free abelian group of rank k, with k free generators $t_1, \dots, t_k$ acting as "translations" on $\hat{M}$ (see Part II, §29.2):

$$t_j\colon \hat{M} \to \hat{M}, \qquad j = 1, \dots, k.$$

Note that the *irrationality index* $\varkappa = (\varkappa_1, \varkappa_2, \dots, \varkappa_k)$ may be regarded as a point of $(k-1)$-dimensional projective space:

$$\varkappa = (\varkappa_1 : \varkappa_2 : \dots : \varkappa_k) \in \mathbb{R}P^{k-1},$$

since the k-tuple $(\varkappa_1, \dots, \varkappa_k)$ is determined only up to multiplication by an arbitrary non-zero constant (provided we do not distinguish between ω and $c\omega$, c constant).

The case $k = 1$ is particularly simple and interesting. Here the form ω (or more precisely some constant multiple of it) can be regarded as determining, via (2) with $k = 1$, an element of the integral cohomology group:

$$[\omega] \in H^1(M; \mathbb{Z}),$$

so that then $\exp(2\pi i S)$ will determine a single-valued function down on M taking complex values of modulus 1, and therefore yielding a map

$$f = \exp(2\pi i S)\colon M \to S^1. \tag{3}$$

Although the problem of constructing an analogue of Morse theory for the critical points of such maps is certainly of classical type, it has until quite recently (1981) never received attention in the literature.

We shall consider to begin with some infinite-dimensional examples of "many-valued functionals" leading naturally to the sort of situation described above. Let W^m be a Riemannian manifold which is complete with respect to its metric $g_{ij}(x)$, and let Ω be a closed 2-form defined on it: $d\Omega = 0$. Let $\{U_\alpha\}$ be an open cover of W^m ($W^m = \bigcup_\alpha U_\alpha$) by regions U_α with the following two properties:

(i) the form Ω is exact on each U_α:

$$\Omega|_{U_\alpha} = d\psi_\alpha \qquad \text{for some 1-form } \psi_\alpha \text{ on } U_\alpha; \tag{4}$$

(ii) corresponding to each smooth map γ of the unit interval (or the circle S^1) into W^m, there should exist a region U_α such that (the image under) γ is wholly contained in U_α.

Let M denote the infinite-dimensional manifold $\Omega(W^m; x_0, x_1)$, whose points are the smooth paths from x_0 to x_1 (or the manifold $\Omega^+(W^m)$ of smooth, closed, directed curves in W^m), and for each α, let N_α denote the subspace of all points γ of M (i.e. paths, or curves) contained in U_α. Each intersection $N_\alpha \cap N_\beta$ can then be represented in the form

$$N_\alpha \cap N_\beta = \bigcup_q N_{\alpha\beta}^{(q)} \quad (q = 1, 2, \ldots),$$

where $N_{\alpha\beta}^{(q)}$ consists of those closed curves in the qth homology class in $H_1(U_\alpha \cap U_\beta; \mathbb{R})$ relative to some enumeration of those classes (and analogously for the homology classes of paths from x_0 to x_1). Consider on each subspace N_α the (single-valued) functional

$$S^{(\alpha)}\{\gamma\} = \int_\gamma dl - \psi_\alpha, \tag{5}$$

where ψ_α is as in (4).

Lemma 1. *For each $q = 1, 2, \ldots$, the difference of functionals $S^{(\alpha)}\{\gamma\} - S^{(\beta)}\{\gamma\}$ is constant on the space $N_{\alpha\beta}^{(q)}$.*

Proof. From (5) we have

$$S^{(\alpha)}\{\gamma\} - S^{(\beta)}\{\gamma\} = \int_\gamma (\psi_\beta - \psi_\alpha). \tag{6}$$

In view of (4) we have $d\psi_\alpha = d\psi_\beta$ on $U_\alpha \cap U_\beta$, so that $\psi_\beta - \psi_\alpha$ is a closed 1-form. Hence the integral in (6) is constant on the homology class of the path $\gamma \in N_\alpha \cap N_\beta$. □

From this it follows that the family of functionals $S^{(\alpha)}$ defines a "many-valued function" S on M, whose "variational differential" δS (the change in S determined by an infinitesimal perturbation $\delta\gamma$; cf. Part I, §37.1) may be considered as a globally defined 1-form on the infinite-dimensional manifold M.

(This situation generalizes naturally as follows: Suppose we are given a sufficiently well-behaved, single-valued functional $S_0\{\gamma\}$ defined on the set of smooth maps $\gamma\colon V^l \to W^m$ between two complete Riemannian manifolds, together with a closed $(l+1)$-form Ω on W^m ($d\Omega = 0$); and a covering $\{U_\alpha\}$ of W^m by regions U_α such that, much as before:

(i) $\Omega|_{U_\alpha} = d\psi_\alpha$ for some l-form ψ_α on U_α;
(ii) for each $\gamma\colon V^l \to W^m$, there exists an index α such that the image under γ is wholly contained in U_α.

Then, analogously to the previous special case, there is defined on the infinite-dimensional manifold M of all smooth maps $V^l \to W^m$ (i.e. of general "chiral fields"; cf. Part II, §32.3) the "many-valued functional" (see §5 of [4])

$$S\{\gamma\} = S_0\{\gamma\} + \int_\gamma \psi_\alpha .)$$

Returning to the case $l = 1$, we remind the reader of our assumption that the Riemannian metric g_{ij} on the manifold W^m, and the given 2-form Ω are such that the Morse index of every stationary point of the functional S is finite, and the "gradient descent" is well defined. Such is the situation for functionals of so-called "Maupertuis–Fermat" type arising in connexion with the investigation of the possible paths of motion of a charged particle in a potential field $u(x)$ and a magnetic field Ω defined on a Riemannian manifold W^m (where here $m = 2$ or 3), under the condition of constant energy E: the paths of motion are those extremizing the functional (cf. Part I, Example 33.2.2(c) and §33.3)

$$\tilde{S}\{\gamma\} = \int_\gamma (d\tilde{l}_E - A_j\,dx^j), \tag{7}$$

where

$$(d\tilde{l}_E)^2 = 2m(E - u(x))g_{ij}\,dx^i\,dx^j, \qquad d(A_j\,dx^j) = \Omega. \tag{8}$$

Note that the 2-form Ω representing the magnetic field is here considered exact (as in (8)). If Ω is allowed to be non-exact, then the situation of present interest arises, namely that of many-valued functionals. Note that the condition (always assumed) that the metric $(d\tilde{l}_E)^2$ be complete is, provided W^m is compact, equivalent to the condition

$$E > \max_{W^m} u(x). \tag{9}$$

For non-simply-connected manifolds W^m (for instance, the m-dimensional torus T^m) it may happen that, independently of the exactness or otherwise of Ω and the details of the above constructon, the 1-form δS turns out *a posteriori* to be exact simply because the path-space M is simply-connected: for exactness of the 1-form δS (and consequent single-valuedness of the functional S) it suffices that the form Ω become exact when pulled back to the universal cover $\hat{W}^m$:

$$q^*\Omega = d\psi, \qquad \text{where} \quad q\colon \hat{W}^m \to W^m$$

is the universal covering map. This occurs precisely if the cohomology class $[\Omega] \in H^2(W^m; \mathbb{R})$ of the form Ω lies in the subgroup determined by the fundamental group:

$$[\Omega] \in H^2(\pi_1; \mathbb{R}) \subset H^2(W^m; \mathbb{R}). \tag{10}$$

(Recall that, by definition $H^2(\pi_1; \mathbb{R}) = H^2(K(\pi_1, 1); \mathbb{R})$; the embedding in (10) then arises from the natural map $W^m \to K(\pi_1, 1)$, constructed essentially as in the proof of Theorem 10.4.)

EXERCISE

1. Find sufficient conditions (on the fundamental group π_1 and the cohomology class $[\Omega] \in H^2(\pi_1; \mathbb{R})$) for the functional S on the space of closed curves in W^m to take on arbitrarily large negative values.

For simply-connected manifolds W^m, on the other hand, this situation does not arise. Here the integrals of the 1-form $\omega = \delta S$ over cycles forming a basis for $H_1(M; \mathbb{Z})$ (cf. (2)) (and the concomitant irrationality degree of ω) are determined by (in fact coincide with) the integrals of the 2-form Ω over appropriate basic 2-cycles in $H_2(W^m; \mathbb{Z})$.

For certain important systems of classical mechanics, the problem of finding the possible trajectories can be recast as extremal problems for functionals of the form (7) (Novikov–Shmeltser):

(i) Kirchhoff's problem concerning the motion of a rigid body in an ideal fluid under the assumption that the flow is determined by a potential fading to zero at infinity;

(ii) the problem of the motion of a rigid body about a fixed point in an axially symmetric (in particular constant) gravitational field (for instance, a spinning top or gyroscope).

Each of these problems can be described mathematically by a system of equations transformable into a Hamiltonian system on the Lie algebra $L = E(3)$ of the group of motions of 3-dimensional Euclidean space, with phase space identified with the dual space L^* (see Part I, §§33, 34 and Part II, §28). In terms of any basis $\{e_i^*\}$ for L^* each element l^* of L^* can be expressed uniquely in the form

$$l^* = \sum l_i e_i^*, \qquad l_i \in \mathbb{R}.$$

We can identify each coefficient l_i with the linear form on L^* whose value at e_j^* is $l_i \delta_i^j$, and hence with an element of $L \simeq (L^*)^*$. Now the *Poisson bracket* of pairs of arbitrary (real-valued) functions on L^*, may be defined by the following conditions (cf. Part I, §34.2 and Part II, §28.2):

(i) The Poisson bracket of two linear functionals l_i, l_j on L^* (i.e. in essence of two elements of L) should coincide with their commutator $[l_i, l_j] = c_{ij}^k l_k$ in L:

$$\{l_i, l_j\} = c_{ij}^k l_k. \tag{11}$$

(ii) The Poisson bracket of any two functionals on L^* (not necessarily linear) should satisfy the following further "bracket" conditions: bilinearity, skew-symmetry, Jacobi's identity, and, finally, Leibniz' product formula, namely

$$\{fg, h\} = \{f, h\}g + \{g, h\}f. \tag{12}$$

(Recall from Part II, §28.2 that, more generally, a Poisson bracket of functions on an arbitrary manifold N^q is defined by a formula of the form (in terms of local co-ordinates on N^q)

$$\{f, g\} = h^{ij}(x)\frac{\partial f}{\partial x^i}\frac{\partial g}{\partial x^j}, \tag{13}$$

where the $h^{ij}(x) = -h^{ij}(x)$ are the components of a skew-symmetric tensor whose inverse (assuming $\det(h^{ij}) \neq 0$) is therefore a 2-form $h = h_{ij}\, dx^i \wedge dx^j$ on N^q ($h_{ij}h^{jk} = \delta_i^k$). The requirement that the bracket (13) satisfy Jacobi's identity then comes down to the closure of this 2-form: $dh \equiv 0$. The simplest case, $h^{ij} = \text{const.}$, is that arising in the "classical" Hamiltonian formalism deriving from the calculus of variations (see Part I, §§33, 34). The next case, namely where the h^{ij} are linear functions of the points $x = (x^1, \ldots, x^q)$, has been intensively investigated in the literature over the last 20 years, essentially for the reason that here $h^{ij}(x) = c_k^{ij}x^k$, where, not surprisingly in view of the defining conditions of a Poisson bracket, the c_k^{ij} are the structural constants of some Lie algebra. The next case up in complexity, namely that of "quadratic" brackets, where $h^{ij} = c_{kl}^{ij}x^k x^l$, also turns out to be of considerable interest, and has been the subject of recent investigations (by Skljanin and Faddeev).)

Returning to the above two problems, and the linear case of a bracket defined (essentially) on pairs of elements of a Lie algebra, more specifically the Lie algebra $L = E(3)$, we now introduce notation for the standard basis for this Lie algebra, namely

$$\{M_1, M_2, M_3, p_1, p_2, p_3\},$$

where the M_i are the approprite "infinitesimal" rotations, and the p_i the "infinitesimal" translations, of Euclidean $\mathbb{R}^3$. According to the above definition (see in particular (11)), the Poisson bracket on $L = E(3)$ is given by commutation in $E(3)$:

$$\begin{aligned} \{M_i, M_j\} &= \sum_k \varepsilon_{ijk} M_k, \qquad \varepsilon_{ijk} = \operatorname{sgn}\begin{bmatrix} 1 & 2 & 3 \\ i & j & k \end{bmatrix}, \\ \{M_i, p_j\} &= \sum_k \varepsilon_{ijk} p_k, \\ \{p_i, p_j\} &= 0. \end{aligned} \tag{14}$$

In the Kirchhoff problem the Hamiltonian $H(M_i, p_i)$, which is just the total energy of the system comprised of the rigid body and the fluid in which it moves, turns out to be a positive definite quadratic form in the variables M_i, p_i, identified with the appropriate linear functionals on L^*:

$$H = \sum a_{ij} M_i M_j + \sum b_{ij} M_i p_j + \sum c_{ij} p_i p_j. \tag{15}$$

(Note that if the rigid body is not simply-connected, then there may occur in addition linear terms.)

In the problem of the motion of a rigid body (such as a top or gyroscope) about a fixed point in an axially symmetric gravitational field with potential

$U(z)$, the Hamiltonian (again on L^*) turns out to be of the form

$$H = \sum a_{ij} M_i M_j + U(d^i p_i), \tag{16}$$

where the d^i are constants determined by the positions relative to one another of the centre of mass and the fixed point. (Here the quadratic form $\Sigma a_{ij} M_i M_j$ is always assumed positive definite.) This system differs further from that arising in the Kirchhoff problem in that, in distinction to the system defined by (15), there are restrictions on the Hamiltonian (16) (in the form of inequalities).

For both problems the equations of motion then have the usual form (see Part I, §34.2)

$$\dot{M}_i = \{H, M_i\}, \qquad \dot{p}_i = \{H, p_i\}. \tag{17}$$

The other "integrals of motion" $f_l(M, p)$ (i.e. conserved expressions in the M_i and p_i), apart from the energy $H = E$, are, in general, those satisfying

$$\{f_l, M_i\} \equiv \{f_l, p_i\} \equiv 0, \tag{18}$$

for $i = 1, 2, 3$, i.e. the elements of the "annihilator" of the Poisson bracket. These expressions, which as it turns out lie in the centre of the "enveloping algebra" of the Lie algebra, reduce essentially to the following two ("Kirchhoff's integrals of motion"):

$$f_1 = \sum p_i^2, \qquad f_2 = \sum M_i p_i. \tag{19}$$

EXERCISE

2. By means of elementary computations, verify that these quantities are indeed conserved along trajectories.

We write p^2 and ps respectively, for the (constant) values of f_1 and f_2 along an (arbitrary) trajectory. (In the spinning-top problem one may assume that $f_1 \equiv 1$ on all trajectories, and in this case $f_2 = s$ is referred to as the "area constant" of the trajectory.) On each joint level surface defined by $f_1 =$ const. $= p^2$, $f_2 =$ const. $= ps$ (a 4-dimensional manifold) the Poisson bracket, given by (14), satisfies $\det(h^{ij}) \neq 0$ provided $p \neq 0$, and therefore defines a "symplectic" 2-form (see Part II, §28.2) given by

$$h = h_{ij}\, dx^i \wedge dx^j, \qquad h_{ij} h^{jk} = \delta_i^k,$$

satisfying $dh = 0$, depending on the particular values of p and s. One has the following important result.

Lemma 2. *On the joint level surface given by $f_1 = p^2 \neq 0$, $f_2 = ps$, consider the new co-ordinates*

$$y^1 = \theta, \qquad y^2 = \varphi, \qquad \xi_1 = p_\theta, \qquad \xi_2 = p_\varphi, \qquad 0 \leq \varphi < 2\pi, \quad -\frac{\pi}{2} \leq \theta \leq \frac{\pi}{2}, \tag{20}$$

defined in terms of the p_i *and* M_i *by*

$$p \sin\theta = p_3, \qquad M_3 - \frac{s}{p} p_3 = -p_\varphi,$$

$$p \cos\theta \cos\varphi = p_2, \qquad M_2 - \frac{s}{p} p_2 = p_\varphi \tan\theta \sin\varphi + p_\theta \cos\varphi,$$

$$p \cos\theta \sin\varphi = p_2, \qquad M_1 - \frac{s}{p} p_1 = p_\varphi \tan\theta \cos\varphi - p_\theta \sin\varphi.$$

In terms of these co-ordinates the Poisson bracket on the joint level surface takes the form

$$\{y^a, y^b\} = 0, \qquad \{y^a, \xi_b\} = \delta^a_b, \qquad \{\xi_1, \xi_2\} = s\cos\theta, \tag{21}$$

and the corresponding symplectic 2-form h is given by

$$h = \sum_{a=1}^{2} dy^a \wedge d\xi_a + s\cos\theta\, d\theta \wedge d\varphi = h_0 + \Omega,$$

where Ω ($= s\cos\theta\, d\theta \wedge d\varphi$) *is a closed 2-form on the 2-sphere* S^2 *(with spherical co-ordinates* φ, θ*).*

The level surface $f_1 = p^2 \neq 0$, $f_2 = ps$ *is diffeomorphic to* $T^*(S^2)$, *the co-tangent bundle over* S^2. *The integral of the form h (or, equivalently, of* Ω*) over a generating cycle* $[S^2] \in H_2(T^*(S^2)) \simeq \mathbb{Z}$ *has the following value:*

$$\iint_{S^2} h = \iint_{[S^2]} \Omega = 4\pi s = 4\pi f_2 f_1^{-1/2}. \tag{22}$$

The proof of this lemma involves essentially just direct calculation. (In particular, the fact that the level surface $f_1 = p^2$, $f_2 = ps$ is diffeomorphic to $T^*(S^2)$ is almost immediate from the forms of the two integrals f_1 and f_2.)

A Poisson bracket on $T^*(M^n)$ given by a 2-form $h = h_0 + \Omega$ where Ω is a closed 2-form on the base M^n (as in the above lemma, where $M^n = S^2$) may be regarded as arising from a system (with bracket given by h_0) after inclusion of a (formal) magnetic field Ω (see Part I, §33.2, Example (c)). Hence the trajectories in the Kirchhoff and spinning-top problems may be obtained using the "Maupertuis–Fermat principle", i.e. by using a functional of the form (7), which now however may be many-valued since Ω need no longer be exact. In view of (22) this will be the case if $s \neq 0$ (or equivalently if $f_2 \neq 0$). (For the classical gyroscope the "area constant" is in fact non-zero.) The Hamiltonian H on a level surface $f_1 = p^2 \neq 0$, $f_2 = ps$, has, in terms of the new co-ordinates (20), the following form:

$$H = \tfrac{1}{2} g^{ab}(y)\xi_a \xi_b + A^a(y)\xi_a + V(y)$$

(with the Poisson bracket as in (21)). This Hamiltonian system, restricted to any region $U_\alpha = S^2 \setminus \{P_1, P_2\}$, the sphere with two antipodal points P_1, P_2 removed, is equivalent to the Lagrangian system given by the following functional:

$$S^{(\alpha)}\{\gamma\} = \int_\gamma (\tfrac{1}{2}g_{ab}\dot{y}^a\dot{y}^b - U(y) - A_a\dot{y}^a - s(\sin\theta)\dot{\varphi})\,dt, \tag{23}$$

where as usual

$$g_{ab}g^{ac} = \delta_b^c, \qquad A_a g^{ac} = A^c, \qquad y^1 = \theta, \qquad y^2 = \varphi, \qquad U = V - \tfrac{1}{2}g_{ab}A^aA^b.$$

Here the action has the same form as the mechanical action of a charged particle moving on the sphere S^2 with metric g_{ab}, on which there exists a potential field $U(x)$ and a magnetic field $\Omega_{12} = \partial_1 A_2 - \partial_2 A_1$ of a non-trivial "monopole" (since if $s \neq 0$ the formal "magnetic field" is topologically non-trivial in the sense of being non-exact). The covering of S^2 by the regions U_α (where the index α runs over the set of all pairs $\{P_1, P_2\}$ of antipodal points of S^2) satisfies the conditions given earlier (see (4) *et seqq.*), allowing the construction of a many-valued functional S (assuming $s \neq 0$) determined by the action appearing in (23) and depending on p and s.

At each fixed energy level E the trajectories of this system may be obtained using the appropriate Maupertuis–Fermat functional $\tilde{S}$ (see (7)), also many-valued, with associated closed, infinite-dimensional 1-form $\delta\tilde{S}$:

$$\begin{aligned} \tilde{S}^{(\alpha)} &= \int_\gamma (d\tilde{l}_E - A_a^{(\alpha)}\,dy^a), \\ d\tilde{l}_E &= \sqrt{2(E-U)g_{ab}\dot{y}^a\dot{y}^b}. \end{aligned} \tag{24}$$

As noted earlier, provided $E > \max_{S^2} U(y)$, the metric $\tilde{l}_E$ is complete.

We single out an important property of single-valued as well as multiple-valued functionals of Maupertuis–Fermat type (7) (in the case of the sphere S^2 by way of exemplification): In the space $M = \Omega^+(S^2)$ of closed, directed curves on S^2, the totality of one-point curves is a non-degenerate critical submanifold of local minima (see §20); it turns out that one can normalize the functional $\tilde{S}$ on the infinite-sheeted covering space $\hat{M}$ of M (on which $\tilde{S}$ is single-valued) so that on one sheet (call it the zeroth S_0^2) of the complete inverse image $p^{-1}(S^2) = \bigcup_n S_n^2$ of the critical submanifold ($\cong S^2$) of one-point curves, the functional $\tilde{S}$ takes on the value zero and, more generally, the value $4\pi ns$ on S_n^2:

$$\begin{aligned} \tilde{S}(\gamma) &= 0 \qquad \text{for all} \quad \gamma \in S_0^2, \\ \tilde{S}(\gamma) &= n\iint_{S^2} \Omega = 4\pi ns \qquad \text{for all} \quad \gamma \in S_n^2. \end{aligned} \tag{25}$$

One can exploit this property of the space M of directed closed curves on S^2 in the following way: Let $I[0, 1]$ be a path in $\hat{M}$ joining the two components S_0^2 and S_1^2 of $p^{-1}(S^2)$; thus 0 denotes a point of $S_0^2 \subset \hat{M}$, and 1 a point of $S_1^2 \subset \hat{M}$. If one now deforms this path monotonically in the direction on $\hat{M}$ in which $\tilde{S}$ decreases most rapidly (as determined by the gradient of $\tilde{S}$), obtaining thereby a continuous family of paths I_τ, $\tau \geq 0$, with $I_0 = I$, one sees that (since

$\tilde{S}$ has local minima at the end-points of I)

(i) the end-points of the paths must remain fixed for all τ;
(ii) $\max_{\gamma \in I_\tau} \tilde{S}(\gamma) \geq 4\pi s$ for each τ.

These properties, together with the appropriate "maximum principle", imply the existence of a critical (saddle) point, having index 1 in the non-degenerate case. Since such critical points correspond to trajectories of the original system, one has the following

Theorem 1 (Novikov). *For all values of the parameters E, p, s satisfying condition* (9), *the systems arising in the Kirchhoff and spinning-top problems have integral trajectories which are periodic relative to a co-ordinate frame moving with the rigid body.*

Remarks

1. By means of perturbation theory several applied mathematicians had earlier obtained in more explicit form families of such periodic solutions for values of the parameters close to those of the integrable cases. However, the possibility of obtaining such solutions corresponding to parameter values far from those allowing integrability had not previously been established.

2. It follows from the "Maupertuis–Fermat principle" (see (7)) that in the case of zero area constant, $s = 0$, one obtains in the spinning-top problem a single-valued functional on S^2, arising from the metric alone. This result was obtained earlier (by Kozlov and Kharlamov) using different methods: in the case $E > \max U(x)$, known results of Lyusternik and Shnirelman may be exploited; the case $E \leq \max U(x)$ is due to Kozlov.

We now turn to the purely topological problem of constructing a Morse theory for closed 1-forms ω on smooth, closed, finite-dimensional manifolds $M = M^n$. As noted earlier (see (3)), in the simplest case, when ω represents an integral cohomology class: $[\omega] \in H^1(M^n; \mathbb{Z})$, we obtain a map f from M^n to the circle

$$f = \exp(2\pi i S)\colon M^n \to S^1, \qquad S = \hat{f}\colon \hat{M} \to \mathbb{R}.$$

We now further examine this situation. In particular, if S has no critical points then the map f is the projection map of a smooth fibre bundle (M^n) with base $B = S^1$. In general (in the situation we are considering), a cyclic $\mathbb{Z}$-covering $\hat{M} \xrightarrow{p} M^n$ may be constructed in the following way: Let N^{n-1} be a submanifold of M^n of dimension $n-1$ representing the cycle $D[\omega] \in H_{n-1}(M^n; \mathbb{Z})$, where D denotes, as usual, the Poincaré-duality operator. (That $D[\omega]$ can be so realized is guaranteed by Corollary 27.11(ii).) Proceeding as in the construction given just prior to Lemma 27.17, we cut the manifold M^n along its submanifold N^{n-1}, obtaining thereby a film W^n with boundary the union of two disjoint parts each diffeomorphic to N^{n-1}: $\partial W = N_0^{n-1} \cup N_1^{n-1}$. Taking copies $W_i \cong W$ of this film, one for each integer i, with boundary $\partial W_i =$

$N_{i,0} \cup N_{i,1}$ (where $N_{i,0}$, $N_{i,1}$ are each diffeomorphic to N^{n-1}), we glue each W_{i-1} to W_i along one of these bounding manifolds to obtain $\hat{M}$:

$$\hat{M} = \bigcup_{-\infty<i<\infty} W_i, \qquad N_{i-1,1} = N_{i,0}, \qquad -\infty < i < \infty.$$

We may in fact suppose that N^{n-1} is a "level surface" of the function S (as a many-valued function on M) or, equivalently, the complete inverse image in M of a point of S^1 under the map $f = \exp(2\pi i S)$. A generator $t\colon \hat{M} \to \hat{M}$ of the monodromy group ($\simeq \mathbb{Z}$) acts on $\hat{M}$ as follows:

$$t\colon W_i \to W_{i+1}, \qquad N_{i,0} \to N_{i,1} = N_{i+1,0}. \tag{26}$$

On the basis of the general argument chracteristic of Morse theory, one might now expect that the functions S on $\hat{M}$ should determine a cellular decomposition (essentially) of $\hat{M}$. However, in the present situation, an important condition fails to hold, on which the usual Morse theory crucially depended (in both its finite- and infinite-dimensional versions), namely the relative compactness of the regions of the form $\{x|S(x) \leq a\}$. On the other hand, one still has, in the present context of the function S on $\hat{M}$, a "surface of steepest descent" of dimension i issuing from each critical point of index i and descending through the various lower levels, which it is natural to regard as a "cell" (or at least if not this surface then some small perturbation of it). However, this "cell" may extend through all lower level surfaces defined by S to $-\infty$, and its algebraic boundary may then involve infinitely many similar such "cells" of dimension $i - 1$. These considerations, together with the fact that under the generating translation $t\colon \hat{M} \to \hat{M}$ the function S is transformed to itself with a constant added (so that the set of critical points is preserved), lead to the following conclusions:

(i) each critical point of S gives rise to a free generator in the complex obtained as indicated above;
(ii) the boundary of a "cell" of this complex may be an infinite linear combination of cells of the complex arising from "lower down" with respect to the values of S (and therefore extending to "infinity" only in that direction in $\hat{M}$ in which S decreases);
(iii) all of these "cells" are obtained from a finite number of "basic" ones via the action of the elements t^j of the monodromy group ($\simeq \mathbb{Z}$) on $\hat{M}$.

Denote by $K = \hat{\mathbb{Z}}[t, t^{-1}]$ the ring of formal Laurent series

$$\sum_{-\infty<\text{const.}<j} m_j t^j \tag{27}$$

with integer coefficients m_j, vanishing for j less than some constant (depending on the series). From the "cell" complex determined as above by S (as a single-valued function on the covering space $\hat{M}$, or, equivalently, a many-valued function on M), we construct in the usual way a chain complex C of K-modules (finitely generated in view of (iii)). The complex C will have the form

$$0 \to C_n \xrightarrow{\partial} C_{n-1} \to \cdots \to C_1 \xrightarrow{\partial} C_0 \to 0,$$

where the boundary operator ∂ is a K-module homomorphism. Note that, in distinction to conventional Morse theory, it is possible here that $C_0 = 0$, $C_n = 0$; in fact, it can be shown that M^n may be endowed with a closed 1-form representing any non-trivial cohomology class in $H^1(M^n; \mathbb{Z})$, with the property that the many-valued function it determines has no local maximum or minimum (i.e. $C_0 = 0$, $C_n = 0$). Note also that on a fibre bundle M^n over S^1 a closed 1-form ω can be defined, whose corresponding S has no critical points, so that $C_n = C_{n-1} = \cdots = C_0 = 0$.

Lemma 3. *The homology groups of the complex C of K-modules, determined as above by any smooth, closed 1-form ω on M^n, are homotopy invariants.*

This lemma (whose proof, though not difficult, we omit) allows us to obtain, in terms of certain numerical invariants of these homology groups, analogues of the Morse inequalities in the present situation of a many-valued function S determining a map from M^n to the circle:

$$\exp(2\pi i S): M^n \to S^1.$$

Since the ring K has cohomological dimension 1 (in view of the fact that if the coefficients are allowed to range over a field, e.g. $\mathbb{Q}$, then K itself becomes a field), it follows that submodules of free K-modules will again be free K-modules. Hence the submodules of "cycles": $Z_k = \operatorname{Ker} \partial \subset C_k$, and of "boundaries": $B_k \subset Z_k$, both have (finite) free bases, and therefore well-defined ranks; we call the difference in these ranks the kth *Betti number*, denoting it by $b_k(M^n, [\omega])$:

$$b_k(M^n, [\omega]) = \operatorname{rank} Z_k - \operatorname{rank} B_k.$$

The analogue $q_k(M^n, [\omega])$ of the kth "torsion number" (i.e. the number of "torsion coefficients" or "invariant factors" of the kth integral homology group of a space) is defined as follows: It can be shown that there exist free bases $\{e_1, \ldots, e_N\}$ for the module Z_k, and $\{e'_1, \ldots, e'_L\}$ for its submodule B_k (where $N - L = b_k$), such that

$$e'_j = \left(n_j + \sum_{k \geq 1} n_{jk} t^k\right) e_j + \sum_{i > L} q_{ij}(t) e_i,$$

where: (i) the integer n_j is divisible by n_{j+1}; (ii) the Laurent series $q_{ij}(t)$ do not involve negative powers of t; and (iii) for all i, j for which $q_{ij}(t) \not\equiv 0$, the integers $q_{ij}(0)$ are non-zero and divisible by n_j. We now define the *torsion number* $q_k(M^n, [\omega])$ to be the number of indices j for which $n_j \neq 1$. The number $q_k + b_k$ is then the least number of generators of the quotient module $H_k = Z_k/B_k$. The following result may be established without great difficulty on the basis of the foregoing discussion.

Theorem 2. *For each i the following analogue of the* ith *Morse inequality (see* §16(1)) *holds for the number* $\mu_i(S)$ (*or* $\mu_i(\omega)$) *of critical points of index i of the map* $\exp(2\pi i S)$ *of* M^n *to the circle (or the closed* 1-*form* ω, $[\omega] \in H^1(M^n; \mathbb{Z})$):

$$\mu_i(S) \geq b_i(M^n, [\omega]) + q_i(M^n, [\omega]) - q_{i-1}(M^n, [\omega]). \qquad (28)$$

Note that although these inequalities are indeed analogous to the "classical" Morse inequalities, the invariants figuring in them have a more complex geometric significance.

For manifolds M^n with $\pi_1(M^n) \simeq \mathbb{Z}$ it makes sense to ask whether the inequalities (28) are best possible (by analogy with a well-known result of Smale concerning single-valued functions on simply-connected manifolds). One may without great difficulty find a level surface $N^{n-1} \subset M^n$ which represents the Poincaré dual of the class $[\omega] \in H^1(M^n; \mathbb{Z})$, and is connected and simply-connected (at least in dimensions $n \geq 5$). By using a Smale function on the film W^n with two boundary components ($\partial W^n = N_0^{n-1} \cup N_1^{n-1}$), obtained by cutting M^n along its submanifold N^{n-1}, one may then in a canonical "minimal" manner construct a closed 1-form ω on M^n (and consequent function S on the covering space $\hat{M}$ of M^n) by extending its definition from the level surface N^{n-1} to the whole of M^n. However, the resulting form ω (or many-valued function on M^n) may be far from having the least number of critical points consistent with (28). In order to obtain in this way a form ω with the number of critical points of each index i equal to the right-hand side of (28), it is necessary to choose the initial submanifold $N^{n-1} \subset M^n$ to be "minimal" in a certain specific sense, if possible. The question of whether this could be achieved was settled affirmatively by Farber in 1984, thus showing that the inequalities (28) are indeed exact.

We conclude with a few remarks concerning the more complex situation where $k > 1$, i.e. where ω has two or more rationally independent integrals over one-dimensional cycles (see (2)):

$$\varkappa_i = \int_{\gamma_i} \omega, \qquad i = 1, \ldots, N; \quad \varkappa_i = 0 \text{ for } i \geq k+1;$$

$$\varkappa_1 \neq 0, \ldots, \varkappa_k \neq 0; \qquad \sum_{i=1}^{k} m_i \varkappa_i \neq 0 \qquad \text{if the } m_i \in \mathbb{Z} \text{ are not all zero,}$$

where $\{\gamma_1, \ldots, \gamma_k, \gamma_{k+1}, \ldots, \gamma_N\}$ is a suitable basis for $H_1(M^n; \mathbb{Z})$. As noted earlier there is a covering map $p: \hat{M} \to M^n$ such that $p^*\omega = dS$, with free abelian monodromy group. We introduce the ring $K_\varkappa$ consisting of all series in k variables of the form

$$b = \sum_{m=(m_1, \ldots, m_k)} b_m t_1^{m_1} \ldots t_k^{m_k}, \qquad b_m \in \mathbb{Z},$$

such that:

(i) $b_m = 0$ whenever $\sum m_i \varkappa_i$ is negative and sufficiently large in absolute value (i.e. sufficiently large negative); and

(ii) the following "stability" property holds with respect to $\varkappa = (\varkappa_1, \dots, \varkappa_k)$: for each series there should exist $\varepsilon > 0$ and $N > 0$ such that $b_m = 0$ whenever

$$\sum_{i=1}^{k} m_i \varkappa_i' < -N \text{ for every choice of numbers } \varkappa_i' \text{ satisfying } \sum_{i=1}^{k} |\varkappa_i' - \varkappa_i| < \varepsilon.$$

Analogously to the case $k = 1$ considered above, a closed 1-form ω on M for which $k > 1$, gives rise to a chain complex of $K_\varkappa$-modules. Again the homology groups of this chain complex are homotopy invariants, and therefore lend themselves to the formulation of inequalities of the Morse type. Here it is of interest to investigate the dependence on $\varkappa$ of this chain complex and its homology groups, under small changes in ω which leave the critical points essentially as before.

If the form ω is without critical points then the manifold M^n has the form

$$M^n \cong \hat{M}/\mathbb{Z}^k \simeq (\hat{N} \times \mathbb{R})/\mathbb{Z}^k,$$

where $\hat{N}$ is diffeomorphic to a leaf of the foliation defined by $\omega = 0$ (see Part II, §§29.1, 29.2). (Note that the leaves of this foliation are all mutually diffeomorphic.) By means of a succession of approximations of ω by forms $\omega_j \to \omega$ also without critical points, and having rational integrals over the basic cycles γ_i, it can be seen that the manifold M^n is a fibre bundle with base the circle. The fibres of this bundle are compact manifolds N_j^{n-1} which are "factors" of $\hat{N}$ in the sense that $\hat{N}$ is a regular covering space for each N_j^{n-1}, with monodromy group $\mathbb{Z}^{k-1}$.

References

1. Novikov, S. P. and Shmeltser, I., 1981. *Funkt. Analiz* **15**, No. 3, p. 54.
2. Novikov, S. P., 1981. *Funkt. Analiz* **15**, No. 4, pp. 37–52.
3. Novikov, S. P., 1981. *DAN SSSR* **260**, No. 1, p. 31.
4. Novikov, S. P., 1982. 'Hamiltonian formalism and a multi-valued analogue of Morse theory'. *UMN* **37**, No. 5, pp. 3–49.
5. Novikov, S. P. and Taĭmanov, I. A., 1984. *DAN SSSR* **274**, No. 1, p. 26. (This article contains corrections of certain inaccuracies in the survey [4].)

APPENDIX 2

Plateau's Problem. Spectral Bordisms and Globally Minimal Surfaces in Riemannian Manifolds

by A. T. Fomenko

1. Locally Minimal Surfaces

As was noted in Part I, §37.5, soap films adhering to fixed wire "contours" (in 3-dimensional Euclidean space) provide good, graphic, physical models of 2-dimensional minimal surfaces. We remind the reader of the definition of the functional of volume in several dimensions: Let V^k be a smooth, compact submanifold of a Riemannian manifold M^n, let $D \subset V$ be a compact region of this submanifold, and denote by g_{ij} the metric induced on V^k from the given Riemannian metric on M^n; the *k-dimensional volume* $\mathrm{vol}_k D$ of the region D with respect to the metric g_{ij} is then defined as the integral over D of the "volume element" of V, given locally by $\sqrt{g}\, dx^1 \ldots dx^k$, where $g = \det(g_{ij})$. Thus, in particular, on the set of k-dimensional compact submanifolds V of M^n the correspondence $V \mapsto \mathrm{vol}_k V$, defines a functional (the "functional of Riemannian volume"). The extremals, or stationary points of this functional, whether or not they actually minimize its value locally, are called "locally minimal surfaces". Recall by way of example that in Part I, §37.5, we obtaind the Euler–Lagrange equations for the extremals of this functional in the case of hypersurfaces in the Euclidean space $\mathbb{R}^n$, and deduced (at least in the case $n = 3$) the following classical result to the effect that the condition of local minimality of a hypersurface $V \subset \mathbb{R}^n$ is equivalent to a condition on a local invariant (the mean curvature) of the embedding of that hypersurface in n-dimensional Euclidean space.

Proposition 1. (*Cf. Theorem* 37.5.2 *of Part I.*) *Let $V^{n-1} \subset \mathbb{R}$ be a smooth hypersurface (possibly with non-empty boundary) in Euclidean n-space. The mean curvature H of V^{n-1} is identically zero if and only if in some neighbourhood of each of its interior points the hypersurface may be represented as the graph* *of a function "extremal" for the volume functional, i.e. of a function satisfying the Euler–Lagrange equation for the locally minimal hypersurfaces.*

It was also proved in Part I, §37.5, that the 2-dimensional minimal surfaces in 3-dimensional Euclidean space admit the following simple analytical part-characterization: Suppose a surface $V^2 \subset \mathbb{R}^3$ is given via a parametrization $r: D(u, v) \to \mathbb{R}^3$, $r = r(u, v)$, where D is a region of the plane with Cartesian co-ordinates u, v; then if u, v furnish conformal co-ordinates on the surface (i.e. if in terms of u, v, the induced metric on V^2 has the form $\lambda(u,v)(du^2 + dv^2)$), then V^2 is locally minimal if and only if the radius vector r is harmonic (with respect to the operator $\partial^2/\partial u^2 + \partial^2/\partial v^2$) in terms of the conformal co-ordinates. (Harmonicity of the radius vector is by itself not sufficient: the surface swept out by a harmonic radius vector need not be locally minimal; see the discussion towards the end of §37.5 of Part I.) The topological structure even of 2-dimensional minimal surfaces can be rather complex; in particular, the evidence afforded by soap films (which incidentally may perhaps be taken as furnishing a practical confirmation of the existence theorem for minimal surfaces bounded by any piecewise smooth closed contour—see below) shows that one does not in general have uniqueness of such surfaces (see Part I, Figure 41), and no results are known as to whether an absolutely minimal surface is in some sense uniquely determined by a given such contour. Moreover, locally minimal surfaces may contain singularities (see Part I, Figure 42).

The phrase "the Plateau problem" is now generally understood as embracing a whole series of problems arising in the study of the extremals and absolute minima of the k-dimensional volume functional on the class of k-dimensional surfaces embedded in an ambient Riemannian manifold and satisfying one or another boundary condition. Various periods in the rich history of the evolution of variational problems of this type differ essentially from one another in their approaches to the basic concepts of "surface", "boundary", "minimization", and their correspondingly different methods of obtaining minimal solutions. The classical Plateau problem for 2-dimensional surfaces-with-boundary in $\mathbb{R}^3$ (and subsequently in $\mathbb{R}^n$) was the first to be posed and solved (J. Douglas [15], T. Rado [20]). This problem may be formulated, in parametric form, as follows:

Let $r(u, v)$ denote the radius vector of a co-ordinate region (or "patch") of a surface V^2 in $\mathbb{R}^n$; thus $r: D \to \mathbb{R}^n$ is a regular map from a region of $\mathbb{R}^2$ to $\mathbb{R}^n$ determining V^2 locally. The area of this patch of the surface is then given by

$$\mathrm{vol}_2 r(D) = \int_D \sqrt{EG - F^2}\, du\, dv,$$

and the total area $\mathrm{vol}_2(V^2)$ is obtained by combining appropriately the areas of the co-ordinate patches comprising V^2 (cf. Part II, §8.2). The question of interest is then as follows: Does there exist a surface V_0^2 (with co-ordinate patches given by maps r_0) with prescribed boundary "contour" (consisting of disjoint embedded circles in $\mathbb{R}^n$) whose area is least among all other surfaces with the same boundary contour?

In addition to this question of the existence of an absolute minimum (in the class of all surfaces with given boundary), the related question was considered of the existence of a minimal surface in a given homotopy class of surfaces with fixed boundary (i.e. of surfaces deformable to one another by means of homotopies of their defining maps). It turns out that in this, the 2-dimensional, case these problems both have affirmative solutions (see the surveys [1] and [2]). (Note however that the minimal surfaces V_0^2 solving these problems may, depending on the shape of the prescribed boundary contour, have self-intersections and other singularities.) The literature concerning these 2-dimensional problems, and related ones, is vast; however, since our present object is rather to give a survey of higher-dimensional problems of Plateau type, we shall limit ourselves to merely referring the reader particularly interested in the 2-dimensional case to the survey articles [5] and [6], and also to [14],]15], [17], [18], [19], [20], [21].

The analysis of the higher-dimensional problem requires certain preliminary concepts culminating in the definition of the "second fundamental form" of a Riemannian manifold.

Let $f: M^k \to W^n$ be a smooth embedding of a smooth manifold $M = M^k$ into a smooth, orientable, connected, closed Riemannian manifold W^n. As usual, we denote by $T_m M$ the tangent space to M at a point $m \in M$, by TM the tangent bundle of M, and by $\langle x, y \rangle$ the scalar product of vectors $x, y \in T_m M$, determined by the metric on M induced from the given Riemannian metric on W. We denote by ∇ the symmetric Riemannian connexion on TW compatible with the given metric, and, for any tensor field P on W, we denote by $\nabla_X P$ the covariant derivative (determined by the connexion ∇) of P along a vector field X on W. If x is the value of the field X at a point $w \in W$ (so that $x \in T_w W$), then the covariant derivative of the field P in the direction of x will be denoted by $\nabla_x P$. (See Part I, §29.)

Identifying M^k with the submanifold $f(M^k) \subset W^n$, we then have, in addition to the tangent bundle TM, the normal bundle NM, since at each point $m \in M$, the subspace N_m^{n-k} of $T_m W$ orthogonal to $T_m M$ is uniquely determined by the given metric on W^n. The Riemannian connexion on TW induces in the natural way Riemannian connexions on TM and NM: if Y is a smooth vector field on the submanifold M and $x \in T_m M$ is any tangent vector to M, we set $\nabla_x^{TM} Y = (\nabla_x Y)^T$, where ∇ is as before the symmetric Riemannian connexion on the ambient manifold W, and $(\ )^T$ denotes the orthogonal projection on the tangent space $T_m M$. It is easily verified that this operation determines a torsion-free (i.e. symmetric) connexion of TM, which in fact coincides with the Riemannian connexion compatible with the induced metric on M. An induced connexion on the normal bundle is defined similarly: take any smooth cross-section V of the bundle NM (i.e. above every point $m \in M$, choose a vector $V(m)$ in TW orthogonal to TM (i.e. $V(m) \in N_m M$) in such a way that $V(m)$ varies smoothly with m; thus V is a smooth (normal) vector field on M). For each $x \in T_m(M)$ we then set $\nabla_x^{NM} V = (\nabla_x V)^N$, where $(\ )^N$ denotes the orthogonal projection of TW on NM. We are now in a position to define the "second fundamental form" of our submanifold M^k (of arbitrary dimension k):

Definition 1. Corresponding to each $v \in N_m M$, choose any smooth tangent vector field V on the manifold W with the property that $V(m) = v$ and V is orthogonal to the submanifold M in some neighbourhood of the point $m \in M$. The linear map $Q^v\colon T_m M \to T_m M$, defined by the formula

$$Q^v(x) = -\nabla_x^{TM} V, \qquad x \in T_m M,$$

determines, via the scalar product $\langle\ ,\ \rangle$, a symmetric bilinear form on $T_m M \times T_m M$ whose value at a pair (x, y) of vectors in $T_m M$ is $\langle Q^v x, y\rangle$, and it is this bilinear form that we call the *second fundamental form* of the submanifold $M \subset W$. (It can be shown that for each $v \in N_m M$, Q^v is well defined, i.e. is independent of the choice of the vector field V on W, and depends smoothly on its arguments.)

We see that we have, in fact, defined here a whole family $Q = \{Q^v\}$ of linear maps $T_m M \to T_m M$ (or, equivalently, forms on $T_m M$) parametrized by $v \in N_m M$. The family Q may equivalently be interpreted as a symmetric bilinear form $Q(x, y)$ on $T_m M$, taking its values in the normal space $N_m M$, by defining the value $Q(x, y) \in N_m M$ of Q at $(x, y) \in T_m M \times T_m M$ by means of the formula

$$\langle Q(x, y), v\rangle = \langle Q^v x, y\rangle \qquad \text{for all} \quad v \in N_m M.$$

Note that by "including" the vector $y \in T_m M$ in a smooth vector field Y on the manifold W, whose restriction to $M \subset W$ is tangential to M, one obtains (via the properties of covariant differentiation listed in Part I, §28.2)

$$Q(x, y) = \nabla_x^{NM} Y.$$

Definition 2. Considering the second fundamental form Q as determining a symmetric bilinear form on the tangent space $T_m M$ at each point $m \in M$, with its values in $N_m M$, and writing $Q(x, y) = (Q_{ij}^l x^i y^j)$ with respect to suitable local co-ordinates on M, we may take in the usual way the "trace" relative to the metric g_{ij} given on W, namely

$$g^{ij} Q_{ij}^l, \qquad g^{ij} g_{jk} = \delta_k^i,$$

again obtaining a vector in $N_m M$; thus the trace of the form Q relative to the Riemannian metric is just a cross-section H of the normal bundle NM. We call this cross-section the *mean curvature* of the embedded manifold $M \subset W$. (In particular, if M is a hypersurfaces in W, then the mean curvature of $M \subset W$ is the scalar (cf. Part I, §37.5)

$$H = \operatorname{tr} G^{-1} \hat{Q},$$

where $G = (g_{ij})$ and $\hat{Q}$ denotes the matrix of Q at each point, i.e. G and $\hat{Q}$ are respectively the matrices of the first and second fundamental forms of $M \subset W$.)

The equivalence in the case of a hypersurface, between the vanishing of the mean curvature and local minimality (see Proposition 1 above), can now (i.e. once we have the above definition of mean curvature) be shown to extend to embeddings in W of manifolds M of arbitrary dimension $<\dim W$.

Proposition 2. *Let M be a compact submanifold of W (as above) and write $v_k(t) = \mathrm{vol}_k f_t(M)$, where $f_t: M \to W$, $0 \le t \le 1$, is an isotopy (or "isotopic variation" of M in W), i.e. f_0 is the inclusion map $M \to W$, and f_t is an embedding for all t. The manifold M is then locally minimal in W, i.e. $dv_k/dt|_{t=0} = 0$ for every local isotopic variation throughout which the boundary ∂M is fixed, precisely if the mean curvature H of $M \subset W$ is everywhere zero.*

Thus, to repeat somewhat, the submanifolds with identically zero mean curvature are precisely those at which the volume functional is stationary, i.e. whose volume remains unchanged to a linear or "first order" approximation under "infinitesimal" (both in extent and "amplitude") variations (to describe in the customary way the intuitive idea associated with vanishing first (variational) derivative). Of course, even if the volume of a submanifold is locally minimal (or maximal), it may decrease (or increase) under finite variations; for instance, while the equator of the standard 2-sphere (in Euclidean $\mathbb{R}^3$) is locally minimal in the sense that small local variations increase its length (it is in fact a "totally geodesic" submanifold), yet it is contractible to a point on the sphere, and is therefore not a "globally" minimal submanifold. (Recall that a "totally geodesic" submanifold is one all of whose geodesics are also geodesics of the ambient manifold. On such a submanifold the second fundamental form is identically zero, so that by Proposition 2 above it will be locally minimal.) The concept of "global" minimality of a submanifold requires for its precise formulation the idea of a "large" variation; the following is the definition of this concept best suited to our purposes.

Definition 3. Let $M^k \subset W^n$ be a compact, orientable, closed submanifold. By a *bordism-deformation* of M^k we shall mean a $(k+1)$-dimensional, smooth, compact, orientable manifold $Z^{k+1} \subset W^n$ with boundary $\partial Z = M \cup (-P)$ where $-P^k$ is a submanifold with the opposite orientation (to that induced from some orientation of W^n). The submanifold P^k is then called a *bordism-variation* of M^k. (In the case where the submanifold $M \subset W$ is non-compact we shall understand a *bordism-deformation* (non-compact version) to be given by a submanifold $P = P^k \subsetneq W$ which coincides with M^k outside some compact region, together with a $(k+1)$-dimensional submanifold Z with piecewise-smooth boundary $\partial Z \subset M \cup (-P)$.)

Recall by way of example that in §33 of Part II it was shown that in Kähler manifolds the globally minimal submanifolds of even dimension are precisely the complex submanifolds.

2. Higher-Dimensional Variational Problems and Bordism Theory

We here give first the "classical" formulation of problems concerning the absolute and "relative" minima (with respect to volume) in classes of surfaces with various prescribed topological properties. We assume given a manifold W^n and a fixed $(k-1)$-dimensional, smooth, closed submanifold A^{k-1}, which we briefly designate as "contour", and consider all possible pairs (M^k, f) consisting of a k-dimensional, smooth, compact manifold M^k with boundary homeomorphic to the contour A, and a continuous (or piecewise-smooth) map $f: M \to W$ identifying the boundary ∂M with A.

PROBLEM 1. Among the pairs (M, f) as above, does there exist a pair (M_0, f_0) such that the map f_0, or the film $X_0 = f_0(M_0)$ (the image of M_0 in W), has reasonable minimality properties? In particular, we require that an inequality of the form $\mathrm{vol}_k\, X_0 \leq \mathrm{vol}_k\, X$ be satisfied, where $X = f(M)$ is any film arising from an admissible pair (M, f), and $\mathrm{vol}_k(\ )$ denotes Riemannian volume or the usual Hausdorff measure. (By way of a "reasonable minimality property" of the film $X_0 = f_0(M_0) \subset W$, one might for instance stipulate (in addition to requiring the inequality $\mathrm{vol}\, X_0 \leq \mathrm{vol}\, X$) that the singular points of X_0 should comprise a nowhere-dense subset Y with the property that around each non-singular point $P \in X_0 \setminus Y$ there is a neighbourhood $U_p \subset W$ such that the intersection $(X_0 \setminus Y) \cap U_p$ is made up of submanifolds V_α of dimensions $\leq k$, which are all locally minimal in the above sense of classical differential geometry, i.e. their mean curvatures should all be identically zero.

PROBLEM 2. Consider pairs (V, g) of orientable, closed, k-dimensional manifolds $V = V^k$, and continuous (or piecewise-smooth) maps $g: V \to W^n$, and write $X = g(V) \subset W^n$. Analogously to the earlier definition, we say that a pair (V', g') is a *bordism-variation* of a pair (V, g) if there exists a compact manifold Z with boundary $\partial Z = V \cup (-V')$ (disjoint union), and a continuous (or piecewise-smooth) map $F: Z \to W$ such that $F|_V = g$, $F|_{V'} = g'$. Does there exist, in the class of bordism-variations of a given pair (V', g'), a pair (V_0, g_0) such that the image $X_0 = g_0(V_0) \subset W$ has reasonable minimality properties, including in particular the property that $\mathrm{vol}_k\, X_0 \leq \mathrm{vol}_k\, X$, $X = g(V) \subset W$, for all pairs (V, g) in the given class of bordism-variations?

Thus in this second type of problem one essentially seeks a film in a given class of bordism-variations which minimizes the volume functional on that class absolutely.

We now formulate the cognate problems where one is interested in "relative" minima (i.e. within just a single homotopy class) rather than minimization of the volume functional over all homotopy classes.

PROBLEM 1′. Let M be a fixed (!) manifold with boundary $\partial M = A$, and let $f': M \to W$ be a fixed continuous (or piecewise-smooth) map. Does there exist a continuous (or piecewise-smooth) map $f_0: M \to W$ homotopic to f' via a homotopy throughout which the boundary A of M remains fixed, such that f_0 (or rather the corresponding film $X_0 = f_0(M)$) has certain desired minimality properties, including the usual one that $\mathrm{vol}_k X_0 \leq \mathrm{vol}_k X$ for all films $X = f(M)$ arising from maps f in the same homotopy class as f'?

Thus, to repeat, in this problem one seeks to minimize the volume functional over a particular homotopy class, rather than all suitable maps as in its prototype.

Problem 2′. Given a fixed closed manifold M^k, and a fixed map $g: M^k \to W^n$, does there exist a map $g_0: M^k \to W^n$, homotopic to g, with appropriate minimality properties (including the property $\mathrm{vol}_k g_0(M) \leq \mathrm{vol}_k g'(M)$) relative to all maps $g': M^k \to W^n$ homotopic to the given initial map g?

For the sake of brevity we shall refer to Problems 1 and 1′ as "contour-spanning", and to Problems 2 and 2′ as "cycle-representing". Surfaces (or maps) which solve these various problems (when solutions exist!) will be termed "globally minimal". Before giving the known existence results (in the next subsection), we describe a major difficulty arising in attempting to establish such results, and some of the means for avoiding or accommodating it.

The difficulty in question consists in the appearance, in some situations inevitably, of "strata" of low dimensions, during the process minimizing the higher-dimensional volume functional. This phenomenon can be accommodated easily in the 2-dimensional case, but becomes critical in dimensions higher than 2. Figure 120 shows a contour A and a film $X_t = f_t(M)$ at various stages of a possible (continuous) process of deformation into a minimal surface in $\mathbb{R}^3$ spanning A. It is clear that at some instant t singularities will appear as the tubular region T is collapsed onto the line segment S. In the present situation (and similarly in the 2-dimensional case generally), this can be remedied by mapping the segment continuously into the 2-dimensional disc spanning the given contour (in the last of the three diagrams comprising Figure 120). The significant point here is that in doing this the parametrization of the film in terms of which the deformation is performed, is not thereby sacrified: the film continues to be the image of an appropriate 2-dimensional manifold-with-boundary.

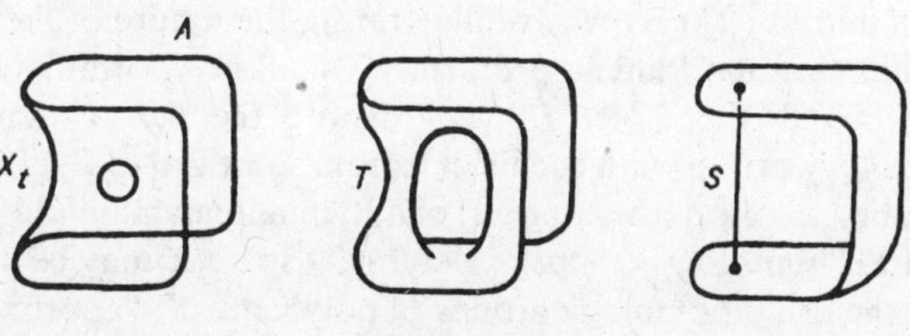

Figure 120

However, in dimensions $k > 2$, the occurrence of situations analogous to that depicted in Figure 120 sharply increases the difficulty of the minimization problem. As the k-dimensional volume of the film $X_t = f_t(M)$ being deformed approaches the minimum value, collapsing may begin to take place in the film, so that the final map $f_1: M \to W$ not only fails to be an embedding or immersion, but sends certain open sets of M into submanifolds of W of dimension $< k$. Thus in the image $X_1 = f_1(M)$ there may exist "pieces" (or "strata") S of dimensions $s \leq k - 1$. In distinction to the 2-dimensional case, it can happen that such "low-dimensional strata" cannot be avoided, or accommodated by mapping them continuously into the "massive portion" (i.e. the k-dimensional part) $X_1^{(k)}$ of the film X_1, without thereby sacrificing the crucial property that the film be a continuous image of some smooth manifold $\tilde{M}$ with boundary A. Since our aim is to find a volume-minimizing film in the class of films of the form $X = f(M)$, i.e. those admitting a continuous parametrization in terms of local co-ordinates on the manifold M, we must ensure that, whatever means are adopted to accommodate or eliminate the "low-dimensional strata", the resulting film $\tilde{X}$ admits as before such a parametrization (possibly in terms of a different manifold). However, as simple examples show, neither by removing the low-dimensional strata nor by attempting to map them into the "massive part" $X^{(k)}$ of the film X (via some continuous map defined on the whole film) can one, in general, arrange that the property of admitting a continuous parametrization be preserved. It might be thought that a simple way around the difficulty would be to concentrate exclusively on the functional $\mathrm{vol}_k(\ )$ whose value on all low-dimensional strata is of course zero, i.e. in effect just to neglect the low-dimensional strata. However, it turns out (see the articles [7]–[9] for the details) that for this simplified approach to succeed one still requires rather extensive information concerning the behaviour of the low-dimensional strata, since they are crucial to the parametrization of the film.

In view of the above-described difficulty arising in the process of minimization of films, it became essential to reformulate the problem in new, cruder, terms, allowing one to nullify the effects of low-dimensional strata. We shall now describe this reformulation, in the language of ordinary homology, of the higher-dimensional Plateau problem.

The appropriate steps towards this reformulation (resulting in the solution of Plateau's problem in terms of homology classes) were taken in the remarkable works [1], [3] and [4]. (Surveys of the basic ideas involved may be found in these papers and in [2].) By way of illustrating the nature of the solution of the higher-dimensional Plateau problem we shall now formulate a result of Reifenberg. To this end, let $H_{k-1}(A)$ denote the $(k-1)$-dimensional "spectral" homology group (with coefficients from some group G) of a $(k-1)$-dimensional submanifold A (the contour) of a Riemannian manifold W. (Note that the "spectral" homology groups of a compact space X may be defined as the "limits" of the usual homology groups of polyhedra X_n "approximating" X in a certain sense.) Let $\{X\}$ denote the class of all k-dimensional "surfaces" in W which contain A, such that under the homomorphism $i_*: H_{k-1}(A) \to H_{k-1}(X)$ induced by the inclusion map $i: A \to X$, the whole of the homology group $H_{k-1}(A)$ is sent to zero, i.e. Ker $i_* = H_{k-1}(A)$. (Both here and in what follows we shall understand the word "surface" to mean a compact, Hausdorff-measurable subset of the Riemannian manifold W.) Write

$$\lambda_k = \inf_{X \in \{X\}} \operatorname{vol}_k X,$$

where, as before, $\operatorname{vol}_k X$ denotes the Riemannian volume of X (if defined) or else the Hausdorff measure of X. It turns out (see [1]–[4] for the details) that there always exists a minimal surface X of the type just defined, i.e. there always exists a k-dimensional compact subset $X_0 \in \{X\}$ such that $\operatorname{vol}_k X_0 = \lambda_k$. Within the context of this approach to minimal surfaces, research has evolved in two directions: an analytic or functional direction (see [1], [4]), and in a more geometric vein ([2], [3]). It has culminated in some remarkable theorems on the existence of absolutely minimal elements in each ordinary homology class, and the regularity almost everywhere of minimal solutions (Federer, Fleming, Almgren, Reifenberg, and others).

In this approach essential use is made of the following fact: if $X \supset Y = \bar{Y}$ where $\dim(\overline{X \setminus Y}) < k$, then $H_k(X) = H_k(Y)$ and $\operatorname{vol}_k X = \operatorname{vol}_k Y$. It is clear from this why the above-described difficulty with low-dimensional strata does not arise in the approach via homology: their effect is negligible both topologically (homologically) and metrically speaking. On the other hand, in using ordinary homology theory as the context for defining the terms "boundary" and "pasting onto (or spanning) a contour", one has become somewhat removed from the classical formulation of the minimal-surface problem (see earlier): For if A is a $(k-1)$-dimensional submanifold of W and X_0 is a minimal "surface" spanning the contour A "homologically" (i.e. in the sense just given), then there may not exist in general a manifold M with $\partial M \cong A$ such that $X_0 = f(M)$ (where $f: M \to W$ identifies ∂M with A). In other words, X_0 need not in general admit a continuous parametrization in terms of a manifold. (See [7]–[9] for the details.)

We now return to that version of Plateau's problem, wherein a minimal solution is sought in the class of surfaces (or rather films) parametrized by manifolds. We wish now to examine the behaviour of such films in all dimensions and not just the maximal dimension k. To realize such a project a more flexible framework than that of ordinary homology is necessary; with this in mind therefore, we introduce certain concepts to be used in setting up this framework.

Definition 4. Let (Y, Z) be pair of compact topological spaces with $Y \supset Z$. By an oriented, *$(k-1)$-dimensional, singular manifold of the pair* (Y, Z), we shall mean a pair (V^{k-1}, f) consisting of a compact, oriented manifold V^{k-1} with boundary ∂V, and a continuous map of pairs $f: (V, \partial V) \to (Y, Z)$, $f(V) \subset Y$, $f(\partial V) \subset Z$. (If $Z = \varnothing$, we assume $\partial V = \varnothing$.) A singular manifold (V, f) is said to be *bordant (or equivalent) to zero* if there exists a compact, oriented manifold M^k and a continuous map $F: M \to Y$, such that:

(i) the manifold V is a (regular) submanifold of the boundary ∂M of M;
(ii) the orientation of V coincides with that induced from the given orientation of M; and
(iii) $F|_V = f$, $F(\partial M \setminus V) \subset Z$.

The operation of forming the *disjoint union of singular manifolds* is induced in the obvious way from the analogous operation on ordinary manifolds. Two singular manifolds (V_1, f_1) and (V_2, f_2) are then said to be *bordant* if their disjoint union $(V_1 \cup V_2, f_1 \cup f_2)$ is bordant to zero. The set of bordism classes of $(k-1)$-dimensional, oriented, singular manifolds of a pair (Y, Z) forms in the usual way (see §27.1) an abelian group $\Omega_{k-1}(Y, Z)$, the $(k-1)$st *group of singular bordisms.* If the condition that the manifolds be oriented (or even orientable) is omitted, the analogous construction yields the $(k-1)$st *group of unoriented singular bordisms* $N_{k-1}(Y, Z)$.

Problems 1 and 2 may now be restated as follows. Let A^{k-1} be a compact, oriented submanifold of a manifold W and $i: A \to X$ an embedding of $A = A^{k-1}$ into a surface X in W.

PROBLEM 1 (reformulated). Among the "surfaces" $X \subset W$ containing A in such a way that the singular bordism (A, i) (where $i: A \to X$ is the inclusion) is equivalent to zero in X, does there exist a "surface" X_0 with the appropriate minimality properties?

Clearly the class of "surfaces" X figuring in this version of Problem 1 is characterized by the property $i_*\sigma = 0$, where $i_*: \Omega_{k-1}(A) \to \Omega_{k-1}(X)$ is the homomorphism induced by the inclusion i, and $\sigma \in \Omega_{k-1}(A)$ is the element determined by the identity map $e: A \to A$.

PROBLEM 2 (reformulated). Does there exist among all singular manifolds (V, g), $g: V \to W$, that are bordant (or "equivalent") to a given singular manifold (V', g'), $g': V' \to W$, a singular manifold (V_0, g_0) such that the surface $X_0 = g_0(V_0)$ has the appropriate minimality properties?

In addition to the groups Ω_{k-1} and N_{k-1}, one needs also the groups Ω^p_{k-1} of singular bordisms reduced modulo p. Each of the three series of groups Ω_{k-1}, N_{k-1}, Ω^p_{k-1}, $k = 1, 2, \ldots$, satisfies the first four of the five Steenrod–Eilenberg axioms listed in §6, by virtue of which it represents an extraordinary (or generalized) homology theory. However, the fifth axiom of ordinary homology theories, to the effect that in positive dimensions the homology groups of the one-point space should all be zero, is not satisfied in general by the bordism groups of dimensions > 0: as already noted in §27.1, these may be non-trivial.

Since generally speaking minimal surfaces may have singularities (with possibly highly complex structures), in order to apply bordism theory to the variational problems we are concerned with here, it is necessary to extend the domain of definition of that theory beyond the class of cell complexes to include the class of surfaces in our present sense (i.e. compact Hausdorff-measurable subsets of a Riemannian manifold W). The process for achieving this is analogous to that whereby spectral homology theory is established on the basis of ordinary homology. Thus one represents the "surface" X as a "limit" of polyhedra X_n, and then defines the spectral bordisms of X as the corresponding "limits" of the usual singular bordisms of the X_n. (For a compact smooth manifold X the spectral bordism groups coincide with the usual bordism groups.)

In what follows, by "bordisms" of surfaces we shall always mean "spectral bordisms". Since the groups N_{k-1} and Ω^p_{k-1} are compact groups (for the class of finite cell complexes), extending their definition to the class of "surfaces" can be carried through unimpeded. On the other hand the extension is less straightforward in the case of the bordism groups Ω_{k-1}; in fact here one needs to consider rather the groups ${}^p\Omega_{k-1} = \Omega_{k-1} \otimes_Z Q_p$, where Q_p is the group of p-adic integers. (See [11] for the details.)

3. Statement of the Existence Theorem for Higher-Dimensional, Globally Minimal Surfaces, at which the Volume Functional Attains an Absolute Minimum

Let W be a smooth, closed, Riemannian manifold, let h denote any one of the spectral bordism theories described above (extended to our variational class B of "surfaces" in W, i.e. compact Hausdorff-measurable subsets), and let A be a fixed surface (the contour) in W. In Problem 1 above one is concerned with those surfaces from the class B which span the contour A in the sense of the spectral bordism theory h, and in Problem 2 with the surfaces from B that represent some prescribed non-trivial element of the appropriate spectral bordism group of W, and one seeks a minimal surface within each of these (restricted) variational classes. Each surface X in the class B has a *stratification*

$$X = A \cup S^{(k)} \cup S^{(k-1)} \cup \cdots,$$

where $S^{(k)}$ is the largest subset of $X \setminus A$ having dimension k at each of its points, and then $S^{(k-1)}$ is the largest subset of $X \setminus (A \cup S^{(k)})$ of dimension $k-1$ at each point, and so on (see [7], [8], [11]). We call the subsets $S^{(i)}$ *strata*. If the strata are measurable then the *stratified volume* $SV(X)$ is defined by

$$SV(X) = (\mathrm{vol}_k S^{(k)}, \mathrm{vol}_{k-1} S^{(k-1)}, \ldots),$$

i.e. as an ordered k-tuple or vector. As we vary X within the class of admissible variations, i.e. remaining always in the variational class B, the vector representing the stratified volume of the surface likewise varies. Our problem then becomes that of finding a surface in B with the least stratified volume

$$SV_B = (d_k, d_{k-1}, \ldots),$$

where the term "least" is to be understood in the following lexicographical sense: We first minimize the initial component of $SV(X)$, i.e. we seek in the class B surfaces X_k satisfying the equality

$$\mathrm{vol}_k\, S^{(k)} = \mathrm{vol}_k(X_k \backslash A) = d_k = \inf_{Y \in B} \mathrm{vol}(Y \backslash A).$$

If there exist such surfaces X_k, we then look to the minimization of the second component among the narrower class of these, i.e. of surfaces X_k such that $\mathrm{vol}_k(X_k \backslash A) = d_k$. Thus at this, second, stage we seek surfaces X_{k-1} among the X_k satisfying

$$\mathrm{vol}_{k-1}[X_{k-1} \backslash (A \cup S^{(k)})] = d_{k-1} = \inf_{\{X_k\}} \mathrm{vol}_{k-1}[X_k \backslash (A \cup S^{(k)})].$$

The first two components of the stratified volume of the X_{k-1} will then be lexicographically least among all surfaces in B. We continue in this way, at the ith stage seeking to minimize the ith component among all surfaces in B with the first $i - 1$ components of their stratified volume lexicographically least. If this process is well defined (and that it is, essentially forms part of the existence theorem given below), then it will terminate at a surface whose stratified volume is globally least among all stratified surfaces in the variational class B. Thus at the root of the reformulation and solution of Plateau's problem lies the concept of stratified volume (due to the present author), and the development of a method of minimizing it in all dimensions (see [7], [8], [9], [11]). A further elaboration of this idea subsequently led to a proof of the existence of globally minimal surfaces in each homotopy class of multivarifolds (see the article [12] by Dao Chong Thi).

Theorem 1 (The Basic Theorem; see [7], [8], [9], [11]). *Let W^n be $\mathbb{R}^n$ or any smooth, closed manifold whose first two homotopy groups are trivial*: $\pi_1(W) = 0$, $\pi_2(W) = 0$, *let $A \subset W$ be a fixed surface (the contour), and let B be any non-empty variational class of surfaces spanning A in the sense of spectral bordism theory (see above). Then there always exists in B a surface X_0 whose statified volume is lexicographically least*:

$$SV(X_0) = (d_k, d_{k-1}, \ldots) = SV_B.$$

This surface has a uniquely defined stratification (*i.e. partition into strata*)

$$X_0 = A \cup S^{(k)} \cup S^{(k-1)} \cup \cdots,$$

such that $\mathrm{vol}_i\, S^{(i)} = d_i$, and each subset $S^{(i)}$ is (except possibly for a subset of zero i-dimensional measure consisting of singular points) a smooth, minimal i-dimensional submanifold of the manifold W (i.e. with identically zero mean curvature).

Corollary 1. *With the same assumptions as in Theorem 1, let B be a variational class appropriate to the spectral version of either Problem 1 or Problem 2 (see above). Then in the class B there exists a globally minimal surface (possibly with singularities, forming sets of measure zero in each stratum) solving Plateau's problem:*

(i) *In the case of Problem 1 this surface is minimal among all surfaces spanning the contour A (in the sense of spectral bordism theory), i.e. admitting continuous parametrizations in terms of spectra of various manifolds with boundary A;*

(ii) *In the case of Problem 2 this surface is minimal among all surfaces representing a given element of the spectral bordism group of the ambient manifold.*

These results are in fact consequences of a significantly more general existence theorem for globally minimal surfaces, established in [7], [8] and [11] in the context of so-called extraordinary (or "generalized") (co)homology theory. We shall not go into the details of this theory here however (since this would require a sizeable addition of material), but shall confine ourselves to the statement of the following instance of the higher-dimensional variational problem (and its solution) formulated in terms of extraordinary spectral cohomology:

Given a "stable", non-trivial vector bundle ξ on a manifold W, consider as our variational class the surfaces $X \subset W$ with the property that the restriction of ξ to X remains "stable" and non-trivial. Then there exists within this class of surfaces a globally minimal one (in the sense of having least stratified volume).

Up until now we have been considering our two problems (the contour-spanning problem and the cycle-representing problem) as distinct. However, it is in a sense more natural to consider the mixed problem, wherein one seeks a minimal surface which simultaneously spans a given contour and represents a given "cycle" in the ambient manifold. We shall now describe briefly how this "mixed" Plateau problem can be solved.

Let h be one of the above spectral bordism theories, let $L = \{L_p\}$ be a fixed collection of subgroups $L_p \subset h_p(A)$ (where the p are integers), and let $L' = \{L'_p\}$ be a fixed collection of subgroups $L'_p \subset h_p(W)$. (Here as usual A is the given contour in the manifold W.)

Definition 5. We denote by $B(A, L, L')$ the class of all "surfaces" X in the manifold W such that:

(i) $A \subset X \subset W$;
(ii) $\bigcup L \subset \operatorname{Ker} i_*$;
(iii) $\bigcup L' \subset \operatorname{Im} j_*$,

where $i: A \to X$ and $j: X \to W$ are the inclusion maps. (The classes $B(A, L, \{0\})$ and $B(\varnothing, \{0\}, L')$ (with appropriate L, L') coincide respectively with the classes B introduced above in connexion with the spectral versions of Problems 1 and 2.)

It turns out that in each class $B = B(A, L, L')$ there always exists a globally minimal surface, i.e. one whose stratified volume is (lexicographically) least among all surfaces in $B(A, L, L')$. We shall now indicate the sequence of steps involved in establishing this result.

With hypotheses as in Theorem 1 (where now B is any non-empty class of the form $B(A, L, L')$), let k be the least integer such that $\mathrm{vol}_k(X) < \infty$ for some $X \in B$; then assuming $k \geq 3$ (our interest lying with the "higher-dimensional" situation), we have $3 \leq k \leq n$, where $n = \dim W$. One can then establish the following successive assertions:

1. There exists a "surface" in B whose k-dimensional volume is globally least. More precisely, if $\{X\}_k$ denotes the class of surfaces $X \in B$ such that

$$\mathrm{vol}_k(X \setminus A) = d_k = \inf_{Y \in B} \mathrm{vol}_k(Y \setminus A),$$

then the class $\{X\}_k$ is non-empty, and $d_k < \infty$. Furthermore, if $d_k > 0$, then each surface $X \in \{X\}_k$ contains a uniquely defined "k-dimensional stratum" $S^{(k)} \subset X \setminus A$ (i.e. having dimension k at each of its points and maximal with respect to this property) such that $A \cup S^{(k)}$ is compact in W, and $S^{(k)}$ contains a (possibly empty) subset Z_k (the set of k-dimensional singular points of X) with the properties that $\mathrm{vol}_k Z_k = 0$ and $S^{(k)} \setminus Z_k$ is a k-dimensional submanifold of W, without boundary and everywhere dense in $S^{(k)}$. (Hence clearly

$$\mathrm{vol}_k S^{(k)} = \mathrm{vol}_k(X \setminus A) = d_k.)$$

If $d_k = 0$, then we set $S^{(k)} = \varnothing$, and say that the surfaces in $\{X\}_k$ have no strata of dimension k.

2. There exist surfaces in $\{X\}_k$ whose $(k-1)$-dimensional volume is least (among all surfaces in $\{X\}_k$), where by "$(k-1)$-dimensional volume" of $X \in \{X\}_k$ is meant, as usual, the $(k-1)$-dimensional volume of the stratum of X of that dimension. More precisely, if $\{X\}_{k-1} (\subset \{X\}_k)$ denotes the class of all surfaces X in B satisfying $\mathrm{vol}_k(X \setminus A) = d_k$ (i.e. $X \in \{X\}_k$), together with the further condition

$$\mathrm{vol}_{k-1}(X \setminus (A \cup S^{(k)})) = d_{k-1} = \inf_{Y \in \{X\}_k} \mathrm{vol}_{k-1}(Y \setminus (A \cup S^{(k)})),$$

then this class is non-empty, and $d_{k-1} < \infty$. Furthermore, in the case $d_{k-1} > 0$, each surface in $\{X\}_{k-1}$ contains a uniquely defined $(k-1)$-dimensional stratum

$$S^{(k-1)} \subset X \setminus (A \cup S^{(k)}),$$

such that $A \cup S^{(k)} \cup S^{(k-1)}$ is compact in W, and $S^{(k-1)}$ contains a (possibly empty) subset Z_{k-1} of $(k-1)$-dimensional measure zero, which when removed from $S^{(k-1)}$ leaves a smooth $(k-1)$-dimensional manifold $S^{(k-1)} \setminus Z_{k-1}$ without boundary and everywhere dense in $S^{(k-1)}$. (Clearly we then have

$$\text{vol}_{k-1} S^{(k-1)} = \text{vol}_{k-1} X \backslash (A \cup S^{(k)}) = d_{k-1}.)$$

If $d_{k-1} = 0$ we take $S^{(k-1)} = \varnothing$.

The succeeding stages are similar: at the ith stage it is established that there exist surfaces in the set $\{X\}_{k-i+1}$, already defined to consist of those surfaces in B with the first $(i-1)$ components of their stratified volumes lexicographically least, with least ith component, i.e. least volume of the ith stratum. This inductive process terminates in a non-empty class $\{X\}$, consisting of the surfaces in B that are globally least in all dimensions, i.e. have least stratified volume, in the natural lexicographic ordering. Moreover, each stratum $S^{(i)}$ is (after removal of its singular points, if any, forming a set of i-dimensional measure zero), a smooth minimal submanifold of dimension i.

We conclude with a few remarks concerning the theorem on the existence of globally minimal surfaces in each homotopy class of multivarifolds. The introduction of the new concept of stratified volume and the methodology developed (in [7], [8], and [11]) for its minimization, actually enable one to solve Plateau's problem in each variational class of multivarifolds in W obtained by means of homotopies from any fixed map $f: V \to W$, whence the result: *In each homotopy class of multivarifolds in W there always exists a globally minimal one* (see [12]). (For the proof of this result, and in order to establish the regularity almost everywhere of minimal solutions, the concepts of strata and stratified volume had to be extended to "multivarifolds".)

References

1. Federer, H., 1969. *Geometric Measure Theory*. Springer-Verlag: Berlin.
2. Morrey, Ch. B., 1966. *Multiple Integrals in the Calculus of Variations*. Springer-Verlag: Berlin.
3. Reifenberg, E. R., 1960. 'Solution of the Plateau problem for m-dimensional surfaces of varying topological type'. *Acta Math.* **104**, pp. 1–92.
4. Almgren, F. J., 1968. 'Existence and regularity almost everywhere of solutions to elliptic variational problems among surfaces of varying topological and singularity structure. *Ann. Math.* **87**, pp. 321–391.
5. Osserman, R., 1969. *A Survey of Minimal Surfaces*. Van Nostrand Reinhold: New York.
6. Osserman, R., 1964. 'Global properties of minimal surfaces in E^3 and E^n'. *Ann. Math.* **80**, pp. 340–364.
7. Fomenko, A. T., 1972. 'The higher-dimensional Plateau problem in Riemannian manifolds'. *Matem. Sb.* **131**, pp. 475–520.
8. Fomenko, A. T., 1972. 'Minimal compact sets in Riemannian manifolds and Reifenberg's conjecture'. *IAN SSSR* **36**, pp. 1049–1080.
9. Fomenko, A. T., 1981. 'Higher-dimensional variational methods in the topology of extremals'. *UMN* **36**, pp. 105–135.
10. Fomenko, A. T., 1971. 'Bott periodicity from the point of view of the higher-dimensional Dirichlet functional'. *IAN SSSR* **35**, pp. 667–681.

11. Fomenko, A. T., 1974, 1978. 'Higher-dimensional Plateau problems on Riemannian manifolds and extraordinary homology and cohomology'. Part I, in: *Proceedings of the Seminar on Vector and Tensor Analysis*, MGU 17, pp. 3–176; Part II, in: *Proceedings of the Seminar on Vector and Tensor Analysis*, MGU 18, pp. 4–93.
12. Dao Chong Thi, 1980. 'Multivarifolds and the classical higher-dimensional Plateau problem'. *IAN SSSR* **44**, pp. 1031–1065.
13. Fomenko, A. T., 1981. 'On the minimal volumes of topological globally minimal surfaces and cobordisms'. *IAN SSSR* **45**, pp. 187–212.
14. Böhme, R. and Tromba, A., 1981. 'The index theorem for classical minimal surfaces'. *Ann. Math.* **113**, pp. 447–499.
15. Douglas, J., 1931. 'Solutions of the problem of Plateau'. *Trans. Amer. Math. Soc.* **33**, pp. 263–321.
16. Fleming, W., 1962. 'On the oriented Plateau problem'. *Rend. Circ. Mat. Palermo* **11**, No. 1, pp. 69–90.
17. Grüter, M., Hildebrandt, S. and Nitsche, C. C., 1981. 'On the boundary behaviour of minimal surfaces with a free boundary, which are not minima of the area'. *Manuscripta Math.* **35**, No. 3, pp. 387–410.
18. Meeks, W. H. and Yau, S. S.-T., 1982. 'The classical Plateau problem and the topology of three-dimensional manifolds. The embedding of the solution given by Dehn's lemma' *Topology* **21**, No. 4, pp. 409–442.
19. Morgan, F., 1976. 'A smooth curve in $\mathbb{R}^4$ bounding a continuum of area minimizing surfaces'. *Duke Math. J.* **43**, pp. 867–870.
20. Rado, T., 1933. *On the Problem of Plateau*. Springer-Verlag: Berlin.
21. Nitsche, J. C. C., 1975. *Vorlesungen über Minimalflächen*. Springer-Verlag: Berlin.

Index

Errata to Part I

Page	Line	
iv	1	Fomenko's address should be as in Part II
3	2	Italicize "world-line".
44	4, 5, 7, 8	The words "matrix" and "matrices" should be replaced by "determinant(s)", and similarly for the corresponding symbols.
49	18	The formula of Exercise 17 should be $\varkappa = \frac{(\dot{r}, \ddot{r}, \dddot{r})}{\lvert[\dot{r}, \ddot{r}]\rvert^2}$.
64	8, 9	Replace "$(x, y) \mapsto (u, v)$" by "$(u, v) \mapsto (x, y)$".
77	5	Replace "d^2l" by "d^2f."
140	2	Insert "be invariant" after "...expression".
146	3	Replace "quantities" by "entities".
222, 299	17, 1	Replace the word "Application" by "Addendum".
224	1	§24.4 should be entitled: "The Invariant Metric on a Transformation Group".
224	2, 4, 7, 10, 11, 14, 22, 23, 25, 29	Replace "Killing metric" by "invariant metric", and "Killing form" by "invariant scalar product" throughout.
	footnote	Delete.
225	15, 17, 26, 27, 30, 32	Same as for p. 224.
232, 233		In Exercises 5, 7, 8, 9, replace "Killing

		metric(s) by "invariant metric(s)", and in Exercise 6 by "invariant scalar product".
264	22	"dx_α" should be "dx^α".
288	9	Delete the minus sign before Γ_{ji}^k.
294	4	For "stain" read "strain".
305	33	"The curvature tensor defined by a Killing metric" should "The curvature tensor of groups of transformations".
306	7, 16, 24, 30, 32, 35	Same as for p. 224.
307	1, 3	Same as for p. 224.
350	14	For "commutation" read "bracket".
377	2	Replace "grade α" by "grad α".
455	10	In reference [6], "Klingenberg" is the correct spelling (and also in Part II).
464	42	For "wedge product", one should look up "exterior algebra".

Errata to Part II

Page	Line	
4	38	"(path)-connected" should be "path-connected" (i.e. delete parentheses).
90	27	Replace "$\mathbb{R}^n$" by "$\mathbb{R}^N$".
125	13	For "collection" read "condition".
129	31	"α," should be "α_1".
169	3	For "...and that.." read "...such that...".
184	6	Replace "(under homeomorphisms)" by "(under homeomorphisms or, more generally, homotopy equivalences)".
	9	Delete the words "somewhat loose".
196	4, 32	Arrowhead needed in each exact sequence.
207	9	Replace "...fundamental groups $\pi_n(S^n)$,..." by "...homotopy groups $\pi_n(S^n)$,...".
216	18	Replace "Theorem 24.1.5" by "Theorem 23.1.5".
233	12	Reference should be to §5.2(f) (not §5.2(e)).
243	11	Delete "(as for all complex line bundles)".
	25	In Exercise 1 replace "$\eta \otimes \ldots \otimes \eta$" by "$\eta \oplus \cdots \oplus \eta$".
252	26	Closing parenthesis needed after "...some $\tau \in g$".
264	4	Replace "$\mathbb{C}^+$" by "$\mathbb{C}^*$".
286	9	In the Remark, replace "(namely those for which $c_2, \chi_2 = 1$)" by "(with $\chi_2 = 1$, c_2 varying)".
343, (344)	36, (1)	Replace "finite-zoned" by "finite-gap".
351	26	Equation (59) should be

$$\dot{u}(x) = \{u(x), H\} = \frac{\partial}{\partial x}\frac{\delta H}{\delta u(x)}$$
$$= -u_{xxx} + 6uu_x + cu_x.$$

Page	Line	Correction
357	19	Replace "…of a semi-simple Lie algebra with Killing metric…" by "…of a Lie algebra with invariant metric…".
364	15	In the second of the equations (14), replace "$\dot{p}$" by "$\dot{\varepsilon}$".
	31	Inside the first set of parentheses of equation (16) replace the term "$\mu' v$" by "$\mu' v'$".
365	3	Replace "$1 - a/r$" by "$-1 + a/r$".
367	2	The equation "$g_{11} = -1/r^4\varepsilon^2$" should be "$g_{11} = (-1/r^4\varepsilon^2) \exp \kappa(R)$".
	4	The formula for g_{11} should have the further factor "$\exp x(R)$".
	6	The equations (20) should be "$-1 + \frac{a}{r} = g^{00}(\dot{r})^2 + g^{11}(r')^2 = (\dot{r}) - (a')^2 \left(\frac{8\pi G}{c^4}\right)^{-2} \exp[-\kappa(R)]$".
	8	In (21) change "$1 - \frac{a(R)}{r}$" to "$-1 + \frac{a(R)}{r}$", and insert as a further factor of the last term under the root sign "$\exp[-\kappa(R)]$".
	12, 13	Delete "(i.e. where $g^{rr} = 1 - a/r > 0$)".
	14	Delete "$\neq 0$", and add after the comma "$g^{rr} < 0$.".
	15, 16	Delete these two lines, i.e. from "so that,…" to "…cannot change.".
	23	In the third of the equations (23) insert minus sign before "$1 - \frac{a}{r}$".
368	2–8	Delete the whole of the sentence "We see also … of the matter.".
	11	In (25) insert minus sign before "$1 - \frac{a}{r}$".
369		In Figure 117 the curve labelled $a(R, 0)$ should leave the origin tangentially to the R-axis (but is correct as it stands for sufficiently large R)
415	9	Replace "one compact" by "are compact".
418	15	For "homology group" read "relative homology group".
421	7	Replace "Wesly" by "Wesley".
426	22	Replace "Möbius bond" by "Möbius band".